Lecture Notes in Computer Science 2724
Edited by G. Goos, J. Hartmanis, and J. van Leeuwen

Springer-Verlag
Berlin Heidelberg GmbH

Erick Cantú-Paz   James A. Foster
Kalyanmoy Deb   Lawrence David Davis
Rajkumar Roy   Una-May O'Reilly
Hans-Georg Beyer   Russell Standish
Graham Kendall   Stewart Wilson
Mark Harman   Joachim Wegener
Dipankar Dasgupta   Mitch A. Potter
Alan C. Schultz   Kathryn A. Dowsland
Natasha Jonoska Julian Miller (Eds.)

Genetic and Evolutionary Computation – GECCO 2003

Genetic and Evolutionary Computation Conference
Chicago, IL, USA, July 12-16, 2003
Proceedings, Part II

Springer

Series Editors

Gerhard Goos, Karlsruhe University, Germany
Juris Hartmanis, Cornell University, NY, USA
Jan van Leeuwen, Utrecht University, The Netherlands

Main Editor

Erick Cantú-Paz
Center for Applied Scientific Computing (CASC)
Lawrence Livermore National Laboratory
7000 East Avenue, L-561, Livermore, CA 94550, USA
E-mail: cantupaz@llnl.gov

Cataloging-in-Publication Data applied for

A catalog record for this book is available from the Library of Congress

Bibliographic information published by Die Deutsche Bibliothek
Die Deutsche Bibliothek lists this publication in the Deutsche Nationalbibliografie;
detailed bibliographic data is available in the Internet at <http://dnb.ddb.de>.

CR Subject Classification (1998): F.1-2, D.1.3, C.1.2, I.2.6, I.2.8, I.2.11, J.3

ISSN 0302-9743
ISBN 978-3-540-40603-7   ISBN 978-3-540-45110-5 (eBook)
DOI 10.1007/978-3-540-45110-5

This work is subject to copyright. All rights are reserved, whether the whole or part of the material is
concerned, specifically the rights of translation, reprinting, re-use of illustrations, recitation, broadcasting,
reproduction on microfilms or in any other way, and storage in data banks. Duplication of this publication
or parts thereof is permitted only under the provisions of the German Copyright Law of September 9, 1965,
in its current version, and permission for use must always be obtained from Springer-Verlag. Violations are
liable for prosecution under the German Copyright Law.

http://www.springer.de

© Springer-Verlag Berlin Heidelberg 2003
Originally published by Springer-Verlag Berlin Heidelberg New York in 2003.

Typesetting: Camera-ready by author, data conversion by PTP Berlin GmbH
Printed on acid-free paper   SPIN 10929001   06/3142   5 4 3 2 1 0

Volume Editors

James A. Foster
University of Idaho
Department of Computer Science
P.O. Box 441010
Moscow, ID 83844-1010, USA
E-mail: foster@uidaho.edu

Kalyanmoy Deb
Kanpur Genetic Algorithms Laboratory
Department of Mechanical Engineering
Indian Institute of Technology
Kanpur, PIN 208 016, India
E-mail: deb@iitk.ac.in

Lawrence David Davis
NuTech Solutions, Inc.
28 Green Street, Newbury, MA 01951, USA
E-mail: david.davis@nutechsolutions.com

Rajkumar Roy
Enterprise Integration
Cranfield University
Bedford MK43 0AL, UK
E-mail: r.roy@cranfield.ac.uk.

Una-May O'Reilly
Artificial Intelligence Lab.
MIT 20
200 Technology SQ, Rm 933
Cambridge, MA 02139, USA
E-mail: unamay@ai.mit.edu

Hans-Georg Beyer
University of Dortmund
Dept. of Computer Science XI
44221 Dortmund, Germany
E-mail: beyer@Ls11.cs.uni-dortmund.de

Russell Standish
School of Mathematics
University of New South Wales
Sydney 2052, Australia
E-mail: R.Standish@unsw.edu.au

Graham Kendall
University of Nottingham
School of Computer Science and IT
Jubilee Campus, Wollaton Road
Nottingham NG8 1BB, UK
E-mail: gxk@cs.nott.ac.uk

Stewart W. Wilson
Prediction Dynamics
Concord, MA 01742, USA
E-mail: wilson@prediction-dynamics.com

Mark Harman
Brunel University
Dept. of Information Systems
and Computing
Uxbridge, Middlesex, 11B8 3PH, UK
E-mail: Mark.Harman@brunel.ac.uk

Joachim Wegener
DaimlerChrysler AG
Research and Technology
Alt-Mohabit 96 A, 10559 Berlin
Germany
E-mail: joachim.wegener@daimlerchrysler.com

Dipankar Dasgupta
University of Memphis
Division of Computer Science
Memphis, TN 38152, USA
E-mail: dasgupta@memphis.edu

Mitchell A. Potter
Naval Research Laboratory
Washington, DC, USA
E-mail: mpotter@aic.nrl.navy.mil

Alan C. Schultz
Navy Center for Applied Research in
Artificial Intelligence
Naval Research Laboratory
4555 Overlook Ave. S.W., Washington
DC 20375
E-mail: schultz@aic.nrl.navy.mil

Kathryn A. Dowsland
Gower Optimal Algorithms Ltd
5 Whitstone Lane, Newton
Swansea SA3 4UH, UK
E-mail: k.a.dowsland@btconnect.com

Natasha Jonoska
University of South Florida
Department of Mathematics
4202 E. Fowler Av. PHY 114
Tampa FL 33559, USA
E-mail: jonoska@math.usf.edu

Julian F. Miller
School of Computer Science
University of Birmingham
Birmingham, B15 2TT, UK
E-mail: j.miller@cs.bham.ac.uk

Preface

These proceedings contain the papers presented at the 5th Annual Genetic and Evolutionary Computation Conference (GECCO 2003). The conference was held in Chicago, USA, July 12–16, 2003.

A total of 417 papers were submitted to GECCO 2003. After a rigorous doubleblind reviewing process, 194 papers were accepted for full publication and oral presentation at the conference, resulting in an acceptance rate of 46.5%. An additional 92 submissions were accepted as posters with two-page extended abstracts included in these proceedings.

This edition of GECCO was the union of the 8th Annual Genetic Programming Conference (which has met annually since 1996) and the 12th International Conference on Genetic Algorithms (which, with its first meeting in 1985, is the longest running conference in the field). Since 1999, these conferences have merged to produce a single large meeting that welcomes an increasingly wide array of topics related to genetic and evolutionary computation.

Possibly the most visible innovation in GECCO 2003 was the publication of the proceedings with Springer-Verlag as part of their Lecture Notes in Computer Science series. This will make the proceedings available in many libraries as well as online, widening the dissemination of the research presented at the conference.

Other innovations included a new track on Coevolution and Artificial Immune Systems and the expansion of the DNA and Molecular Computing track to include quantum computation.

In addition to the presentation of the papers contained in these proceedings, the conference included 13 workshops, 32 tutorials by leading specialists, and presentation of late-breaking papers.

GECCO is sponsored by the International Society for Genetic and Evolutionary Computation (ISGEC). The ISGEC by-laws contain explicit guidance on the organization of the conference, including the following principles:

(i) GECCO should be a broad-based conference encompassing the whole field of genetic and evolutionary computation.

(ii) Papers will be published and presented as part of the main conference proceedings only after being peer-reviewed. No invited papers shall be published (except for those of up to three invited plenary speakers).

(iii) The peer-review process shall be conducted consistently with the principle of division of powers performed by a multiplicity of independent program committees, each with expertise in the area of the paper being reviewed.

(iv) The determination of the policy for the peer-review process for each of the conference's independent program committees and the reviewing of papers for each program committee shall be performed by persons who occupy their positions by virtue of meeting objective and explicitly stated qualifications based on their previous research activity.

(v) Emerging areas within the field of genetic and evolutionary computation shall be actively encouraged and incorporated in the activities of the conference by providing a semiautomatic method for their inclusion (with some procedural flexibility extended to such emerging new areas).

(vi) The percentage of submitted papers that are accepted as regular full-length papers (i.e., not posters) shall not exceed 50%.

These principles help ensure that GECCO maintains high quality across the diverse range of topics it includes.

Besides sponsoring the conference, ISGEC supports the field in other ways. ISGEC sponsors the biennial Foundations of Genetic Algorithms workshop on theoretical aspects of all evolutionary algorithms. The journals *Evolutionary Computation* and *Genetic Programming and Evolvable Machines* are also supported by ISGEC. All ISGEC members (including students) receive subscriptions to these journals as part of their membership. ISGEC membership also includes discounts on GECCO and FOGA registration rates as well as discounts on other journals. More details on ISGEC can be found online at http://www.isgec.org.

Many people volunteered their time and energy to make this conference a success. The following people in particular deserve the gratitude of the entire community for their outstanding contributions to GECCO:

James A. Foster, the General Chair of GECCO for his tireless efforts in organizing every aspect of the conference.
David E. Goldberg and John Koza, members of the Business Committee, for their guidance and financial oversight.
Alwyn Barry, for coordinating the workshops.
Bart Rylander, for editing the late-breaking papers.
Past conference organizers, William B. Langdon, Erik Goodman, and Darrell Whitley, for their advice.
Elizabeth Ericson, Carol Hamilton, Ann Stolberg, and the rest of the AAAI staff for their outstanding efforts administering the conference.
Gerardo Valencia and Gabriela Coronado, for Web programming and design.
Jennifer Ballentine, Lee Ballentine and the staff of Professional Book Center, for assisting in the production of the proceedings.
Alfred Hofmann and Ursula Barth of Springer-Verlag for helping to ease the transition to a new publisher.

Sponsors who made generous contributions to support student travel grants:

Air Force Office of Scientific Research
DaimlerChrysler
National Science Foundation
Naval Research Laboratory
New Light Industries
Philips Research
Sun Microsystems

The track chairs deserve special thanks. Their efforts in recruiting program committees, assigning papers to reviewers, and making difficult acceptance decisions in relatively short times, were critical to the success of the conference:

A-Life, Adaptive Behavior, Agents, and Ant Colony Optimization, Russell Standish
Artificial Immune Systems, Dipankar Dasgupta
Coevolution, Graham Kendall
DNA, Molecular, and Quantum Computing, Natasha Jonoska
Evolution Strategies, Evolutionary Programming, Hans-Georg Beyer
Evolutionary Robotics, Alan Schultz, Mitch Potter
Evolutionary Scheduling and Routing, Kathryn A. Dowsland
Evolvable Hardware, Julian Miller
Genetic Algorithms, Kalyanmoy Deb
Genetic Programming, Una-May O'Reilly
Learning Classifier Systems, Stewart Wilson
Real-World Applications, David Davis, Rajkumar Roy
Search-Based Software Engineering, Mark Harman, Joachim Wegener

The conference was held in cooperation and/or affiliation with:

American Association for Artificial Intelligence (AAAI)
Evonet: the Network of Excellence in Evolutionary Computation
5th NASA/DoD Workshop on Evolvable Hardware
Evolutionary Computation
Genetic Programming and Evolvable Machines
Journal of Scheduling
Journal of Hydroinformatics
Applied Soft Computing

Of course, special thanks are due to the numerous researchers who submitted their best work to GECCO, reviewed the work of others, presented a tutorial, organized a workshop, or volunteered their time in any other way. I am sure you will be proud of the results of your efforts.

May 2003

Erick Cantú-Paz
Editor-in-Chief GECCO 2003
Center for Applied Scientific Computing
Lawrence Livermore National Laboratory

GECCO 2003 Conference Organization

Conference Committee

General Chair: James A. Foster
Proceedings Editor-in-Chief: Erick Cantú-Paz
Business Committee: David E. Goldberg, John Koza, J.A. Foster
Chairs of Program Policy Committees:
  A-Life, Adaptive Behavior, Agents, and Ant Colony Optimization, Russell Standish
  Artificial Immune Systems, Dipankar Dasgupta
  Coevolution, Graham Kendall
  DNA, Molecular, and Quantum Computing, Natasha Jonoska
  Evolution Strategies, Evolutionary Programming, Hans-Georg Beyer
  Evolutionary Robotics, Mitchell A. Potter and Alan C. Schultz
  Evolutionary Scheduling and Routing, Kathryn A. Dowsland
  Evolvable Hardware, Julian Miller
  Genetic Algorithms, Kalyanmoy Deb
  Genetic Programming, Una-May O'Reilly
  Learning Classifier Systems, Stewart Wilson
  Real-World Applications, David Davis, Rajkumar Roy
  Search-Based Software Engineering, Mark Harman and Joachim Wegener
Workshops Chair: Alwyn Barry
Late-Breaking Papers Chair: Bart Rylander

Workshop Organizers

Biological Applications for Genetic and Evolutionary Computation (Bio GEC 2003), Wolfgang Banzhaf, James A. Foster
Application of Hybrid Evolutionary Algorithms to NP-Complete Problems, Francisco Baptista Pereira, Ernesto Costa, Günther Raidl
Evolutionary Algorithms for Dynamic Optimization Problems, Jürgen Branke
Hardware Evolutionary Algorithms and Evolvable Hardware (HEAEH 2003), John C. Gallagher
Graduate Student Workshop, Maarten Keijzer, Sean Luke, Terry Riopka
Workshop on Memetic Algorithms 2003 (WOMA-IV), Peter Merz, William E. Hart, Natalio Krasnogor, Jim E. Smith
Undergraduate Student Workshop, Mark M. Meysenburg
Learning, Adaptation, and Approximation in Evolutionary Computation, Sibylle Mueller, Petros Koumoutsakos, Marc Schoenauer, Yaochu Jin, Sushil Louis, Khaled Rasheed
Grammatical Evolution Workshop (GEWS 2003), Michael O'Neill, Conor Ryan
Interactive Evolutionary Search and Exploration Systems, Ian Parmee

Analysis and Design of Representations and Operators (ADoRo 2003), Franz Rothlauf, Dirk Thierens

Challenges in Real-World Optimisation Using Evolutionary Computing, Rajkumar Roy, Ashutosh Tiwari

International Workshop on Learning Classifier Systems, Wolfgang Stolzmann, Pier-Luca Lanzi, Stewart Wilson

Tutorial Speakers

Parallel Genetic Algorithms, Erick Cantú-Paz
Using Appropriate Statistics, Steffan Christiensen
Multiobjective Optimization with EC, Carlos Coello
Making a Living with EC, Yuval Davidor
A Unified Approach to EC, Ken DeJong
Evolutionary Robotics, Dario Floreano
Immune System Computing, Stephanie Forrest
The Design of Innovation & Competent GAs, David E. Goldberg
Genetic Algorithms, Robert Heckendorn
Evolvable Hardware Applications, Tetsuya Higuchi
Bioinformatics with EC, Daniel Howard
Visualization in Evolutionary Computation, Christian Jacob
Data Mining and Machine Learning, Hillol Kargupta
Evolvable Hardware, Didier Keymeulen
Genetic Programming, John Koza
Genetic Programming Theory I & II, William B. Langdon, Riccardo Poli
Ant Colony Optimization, Martin Middendorf
Bionics: Building on Biological Evolution, Ingo Rechenberg
Grammatical Evolution, C. Ryan, M. O'Neill
Evolution Strategies, Hans-Paul Schwefel
Quantum Computing, Lee Spector
Anticipatory Classifier Systems, Wolfgang Stolzmann
Mathematical Theory of EC, Michael Vose
Computational Complexity and EC, Ingo Wegener
Software Testing via EC, J. Wegener, M. Harman
Testing & Evaluating EC Algorithms, Darrell Whitley
Learning Classifier Systems, Stewart Wilson
Evolving Neural Network Ensembles, Xin Yao
Neutral Evolution in EC, Tina Yu
Genetics, Annie S. Wu

Keynote Speakers

John Holland, "The Future of Genetic Algorithms"
Richard Lenski, "How the Digital Leopard Got His Spots: Thinking About Evolution Inside the Box"

Members of the Program Committee

Hussein Abbass
Adam Adamopoulos
Alexandru Agapie
José Aguilar
Jesús Aguilar
Hernán Aguirre
Chang Wook Ahn
Uwe Aickelin
Enrique Alba
Javier Alcaraz Soria
Dirk Arnold
Tughrul Arslan
Atif Azad
Meghna Babbar
Vladan Babovic
B.V. Babu
Thomas Bäck
Julio Banga
Francisco Baptista Pereira
Alwyn Barry
Cem Baydar
Thomas Beielstein
Theodore Belding
Fevzi Belli
Ester Bernado-Mansilla
Tom Bersano-Begey
Hugues Bersini
Hans-Georg Beyer
Filipic Bogdan
Andrea Bonarini
Lashon Booker
Peter Bosman
Terry Bossomaier
Klaus Bothe
Leonardo Bottaci
Jürgen Branke
Wilker Bruce
Peter Brucker
Anthony Bucci
Dirk Bueche
Magdalena Bugajska
Larry Bull
Edmund Burke
Martin Butz

Stefano Cagnoni
Xiaoqiang Cai
Erick Cantú-Paz
Uday Chakraborty
Weng-Tat Chan
Alastair Channon
Ying-Ping Chen
Shu-Heng Chen
Junghuei Chen
Prabhas Chongstitvatana
John Clark
Lattaud Claude
Manuel Clergue
Carlos Coello Coello
David Coley
Philippe Collard
Pierre Collet
Clare Bates Congdon
David Corne
Ernesto Costa
Peter Cowling
Bart Craenen
Jose Cristóbal Riquelme Santos
Keshav Dahal
Paul Darwen
Dipankar Dasgupta
Lawrence Davis
Anthony Deakin
Kalyanmoy Deb
Ivanoe De Falco
Hugo De Garis
Antonio Della Cioppa
A. Santos Del Riego
Brahma Deo
Dirk Devogelaere
Der-Rong Din
Phillip Dixon
Jose Dolado Cosin
Marco Dorigo
Keith Downing
Kathryn Dowsland
Gerry Dozier
Rolf Drechsler

Stefan Droste
Marc Ebner
R. Timothy Edwards
Norberto Eiji Nawa
Aniko Ekart
Christos Emmanouilidis
Hector Erives
Felipe Espinoza
Matthew Evett
Zhun Fan
Marco Farina
Robert Feldt
Francisco Fernández
Sevan Ficici
Peter John Fleming
Stuart Flockton
Dario Floreano
Cyril Fonlupt
Carlos Fonseca
Stephanie Forrest
Alex Freitas
Clemens Frey
Chunsheng Fu
Christian Gagne
M. Gargano
Ivan Garibay
Josep Maria Garrell i Guiu
Alessio Gaspar
Michel Gendreau
Zhou Gengui
Pierre Gérard
Andreas Geyer-Schulz
Tushar Goel
Fabio Gonzalez
Jens Gottlieb
Kendall Graham
Buster Greene
John Grefenstette
Darko Grundler
Dongbing Gu
Steven Gustafson
Charles Guthrie
Pauline Haddow
Hani Hagras

Hisashi Handa
Georges Harik
Mark Harman
Emma Hart
William Hart
Inman Harvey
Michael Herdy
Jeffrey Hermann
Arturo Hernández
  Aguirre
Francisco Herrera
Jürgen Hesser
Robert Hierons
Mika Hirvensalo
John Holmes
Tadashi Horiuchi
Daniel Howard
William Hsu
Jianjun Hu
Jacob Hurst
Hitoshi Iba
Kosuke Imamura
Iñnaki Inza
Christian Jacob
Thomas Jansen
Segovia Javier
Yaochu Jin
Bryan Jones
Natasha Jonoska
Hugues Juille
Bryant Julstrom
Mahmoud Kaboudan
Charles Karr
Balakrishnan Karthik
Sanza Kazadi
Maarten Keijzer
Graham Kendall
Didier Keymeulen
Michael Kirley
Joshua Knowles
Gabriella Kokai
Arthur Kordon
Bogdan Korel
Erkan Korkmaz
Tim Kovacs
Natalio Krasnogor

Kalmanje Krishnakumar
Renato Krohling
Sam Kwong
Gary Lamont
William Langdon
Pedro Larrañaga
Jesper Larse
Marco Laumanns
Paul Layzell
Martin Lefley
Claude Le Pape
Kwong Sak Leung
Warren Liao
Derek Linden
Michael Littman
Xavier Llora
Fernando Lobo
Jason Lohn
Michael Lones
Sushil Louis
Manuel Lozano
Jose Antonio Lozano
Jose Lozano
Pier Luca Lanzi Sean Luke
John Lusth
Evelyne Lutton
Nicholas Macias
Ana Madureira
Spiros Mancoridis
Martin Martin
Pete Martin
Arita Masanori
Iwata Masaya
Keith Mathias
Dirk Mattfeld
Giancarlo Mauri
David Mayer
Jon McCormack
Robert McKay
Nicholas McPhee
Lisa Meeden
Jörn Mehnen
Karlheinz Meier
Ole Mengshoel
Mark Meysenburg
Zbigniew Michalewicz

Martin Middendorf
Risto Miikkulainen
Julian Miller
Brian Mitchell
Chilukuri Mohan
David Montana
Byung-Ro Moon
Frank Moore
Alberto Moraglio
Manuel Moreno
Yunjun Mu
Sibylle Mueller
Masaharu Munetomo
Kazuyuki Murase
William Mydlowec
Zensho Nakao
Tomoharu Nakashima
Olfa Nasraoui
Bart Naudts
Mark Neal
Chrystopher Nehaniv
David Newth
Miguel Nicolau
Nikolay Nikolaev
Fernando Nino
Stefano Nolfi
Peter Nordin
Bryan Norman
Wim Nuijten
Leandro Nunes De Castro
Gabriela Ochoa
Victor Oduguwa
Charles Ofria
Gustavo Olague
Markus Olhofer
Michael O'Neill
Una-May O'Reilly
Franz Oppacher
Jim Ouimette
Charles Palmer
Liviu Panait
Gary Parker
Anil Patel
Witold Pedrycz
Martin Pelikan
Marek Perkowski

Sanja Petrovic
Hartmut Pohlheim
Riccardo Poli
Tom Portegys
Reid Porter
Marie-Claude Portmann
Mitchell A. Potter
Walter Potter
Jean-Yves Potvin
Dilip Pratihar
Alexander Pretschner
Adam Prügel-Bennett
William Punch
Günther Raidl
Khaled Rasheed
Tom Ray
Tapabrata Ray
Victor Rayward-Smith
Patrick Reed
John Reif
Andreas Reinholz
Rick Riolo
Jose Riquelme
Denis Robilliard
Katya Rodriguez-Vazquez
Marc Roper
Brian Ross
Franz Rothlauf
Jon Rowe
Rajkumar Roy
Günter Rudolph
Thomas Runarsson
Conor Ryan
Bart Rylander
Kazuhiro Saitou
Ralf Salomon
Eugene Santos
Kumara Sastry
Yuji Sato
David Schaffer
Martin Schmidt
Thorsten Schnier
Marc Schoenauer
Sonia Schulenburg
Alan C. Schultz

Sandip Sen
Bernhard Sendhoff
Kisung Seo
Franciszek Seredynski
Jane Shaw
Martin Shepperd
Alaa Sheta
Robert Shipman
Olivier Sigaud
Anabela Simões
Mark Sinclair
Abhishek Singh
Andre Skusa
Jim Smith
Robert Smith
Donald Sofge
Alan Soper
Terence Soule
Lee Spector
Andreas Spillner
Russell Standish
Harmen Sthamer
Adrian Stoica
Wolfgang Stolzmann
Matthew Streeter
V. Sundararajan
Gil Syswerda
Walter Tackett
Keiki Takadama
Uwe Tangen
Alexander Tarakanov
Ernesto Tarantino
Gianluca Tempesti
Hugo Terashima-Marin
Sam Thangiah
Scott Thayer
Lothar Thiele
Dirk Thierens
Adrian Thompson
Jonathan Thompson
Jonathan Timmis
Ashutosh Tiwari
Marco Tomassini
Andy Tomlinson
Jim Torresen

Paolo Toth
Michael Trick
Shigeyoshi Tsutsui
Andy Tyrrell
Jano Van Hemert
Clarissa Van Hoyweghen
Leonardo Vanneschi
David Van Veldhuizen
Robert Vanyi
Manuel Vazquez-
  Outomuro
Oswaldo Vélez-Langs
Hans-Michael Voigt
Roger Wainwright
Matthew Wall
Jean-Paul Watson
Ingo Wegener
Joachim Wegener
Karsten Weicker
Peter Whigham
Ronald While
Darrell Whitley
R. Paul Wiegand
Kay Wiese
Dirk Wiesmann
Janet Wile
Janet Wiles
Wendy Williams
Stewart Wilson
Mark Wineberg
Alden Wright
Annie Wu
Zheng Wu
Chia-Hsuan Yeh
Ayse Yilmaz
Tian-Li Yu
Tina Yu
Hongnian Yu
Ricardo Zebulum
Andreas Zell
Byoung-Tak Zhang
Lyudmila A. Zinchenko

A Word from the Chair of ISGEC

You may have just picked up your proceedings, in hard copy and CD-ROM, at GECCO 2003, or purchased it after the conference. You've doubtless already noticed the new format – publishing our proceedings as part of Springer's Lecture Notes in Computer Science (LNCS) series will make them available in many more libraries, broadening the impact of the GECCO conference dramatically!

If you attended GECCO 2003, we, the organizers, hope your experience was memorable and productive, and you have found these proceedings to be of continuing value. The opportunity for first-hand interaction among authors and other participants in GECCO is a big part of what makes it exciting, and we all hope you came away with many new insights and ideas.

If you were unable to come to GECCO 2003 in person, I hope you'll find many stimulating ideas from the world's leading researchers in evolutionary computation reported in these proceedings, and that you'll be able to participate in a future GECCO – for example, next year, in Seattle!

The International Society for Genetic and Evolutionary Computation, the sponsoring organization of the annual GECCO conferences, is a young organization, formed through the merger of the International Society for Genetic Algorithms (sponsor of the ICGA conferences) and the organization responsible for the annual Genetic Programming conferences. It depends strongly on the voluntary efforts of many of its members. It is designed to promote not only the exchange of ideas among innovators and practitioners of well-known methods such as genetic algorithms, genetic programming, evolution strategies, evolutionary programming, learning classifier systems, etc., but also the growth of newer areas such as artificial immune systems, evolvable hardware, agentbased search, and others. One of the founding principles is that ISGEC operates as a confederation of groups with related but distinct approaches and interests, and their mutual prosperity is assured by their representation in the program committees, editorial boards, etc., of the conferences and journals with which ISGEC is associated. This also insures that ISGEC and its functions continue to improve and evolve with the diversity of innovation that has characterized our field.

ISGEC has seen many changes this year, in addition to its growth in membership. We have completed the formalities for recognition as a tax-exempt charitable organization. We have created the new designations of Fellow and Senior Fellow of ISGEC to recognize the achievements of leaders in the field, and by the time you read this, we expect to have elected the first cohort. Additional Fellows and Senior Fellows will be added annually. GECCO continues to be subject to dynamic development – the many new tutorials, workshop topics, and tracks will evolve again next year, seeking to follow and encourage the developments of the many fields represented at GECCO. The best paper awards were presented for the second time at GECCO 2003, and we hope many of you participated in the balloting. This year, for the first time, most presentations at GECCO were electronic, displayed with the LCD projectors that ISGEC has recently purchased. Our journals, *Evolutionary Computation* and *Genetic Programming and Evolvable Machines*, continue to prosper, and we are exploring ways to make them even more widely available. The inclusion of the proceedings in Springer's Lecture Notes in Computer Science series, making them available in many more libraries worldwide, should have a strong positive impact on our field.

ISGEC is your society, and we urge you to become involved or continue your involvement in its activities, to the mutual benefit of the whole evolutionary computation community. Three members were elected to new five-year terms on the Executive Board at GECCO 2002 – Wolfgang Banzhaf, Marco Dorigo, and Annie Wu.

Since that time, ISGEC has been active on many issues, through actions of the Board and the three Councils – the Council of Authors, Council of Editors, and Council of Conferences.

The organizers of GECCO 2003 are listed in this frontmatter, but special thanks are due to James Foster, General Chair, and Erick Cantú-Paz, Editor-in-Chief of the Proceedings, as well as to John Koza and Dave Goldberg, the Business Committee. All of the changes this year, particularly in the publication of the proceedings, have meant a lot of additional work for this excellent team, and we owe them our thanks for a job well done.

Of course, we all owe a great debt to those who chaired or served on the various core and special program committees that reviewed all of the papers for GECCO 2003. Without their effort it would not have been possible to put on a meeting of this quality.

Another group also deserves the thanks of GECCO participants and ISGEC members – the members of the ISGEC Executive Board and Councils, who are listed below. I am particularly indebted to them for their thoughtful contributions to the organization and their continuing demonstrations of concern for the welfare of ISGEC.

I invite you to communicate with me (goodman@egr.msu.edu) if you have questions or suggestions for ways ISGEC can be of greater service to its members, or if you would like to get more involved in ISGEC and its functions.

Don't forget about the 8th Foundations of Genetic Algorithms (FOGA) workshop, also sponsored by ISGEC, the biennial event that brings together the world's leading theorists on evolutionary computation, which will be held in 2004.

Finally, I hope you will join us at GECCO 2004 in Seattle. Get your ideas to Ricardo Poli, the General Chair of GECCO 2004, when you see him at GECCO 2003, and please check the ISGEC Website, www.isgec.org, regularly for details as the planning for GECCO 2004 continues.

Erik D. Goodman

ISGEC Executive Board

Erik D. Goodman (Chair)
David Andre
Wolfgang Banzhaf
Kalyanmoy Deb
Kenneth De Jong
Marco Dorigo
David E. Goldberg
John H. Holland
John R. Koza
Una-May O'Reilly
Ingo Rechenberg
Marc Schoenauer
Lee Spector
Darrell Whitley
Annie S. Wu

Council of Authors

Erick Cantú-Paz (chair), Lawrence Livermore National Laboratory
David Andre, University of California – Berkeley
Plamen P. Angelov, Loughborough University
Vladan Babovic, Danish Hydraulic Institute
Wolfgang Banzhaf, University of Dortmund
Forrest H. Bennett III, FX Palo Alto Laboratory, Inc.
Hans-Georg Beyer, University of Dortmund
Jergen Branke, University of Karlsruhe
Martin Butz, University of Illinois at Urbana-Champaign
Runwei Cheng, Ashikaga Institute of Technology
David A. Coley, University of Exeter
Marco Dorigo, IRIDIA, Université Libre de Bruxelles
Rolf Drechsler, University of Freiburg
Emanuel Falkenauer, Optimal Design and Brussels University (ULB)
Stephanie Forrest, University of New Mexico
Mitsuo Gen, Ashikaga Institute of Technology
Andreas Geyer-Schulz, Abteilung fuer Informationswirtschaft
David E. Goldberg, University of Illinois at Urbana-Champaign
Jens Gottlieb, SAP, AG
Wolfgang A. Halang, Fernuniversitaet
John H. Holland, University of Michigan and Sante Fe Institute
Hitoshi Iba, University of Tokyo
Christian Jacob, University of Calgary
Robert E. Keller, University of Dortmund
Dimitri Knjazew, SAP, AG

John R. Koza, Stanford University
Sam Kwong, City University of Hong Kong
William B. Langdon, University College, London
Dirk C. Mattfeld, University of Bremen
Pinaki Mazumder, University of Michigan
Zbigniew Michalewicz, University of North Carolina at Charlotte
Melanie Mitchell, Oregon Health and Science University
Ian Parmee, University of North Carolina at Charlotte
Frederick E. Petry, University of North Carolina at Charlotte
Riccardo Poli, University of Essex
Moshe Sipper, Swiss Federal Institute of Technology
William M. Spears, University of Wyoming
Wallace K.S. Tang, Swiss Federal Institute of Technology
Adrian Thompson, University of Sussex
Michael D. Vose, University of Tennessee
Man Leung Wong, Lingnan University

Council of Editors

Erick Cantú-Paz (chair), Lawrence Livermore National Laboratory
Karthik Balakrishnan, Fireman's Fund Insurance Company
Wolfgang Banzhaf, University of Dortmund
Peter Bentley, University College, London
Lance D. Chambers, Western Australian Department of Transport
Dipankar Dasgupta, University of Memphis
Kenneth De Jong, George Mason University
Francisco Herrera, University of Granada
William B. Langdon, University College, London
Pinaki Mazumder, University of Michigan
Eric Michielssen, University of Illinois at Urbana-Champaign
Witold Pedrycz, University of Alberta
Rajkumar Roy, Cranfield University
Elizabeth M. Rudnick, University of Illinois at Urbana-Champaign
Marc Schoenauer, INRIA Rocquencourt
Lee Spector, Hampshire College
Jose L. Verdegay, University of Granada, Spain

Council of Conferences, Riccardo Poli (Chair)

The purpose of the Council of Conferences is to provide information about the numerous conferences that are available to researchers in the field of Genetic and Evolutionary Computation, and to encourage them to coordinate their meetings to maximize our collective impact on science.

ACDM, Adaptive Computing in Design and Manufacture, 2004, Ian Parmee
(Ian.Parmee@uwe.ac.uk)

EuroGP, European Conference on Genetic Programming, Portugal, April 2004,
Ernesto Costa (ernesto@dei.uc.pt)

EvoWorkshops, European Evolutionary Computing Workshops, Portugal, April
2004, Stefano Cagnoni (cagnoni@ce.unipr.it)

FOGA, Foundations of Genetic Algorithms Workshop, 2004

GECCO 2004, Genetic and Evolutionary Computation Conference, Seattle,
June 2004, Riccardo Poli (rpoli@essex.ac.uk)

INTROS, INtroductory TutoRials in Optimization, Search and Decision Support
Methodologies, August 12, 2003, Nottingham, UK, Edmund Burke
(ekb@cs.nott.ac.uk)

MISTA, 1st Multidisciplinary International Conference on Scheduling: Theory
and Applications August 8-12, 2003, Nottingham, UK, Graham Kendall
(gxk@cs.nott.ac.uk)

PATAT 2004, 5th International Conference on the Practice and Theory of Automated Timetabling, Pittsburgh, USA, August 18–20, 2004, Edmund Burke
(ekb@cs.nott.ac.uk)

WSC8, 8th Online World Conference on Soft Computing in Industrial Applications, September 29th - October 10th, 2003, Internet (hosted by University
of Dortmund), Frank Hoffmann (hoffmann@esr.e-technik.uni-dortmund.de)

An up-to-date roster of the Council of Conferences is available online at
http://www.isgec.org/conferences.html.

Please contact the COC chair Riccardo Poli (rpoli@essex.ac.uk) for additions to this list.

Papers Nominated for Best Paper Awards

In 2002, ISGEC created a best paper award for GECCO. As part of the double blind peer review, the reviewers were asked to nominate papers for best paper awards. The chairs of core and special program committees selected the papers that received the most nominations for consideration by the conference. One winner for each program track was chosen by secret ballot of the GECCO attendees after the papers were presented in Chicago. The titles and authors of the winning papers are available at the GECCO 2003 website (www.isgec.org/GECCO-2003).

Finite Population Models of Co-evolution and Their Application to Haploidy versus Diploidy, Anthony M.L. Liekens, Huub M.M. ten Eikelder, and Peter A.J. Hilbers

A Game-Theoretic Memory Mechanism for Coevolution, Sevan G. Ficici and Jordan B. Pollack

A Non-dominated Sorting Particle Swarm Optimizer for Multiobjective Optimization, Xiaodong Li

Emergence of Collective Behavior in Evolving Populations of Flying Agents, Lee Spector, Jon Klein, Chris Perry, and Mark Feinstein

Immune Inspired Somatic Contiguous Hypermutation for Function Optimisation, Johnny Kelsey and Jon Timmis

Efficiency and Reliability of DNA-Based Memories, Max H. Garzon, Andrew Neel, and Hui Chen

Hardware Evolution of Analog Speed Controllers for a DC Motor, D.A. Gwaltney and M.I. Ferguson

Integration of Genetic Programming and Reinforcement Learning for Real Robots, Shotaro Kamio, Hideyuki Mitshuhashi, and Hitoshi Iba

Co-evolving Task-Dependent Visual Morphologies in Predator-Prey Experiments, Gunnar Buason and Tom Ziemke

The Steady State Behavior of $(\mu/\mu_I, \lambda)$-ES on Ellipsoidal Fitness Models Disturbed by Noise, Hans-Georg Beyer and Dirk V. Arnold

On the Optimization of Monotone Polynomials by the (1+1) EA and Randomized Local Search, Ingo Wegener and Carsten Witt

Ruin and Recreate Principle Based Approach for the Quadratic Assignment Problem, Alfonsas Misevicius

Evolutionary Computing as a tool for Grammar Development, Guy De Pauw

Adaptive Elitist-Population Based Genetic Algorithm for Multimodal Function Optimization, Kwong-Sak Leung and Yong Liang

Scalability of Selectorecombinative Genetic Algorithms for Problems with Tight Linkage, Kumara Sastry and David E. Goldberg

Effective Use of Directional Information in Multi-objective Evolutionary Computation, Martin Brown and R.E. Smith

Are Multiple Runs of Genetic Algorithms Better Than One? Erick Cantú-Paz and David E. Goldberg

Selection in the Presence of Noise, Jürgen Branke and Christian Schmidt

Difficulty of Unimodal and Multimodal Landscapes in Genetic Programming, Leonardo Vanneschi, Marco Tomassini, Manuel Clergue, and Philippe Collard

Dynamic Maximum Tree Depth: a Simple Technique for Avoiding Bloat in Tree-Based GP, Sara Silva and Jonas Almeida

Generative Representations for Evolving Families of Designs, Gregory S. Hornby

Identifying Structural Mechanisms in Standard Genetic Programming, Jason M. Daida and Adam M. Hilss

Visualizing Tree Structures in Genetic Programming, Jason M. Daida, Adam M. Hilss, David J. Ward, and Stephen L. Long

Methods for Evolving Robust Programs, Liviu Panait and Sean Luke

Population Implosion in Genetic Programming, Sean Luke, Gabriel Catalin Balan, and Liviu Panait

Designing Efficient Exploration with MACS: Modules and Function Approximation, Pierre Gérard and Olivier Sigaud

Tournament Selection: Stable Fitness Pressure in XCS, Martin V. Butz, Kumara Sastry, and David E. Goldberg

Towards Building Block Propagation in XCS: a Negative Result and Its Implications, Kurian K. Tharakunnel, Martin V. Butz, and David E. Goldberg

Quantum-Inspired Evolutionary Algorithm-Based Face Verification, Jun-Su Jang, Kuk-Hyun Han, and Jong-Hwan Kim

Mining Comprehensive Clustering Rules with an Evolutionary Algorithm, Ioannis Sarafis, Phil Trinder and Ali Zalzala

System-Level Synthesis of MEMS via Genetic Programming and Bond Graphs, Zhun Fan, Kisung Seo, Jianjun Hu, Ronald C. Rosenberg, and Erik D. Goodman

Active Guidance for a Finless Rocket Using Neuroevolution, Faustino J. Gomez and Risto Miikkulainen

Extracting Test Sequences from a Markov Software Usage Model by ACO, Karl Doerner and Walter J. Gutjahr

Modeling the Search Landscape of Metaheuristic Software Clustering Algorithms, Brian S. Mitchell and Spiros Mancoridis

Table of Contents

Volume II

Genetic Algorithms (continued)

Design of Multithreaded Estimation of Distribution Algorithms 1247
Jiri Ocenasek, Josef Schwarz, Martin Pelikan

Reinforcement Learning Estimation of Distribution Algorithm 1259
Topon Kumar Paul, Hitoshi Iba

Hierarchical BOA Solves Ising Spin Glasses and MAXSAT 1271
Martin Pelikan, David E. Goldberg

ERA: An Algorithm for Reducing the Epistasis of SAT Problems 1283
Eduardo Rodriguez-Tello, Jose Torres-Jimenez

Learning a Procedure That Can Solve Hard Bin-Packing Problems:
A New GA-Based Approach to Hyper-heuristics 1295
Peter Ross, Javier G. Marín-Blázquez, Sonia Schulenburg, Emma Hart

Population Sizing for the Redundant Trivial Voting Mapping 1307
Franz Rothlauf

Non-stationary Function Optimization Using Polygenic Inheritance...... 1320
Conor Ryan, J.J. Collins, David Wallin

Scalability of Selectorecombinative Genetic Algorithms for
Problems with Tight Linkage 1332
Kumara Sastry, David E. Goldberg

New Entropy-Based Measures of Gene Significance and Epistasis 1345
Dong-Il Seo, Yong-Hyuk Kim, Byung-Ro Moon

A Survey on Chromosomal Structures and Operators for Exploiting
Topological Linkages of Genes 1357
Dong-Il Seo, Byung-Ro Moon

Cellular Programming and Symmetric Key Cryptography Systems 1369
Franciszek Seredyński, Pascal Bouvry, Albert Y. Zomaya

Mating Restriction and Niching Pressure: Results from Agents and
Implications for General EC 1382
R.E. Smith, Claudio Bonacina

EC Theory: A Unified Viewpoint 1394
　Christopher R. Stephens, Adolfo Zamora

Real Royal Road Functions for Constant Population Size.............. 1406
　Tobias Storch, Ingo Wegener

Two Broad Classes of Functions for Which a No Free Lunch Result
Does Not Hold 1418
　Matthew J. Streeter

Dimensionality Reduction via Genetic Value Clustering 1431
　Alexander Topchy, William Punch III

The Structure of Evolutionary Exploration: On Crossover,
Buildings Blocks, and Estimation-of-Distribution Algorithms 1444
　Marc Toussaint

The Virtual Gene Genetic Algorithm 1457
　Manuel Valenzuela-Rendón

Quad Search and Hybrid Genetic Algorithms 1469
　Darrell Whitley, Deon Garrett, Jean-Paul Watson

Distance between Populations 1481
　Mark Wineberg, Franz Oppacher

The Underlying Similarity of Diversity Measures Used in
Evolutionary Computation 1493
　Mark Wineberg, Franz Oppacher

Implicit Parallelism 1505
　Alden H. Wright, Michael D. Vose, Jonathan E. Rowe

Finding Building Blocks through Eigenstructure Adaptation............ 1518
　Danica Wyatt, Hod Lipson

A Specialized Island Model and Its Application in
Multiobjective Optimization... 1530
　Ningchuan Xiao, Marc P. Armstrong

Adaptation of Length in a Nonstationary Environment 1541
　Han Yu, Annie S. Wu, Kuo-Chi Lin, Guy Schiavone

Optimal Sampling and Speed-Up for Genetic Algorithms on the
Sampled OneMax Problem 1554
　Tian-Li Yu, David E. Goldberg, Kumara Sastry

Building-Block Identification by Simultaneity Matrix 1566
　Chatchawit Aporntewan, Prabhas Chongstitvatana

A Unified Framework for Metaheuristics 1568
   Jürgen Branke, Michael Stein, Hartmut Schmeck

The Hitting Set Problem and Evolutionary Algorithmic Techniques
with ad-hoc Viruses (HEAT-V) 1570
   Vincenzo Cutello, Francesco Pappalardo

The Spatially-Dispersed Genetic Algorithm.......................... 1572
   Grant Dick

Non-universal Suffrage Selection Operators Favor Population
Diversity in Genetic Algorithms.................................... 1574
   Federico Divina, Maarten Keijzer, Elena Marchiori

Uniform Crossover Revisited: Maximum Disruption in
Real-Coded GAs 1576
   Stephen Drake

The Master-Slave Architecture for Evolutionary Computations
Revisited 1578
   Christian Gagné, Marc Parizeau, Marc Dubreuil

Genetic Algorithms – Posters

Using Adaptive Operators in Genetic Search 1580
   Jonatan Gómez, Dipankar Dasgupta, Fabio González

A Kernighan-Lin Local Improvement Heuristic That Solves Some Hard
Problems in Genetic Algorithms 1582
   William A. Greene

GA-Hardness Revisited 1584
   Haipeng Guo, William H. Hsu

Barrier Trees For Search Analysis 1586
   Jonathan Hallam, Adam Prügel-Bennett

A Genetic Algorithm as a Learning Method Based on
Geometric Representations 1588
   Gregory A. Holifield, Annie S. Wu

Solving Mastermind Using Genetic Algorithms....................... 1590
   Tom Kalisker, Doug Camens

Evolutionary Multimodal Optimization Revisited 1592
   Rajeev Kumar, Peter Rockett

Integrated Genetic Algorithm with Hill Climbing for Bandwidth
Minimization Problem 1594
   Andrew Lim, Brian Rodrigues, Fei Xiao

A Fixed-Length Subset Genetic Algorithm for the p-Median Problem.... 1596
    Andrew Lim, Zhou Xu

Performance Evaluation of a Parameter-Free Genetic Algorithm for
Job-Shop Scheduling Problems 1598
    Shouichi Matsui, Isamu Watanabe, Ken-ichi Tokoro

SEPA: Structure Evolution and Parameter Adaptation in
Feed-Forward Neural Networks 1600
    Paulito P. Palmes, Taichi Hayasaka, Shiro Usui

Real-Coded Genetic Algorithm to Reveal Biological Significant
Sites of Remotely Homologous Proteins 1602
    Sung-Joon Park, Masayuki Yamamura

Understanding EA Dynamics via Population Fitness Distributions 1604
    Elena Popovici, Kenneth De Jong

Evolutionary Feature Space Transformation Using Type-Restricted
Generators 1606
    Oliver Ritthoff, Ralf Klinkenberg

On the Locality of Representations 1608
    Franz Rothlauf

New Subtour-Based Crossover Operator for the TSP 1610
    Sang-Moon Soak, Byung-Ha Ahn

Is a Self-Adaptive Pareto Approach Beneficial for Controlling
Embodied Virtual Robots? 1612
    Jason Teo, Hussein A. Abbass

A Genetic Algorithm for Energy Efficient Device Scheduling in
Real-Time Systems 1614
    Lirong Tian, Tughrul Arslan

Metropolitan Area Network Design Using GA Based on Hierarchical
Linkage Identification.. 1616
    Miwako Tsuji, Masaharu Munetomo, Kiyoshi Akama

Statistics-Based Adaptive Non-uniform Mutation for Genetic
Algorithms 1618
    Shengxiang Yang

Genetic Algorithm Design Inspired by Organizational Theory:
Pilot Study of a Dependency Structure Matrix Driven
Genetic Algorithm 1620
    Tian-Li Yu, David E. Goldberg, Ali Yassine, Ying-Ping Chen

Are the "Best" Solutions to a Real Optimization Problem Always
Found in the Noninferior Set? Evolutionary Algorithm for Generating
Alternatives (EAGA) 1622
    Emily M. Zechman, S. Ranji Ranjithan

Population Sizing Based on Landscape Feature 1624
    Jian Zhang, Xiaohui Yuan, Bill P. Buckles

Genetic Programming

Structural Emergence with Order Independent Representations 1626
    R. Muhammad Atif Azad, Conor Ryan

Identifying Structural Mechanisms in Standard Genetic Programming ... 1639
    Jason M. Daida, Adam M. Hilss

Visualizing Tree Structures in Genetic Programming 1652
    Jason M. Daida, Adam M. Hilss, David J. Ward, Stephen L. Long

What Makes a Problem GP-Hard? Validating a Hypothesis of
Structural Causes 1665
    Jason M. Daida, Hsiaolei Li, Ricky Tang, Adam M. Hilss

Generative Representations for Evolving Families of Designs 1678
    Gregory S. Hornby

Evolutionary Computation Method for Promoter Site Prediction
in DNA 1690
    Daniel Howard, Karl Benson

Convergence of Program Fitness Landscapes 1702
    W.B. Langdon

Multi-agent Learning of Heterogeneous Robots by
Evolutionary Subsumption 1715
    Hongwei Liu, Hitoshi Iba

Population Implosion in Genetic Programming 1729
    Sean Luke, Gabriel Catalin Balan, Liviu Panait

Methods for Evolving Robust Programs 1740
    Liviu Panait, Sean Luke

On the Avoidance of Fruitless Wraps in Grammatical Evolution 1752
    Conor Ryan, Maarten Keijzer, Miguel Nicolau

Dense and Switched Modular Primitives for Bond Graph Model Design .. 1764
    Kisung Seo, Zhun Fan, Jianjun Hu, Erik D. Goodman,
    Ronald C. Rosenberg

Dynamic Maximum Tree Depth 1776
  Sara Silva, Jonas Almeida

Difficulty of Unimodal and Multimodal Landscapes in
Genetic Programming 1788
  Leonardo Vanneschi, Marco Tomassini, Manuel Clergue,
  Philippe Collard

Genetic Programming – Posters

Ramped Half-n-Half Initialisation Bias in GP 1800
  Edmund Burke, Steven Gustafson, Graham Kendall

Improving Evolvability of Genetic Parallel Programming Using
Dynamic Sample Weighting 1802
  Sin Man Cheang, Kin Hong Lee, Kwong Sak Leung

Enhancing the Performance of GP Using an Ancestry-Based Mate
Selection Scheme 1804
  Rodney Fry, Andy Tyrrell

A General Approach to Automatic Programming Using Occam's Razor,
Compression, and Self-Inspection 1806
  Peter Galos, Peter Nordin, Joel Olsén, Kristofer Sundén Ringnér

Building Decision Tree Software Quality Classification Models
Using Genetic Programming 1808
  Yi Liu, Taghi M. Khoshgoftaar

Evolving Petri Nets with a Genetic Algorithm 1810
  Holger Mauch

Diversity in Multipopulation Genetic Programming 1812
  Marco Tomassini, Leonardo Vanneschi, Francisco Fernández,
  Germán Galeano

An Encoding Scheme for Generating λ-Expressions in
Genetic Programming 1814
  Kazuto Tominaga, Tomoya Suzuki, Kazuhiro Oka

AVICE: Evolving Avatar's Movement 1816
  Hiromi Wakaki, Hitoshi Iba

Learning Classifier Systems

Evolving Multiple Discretizations with Adaptive Intervals for a
Pittsburgh Rule-Based Learning Classifier System 1818
  Jaume Bacardit, Josep Maria Garrell

Limits in Long Path Learning with XCS 1832
  Alwyn Barry

Bounding the Population Size in XCS to Ensure
Reproductive Opportunities 1844
  Martin V. Butz, David E. Goldberg

Tournament Selection: Stable Fitness Pressure in XCS 1857
  Martin V. Butz, Kumara Sastry, David E. Goldberg

Improving Performance in Size-Constrained Extended
Classifier Systems 1870
  Devon Dawson

Designing Efficient Exploration with MACS: Modules and Function
Approximation 1882
  Pierre Gérard, Olivier Sigaud

Estimating Classifier Generalization and Action's Effect:
A Minimalist Approach 1894
  Pier Luca Lanzi

Towards Building Block Propagation in XCS: A Negative Result and
Its Implications 1906
  Kurian K. Tharakunnel, Martin V. Butz, David E. Goldberg

Learning Classifier Systems – Posters

Data Classification Using Genetic Parallel Programming 1918
  Sin Man Cheang, Kin Hong Lee, Kwong Sak Leung

Dynamic Strategies in a Real-Time Strategy Game 1920
  William Joseph Falke II, Peter Ross

Using Raw Accuracy to Estimate Classifier Fitness in XCS 1922
  Pier Luca Lanzi

Towards Learning Classifier Systems for Continuous-Valued
Online Environments 1924
  Christopher Stone, Larry Bull

Real World Applications

Artificial Immune System for Classification of Gene Expression Data 1926
  Shin Ando, Hitoshi Iba

Automatic Design Synthesis and Optimization of Component-Based
Systems by Evolutionary Algorithms 1938
  P.P. Angelov, Y. Zhang, J.A. Wright, V.I. Hanby, R.A. Buswell

Studying the Advantages of a Messy Evolutionary Algorithm for
Natural Language Tagging 1951
   Lourdes Araujo

Optimal Elevator Group Control by Evolution Strategies 1963
   Thomas Beielstein, Claus-Peter Ewald, Sandor Markon

A Methodology for Combining Symbolic Regression and Design of
Experiments to Improve Empirical Model Building 1975
   Flor Castillo, Kenric Marshall, James Green, Arthur Kordon

The General Yard Allocation Problem 1986
   Ping Chen, Zhaohui Fu, Andrew Lim, Brian Rodrigues

Connection Network and Optimization of Interest Metric for
One-to-One Marketing 1998
   Sung-Soon Choi, Byung-Ro Moon

Parameter Optimization by a Genetic Algorithm for a Pitch
Tracking System 2010
   Yoon-Seok Choi, Byung-Ro Moon

Secret Agents Leave Big Footprints: How to Plant a Cryptographic
Trapdoor, and Why You Might Not Get Away with It 2022
   John A. Clark, Jeremy L. Jacob, Susan Stepney

GenTree: An Interactive Genetic Algorithms System for Designing
3D Polygonal Tree Models 2034
   Clare Bates Congdon, Raymond H. Mazza

Optimisation of Reaction Mechanisms for Aviation Fuels Using a
Multi-objective Genetic Algorithm 2046
   *Lionel Elliott, Derek B. Ingham, Adrian G. Kyne, Nicolae S. Mera,
   Mohamed Pourkashanian, Chritopher W. Wilson*

System-Level Synthesis of MEMS via Genetic Programming and
Bond Graphs 2058
   *Zhun Fan, Kisung Seo, Jianjun Hu, Ronald C. Rosenberg,
   Erik D. Goodman*

Congressional Districting Using a TSP-Based Genetic Algorithm 2072
   Sean L. Forman, Yading Yue

Active Guidance for a Finless Rocket Using Neuroevolution 2084
   Faustino J. Gomez, Risto Miikkulainen

Simultaneous Assembly Planning and Assembly System Design Using
Multi-objective Genetic Algorithms 2096
   Karim Hamza, Juan F. Reyes-Luna, Kazuhiro Saitou

Multi-FPGA Systems Synthesis by Means of
Evolutionary Computation 2109
    J.I. Hidalgo, F. Fernández, J. Lanchares, J.M. Sánchez, R. Hermida,
    M. Tomassini, R. Baraglia, R. Perego, O. Garnica

Genetic Algorithm Optimized Feature Transformation –
A Comparison with Different Classifiers 2121
    Zhijian Huang, Min Pei, Erik Goodman, Yong Huang, Gaoping Li

Web-Page Color Modification for Barrier-Free Color Vision with
Genetic Algorithm 2134
    Manabu Ichikawa, Kiyoshi Tanaka, Shoji Kondo, Koji Hiroshima,
    Kazuo Ichikawa, Shoko Tanabe, Kiichiro Fukami

Quantum-Inspired Evolutionary Algorithm-Based Face Verification 2147
    Jun-Su Jang, Kuk-Hyun Han, Jong-Hwan Kim

Minimization of Sonic Boom on Supersonic Aircraft Using an
Evolutionary Algorithm 2157
    Charles L. Karr, Rodney Bowersox, Vishnu Singh

Optimizing the Order of Taxon Addition in Phylogenetic Tree
Construction Using Genetic Algorithm 2168
    Yong-Hyuk Kim, Seung-Kyu Lee, Byung-Ro Moon

Multicriteria Network Design Using Evolutionary Algorithm 2179
    Rajeev Kumar, Nilanjan Banerjee

Control of a Flexible Manipulator Using a Sliding Mode Controller
with Genetic Algorithm Tuned Manipulator Dimension 2191
    N.M. Kwok, S. Kwong

Daily Stock Prediction Using Neuro-genetic Hybrids 2203
    Yung-Keun Kwon, Byung-Ro Moon

Finding the Optimal Gene Order in Displaying Microarray Data 2215
    Seung-Kyu Lee, Yong-Hyuk Kim, Byung-Ro Moon

Learning Features for Object Recognition 2227
    Yingqiang Lin, Bir Bhanu

An Efficient Hybrid Genetic Algorithm for a Fixed Channel
Assignment Problem with Limited Bandwidth 2240
    Shouichi Matsui, Isamu Watanabe, Ken-ichi Tokoro

Using Genetic Algorithms for Data Mining Optimization in an
Educational Web-Based System 2252
    Behrouz Minaei-Bidgoli, William F. Punch III

Improved Image Halftoning Technique Using GAs with Concurrent
Inter-block Evaluation 2264
    Emi Myodo, Hernán Aguirre, Kiyoshi Tanaka

Complex Function Sets Improve Symbolic Discriminant Analysis of
Microarray Data 2277
    *David M. Reif, Bill C. White, Nancy Olsen, Thomas Aune,
Jason H. Moore*

GA-Based Inference of Euler Angles for Single Particle Analysis 2288
    *Shusuke Saeki, Kiyoshi Asai, Katsutoshi Takahashi, Yutaka Ueno,
Katsunori Isono, Hitoshi Iba*

Mining Comprehensible Clustering Rules with an
Evolutionary Algorithm.. 2301
    Ioannis Sarafis, Phil Trinder, Ali Zalzala

Evolving Consensus Sequence for Multiple Sequence Alignment with
a Genetic Algorithm... 2313
    Conrad Shyu, James A. Foster

A Linear Genetic Programming Approach to Intrusion Detection........ 2325
    Dong Song, Malcolm I. Heywood, A. Nur Zincir-Heywood

Genetic Algorithm for Supply Planning Optimization under
Uncertain Demand 2337
    Tezuka Masaru, Hiji Masahiro

Genetic Algorithms: A Fundamental Component of an Optimization
Toolkit for Improved Engineering Designs 2347
    Siu Tong, David J. Powell

Spatial Operators for Evolving Dynamic Bayesian Networks from
Spatio-temporal Data 2360
    Allan Tucker, Xiaohui Liu, David Garway-Heath

An Evolutionary Approach for Molecular Docking 2372
    Jinn-Moon Yang

Evolving Sensor Suites for Enemy Radar Detection.................... 2384
    *Ayse S. Yilmaz, Brian N. McQuay, Han Yu, Annie S. Wu,
John C. Sciortino, Jr.*

Real World Applications – Posters

Optimization of Spare Capacity in Survivable WDM Networks 2396
    H.W. Chong, Sam Kwong

Partner Selection in Virtual Enterprises by Using Ant
Colony Optimization in Combination with the Analytical
Hierarchy Process 2398
   Marco Fischer, Hendrik Jähn, Tobias Teich

Quadrilateral Mesh Smoothing Using a Steady State
Genetic Algorithm 2400
   Mike Holder, Charles L. Karr

Evolutionary Algorithms for Two Problems from the Calculus of
Variations.. 2402
   Bryant A. Julstrom

Genetic Algorithm Frequency Domain Optimization of an
Anti-Resonant Electromechanical Controller 2404
   Charles L. Karr, Douglas A. Scott

Genetic Algorithm Optimization of a Filament Winding Process 2406
   Charles L. Karr, Eric Wilson, Sherri Messimer

Circuit Bipartitioning Using Genetic Algorithm 2408
   Jong-Pil Kim, Byung-Ro Moon

Multi-campaign Assignment Problem and Optimizing
Lagrange Multipliers 2410
   Yong-Hyuk Kim, Byung-Ro Moon

Grammatical Evolution for the Discovery of Petri Net Models of
Complex Genetic Systems 2412
   Jason H. Moore, Lance W. Hahn

Evaluation of Parameter Sensitivity for Portable Embedded Systems
through Evolutionary Techniques 2414
   James Northern III, Michael Shanblatt

An Evolutionary Algorithm for the Joint Replenishment of
Inventory with Interdependent Ordering Costs 2416
   Anne Olsen

Benefits of Implicit Redundant Genetic Algorithms for Structural
Damage Detection in Noisy Environments 2418
   Anne Raich, Tamás Liszkai

Multi-objective Traffic Signal Timing Optimization Using
Non-dominated Sorting Genetic Algorithm II 2420
   Dazhi Sun, Rahim F. Benekohal, S. Travis Waller

Exploration of a Two Sided Rendezvous Search Problem Using
Genetic Algorithms 2422
   T.Q.S. Truong, A. Stacey

Taming a Flood with a T-CUP – Designing Flood-Control Structures
with a Genetic Algorithm 2424
   Jeff Wallace, Sushil J. Louis

Assignment Copy Detection Using Neuro-genetic Hybrids 2426
   Seung-Jin Yang, Yong-Geon Kim, Yung-Keun Kwon, Byung-Ro Moon

Search Based Software Engineering

Structural and Functional Sequence Test of Dynamic and
State-Based Software with Evolutionary Algorithms 2428
   André Baresel, Hartmut Pohlheim, Sadegh Sadeghipour

Evolutionary Testing of Flag Conditions 2442
   Andr#e Baresel, Harmen Sthamer

Predicate Expression Cost Functions to Guide Evolutionary Search
for Test Data 2455
   Leonardo Bottaci

Extracting Test Sequences from a Markov Software Usage Model
by ACO 2465
   Karl Doerner, Walter J. Gutjahr

Using Genetic Programming to Improve Software Effort Estimation
Based on General Data Sets 2477
   Martin Lefley, Martin J. Shepperd

The State Problem for Evolutionary Testing 2488
   Phil McMinn, Mike Holcombe

Modeling the Search Landscape of Metaheuristic Software
Clustering Algorithms 2499
   Brian S. Mitchell, Spiros Mancoridis

Search Based Software Engineering – Posters

Search Based Transformations 2511
   Deji Fatiregun, Mark Harman, Robert Hierons

Finding Building Blocks for Software Clustering 2513
   Kiarash Mahdavi, Mark Harman, Robert Hierons

Author Index

Volume I

A-Life, Adaptive Behavior, Agents, and Ant Colony Optimization

Swarms in Dynamic Environments 1
  T.M. Blackwell

The Effect of Natural Selection on Phylogeny
Reconstruction Algorithms 13
  Dehua Hang, Charles Ofria, Thomas M. Schmidt, Eric Torng

AntClust: Ant Clustering and Web Usage Mining 25
  Nicolas Labroche, Nicolas Monmarché, Gilles Venturini

A Non-dominated Sorting Particle Swarm Optimizer for
Multiobjective Optimization.. 37
  Xiaodong Li

The Influence of Run-Time Limits on Choosing Ant
System Parameters.. 49
  Krzysztof Socha

Emergence of Collective Behavior in Evolving Populations of
Flying Agents 61
  Lee Spector, Jon Klein, Chris Perry, Mark Feinstein

On Role of Implicit Interaction and Explicit Communications in
Emergence of Social Behavior in Continuous Predators-Prey
Pursuit Problem 74
  Ivan Tanev, Katsunori Shimohara

Demonstrating the Evolution of Complex Genetic Representations:
An Evolution of Artificial Plants 86
  Marc Toussaint

Sexual Selection of Co-operation 98
  M. Afzal Upal

Optimization Using Particle Swarms with Near Neighbor Interactions ... 110
  Kalyan Veeramachaneni, Thanmaya Peram, Chilukuri Mohan,
  Lisa Ann Osadciw

Revisiting Elitism in Ant Colony Optimization 122
  Tony White, Simon Kaegi, Terri Oda

A New Approach to Improve Particle Swarm Optimization 134
  Liping Zhang, Huanjun Yu, Shangxu Hu

A-Life, Adaptive Behavior, Agents, and Ant Colony Optimization – Posters

Clustering and Dynamic Data Visualization with Artificial Flying Insect 140
  S. Aupetit, N. Monmarché, M. Slimane, C. Guinot, G. Venturini

Ant Colony Programming for Approximation Problems 142
  Mariusz Boryczka, Zbigniew J. Czech, Wojciech Wieczorek

Long-Term Competition for Light in Plant Simulation 144
  Claude Lattaud

Using Ants to Attack a Classical Cipher 146
  Matthew Russell, John A. Clark, Susan Stepney

Comparison of Genetic Algorithm and Particle Swarm Optimizer When Evolving a Recurrent Neural Network 148
  Matthew Settles, Brandon Rodebaugh, Terence Soule

Adaptation and Ruggedness in an Evolvability Landscape 150
  Terry Van Belle, David H. Ackley

Study Diploid System by a Hamiltonian Cycle Problem Algorithm 152
  Dong Xianghui, Dai Ruwei

A Possible Mechanism of Repressing Cheating Mutants in Myxobacteria 154
  Ying Xiao, Winfried Just

Tour Jeté, Pirouette: Dance Choreographing by Computers 156
  Tina Yu, Paul Johnson

Multiobjective Optimization Using Ideas from the Clonal Selection Principle 158
  Nareli Cruz Cortés, Carlos A. Coello Coello

Artificial Immune Systems

A Hybrid Immune Algorithm with Information Gain for the Graph Coloring Problem 171
  Vincenzo Cutello, Giuseppe Nicosia, Mario Pavone

MILA – Multilevel Immune Learning Algorithm 183
  Dipankar Dasgupta, Senhua Yu, Nivedita Sumi Majumdar

The Effect of Binary Matching Rules in Negative Selection 195
  Fabio González, Dipankar Dasgupta, Jonatan Gómez

Immune Inspired Somatic Contiguous Hypermutation for
Function Optimisation 207
    Johnny Kelsey, Jon Timmis

A Scalable Artificial Immune System Model for Dynamic
Unsupervised Learning 219
    Olfa Nasraoui, Fabio Gonzalez, Cesar Cardona, Carlos Rojas,
    Dipankar Dasgupta

Developing an Immunity to Spam 231
    Terri Oda, Tony White

Artificial Immune Systems – Posters

A Novel Immune Anomaly Detection Technique Based on
Negative Selection 243
    F. Niño, D. Gómez, R. Vejar

Visualization of Topic Distribution Based on Immune Network Model ... 246
    Yasufumi Takama

Spatial Formal Immune Network 248
    Alexander O. Tarakanov

Coevolution

Focusing versus Intransitivity (Geometrical Aspects of Co-evolution) 250
    Anthony Bucci, Jordan B. Pollack

Representation Development from Pareto-Coevolution 262
    Edwin D. de Jong

Learning the Ideal Evaluation Function 274
    Edwin D. de Jong, Jordan B. Pollack

A Game-Theoretic Memory Mechanism for Coevolution 286
    Sevan G. Ficici, Jordan B. Pollack

The Paradox of the Plankton: Oscillations and Chaos in
Multispecies Evolution 298
    Jeffrey Horn, James Cattron

Exploring the Explorative Advantage of the Cooperative
Coevolutionary (1+1) EA 310
    Thomas Jansen, R. Paul Wiegand

PalmPrints: A Novel Co-evolutionary Algorithm for Clustering
Finger Images 322
    Nawwaf Kharma, Ching Y. Suen, Pei F. Guo

Coevolution and Linear Genetic Programming for Visual Learning 332
  Krzysztof Krawiec, Bir Bhanu

Finite Population Models of Co-evolution and Their Application to
Haploidy versus Diploidy 344
  Anthony M.L. Liekens, Huub M.M. ten Eikelder, Peter A.J. Hilbers

Evolving Keepaway Soccer Players through Task Decomposition 356
  Shimon Whiteson, Nate Kohl, Risto Miikkulainen, Peter Stone

Coevolution – Posters

A New Method of Multilayer Perceptron Encoding 369
  Emmanuel Blindauer, Jerzy Korczak

An Incremental and Non-generational Coevolutionary Algorithm 371
  Ramón Alfonso Palacios-Durazo, Manuel Valenzuela-Rendón

Coevolutionary Convergence to Global Optima 373
  Lothar M. Schmitt

Generalized Extremal Optimization for Solving Complex Optimal
Design Problems 375
  Fabiano Luis de Sousa, Valeri Vlassov, Fernando Manuel Ramos

Coevolving Communication and Cooperation for Lattice
Formation Tasks 377
  *Jekanthan Thangavelautham, Timothy D. Barfoot,
  Gabriele M.T. D'Eleuterio*

DNA, Molecular, and Quantum Computing

Efficiency and Reliability of DNA-Based Memories 379
  Max H. Garzon, Andrew Neel, Hui Chen

Evolving Hogg's Quantum Algorithm Using Linear-Tree GP 390
  André Leier, Wolfgang Banzhaf

Hybrid Networks of Evolutionary Processors 401
  *Carlos Martín-Vide, Victor Mitrana, Mario J. Pérez-Jiménez,
  Fernando Sancho-Caparrini*

DNA-Like Genomes for Evolution *in silico* 413
  Michael West, Max H. Garzon, Derrel Blain

DNA, Molecular, and Quantum Computing – Posters

String Binding-Blocking Automata 425
  M. Sakthi Balan

On Setting the Parameters of QEA for Practical Applications: Some
Guidelines Based on Empirical Evidence 427
  Kuk-Hyun Han, Jong-Hwan Kim

Evolutionary Two-Dimensional DNA Sequence Alignment 429
  Edgar E. Vallejo, Fernando Ramos

Evolvable Hardware

Active Control of Thermoacoustic Instability in a Model Combustor
with Neuromorphic Evolvable Hardware 431
  John C. Gallagher, Saranyan Vigraham

Hardware Evolution of Analog Speed Controllers for a DC Motor 442
  David A. Gwaltney, Michael I. Ferguson

Evolvable Hardware – Posters

An Examination of Hypermutation and Random Immigrant Variants of
mrCGA for Dynamic Environments 454
  Gregory R. Kramer, John C. Gallagher

Inherent Fault Tolerance in Evolved Sorting Networks 456
  Rob Shepherd, James Foster

Evolutionary Robotics

Co-evolving Task-Dependent Visual Morphologies in Predator-Prey
Experiments... 458
  Gunnar Buason, Tom Ziemke

Integration of Genetic Programming and Reinforcement Learning for
Real Robots... 470
  Shotaro Kamio, Hideyuki Mitsuhashi, Hitoshi Iba

Multi-objectivity as a Tool for Constructing Hierarchical Complexity 483
  Jason Teo, Minh Ha Nguyen, Hussein A. Abbass

Learning Biped Locomotion from First Principles on a Simulated
Humanoid Robot Using Linear Genetic Programming.................. 495
  Krister Wolff, Peter Nordin

Evolutionary Robotics – Posters

An Evolutionary Approach to Automatic Construction of
the Structure in Hierarchical Reinforcement Learning 507
  Stefan Elfwing, Eiji Uchibe, Kenji Doya

Fractional Order Dynamical Phenomena in a GA 510
  E.J. Solteiro Pires, J.A. Tenreiro Machado, P.B. de Moura Oliveira

Evolution Strategies/Evolutionary Programming

Dimension-Independent Convergence Rate for
Non-isotropic $(1, \lambda) - ES$ 512
 Anne Auger, Claude Le Bris, Marc Schoenauer

The Steady State Behavior of $(\mu/\mu_I, \lambda)$-ES on Ellipsoidal Fitness
Models Disturbed by Noise 525
 Hans-Georg Beyer, Dirk V. Arnold

Theoretical Analysis of Simple Evolution Strategies in Quickly
Changing Environments 537
 Jürgen Branke, Wei Wang

Evolutionary Computing as a Tool for Grammar Development 549
 Guy De Pauw

Solving Distributed Asymmetric Constraint Satisfaction Problems
Using an Evolutionary Society of Hill-Climbers 561
 Gerry Dozier

Use of Multiobjective Optimization Concepts to Handle Constraints
in Single-Objective Optimization 573
 Arturo Hernández Aguirre, Salvador Botello Rionda,
 Carlos A. Coello Coello, Giovanni Lizárraga Lizárraga

Evolution Strategies with Exclusion-Based Selection Operators
and a Fourier Series Auxiliary Function 585
 Kwong-Sak Leung, Yong Liang

Ruin and Recreate Principle Based Approach for the Quadratic
Assignment Problem 598
 Alfonsas Misevicius

Model-Assisted Steady-State Evolution Strategies 610
 Holger Ulmer, Felix Streichert, Andreas Zell

On the Optimization of Monotone Polynomials by the (1+1) EA and
Randomized Local Search 622
 Ingo Wegener, Carsten Witt

Evolution Strategies/Evolutionary Programming – Posters

A Forest Representation for Evolutionary Algorithms Applied to
Network Design 634
 A.C.B. Delbem, Andre de Carvalho

Solving Three-Objective Optimization Problems Using Evolutionary
Dynamic Weighted Aggregation: Results and Analysis 636
  Yaochu Jin, Tatsuya Okabe, Bernhard Sendhoff

The Principle of Maximum Entropy-Based Two-Phase Optimization of
Fuzzy Controller by Evolutionary Programming....................... 638
  Chi-Ho Lee, Ming Yuchi, Hyun Myung, Jong-Hwan Kim

A Simple Evolution Strategy to Solve Constrained
Optimization Problems 640
  Efrén Mezura-Montes, Carlos A. Coello Coello

Effective Search of the Energy Landscape for Protein Folding 642
  Eugene Santos Jr., Keum Joo Kim, Eunice E. Santos

A Clustering Based Niching Method for Evolutionary Algorithms 644
  Felix Streichert, Gunnar Stein, Holger Ulmer, Andreas Zell

Evolutionary Scheduling Routing

A Hybrid Genetic Algorithm for the Capacitated Vehicle
Routing Problem 646
  Jean Berger, Mohamed Barkaoui

An Evolutionary Approach to Capacitated Resource Distribution by
a Multiple-agent Team.. 657
  *Mudassar Hussain, Bahram Kimiaghalam, Abdollah Homaifar,
  Albert Esterline, Bijan Sayyarodsari*

A Hybrid Genetic Algorithm Based on Complete Graph
Representation for the Sequential Ordering Problem 669
  Dong-Il Seo, Byung-Ro Moon

An Optimization Solution for Packet Scheduling: A Pipeline-Based
Genetic Algorithm Accelerator...................................... 681
  Shiann-Tsong Sheu, Yue-Ru Chuang, Yu-Hung Chen, Eugene Lai

Evolutionary Scheduling Routing – Posters

Generation and Optimization of Train Timetables Using Coevolution 693
  Paavan Mistry, Raymond S.K. Kwan

Genetic Algorithms

Chromosome Reuse in Genetic Algorithms 695
  Adnan Acan, Yüce Tekol

Real-Parameter Genetic Algorithms for Finding Multiple Optimal
Solutions in Multi-modal Optimization 706
  Pedro J. Ballester, Jonathan N. Carter

An Adaptive Penalty Scheme for Steady-State Genetic Algorithms 718
   Helio J.C. Barbosa, Afonso C.C. Lemonge

Asynchronous Genetic Algorithms for Heterogeneous Networks
Using Coarse-Grained Dataflow 730
   John W. Baugh Jr., Sujay V. Kumar

A Generalized Feedforward Neural Network Architecture and Its
Training Using Two Stochastic Search Methods 742
   Abdesselam Bouzerdoum, Rainer Mueller

Ant-Based Crossover for Permutation Problems 754
   Jürgen Branke, Christiane Barz, Ivesa Behrens

Selection in the Presence of Noise 766
   Jürgen Branke, Christian Schmidt

Effective Use of Directional Information in Multi-objective
Evolutionary Computation 778
   Martin Brown, R.E. Smith

Pruning Neural Networks with Distribution Estimation Algorithms.... 790
   Erick Cantú-Paz

Are Multiple Runs of Genetic Algorithms Better than One? 801
   Erick Cantú-Paz, David E. Goldberg

Constrained Multi-objective Optimization Using Steady State
Genetic Algorithms 813
   Deepti Chafekar, Jiang Xuan, Khaled Rasheed

An Analysis of a Reordering Operator with Tournament Selection on
a GA-Hard Problem 825
   Ying-Ping Chen, David E. Goldberg

Tightness Time for the Linkage Learning Genetic Algorithm.......... 837
   Ying-Ping Chen, David E. Goldberg

A Hybrid Genetic Algorithm for the Hexagonal Tortoise Problem 850
   Heemahn Choe, Sung-Soon Choi, Byung-Ro Moon

Normalization in Genetic Algorithms 862
   Sung-Soon Choi and Byung-Ro Moon

Coarse-Graining in Genetic Algorithms: Some Issues and Examples.... 874
   *Andrés Aguilar Contreras, Jonathan E. Rowe,
   Christopher R. Stephens*

Building a GA from Design Principles for Learning Bayesian Networks... 886
   Steven van Dijk, Dirk Thierens, Linda C. van der Gaag

A Method for Handling Numerical Attributes in GA-Based Inductive
Concept Learners 898
  Federico Divina, Maarten Keijzer, Elena Marchiori

Analysis of the (1+1) EA for a Dynamically Bitwise
Changing ONEMAX 909
  Stefan Droste

Performance Evaluation and Population Reduction for a Self
Adaptive Hybrid Genetic Algorithm (SAHGA) 922
  Felipe P. Espinoza, Barbara S. Minsker, David E. Goldberg

Schema Analysis of Average Fitness in Multiplicative Landscape 934
  Hiroshi Furutani

On the Treewidth of NK Landscapes 948
  Yong Gao, Joseph Culberson

Selection Intensity in Asynchronous Cellular Evolutionary Algorithms ... 955
  Mario Giacobini, Enrique Alba, Marco Tomassini

A Case for Codons in Evolutionary Algorithms 967
  Joshua Gilbert, Maggie Eppstein

Natural Coding: A More Efficient Representation for
Evolutionary Learning 979
  Raúl Giráldez, Jesús S. Aguilar-Ruiz, José C. Riquelme

Hybridization of Estimation of Distribution Algorithms with a
Repair Method for Solving Constraint Satisfaction Problems 991
  Hisashi Handa

Efficient Linkage Discovery by Limited Probing 1003
  Robert B. Heckendorn, Alden H. Wright

Distributed Probabilistic Model-Building Genetic Algorithm 1015
  Tomoyuki Hiroyasu, Mitsunori Miki, Masaki Sano, Hisashi Shimosaka,
  Shigeyoshi Tsutsui, Jack Dongarra

HEMO: A Sustainable Multi-objective Evolutionary
Optimization Framework 1029
  Jianjun Hu, Kisung Seo, Zhun Fan, Ronald C. Rosenberg,
  Erik D. Goodman

Using an Immune System Model to Explore Mate Selection in
Genetic Algorithms 1041
  Chien-Feng Huang

Designing A Hybrid Genetic Algorithm for the Linear
Ordering Problem 1053
   Gaofeng Huang, Andrew Lim

A Similarity-Based Mating Scheme for Evolutionary
Multiobjective Optimization 1065
   Hisao Ishibuchi, Youhei Shibata

Evolutionary Multiobjective Optimization for Generating an
Ensemble of Fuzzy Rule-Based Classifiers 1077
   Hisao Ishibuchi, Takashi Yamamoto

Voronoi Diagrams Based Function Identification 1089
   Carlos Kavka, Marc Schoenauer

New Usage of SOM for Genetic Algorithms 1101
   Jung-Hwan Kim, Byung-Ro Moon

Problem-Independent Schema Synthesis for Genetic Algorithms 1112
   Yong-Hyuk Kim, Yung-Keun Kwon, Byung-Ro Moon

Investigation of the Fitness Landscapes and Multi-parent
Crossover for Graph Bipartitioning 1123
   Yong-Hyuk Kim, Byung-Ro Moon

New Usage of Sammon's Mapping for Genetic Visualization 1136
   Yong-Hyuk Kim, Byung-Ro Moon

Exploring a Two-Population Genetic Algorithm 1148
   Steven Orla Kimbrough, Ming Lu, David Harlan Wood, D.J. Wu

Adaptive Elitist-Population Based Genetic Algorithm for
Multimodal Function Optimization 1160
   Kwong-Sak Leung, Yong Liang

Wise Breeding GA via Machine Learning Techniques for
Function Optimization 1172
   Xavier Llorà, David E. Goldberg

Facts and Fallacies in Using Genetic Algorithms for Learning
Clauses in First-Order Logic 1184
   Flaviu Adrian Mărginean

Comparing Evolutionary Computation Techniques via
Their Representation 1196
   Boris Mitavskiy

Dispersion-Based Population Initialization 1210
   Ronald W. Morrison

A Parallel Genetic Algorithm Based on Linkage Identification 1222
   Masaharu Munetomo, Naoya Murao, Kiyoshi Akama

Generalization of Dominance Relation-Based Replacement Rules for
Memetic EMO Algorithms 1234
   Tadahiko Murata, Shiori Kaige, Hisao Ishibuchi

Author Index

Connection Network and Optimization of Interest Metric for One-to-One Marketing

Sung-Soon Choi and Byung-Ro Moon

School of Computer Science and Engineering
Seoul National University
Seoul, 151-742 Korea
{sschoi,moon}@soar.snu.ac.kr

Abstract. With the explosive growth of data in electronic commerce, rule finding becomes a crucial part in marketing. In this paper, we discuss the essential limitations of the existing metrics to quantify the interests of rules, and present the need of optimizing the interest metric. We describe the construction of the connection network that represents the relationships between items and propose a natural marketing model using the network. Although simple interest metrics were used, the connection network model showed stable performance in the experiment with field data. By constructing the network based on the optimized interest metric, the performance of the model was significantly improved.

1 Introduction

The progress in modern technologies made it possible for finance and retail organizations to collect and store a massive amount of data. In consequence, it has attracted great attention to identify systems that explain the data, particularly in the data mining area. An early representative problem is the "market-basket" problem [1]. In the problem, we are given a set of items and a collection of transactions each of which is a subset (basket) of items purchased together by a customer in a visit. The objective is "mining" relationships between items from the baskets data.

First of all, the attraction of the problem arises in a great variety of applications [2]. A typical example (from which the problem got its name) is the customers' shopping behavior in a supermarket. In the example, the items are products and the transactions are customer purchases at the checkout. Determining what products customers are likely to buy together is useful for display and marketing. There are many other applications with different data characteristics. Some examples are student enrollment in classes [2], copy detection (identifying identical or similar documents or web pages) [3] [4], clustering (identifying similar vectors in high-dimensional spaces) [5] [6], etc.

With the rapid growth of the electronic commerce (e-commerce) field, the problem becomes increasingly important. Various solutions to the problem were used for marketing in the e-commerce field [7] [8]. Among them, collaborative filtering (tracking user behavior and providing recommendations to individuals based on the similarities of their preferences) has been playing as one of the core marketing strategies [9] [10] [11].

In this paper, we suggest a genetic approach to optimize the interest metric that measures the relationships between items. The optimized metric is used to construct a weighted relationship network of the items. Then, we propose a natural marketing model using the network. Experimental results showed that our model was more stable and better than collaborative filtering.

The rest of this paper is organized as follows. In Sect. 2, we describe interest metrics introduced so far that measure the degree of connection between items. We also give a brief description of methods for one-to-one marketing. In Sect. 3, we present our theoretical view of the interest metrics described in Sect. 2 and mention the need of optimizing the interest metric. In Sect. 4, we explain the connection network that represents the relationships among the items, and describe the marketing model based on the network. The genetic framework for optimizing the interest metric is provided in Sect. 5. Section 6 provides the experimental results on real-world data sets. Finally, we make our conclusions in Sect. 7.

2 Preliminaries

2.1 Interest Metrics of Rules

Recently, there were many studies to find the meaningful rules between items based on the interest metrics that measure the degree of connection between items. A rule is denoted by $X \rightarrow Y$, for two item sets X and Y, which means that the occurrence of X implies the occurrence of Y.

Agrawal et al. [1] proposed the interest metrics called *support* and *confidence* and used them to build rules between items. In particular, they called such rules *association rules*. The support of an item set X is defined to be the number of transactions in which the item set occurs (the number of occurrences of X). For a rule $X \rightarrow Y$, the confidence of this rule is the fraction of transactions containing Y among those containing X. In order that a rule $X \rightarrow Y$ becomes an association rule, the support of the item set $X \cup Y$ and the confidence of the rule $X \rightarrow Y$ must exceed given thresholds θ_s and θ_c, respectively.

Note that the support of the item set $X \cup Y$ corresponds to the number of transactions in which the item sets X and Y occur together (the number of co-occurrences of X and Y). And the confidence of the rule $X \rightarrow Y$ corresponds to the number of co-occurrences of X and Y over the number of occurrences of X. We denote the number of occurrences of an item set X by $n(X)$ and the number of co-occurrences of item sets X and Y by $n(X,Y)$. Then, from the definitions of support and confidence, we have

$$sup(X \rightarrow Y) = n(X,Y) \text{ and } conf(X \rightarrow Y) = \frac{n(X,Y)}{n(X)} . \quad (1)$$

The association rule $X \to Y$ is said to *hold* if and only if

$$sup(X \to Y) = n(X,Y) > \theta_s \text{ and } conf(X \to Y) = \frac{n(X,Y)}{n(X)} > \theta_c. \quad (2)$$

Cohen et al. [12] pointed out a problem of association rules which require high support: Most rules with high supports are obvious and well-known and it is the rules of low supports that provide interesting new insights. Besides, several recent papers mentioned that it is not reasonable to use confidence as the interest measure of rules [2] [13] [14]. In this context, a number of different metrics to quantify "interestingness" or "goodness" of rules were proposed [15]. Among them are *gain* [16], proposed by Piatetsky-Shapiro [17], *variance* and *chi-squared value* [18], *entropy gain* [18] [19], *gini* [19], *laplace* [20] [21], *lift* [22] (also known as *interest* [2] or *strength* [23]), *conviction* [2], and *similarity* [12].

Except *similarity*, Bayardo and Agrawal [15] expressed the definitions of all these metrics with the supports of the related item sets and the confidence of the rule. It is also possible to express *similarity* in the same way. This indicates that all these metrics can be described with the numbers of occurrences and co-occurrences of the related item sets. If we denote by T the set of all transactions, the *laplace*, *gain*, Piatetsky-Shapiro's metric (*p-s*), *conviction*, *lift*, and *similarity* values for a rule $X \to Y$ are expressed as follows:[1]

$$laplace(X \to Y) = \frac{n(X,Y)+1}{n(X)+k},$$

$$gain(X \to Y) = n(X,Y) - c \cdot n(X),$$

$$p\text{-}s(X \to Y) = n(X,Y) - \frac{n(X) \cdot n(Y)}{|T|},$$

$$conviction(X \to Y) = \frac{|T| \cdot n(X) - n(X) \cdot n(Y)}{|T| \cdot (n(X) - n(X,Y))},$$

$$lift(X \to Y) = \frac{|T| \cdot n(X,Y)}{n(X) \cdot n(Y)}, \text{ and}$$

$$similarity(X \to Y) = \frac{n(X,Y)}{n(X)+n(Y)-n(X,Y)}. \quad (3)$$

The rest of the above metrics are also able to be expressed in the same way.

It is difficult to come up with a single metric among the above metrics. It is, however, clear that all the metrics consider only the numbers of occurrences and co-occurrences of item sets. This fact is utilized in modeling the interest metric and optimizing it. And note that these metrics numerically evaluate the relationships between item sets. So, these metrics can be used in quantifying the strengths of the connections between item sets.

---

[1] k in *laplace* is an integer greater than 1 and c in *gain* is a fractional constant between 0 and 1.

2.2 One-to-One Marketing

Personalization is a sharply growing issue in modern marketing. It helps boost customers' loyalty by providing the most attractive contents to each customer or by locating the most proper set of customers for an arbitrary advertisement [24]. As mentioned before, personal information and huge activity logs are accumulated due to the progress of modern technologies. These data implicitly contain valuable trends and patterns which are useful to improve business decisions and efficiency.

Diverse data mining tools to discover implicit knowledge hidden in a large database have been studied with various tools including neural networks, decision trees, rule induction, Bayesian belief networks, evolutionary algorithms, fuzzy sets, clustering, association rules, and collaborative filtering [25] [26]. Among them, collaborative filtering, which tracks user behavior and makes recommendations to individuals on the basis of the similarities of their preferences, is known to be a standard for personalized recommendations [10] [27]. Most of the personalized marketing tools including collaborative filtering utilize customer profiles [7] [8] [28] [29] [30].

These strategies often have the data-sparsity problem, which greatly undermines the quality of recommendations, as it is in general difficult to collect customers' personal information or preferences [31] [32].

3 Rule Space and Optimized Interest Metric

We saw that $sup(X \rightarrow Y) = n(X,Y)$ and $conf(X \rightarrow Y) = \frac{n(X,Y)}{n(X)}$ for a rule $X \rightarrow Y$ in the previous section. Now if we replace $n(X)$, $n(Y)$, and $n(X,Y)$ with independent variables x, y, and z, respectively, the necessary and sufficient condition for an association rule $X \rightarrow Y$ to hold are as follows:

$$z > \theta_s \quad \text{and} \quad \frac{z}{x} > \theta_c . \quad (x > 0, y \geq 0, z \geq 0) \quad (4)$$

This means that the values of x, y, and z for the rule $X \rightarrow Y$ to hold correspond to the points in three dimensions that satisfy the above inequality (4). At this time, the rule space consisting of x, y, and z axes is partitioned by the plane corresponding to the support condition, $f_s(x,y,z) = z = \theta_s$, and the plane corresponding to the confidence condition, $f_c(x,y,z) = \frac{z}{x} = \theta_c$. (e.g., Figure 1(a))

It was mentioned that, for a rule $X \rightarrow Y$, the metrics introduced in the previous section are able to be described with $n(X)$, $n(Y)$, and $n(X,Y)$. Therefore, it is also possible to express all the metrics of the previous section as the formulas of x, y, and z. For example, $conviction(X \rightarrow Y)$ for a rule $X \rightarrow Y$ can be described as follows:

$$conviction(X \rightarrow Y) = \frac{|T| \cdot n(X) - n(X) \cdot n(Y)}{|T|(n(X) - n(X,Y))} = \frac{|T|x - xy}{|T|(x - z)} . \quad (5)$$

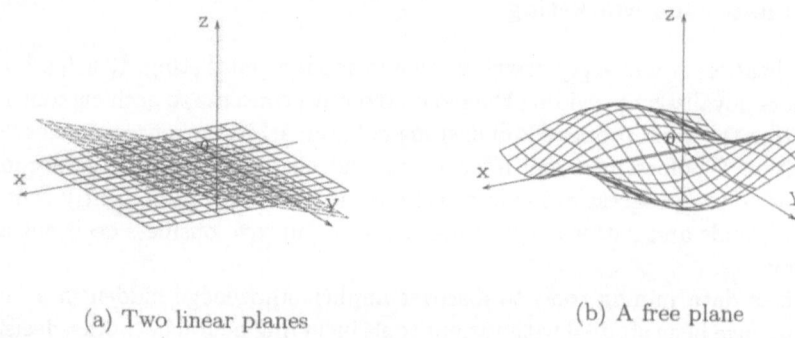

(a) Two linear planes   (b) A free plane

Fig. 1. A rule space partitioned by planes

After all, given a threshold θ, each of the metrics mentioned in the previous section is a plane, $f(x, y, z) = \theta$, which partitions the rule space. The selection of interest metrics allows the space partition of a fixed shape regardless of the characteristics of the data set. We suspect that the optimal shapes of the partition planes are different depending on the data sets. In this context, we guarantee the degree of freedom for the partition plane to the utmost, then find the optimized plane for the given data set, namely the optimized interest metric. (Figure 1(b))

To maximize the degree of freedom for the plane, it is desirable to assume that the plane $f(x, y, z) = \theta$ is a free plane. In this case, however, the search space becomes so huge that the learning time increases excessively. We restrict the shape of the plane to help perform the learning. We set a model of $f(x, y, z)$ as follows:

$$f(x, y, z) = (a_x x^{e_x} + b_x)(a_y y^{e_y} + b_y)(a_z z^{e_z} + b_z)$$

where

$$\begin{cases} 0 < a_x, a_y, a_z \leq 10 \\ -1 \leq e_x, e_y, e_z \leq 1 \\ 0 \leq b_x, b_y, b_z \leq 10 \,. \end{cases} \quad (6)$$

We use a genetic algorithm to search the optimal coefficients and exponents of $f(x, y, z)$ for the data set. The details of optimization are described in Sect. 5.

4 Personalized Marketing Model Using Connection Networks

In the previous section, we made a model of the metric $f(X \to Y)$ $(= f(x, y, z))$ that evaluates the strength of a rule $X \to Y$. Intuitively the value of $f(X \to Y)$ indicates the strength of the connection between the item sets X and Y. The optimized metric is used to measure the strength of connection between item sets. Here, we only consider the case that each item set has just one item. Then,

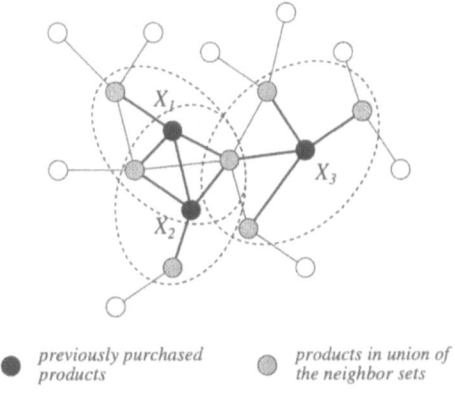

Fig. 2. A connection network

the value of $f(X \rightarrow Y)$ represents the strength for the item X to imply the item Y.

Now we are ready to construct a connection network. We set a vertex for each item. We put an arc from the vertex X to the vertex Y with the weight $f(X \rightarrow Y)$. We have a directed graph $G = (V, A)$, where V is the vertex set and A is the arc set.

We perform one-to-one marketing using the connection network. Suppose that a customer purchased the products $X_1, X_2, \ldots, X_k$ so far. Let $N(X_i)$ be the set of neighbor vertices of X_i in the connection network ($1 \leq i \leq k$). We define a score function for recommendation, $s : V \longrightarrow \mathbb{R}$, as follows ($\mathbb{R}$: the set of real numbers):

$$s(Y) = \begin{cases} \{\sum_{1 \leq i \leq k} f(X_i \rightarrow Y)\}/(a_k k^{e_k} + b_k) \text{ , if } Y \in \bigcup_{1 \leq i \leq k} N(X_i) \\ 0 \qquad\qquad\qquad\qquad\qquad\qquad\text{, otherwise .} \end{cases} \quad (7)$$

We recommend the products of high scores to the customer. The value of the score function $s(Y)$ for a product Y is proportional to the weight sum of the arcs from the previously purchased products to the product Y. Figure 2 shows an example connection network. We divide the sum of weights by a function of k (the number of previously purchased products). This prevents the recommendations from flowing in upon a few customers that purchased excessively many products before. a_k, e_k, and b_k are optimized in the following ranges,

$$0 < a_k \leq 10, \quad -1 \leq e_k \leq 1, \quad 0 \leq b_k \leq 10, \quad (8)$$

and the details are described in the next section.

Such a recommendation strategy is different from the existing ones based on the customers' profiles in that it performs recommendations just by the relationships between products regardless of the customers' profiles. We need not handle the customer profile vectors of high dimensions nor quantify each field

```
create initial population of a fixed size;
do {
    choose parent1 and parent2 from population;
    offspring = crossover(parent1, parent2);
    mutation(offspring);
    replace(population, offspring);
} until (stopping condition);
return the best individual;
```

Fig. 3. The outline of the genetic algorithm

of the customer profiles. We merely optimize the quantitative representation of topology among items and perform recommendation on the basis of this. The computational cost for the recommendation is significantly low compared with existing recommendation strategies. In addition, when reflecting new data into the established network, the cost of updating the network is also fairly low. Furthermore, such a strategy is less sensitive to the data sparsity problem that greatly undermines the quality of recommendations for the existing ones.

5 Genetic Algorithm

The selection of an interest metric and a score function has a great effect on the quality of recommendations. The problem is to find the best set of coefficients related to the interest metric and score function that maximizes the *response rate*, which is defined to be

$$response\ rate = \frac{\#\ of\ purchases}{\#\ of\ recommended\ items}. \tag{9}$$

We used a steady-state genetic algorithm to optimize the nine parameters related to the interest metric (in Sect. 3) and the three parameters related to the score function (in Sect. 4). Figure 3 shows the outline of the genetic algorithm. The details are described in the following.

- **Encoding:** Each solution is a set of 12 real values. A solution is represented by a chromosome; a chromosome is a real array of 12 elements. Figure 4 shows the structure of chromosomes. Each element of the array is called a gene and we restrict the range of each gene as mentioned before.
- **Initialization:** We set the population size to be 100. For each gene in a chromosome, we randomly generate a real number in the restricted range.
- **Parent Selection:** The fitness value F_i of chromosome i is assigned as follows:

$$F_i = (R_i - R_w) + (R_b - R_w)/4 \tag{10}$$

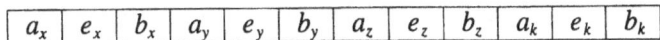

Fig. 4. The structure of chromosomes

where

R_w : the response rate of the worst,
R_b : the response rate of the best, and
R_i : the response rate of chromosome i.

Each chromosome is selected as a parent with a probability proportional to its fitness value. This is a typical proportional selection scheme.
- **Crossover:** We use traditional one-point crossover.
- **Mutation:** We randomly select each gene with a low probability (P=0.1) and perform the non-uniform mutation in which the perturbation degree decreases over time. Non-uniform mutation was first proposed by Michalewicz [33].
- **Replacement:** We replace the inferior of the two parents if the offspring is not worse than both parents. Otherwise, we replace the worst member of the population. This scheme is a compromise between preselection [34] and GENITOR-style replacement [35].

6 Experimental Results

We conducted experiments with two different types of data sets to evaluate the performance of the optimized interest metric and the marketing model using the connection network.

First, we conducted experiments with a massive amount of purchase data set from June 1996 to August 2000 of a representative e-commerce company in Korea. In this data set, the items are products and a transaction is a customer's purchase of one or more items with a time stamp. We first divided the whole data set into two disjoint sets in terms of dates. Then we predicted the purchases of customers in the latter set, based on the data of the former set.

In this experiment, a number of different recommendation models were used: collaborative filtering (CF), a few plain connection networks (PCNs), and optimized connection network (OCN). As mentioned before, collaborative filtering is a proven standard for personalized recommendations [9] [10] [27]. PCNs are our early recommendation models (commercialized by Optus Inc.) in which we construct the networks based on the interest metrics mentioned in Sect. 2 (such as *laplace*, *gain*, and so on) and then recommend attractive products based on the prescribed thresholds. OCN is the recommendation model in which we construct the network on the basis of the optimized interest metric and then recommend the products according to the score-function values.

Table 1. Comparison of experiments with purchase data set

Dates	CF	PCN-*laplace*	PCN-*gain*	PCN-*p-s*	OCN
May 31	.0052 ($\frac{26}{5018}$)	.0032 ($\frac{14}{4310}$)	.0046 ($\frac{23}{5003}$)	.0030 ($\frac{15}{5011}$)	.0052 ($\frac{26}{4994}$)
June 7	.0061 ($\frac{32}{5210}$)	.0034 ($\frac{15}{4411}$)	.0046 ($\frac{24}{5183}$)	.0043 ($\frac{23}{5390}$)	.0087 ($\frac{45}{5162}$)
June 14	.0054 ($\frac{30}{5513}$)	.0028 ($\frac{13}{4565}$)	.0046 ($\frac{25}{5415}$)	.0045 ($\frac{25}{5602}$)	.0092 ($\frac{52}{5634}$)
June 21	.0063 ($\frac{36}{5695}$)	.0032 ($\frac{18}{5742}$)	.0053 ($\frac{29}{5448}$)	.0041 ($\frac{23}{5627}$)	.0099 ($\frac{58}{5859}$)
June 28	.0062 ($\frac{37}{5947}$)	.0034 ($\frac{17}{4937}$)	.0051 ($\frac{31}{6094}$)	.0043 ($\frac{25}{5767}$)	.0088 ($\frac{55}{6229}$)
July 5	.0055 ($\frac{34}{6138}$)	.0027 ($\frac{15}{5459}$)	.0046 ($\frac{28}{6135}$)	.0044 ($\frac{28}{6343}$)	.0090 ($\frac{55}{6117}$)
July 12	.0041 ($\frac{26}{6327}$)	.0043 ($\frac{27}{6289}$)	.0068 ($\frac{41}{6036}$)	.0058 ($\frac{38}{6582}$)	.0084 ($\frac{52}{6192}$)
July 19	.0043 ($\frac{28}{6571}$)	.0063 ($\frac{40}{6314}$)	.0089 ($\frac{56}{6324}$)	.0084 ($\frac{55}{6511}$)	.0084 ($\frac{54}{6458}$)
July 26	.0038 ($\frac{26}{6879}$)	.0069 ($\frac{56}{8111}$)	.0088 ($\frac{59}{6695}$)	.0085 ($\frac{59}{6982}$)	.0112 ($\frac{79}{7036}$)
August 2	.0032 ($\frac{23}{7130}$)	.0069 ($\frac{49}{7067}$)	.0078 ($\frac{58}{7427}$)	.0078 ($\frac{56}{7190}$)	.0079 ($\frac{58}{7372}$)
August 9	.0027 ($\frac{20}{7386}$)	.0045 ($\frac{29}{6385}$)	.0053 ($\frac{39}{7315}$)	.0053 ($\frac{39}{7330}$)	.0061 ($\frac{43}{7047}$)

Table 2. Comparison of experiments with mobile-content data set

CF	PCN-*laplace*	PCN-*gain*	PCN-*similarity*	OCN
.0241 ($\frac{4886}{203054}$)	.0256 ($\frac{5215}{203379}$)	.0269 ($\frac{5522}{205061}$)	.0247 ($\frac{5096}{206474}$)	.0291 ($\frac{5535}{190417}$)

For experiments, we selected eleven pivot dates over the weeks from May 2000 to August 2000. For all the models, the data before the pivot date were used as the training set; the data after the date were used as the test set. In the case of OCN, the training set was further divided into the real training set and the validation set, to optimize the interest metric and the score function.

Table 1 shows the response rates (and the numbers of purchases over the numbers of recommended products) of CF, PCNs, and OCN, respectively. We omitted the experimental results of PCNs based on *conviction*, *lift*, and *similarity* since their performances were not so comparable. Here, the numbers of recommendations were adjusted to be comparable for fair comparison. PCNs showed comparable performance with CF although the networks were constructed based on the general metrics. We consider this to be an evidence of the suitability of the connection network as a recommendation model. The optimized model, OCN, significantly improved all the PCN models. OCN showed on average 76 % and 40 % better results than CF and PCN-*gain* (the best among PCNs), respectively.

Next, we conducted similar experiments with a massive amount of Internet contents access data from August 2001 to January 2002 of a representative contents service company in Korea. In this data set, the items are contents provided by the company and a transaction is composed of the contents that a customer used in a certain period of time. We divided the whole data set into two disjoint sets in terms of dates in the ratio of three to one.

Table 2 shows the response rates of CF, PCNs, and OCN, respectively. The response rate is the number of uses (hits) over the number of recommended contents as in the parentheses. In the table, the experimental results of PCNs based on *p-s*, *conviction*, and *lift* were omitted since their performances were not so comparable. Similarly to the experiments with the purchase data set, PCNs showed comparable performance with CF, and OCN performed significantly better than the other models.

7 Conclusion

We discussed the essential limitations of existing rule-interest metrics and raised the need of optimizing the interest metric. We proposed a novel marketing model using connection networks. We maximized the performance of the proposed model by constructing the network with the optimized interest metric.

Although the suggested method performed impressively, we consider that there remains room for further improvement. More elaborate modeling of the interest metric and the score function is a candidate. Other function models such as neural network and relevant optimization are also worth trying.

The optimized interest metric and the connection network model is not restricted to the personalized marketing only. We believe that they are applicable to various other problems; so far, we found that they are applicable to a few practical problems such as e-mail auto-response system and personalized search engine.

Acknowledgement. The authors thank Yung-Keun Kwon for insightful discussions about data mining techniques. This work was partly supported by Optus Inc. and Brain Korea 21 Project. The RIACT at Seoul National University provided research facilities for this study.

References

1. R. Agrawal, T. Imilienski, and A. Swami. Mining association rules between sets of items in large databases. In *Proc. of the ACM SIGMOD Int'l Conf. on Management of Data*, pages 207–216, 1993.
2. S. Brin, R. Motwani, J.D. Ullman, and S. Tsur. Dynamic itemset counting and implication rules for market basket data. In *Proc. of the ACM SIGMOD Int'l Conf. on Management of Data*, pages 255–264, 1997.
3. N. Shivakumar and H. Garcia-Molina. Building a scalable and accurate copy detection mechanism. In *Proc. Third Int'l Conf. Theory and Practice of Digital Libraries*, 1996.
4. A. Broder. On the resemblance and containment of documents. In *Proc. Compression and Complexity of Sequences Conf. (SEQUENCES'97)*, pages 21–29, 1998.
5. R. O. Duda and P.E. Hart. *Pattern Classification and Scene Analysis*. Wiley InterScience, New York, 1973.
6. S. Guha, R. Rastogi, and K. Shim. CURE – an efficient clustering algorithm for large databases. In *Proc. of the ACM SIGMOD Int'l Conf. on Management of Data*, pages 73–84, 1998.

7. D. Greening. *Building Consumer Trust with Accurate Product Recommendations.* LikeMinds White Paper LMWSWP-210-6966, 1997.
8. C.C. Aggarwal, J.L. Wolf, K. Wu, and P.S. Yu. Horting hatches an egg: A new graph-theoretic approach to collaborative filtering. In *Proc. of the fifth ACM SIGKDD Int'l Conf. on Knowledge Discovery and Data Mining*, pages 201–212, 1999.
9. D. Goldberg, D. Nichols, B.M. Oki, and D. Terry. Using collaborative filtering to weave an information tapestry. *Comm. of the ACM*, 35(12):61–70, 1992.
10. J.A. Konstan et al. Grouplens: Applying collaborative filtering to usenet news. *Comm. of the ACM*, 40:77–87, 1997.
11. H.R. Varian and P. Resnick. CACM special issue on recommender systems. *Comm. of the ACM*, 40, 1997.
12. E. Cohen et al. Finding interesting associations without support pruning. *IEEE Trans. on Knowledge and Data Engineering*, 13(1):64–78, 2001.
13. C. Silverstein, S. Brin, and R. Motwani. Beyond market baskets: Generalizing association rules to dependence rules. *Data Mining and Knowledge Discovery*, 2:69–96, 1998.
14. C. Silverstein, S. Brin, R. Motwani, and J.D. Ullman. Scalable techniques for mining casual structures. In *Proc. 24th Int'l Conf. Very Large Data Bases*, pages 594–605, 1998.
15. R.J. Bayardo Jr. and R. Agrawal. Mining the most interesting rules. In *Proc. of the fifth ACM SIGKDD Int'l Conf. on Knowledge Discovery and Data Mining*, pages 145–154, 1999.
16. T. Fukuda, Y. Morimoto, S. Morishita, and T. Tokuyama. Data mining using two-dimensional optimized association rules: Scheme, algorithms, and visualization. In *Proc. of the ACM SIGMOD Int'l Conf. on Management of Data*, pages 13–23, 1996.
17. G. Piatetsky-Shapiro. *Discovery, Analysis, and Presentation of Strong Rules.* AAAI/MIT Press, 1991. Chapter 13 of Knowledge Discovery in Databases.
18. S. Morishita. On classification and regression. In *Proc. First Int'l Conf. on Discovery Science – Lecture Notes in Artificial Intelligence 1532*, pages 40–57, 1998.
19. Y. Morimoto, T. Fukuda, H. Matsuzawa, T. Tokuyama, and K. Yoda. Algorithms for mining association rules for binary segmentations of huge categorical databases. In *Proc. of the 24th Very Large Data Bases Conf.*, pages 380–391, 1998.
20. P. Clark and P. Boswell. Rule induction with CN2: Some recent improvements. In *Machine Learning: Proc. of the Fifth European Conference*, pages 151–163, 1991.
21. G.I. Webb. OPUS: An efficient admissible algorithm for unordered search. *Journal of Artificial Intelligence Research*, 3:431–465, 1995.
22. International Business Machines. *IBM Intelligent Miner User's Guide*, 1996. Version 1, Release 1.
23. V. Dhar and A. Tuzhilin. Abstract-driven pattern discovery in databases. *IEEE Trans. on Knowledge and Data Engineering*, 5(6), 1993.
24. R. Dewan, B. Jing, and A. Seidmann. One-to-one marketing on the internet. In *Proc. of the 20th Int'l Conf. on Information Systems*, pages 93–102, 1999.
25. M.S. Chen, P.S. Han, and J.Yu. Data mining: An overview from a database perspective. *IEEE Trans. on Knowledge and Data Engineering*, 8(6):866–883, 1996.
26. M. Goebel and L. Gruenwald. A survey of data mining and knowledge discovery software tools. *SIGKDD Explorations*, 1:20–33, 1999.

27. J.L. Herlocker, J.A. Konstan, A. Borchers, and J. Riedl. An algorithmic framework for performing collaborative filtering. In *Proc. of the 22nd Annual Int'l ACM SIGIR Conf. on Research and Development in Information Retrieval*, pages 230–237, 1999.
28. V. Venugopal and W. Baets. Neural networks and their applications in marketing management. *Journal of Systems Management*, 45(9):16–21, 1994.
29. S. Park. Neural networks and customer grouping in e-commerce: A framework using fuzzy ART. In *Proc. Academia/Industry Working Conf. on Research Challenges '00*, pages 331–336, 2000.
30. Y.K. Kwon and B.R. Moon. Personalized email marketing with a genetic programming circuit model. In *Genetic and Evolutionary Computation Conference*, pages 1352–1358, 2001.
31. M.J. Culnan. "How did they get my name ?": An exploratory investigation of customer attitudes toward secondary information use. *MIS Quarterly*, 17(3):341–363, 1993.
32. J. III Hagel and J.F. Rayport. The coming battle for customer information. *Havard Business Review*, 75(1):53–65, 1997.
33. Z. Michalewicz. *Genetic Algorithms + Data Structures = Evolutionary Programs*. Springer, 1992.
34. D. Cavicchio. *Adaptive Search Using Simulated Evolution*. PhD thesis, University of Michigan, Ann Arbor, MI, 1970. Unpublished.
35. D. Whitley and J. Kauth. Genitor: A different genetic algorithm. In *Rocky Mountain Conf. on Artificial Intelligence*, pages 118–130, 1988.

Parameter Optimization by a Genetic Algorithm for a Pitch Tracking System

Yoon-Seok Choi and Byung-Ro Moon

School of Computer Science & Engineering, Seoul National University
Sillim-dong, Gwanak-gu, Seoul, 151-742 Korea
{mickey, moon}@soar.snu.ac.kr

Abstract. The emergence of multimedia data in databases requires adequate methods for information retrieval. In a music data retrieval system by humming, the first stage is to extract exact pitch periods from a flow of signals. Due to the complexity of speech signals, it is difficult to make a robust and practical pitch tracking system. We adopt genetic algorithm in optimizing the control parameters for note segmentation and pitch determination. We applied the results to *HumSearch*, a commercialized product, as a pitch tracking engine. Experimental results showed that the proposed engine notably improved the performance of the existing engine in *HumSearch*.

1 Introduction

As information systems advance, databases include multimedia data such as images, musics, and movies. For music databases, there are many search methods based on melodic contours, authors, singers, or lyrics. People want to find music with a few notes that are entered by a convenient method like humming. We aim to develop a pitch tracking engine for a music retrieval system that supports queries by humming.

A number of music search algorithms have been proposed. Handel [1] emphasized that the melodic contour is the most critical factor that listeners use to distinguish a song from the others. In other words, similar melodic contours make listeners consider songs as the same. Music search by humming matches songs and input queries according to properties such as melodic contours and rhythmic variations. Although being attractive, it contains some difficult processes such as melodic contour representation, note segmentation, and pitch period determination. Chou et al. [2] proposed a chord decision algorithm which transforms songs and queries into *chord strings*. Ghias et al. [3] used *melodic contours* of hummed queries, which consist of a sequence of relative differences in pitch between successive notes. They used an alphabet of three possible relationships between pitches: "U", "D", and "S". Each alphabet represents the situation that the current note is above ("U"), below ("D"), or the same ("S"), respectively, as the previous one.

To represent a melody by a melodic contour, a high-precision pitch tracking process is required. Pitch tracking is a process to determine the pitch period of each note in a melody. It involves two main processes: *note segmentation* and *pitch determination*.

Note segmentation determines where notes begin and terminate, and extracts the notes from a flow of signals. Rodger et al. [4] performed note segmentation in two ways: One based on amplitude and the other on pitch. Ahmadi et al. [5] presented an improved system for voiced/unvoiced classification from hummed queries based on statistical analysis of cepstral peaks, zero-crossing rates, and energy magnitudes of short-time speech segments. Mingyang et al. [6] proposed a system for the pitch tracking of noisy speech with statistical anticipation.

Pitch determination calculates pitch period of a segmented note. There are several approaches to calculate the basic frequency such as autocorrelation [7], AMDF [8], and cepstrum analysis [9]. There are also a number of studies under noisy environments. Shimomuro et al. [10] used weighted autocorrelation for pitch extraction of noisy speech. Kunieda et al. [11] calculated the fundamental frequency by applying autocorrelation in order to extract periodic features.

Few people usually have perfect pitch and they cannot remember all the exact pitches of their favorite songs either, even when they sing the songs very often. There may be errors as well in the melodic contours extracted from hummed queries. For this reason, we need approximate pattern matching. Our engine is utilized in *HumSearch* which supports queries by humming with an advanced approximate pattern matching.

In this paper, we suggest a genetic algorithm for pitch tracking that transcribes a hummed query into a melodic contour under noisy environments. We not only used the classical methods for pitch tracking such as the energy of short-time speech segment, zero-crossing rate, and cepstrum analysis, but also designed a contour analysis model. While the classical algorithms determined the threshold for note segmentation using a statistical method [5] [6] or an adaptive method [4], we optimize the control parameters of the methods by combining the classical methods with the genetic framework to enhance the performance of the pitch tracking engine.

The remainder of this paper is organized as follows. In Section 2, we summarize preliminaries for pitch tracking. In Section 3, we explain our additional methods such as contour analysis for note segmentation and describe our system in Section 4. We perform experiments and compare the results with an existing pitch tracking engine in Section 5. Finally, we make our conclusions in Section 6.

2 Preliminaries

2.1 Note Segmentation

In pitch tracking systems, note segmentation is the most important process. To transcribe a hummed query into a melodic contour, we need to compute the number of notes contained in the query. Figure 2 shows an energy diagram of

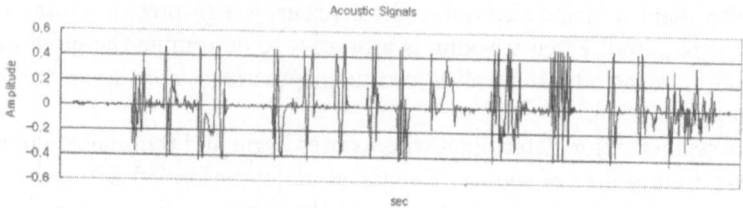

Fig. 1. Original vocal waveform (spaced by 0.01s)

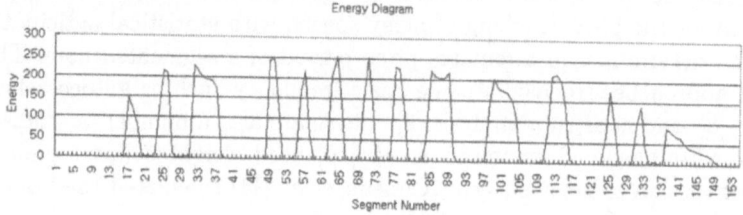

Fig. 2. Energy diagram of short-time speech segment

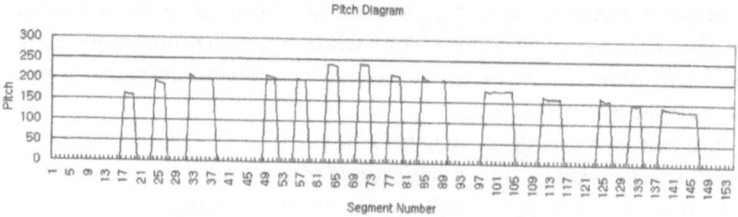

Fig. 3. Pitch diagram of speech segment

a hummed query. Figure 3 is the result of pitch determination from the energy diagram.

Speech Segment. Because of a temporary varying occurrence factor, speech processing is naturally a non-stationary process. However, we assume that the speech signal is short-time stationary for simple speech processing. We divide the whole sound signal into small speech segments and try to determine pitches on these small blocks. It is accomplished by multiplying the signal by a window function, w_n, whose value is zero outside some defined range. The rectangular window is defined as follows:

$$w_n = \begin{cases} 1, \text{ if } 0 \leq n < N \\ 0, \text{ } otherwise. \end{cases}$$

where n is the time point of speech samples and N is the window size. Note that if the size of the window is too large, our assumption becomes unreasonable.

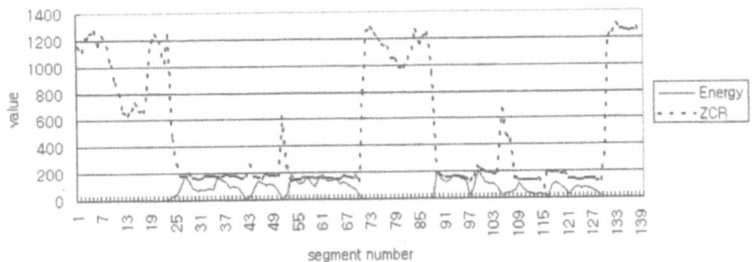

Fig. 4. Note segmentation based on zero-crossing rate

In contrast, if the size of the window is too small, we lack enough information for delimitating a pitch. The size of the window is thus an important factor of determining pitch period.

Energy of Speech Segment. The energy of a speech segment is defined to be the sum of amplitudes of the input wave as follows:

$$E = \sum_{i=0}^{N} |x(i)|$$

where $x(i)$ is the amplitude of original waveform signals and N is the size of a speech segment [12]. A statistical metric such as average or median is used to determine a threshold which classifies segments with or without voice. To make classification simpler, a user sings by "da" or "ta" because the consonants cause a drop in amplitude at each note boundary [3].

Zero-Crossing Rate. Zero-crossing occurs when neighboring points in a wave have different signs. The zero-crossing rate measures the number of zero-crossings in a given time interval [5]. The zero-crossing rate corresponding to the ith speech segment is computed as follows:

$$ZCR_i = \sum_{n=1}^{N-1} |sgn[x_i(n)] - sgn[x_i(n-1)]|$$

where $x_i(n)$ is the value at position n in the ith speech segment and N denotes the size of the speech segment, $x_i(n)$. A segment with a high zero-crossing rate is judged to be unvoiced or sibilant sounds; a segment with a low number is judged to be voiced intervals (See Figure 4).

2.2 Pitch Determination

Algorithms for determining the pitch on a vocal signal are roughly classified into two types: time-domain methods and frequency-domain methods. The time-domain methods such as autocorrelation and AMDF examine the structure of the sampled waveform, and the frequency-domain methods such as cepstrum analysis perform the Fourier transform and examine the resulting cepstrum.

Autocorrelation. Autocorrelation is one of the oldest classical algorithms for pitch determination. It shows the similarity in phase between two values of the speech signal at times x_n and x_{n+k}. The autocorrelation function $R(k)$ is calculated from a numerical sequence as

$$R(k) = \sum_{n=1}^{N-k} x(n) \cdot x(n+k)$$

where k is the correlation distance in the time sequence, N is the length of the sequence, and $x(n)$ is the value at position n in the time sequence. The function value with respect to k expresses the average correlation between numbers separated by distance k in the sequence. The correlation distance k with the maximum function value corresponds to the pitch period of the input signal.

Cepstrum Analysis. The cepstrum is defined to be the real part of the inverse Fourier transform of the log-power spectrum of $x(n)$ which is the value of acoustic signal in time sequence. Since cepstrum analysis requires a high computation cost, we use the *Fast Fourier Transformation* (FFT). For the FFT, we restrict the window size of the speech segment to be a dyadic number. Figure 7 shows the pitch period extracted from the vocal waveform of Figure 6.

2.3 The MIDI Notes Presentation

Since musical units such as *octaves or cents* are relative measures, we use the MIDI notation for note presentation. MIDI is a standard for communicating with electronic musical instruments and also a standard representation of the western musical scale. It is thus appropriate for representing a *melodic contour* of song or hummed input. MIDI assigns an integer in [0, 127] to each note. Middle C (C4) is assigned 60, the note just above is 61, and that below is 59. MIDI note 0 corresponds to 8.1279 Hz and the highest defined note, 127, corresponds to 13,344 Hz in frequency. The output of the pitch tracking system is eventually stored in MIDI form. Figure 12 contains a sequence of 12 MIDI notes from a hummed input.

3 Enhanced Methods for Pitch Tracking System

Since a user hums at a slide, the amplitude of an unvoiced sound segment does not drop. Thus the repeated notes may not be segmented successfully using the

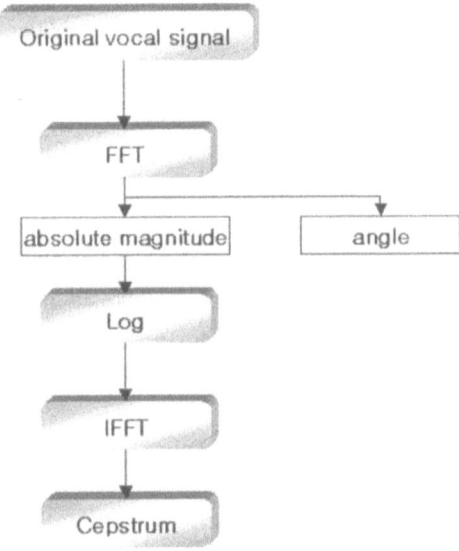

Fig. 5. Cepstrum analysis

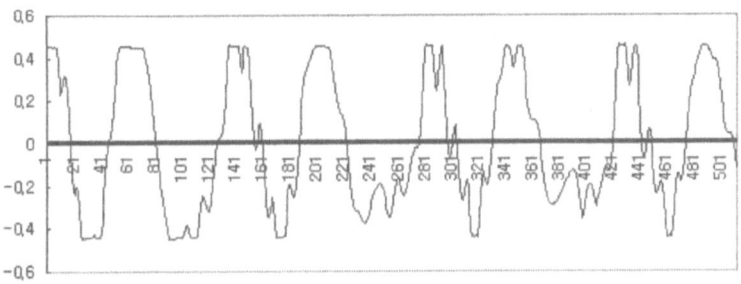

Fig. 6. Original waveform signal

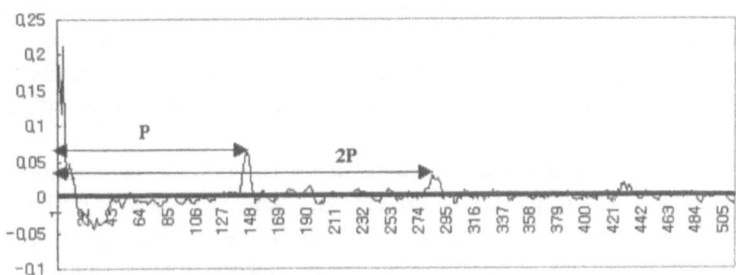

Fig. 7. Cepstrum pitch determination (p = pitch period)

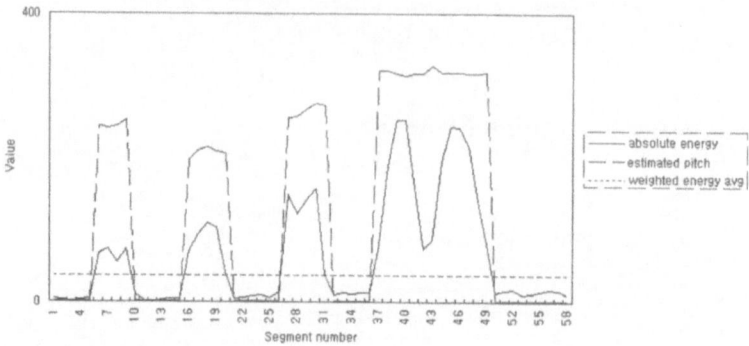

Fig. 8. Note segmentation based on energy of short-time speech segment

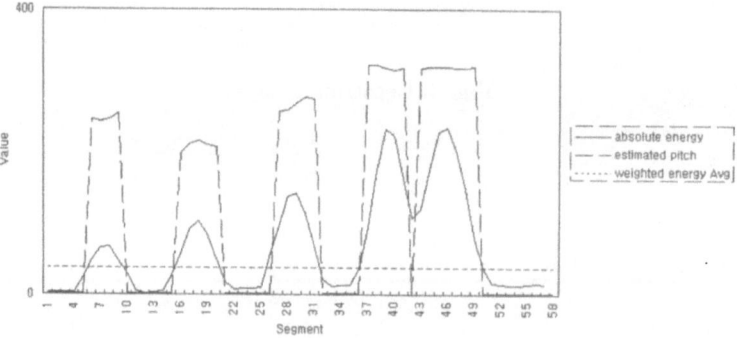

Fig. 9. Note segmentation based on smoothed energy diagram for detecting a valley

methods such as energy of the speech segment and zero-crossing rate. Unsegmented notes should separate into two notes (See Figure 8). We try to separate the successive notes with contour analysis of energy diagram. We detect the valley in the energy diagram and adopt it as a new feature for note segmentation. We use a moving average method and make an energy diagram flatten to remove the small valleys. This method divides the successive notes into two notes (Figure 9).

In our system, we divided the resolution of the relative pitch differences into 15 levels (U1 ∼ U7 (Up), D1 ∼ D7 (Down), and R (Same)). If the difference is beyond the range, it is assigned one of the boundary levels, U7 or D7. More accurate representation of pitch differences in successive notes is possible by this fine-grained strategy.

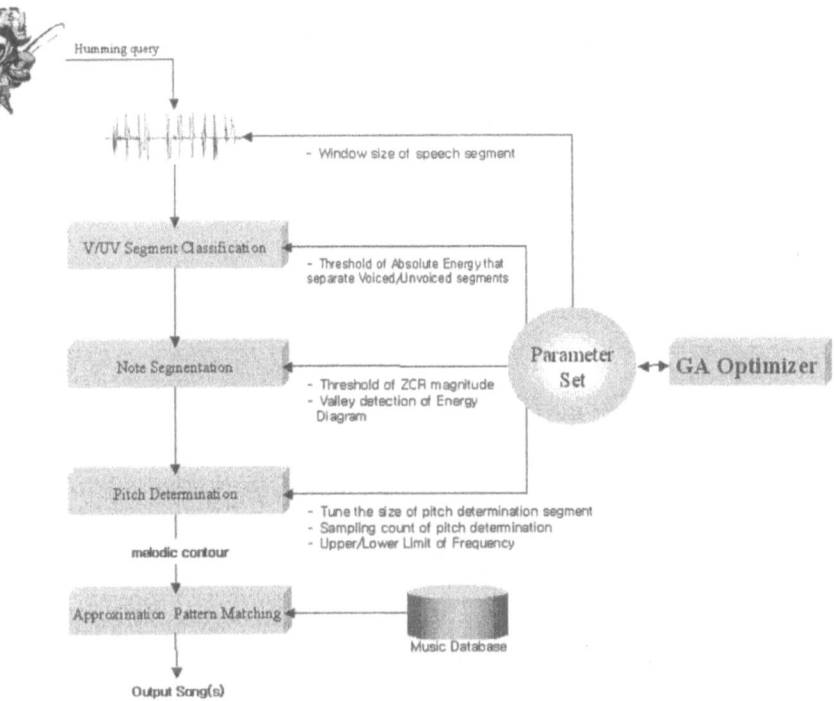

Fig. 10. Optimized pitch tracking system with GA

4 Optimized Pitch Tracking System with GA: GAPTS

4.1 System Architecture

The system architecture is shown in Figure 10. A hummed query fed into the system is evaluated through the pitch tracking and query systems. The efficiency of the pitch tracking module depends on the control parameters such as weighted value, window size, etc. There are three main modules in the system : *voiced/unvoiced classification*, *note segmentation*, and *pitch determination*.

4.2 GA Framework

Encoding. A chromosome consists of 11 genes. Each gene corresponds to a feature that affects controlling the pitch tracking system. The function of each gene is explained in Table 1.

Initialization. Initial solutions are generated at random. We set the population size to be 100.

```
Create initial population;
Evaluation all chromosomes ;
do
{
    Choose parent1 and parent2 from population ;
    offspring = crossover (parent1, parent2) ;
    mutation(offspring) ;
    evaluation (offspring) ;
    replace(parent1, parent2, offspring);
} until (stopping condition) ;
return the best solution;
```

Fig. 11. Genetic algorithm framework

Table 1. The functions of genes

Genes	Description
0	The window size of speech segment
1	A weighted value for determining the threshold of short-time energy which divides voiced/unvoiced segments
2	The upper bound of frequency considered as voiced pitch
3	The lower bound of frequency considered as voiced pitch
4	Weighted value for determining the threshold of zero-crossing rate which divides voiced/unvoiced segments
5	Boundary value for deciding a sampling count to determine pitch periods
6	The window size of analysis section for pitch determining
7	A window size of the moving average method for smoothing the contour of a short-time energy diagram. If the size is set to 0, the method is not applied to the system
8	The window size of the moving average method for smoothing the contour of a zero-crossing rate diagram. if the size is set to 0, the method is not applied to the system
9	Sampling count for pitch determining
10	Weighted value to determine the threshold that cuts a note into two or more for note segmentation with contour analysis

Parent Selection. We assign to each chromosome a fitness value. A fitness value is defined to be the sum of edit distances between the melodic contours of the hummed query and the original song. Let $SED(t, q)$ be the string edit distance between a hummed query q and the target contour t. The fitness value F_i of chromosome i is defined as follows:

$$F_i = \sum_{i=1}^{N} SED(Q(i), T(i))$$

where $Q(i)$ is the ith hummed query, $T(i)$ is the ith target contour, and N is the number of the hummed queries. We use the tournament selection method [13]. The tournament size is 2.

Crossover. We use the uniform crossover [14]. The relatively high disruptivity of the uniform crossover helped our algorithm escape from local optima and converge to better solutions.

Table 2. sets of parameters

gene number	Optus PTS	GAPTS
0	11	11
1	0.4	0.47
2	530.0	454.91
3	90.0	64.28
4	0.4	0.5
5	6	10
6	2	3
7	0	1
8	0	0
9	1.5	5.00
10	1.5	0.68

Mutation. We randomly select each genes with a probability ($= 0.5$) and mutate the gene values within its admitted range.

Replacement. After generating an offspring, GAPTS replaces a member of the population by the offspring. We replace the inferior of the two parents with the offspring if the offspring is not worse than both parents. Otherwise, we replace the worst member of the population by the offspring. This scheme is a compromise between preselection [15] and GENITOR-style replacement [16].

Stopping Condition. The GA stops if the generation count reaches a predetermined number, or it shows 2000 times of consecutive fails to replace one of the parents.

5 Experimental Results

The melodic contours extracted by GAPTS are evaluated by SED. SED is the error in the string edit distance as mentioned in Section 4.2. Ten people hummed 95 songs for test. Following the convention for clear segmentation, they sang by "*ta*" or "*da*" into microphone under a usual condition with noise. All programs were written in $C++$ language and run on PentiumIII 866MHz with the Linux operating system. We did not use any hardware device for acoustic signal processing except the microphone.

For robust comparison between Optus PTS and GAPTS (combined with GA), we followed the 5-fold cross-validation approach [17] [18]. We randomly divided the entire hummed query set D into 5 mutually exclusive subsets of approximately equal size. The GAPTS was trained and tested 5 times. Table 2 shows the sets of parameters; the first is the set of parameters that has been used in *HumSearch*, a commercial query-by-humming product of Optus Inc., and the second is the set we have found by the GA.

The original SED of *HumSearch* was 443. The GAPTS improved it to 424, which is a notable improvement. When we replaced the existing set of parameters in *HumSearch* by the new set, *HumSearch* also showed improvement. Out of the

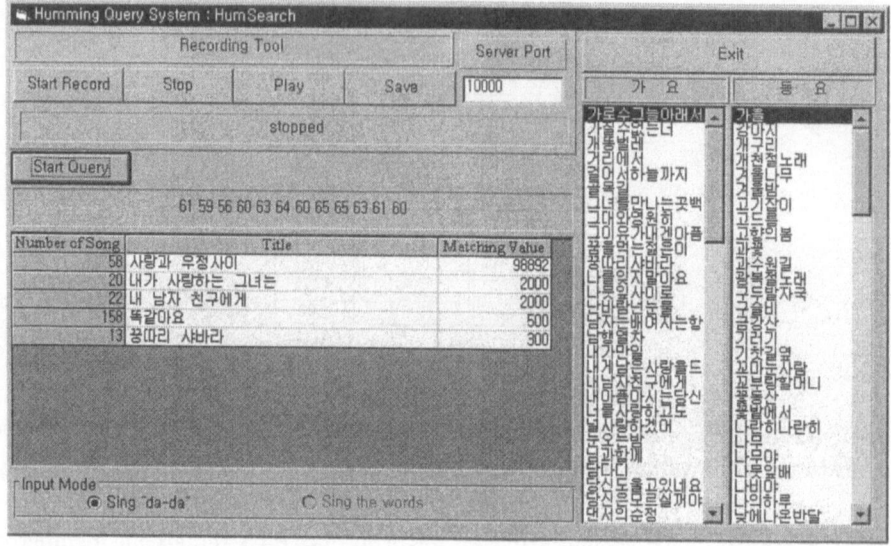

Fig. 12. A snapshot of our hummed query system

95 songs, we found 10 songs ranked up and 8 songs ranked down. We should note that the parameter set for the *HumSearch* has been tuned for a long time to be a commercial product. Considering this, the improvement in SED and the change of ranks is notable. Figure 12 shows a snapshot of our interactive hummed query system.

6 Conclusions

We extracted the parameters that control the note segmentation and pitch determination. We optimized them by combining the classical methods such as short-time energy diagram, zero-crossing rate, contour analysis of short-time energy diagram for note segmentation and cepstrum analysis for pitch determination. Our pitch tracking system worked well regardless of testing environments under various conditions, e.g., sexuality, noise, and input devices (mobile phone or microphone).

Acknowledgments. This work was partly supported by Optus Inc. and Brain Korea 21 Project. The RIACT at Seoul National University provided research facilities for this study.

References

1. Stephen Handel. Listening : An introduction to the perception of auditory events. In *The MIT Press*, 1989.

2. Ta Chou, Arbee L.P., and C.C. Liu Chen. Music database : indexing technique and implementation. In *Proc. of IEEE, International Workshop on Multimedia data base Menagement System*, 1996.
3. A. Ghais, J. Logan, D. Chamberlin, and B.C. Smith. Query by humming. In *Proceeding ACM Multimedia 95*, 1995.
4. Rodger J. McNab, Lloyd A. Smith, Ian H. Witten, Cclar L. Henderson, and Sally J. Cunningham. Towards the digital music library: Tune retrieval from acoustic input. In *Proc. ACM Digital Libraries*, 1996.
5. Sassan Ahmadi and Andreas S. Spanias. Cepstrum-based pitch detection using a new statistical v/uv classification algorithm. *IEEE Trans. Speech and Audio Processing*, 7(3):333–338, 1999.
6. Mingyang Wu, DeLiang Wang, and G.J. Brown. Pitch tracking based on statistical anticipation. In *Proceedings. IJCNN '01. International Joint Conference*, volume 2, pages 866–871, 2001.
7. L.R. Rabiner. On the use of autocorrelation analysis for pitch detection. *IEEE Trans. Acoust,, Speech, Signal Processing*, ASSP-25(1):24–33, 1977.
8. Ross. M, J, H.L. Shaffer, A. Cohen, R. Freudberg, and H. J. Manley. Average magnitude difference function pitch extractor. *IEEE Trans. Acoust,, Speech, Signal Processing*, ASSP-22(1):353–362, 1977.
9. Noll A.M. Cepstrum pitch determination. *Journal of Acoust, Soc, Amer.*, 41:293–309, 1967.
10. T. Shimomuro and H. Kobayashi. Weighted autocorrelation for pitch extraction of noisy speech. *IEEE Trans. Speech and Audio Processing*, 9(7):727–730, 2001.
11. Shimamura T. Kunieda N. and Suzuki J. Robust method of measurement of fundamental frequency by aclos: autocorrelation of log spectrum. In *IEEE International Conference Acoustics, Speech, and Signal Processing*, pages 232–235, 1996.
12. J.K. Kaiser. On a simple algorithm to calculate the 'energy' of a signal. In *Proceeding IEEE ICASSP*, pages 381–384, 1990.
13. D. E. Goldberg, K. Deb, and B. Korb. Do not worry, be messy. In *Proceedings of the Fourth International Conference on Genetic Algorithms*, pages 24–30, 1991.
14. Syswerda. Uniform crossover in genetic algorithms. In *International Conference on Genetic Algorithms*, pages 2–9, 1989.
15. D. Cavicchio. *Adaptive Search Using Simulated Evolution*. PhD thesis, University of Michigan, Ann Arbor, MI, 1970. Unpublished.
16. D. Whitley and J. Kauth. GENITOR: A different genetic algorithm. In *Proceedings of Rocky Mountain Conference on Artificial Intelligence*, pages 118–130, 1988.
17. B. Efron an dR.Tibshirani. Cross-validation and the bootstrap: Estimating the error rate of a prediction rule. In *Technical Report(TR-477), Dept. of Statistics, Stanford University.*, 1995.
18. R. Kohavi. A study of cross-validation and bootstrap of accuracy estimation and model selection. In *Proceedings of hte Rourteenth International Joint Conference on Artificial Intelligence*, pages 1137–1143, 1995.

Secret Agents Leave Big Footprints: How to Plant a Cryptographic Trapdoor, and Why You Might Not Get Away with It

John A. Clark, Jeremy L. Jacob, and Susan Stepney

Department of Computer Science, University of York
Heslington, York, YO10 5DD, UK
{jac,jeremy,susan}@cs.york.ac.uk

Abstract. This paper investigates whether optimisation techniques can be used to evolve artifacts of cryptographic significance which are apparently secure, but which have hidden properties that may facilitate cryptanalysis. We show how this might be done and how such sneaky use of optimisation may be detected.

1 Introduction

The issue of optimisation is fundamental to cryptography. Pseudo-random bit streams should appear as random as possible, cryptanalysts try to extract the maximum amount of useful information from available data and designers wish to make attacks as difficult as possible. It is not surprising that some security researchers have begun to consider the use of heuristic optimisation algorithms. Both simulated annealing and genetic algorithms have been used to decrypt simple substitution and transposition ciphers [6,8,13,14], and there have been some recent and successful attempts to attack modern day crypto protocols based on NP-hard problems [3]. Recent research has used optimisation for design synthesis, evolving Boolean functions for cryptographic products, most notably using hill climbing [10] and genetic algorithms [9]. Other recent work has applied search to the synthesis of secure protocols [1].

Some cryptographic algorithms have been shown to be insecure. Others are thought to be secure, but subsequent discoveries may prove otherwise. Cryptanalysts live in hope that the ciphers they investigate have as yet undiscovered features that will allow them to be broken, 'trapdoors' if you like, hidden ways of opening up the algorithms to let the secrets of those using them fall out! Discovering such trapdoors is very hard work. There is, however, a somewhat sneakier approach — deliberately create an algorithm with a subtle trapdoor in it but which looks secure according to currently accepted criteria. Users of the algorithm may communicate secrets, but the analyst who knows the trapdoor will be able to access them. If users subsequently find out that a trapdoor had been deliberately engineered into the algorithm, they may not be very happy. The trapdoor should be so subtle that they have little chance of discovering it.

This paper investigates whether optimisation techniques can be used to surreptitiously plant such trapdoors.

This issue goes to the heart of cryptosystem design in the past. The debate about whether the Data Encryption Standard [11] has a secret trapdoor in its substitution boxes [1] has raged since its publication [12]. A major factor in this debate is that initially the design criteria were not made public (and the full criteria are still unknown). There was a distinct suspicion that security authorities had designed in a trapdoor or otherwise knew of one. [2] We contend that combinatorial and other numerical optimisation techniques can be *demonstrably open* or *honest*. Goals are overt (that is, maximising or minimising a specified design function). The means of achieving those ends is entirely clear — the algorithm for design is the search technique and its initialisation data coupled with the specified design function to be minimised. An analyst who doubts the integrity of the design process can simply replay the search to obtain the same result. Furthermore, the approach permits some surprising analyses.

2 Public and Hidden Design Criteria

The set of all possible designs forms a design space D. From within the design space we generally seek one instantiation whose exhibited properties are "good" in some sense. (In the case of DES, one design space may be thought of as the set of legitimate S-box configurations. The standard uses a particular configuration but many others are possible.) When designing a cryptosystem, we may require that certain stringent public properties P be met. Generally only small fraction of possible designs satisfy these properties. Consider now a trapdoor property T. If a design has property T then cryptanalysis may be greatly facilitated. Some designs may simultaneously satisfy P and T. There two possibilities here:

- All designs (or a good number) satisfying P also satisfy T. This suggests that the basic design family is flawed (though the flaw may not be publicly known).
- Only a small fraction of designs satisfying P also satisfy T. Here the trapdoor planter must drive the design in such a way that T is achieved simultaneously with P.

---

[1] DES is a 16-round cipher. Each round is a mini-encryption algorithm with the outputs of one round forming the inputs of the next. Each round application further obscures the relationship to the initial inputs. At the end of 16 rounds, there should be a *seemingly* random relationship between the initial inputs and the final outputs. (Of course, the relationship is not random, since if you have the secret key you can recover the plaintext inputs) Within each round there are several substitution boxes (S-boxes). These take 6 input bits and produce 4-bit outputs. The details need not concern us here. Suffice it to say that S-boxes are designed to give a complex input-output mapping to prevent particular types of attacks.

[2] As it happens, it seems they intervened to make the algorithm secure against differential cryptanalysis, a form of attack that would be discovered by academics only in the late 1980s.

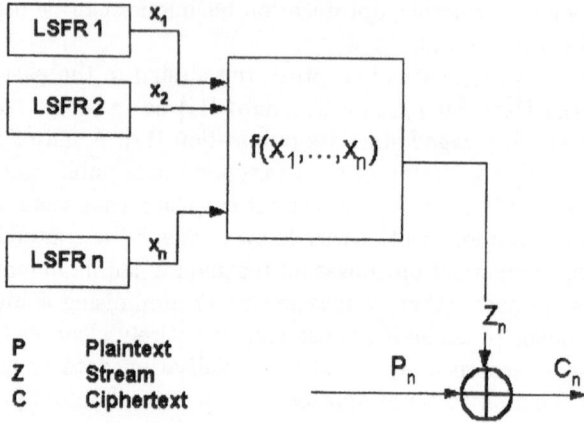

Fig. 1. Combining bit streams

It may be computationally very hard to locate good designs. Heuristic optimisation offers a potential solution. Suppose a cost function motivated solely by the goal of evolving a design (from design space D) to satisfy P is given by $h : D \to \Re$. (We denote this cost function by h since its choice is *honest* — free from hidden motives.) Suppose also that a cost function aimed at evolving a design satisfying the trapdoor property T is given by $t : D \to \Re$ (t for trapdoor). If cryptographic functionality is developed using only trapdoor criteria T it is highly likely that someone will notice! Indeed, reasonable performance of the resulting artifacts when judged against public properties P would be accidental. Similarly, an artifact developed using only the honest (public) criteria P will generally have poor trapdoor performance. There is a tradeoff between malicious effectiveness (i.e. attaining good trapdoor properties) and being caught out. The more blunt the malice, the more likely the detection. We capture the degree of bluntness via a parameter λ, which we term the *malice factor*. The design problem for malice factor $\lambda \in [0.0, 1.0]$ is to minimise a cost function of the form

$$cost(d) = (1 - \lambda)h(d) + \lambda t(d) \qquad (1)$$

At the extremes we have the honest design problem, $\lambda = 0$, and unrestrained malice, $\lambda = 1$. In between we have a range of tradeoff problems. We now illustrate how optimisation can be used and abused in the development of perhaps the simplest cryptographically relevant element, namely a Boolean function.

3 Cryptographic Design with Optimisation

3.1 Boolean Functions

Boolean functions play a critical role in cryptography. Each output from an S-box can be regarded as a Boolean function in its own right. Furthermore, in stream ciphers Boolean functions may be used to combine the outputs of several bit streams. Figure 1 illustrates the classic stream cipher model. The plaintext stream $\{P_i\}$ of bits is XOR-ed with a pseudo-random bit stream $\{Z_i\}$ to give a cipher text stream $\{C_i\}$. The plaintext is recovered by the receiver by XOR-ing the cipherstream with the same pseudo-random stream. The pseudo-random stream is formed from several bit streams generated by Linear Feedback Shift Registers (LFSRs) using a suitable combining function. The initial state of the registers forms the secret key. The function f must be sufficiently complex that cryptanalysts will not be able to determine the initial state of the registers, even when they know what the plaintext is (and so can recover the pseudo-random stream $\{Z_i\}$). Two desirable features are that the function should be highly non-linear and posses low auto-correlation. These are explained below, together with some further preliminaries.

We denote the *binary truth table* of a Boolean function by $f : \{0,1\}^n \to \{0,1\}$ mapping each combination of n binary variables to some binary value. If the number of combinations mapping to 0 is the same as the number mapping to 1 then the function is said to be *balanced*. A cryptographic designer may require a Boolean function to have this property. The *polarity truth table* is a particularly useful representation for our purposes. It is defined by $\hat{f}(x) = (-1)^{f(x)}$. Two functions f, g are said to be *uncorrelated* when $\sum_x \hat{f}(x)\hat{g}(x) = 0$. If so, if you approximate f by using g you will be right half the time and wrong half the time.

An area of particular importance for cryptanalysts is the ability to approximate a function f by a simple linear function, that is, one of the form $L_\omega(x) = \omega_1 x_1 \oplus \omega_2 x_2 \ldots \oplus \omega_n x_n$ (where $\oplus$ is exclusive or). For example, $x_1 \oplus x_3 \oplus x_4$ is a linear function (with $\omega = 1011$), whereas $x_1 x_2 \oplus x_3 \oplus x_4$ is not: it has a quadratic term. One of the cryptosystem designer's tasks is to make such approximation as difficult as possible; the function f should exhibit high non-linearity [3]. In terms of the nomenclature given above, the formal definition of non-linearity is given by

$$NL(f) = \frac{1}{2}\left(2^n - \max_\omega \left|\sum_x \hat{f}(x)\hat{L}_\omega(x)\right|\right) \qquad (2)$$

The 2^n functions $\{\hat{L}_\omega(x), \omega : 0..(2^{n-1})\}$ span the space of Boolean functions on n variables. The term $\sum_x \hat{f}(x)\hat{L}_\omega(x)$ (usually referred to as the Walsh value at ω) can be thought of as the dot product of $\hat{f}$ with $\hat{L}_\omega$. Thus, we are trying to minimise the maximum absolute value of a projection onto a linear function.

---

[3] If it doesn't then various forms of attack become possible, e.g. the Best Affine Attack. Loosely, linear functions are not very complex input-output mappings. If the function $\hat{f}$ can be approximated by a linear function, it is just too predictable to resist attack. See [5] for details.

Low *auto-correlation* is another important property that cryptographically strong Boolean functions should possess. Auto-correlation is defined by

$$AC_{max} = \max_{s \neq 0} \left| \sum_x \hat{f}(x)\hat{f}(x \oplus s) \right| \qquad (3)$$

It is essentially a property about dependencies between periodically spaced elements. The details need not concern us here. What is important is that non-linearity and autocorrelation of a given Boolean function f can be measured. For the purpose of this paper we simply identify balancedness, high non-linearity and low autocorrelation to serve as our desirable publicly stated properties P.

3.2 An Honest Cost Function

We aim to provide balanced Boolean functions of 8 variables with high non-linearity and low autocorrelation. We adopt a search strategy that starts with a balanced (but otherwise random) function, and explores the design space by moving between neighbouring functions until an appropriate solution is found. A function $\hat{f}$ can be represented by a vector of length 256 with elements equal to 1 or -1. The first element is $\hat{f}(00000000)$ and the last $\hat{f}(11111111)$ etc. We define the neighbourhood of a balanced function $\hat{f}$ to be all functions $\hat{g}$ obtained from $\hat{f}$ by negating any two dissimilar elements in the vector (that is, changing a 1 to a -1 and -1 to a 1). Provided the initial function was balanced, the search maintains that balance.

To use heuristic optimisation techniques we need a cost function whose minimisation gives good values of non-linearity and low values for autocorrelation. We use the (honest) cost function

$$h(\hat{f}) = \sum_\omega \left| \left| \sum_x \hat{f}(x)\hat{L}_\omega(x) \right| - 12 \right|^3 \qquad (4)$$

where ω, x range over $\{0, 1\}^8$. The details of this cost function and the rationale for it can be found in [2]. It gives very good results for Boolean functions of 8 variables; but we admit it is not intuitively clear. There is a tradeoff between a "clearly" honest cost function, and one with good performance.

The primary search technique used was the simulated annealing algorithm [7] with the honest cost function of Equation 4. This was followed by hill-climbing (attempting to maximise directly the non-linearity as defined in Equation 2) from the resulting function $\hat{f}$. Figure 2 shows the results from 400 runs of the technique using $\lambda = 0$, that is, total honesty. 115 of the 400 runs gave rise to functions with non-linearity of 116 and autocorrelation of 32. The highest non-linearity ever attained by *any* technique (ignoring auto-correlation) is 116. Until recently the lowest autocorrelation possible was thought to be 24. Recent optimisation based work [4] generated functions with autocorrelation of 16 (but with nonlinearity value of 112). No function has ever been demonstrated with nonlinearity 116 and autocorrelation 16.

AC_{max}	Non-linearity			
	110	112	114	116
56	0	0	0	0
48	0	2	5	13
40	0	9	68	80
32	0	29	74	115
24	0	1	1	3

Fig. 2. Honest Evolution, sample size = 400. Simultaneous high non-linearity and low autocorrelation (bottom right corner) are desirable.

Our approach appears to give excellent results for both criteria simulataneously. Thus the technique can be used effectively by honest researchers to achieve honest goals. The search proceeded from a random starting function to its final goals guided by the honest cost function at each stage. Now we choose to guide the search in a malicious way.

3.3 Evolving a Trapdoor

We now seek designs that perform well when judged against the public criteria P but also possess some secret trapdoor property T. For illustrative purposes only, we choose to define a trapdoor property T as sufficient closeness to some arbitrary but fixed Boolean function represented by g. We take $\left|\sum_x \hat{f}(x)\hat{g}(x)\right|$ as a measure of closeness. The maximum value 256 occurs when f and g are the same function (or when one is the logical negation of the other). Since our design framework is based around minimising cost, the closer the value to 256 the smaller should be the cost. This gives rise to a trapdoor cost function

$$t(\hat{f}) = \left|256 - \left|\sum_x \hat{f}(x)\hat{g}(x)\right|\right|^3 \tag{5}$$

The exponent 3 is adopted for comparability with the honest cost function. It serves to punish poorly performing functions very highly and affects the rate at which the search converges. We now combine the honest and dishonest functions in a single cost function *cost* as given in Equation 1 (with the function $\hat{f}$ playing the role of design d).

We applied the technique for various values of λ in the range $[0, 1]$. For a randomly generated trapdoor g, 30 runs were carried out for each of $\lambda \in \{0.0, 0.2, 0.4, 0.6, 0.8, 1.0\}$. The same trapdoor was used in all runs. Figure 3 records the results. Thus, for $\lambda = 0.0$, 12 of the 30 runs give rise to functions with nonlinearity of 116 and autocorrelation of 32. The average closeness to the trapdoor g (given by $\left|\sum_x \hat{f}(x)\hat{g}(x)\right|$) for the 30 runs in this case is 12.8.

The results show that optimisation techniques can be used effectively to obtain artifacts that provide good public performance measures and good trapdoor functionality. The figures show that as λ increases the performance of the evolved

AC_{max}	$\lambda = 0.0$ Non-linearity				$\lambda = 0.2$ Non-linearity				$\lambda = 0.4$ Non-linearity			
	110	112	114	116	110	112	114	116	110	112	114	116
64	0	0	0	0	0	0	0	0	0	0	1	0
56	0	0	0	0	0	0	1	0	0	0	2	0
48	0	0	0	1	0	0	7	0	0	1	6	0
40	0	0	3	4	0	0	16	0	0	2	17	0
32	0	2	7	12	0	0	6	0	0	0	1	0
24	0	0	0	1	0	0	0	0	0	0	0	0
	Mean Trap = 12.8				Mean Trap = 198.9				Mean Trap = 213.1			

AC_{max}	$\lambda = 0.6$ Non-linearity				$\lambda = 0.8$ Non-linearity				$\lambda = 1.0$ Non-linearity			
	110	112	114	116	110	112	114	116	110	112	114	116
80	0	0	0	0	0	0	0	0	2	0	0	0
72	0	0	0	0	0	0	0	0	4	1	0	0
64	0	0	1	0	0	1	0	0	10	6	0	0
56	0	1	2	0	0	4	1	0	2	5	0	0
48	0	5	7	0	0	19	1	0	0	0	0	0
40	0	2	12	0	0	3	1	0	0	0	0	0
32	0	0	0	0	0	0	0	0	0	0	0	0
24	0	0	0	0	0	0	0	0	0	0	0	0
	Mean Trap = 222.1				Mean Trap = 232.3				Mean Trap = 242.7			

Fig. 3. Performance for various malice factors λ. Sample size = 30 for all experiments. For the $\lambda = 1$ result, any remaining non-linearity and auto-correlation arises from the dishonest g and the final hill-climb on non-linearity.

designs with respect to the public criteria worsens (i.e their non-linearity gets smaller and autocorrelation gets bigger). Of course, the closeness of the evolved designs to the trapdoor property gets radically better. An honest sample (that is, $\lambda = 0$) has an average closeness of 12.8. With $\lambda = 0.2$ the average solution obtained has a closeness value of 198.9. With $\lambda = 1.0$ the average is 242.7[4].

4 Detection

If you know what the trapdoor is then you can of course verify that the artifact has it. But what if you don't know what the trapdoor is?

4.1 Telling Truth from Fiction

We wish to distinguish honestly derived functions from maliciously derived ones. We would like to develop discriminating techniques with wide ranging applicability. We observe that functions (and indeed designs more generally) have some

---

[4] Recall that after optimising with respect to the indicated cost function, we then hill-climb (maximise) with respect to non-linearity alone. This typically serves to reduce the trapdoor value attained - it is simply measured at the end of the hill-climb. It plays no part during the hill-climb.

concrete binary representation. Our current function representation is enumerative, but *programs* that implement particular functions also have binary representations (the vector of object code bits being a particular one). A function within a particular family may be defined by some set of configuration data (e.g. S-Boxes). It is the properties of such vector representations of designs that we propose to use as the basis of detection. This has the benefit of being more problem independent.

We illustrate one such way in which families of designs can be discriminated. Suppose we have two groups of designs G_1, G_2. Each group comprises a number of designs represented by vectors. For each group we calculate the mean vector for the vectors in that group. Denote these mean vectors by m_1, m_2. Re-express $m_1 = \mu m_2 + z$, where z is orthogonal to m_2. The first component μm_2 is the projection of m_1 onto m_2. We now project each design onto this residual z and measure the length of such projections. In general if the groups G_1 and G_2 are different we would expect those in G_1 to give rise to projections of large magnitude and those in G_2 to give rise to vectors of small magnitude. More formally

- Let d_{1j} and d_{2j} be the jth designs of G_1 and G_2 respectively. Let d_{1j} be represented by a vector of N bit values $(d_{1jk}), k = 1..N$. Similarly for d_{2j}.
- For each group form the mean vector $m_i = (m_{ik})$, where $m_{ik} = \frac{1}{S}\sum_{j=1}^{D} d_{ijk}$
- Calculate the residual vector when m_1 is projected on m_2. For vectors x, y the residual of x after projection on y is given by $resid(x, y) = x - \frac{(x,y)}{(y,y)} y$, where (x, y) is the 'dot product' for vectors x and y.
- for G_1, G_2 find the dot product of each of its designs onto this residual vector and find its absolute value.

4.2 Results

Figure 4 shows the the summary statistics (with the honest group as the reference sample) for each of the λ values, for 30 runs. The aim is to show the degree to which the summary statistics for each malicious sample overlaps with the set of values from the honest one. For $\lambda = 0$, every value is ≤ 0, by construction. By contrast, for other values of λ, very few results are ≤ 0. For $\lambda = 0.6$ only one design has a result in the range of honest results; for other values of λ the overlap is greater.

We see that the summary statistic provides a reasonable, and in some cases good, way of discriminating between designs developed using different cost functions. In some cases the discrimination capability is not as good as we should like, because the calculated statistic loses a great deal of information. However, even with this gross loss of information we can still detect a discernible difference in malicious and honest designs. Clearly more sophisticated tests are needed.

The above is intended purely as an illustration that families of designs give rise to their own distributions of summary statistics. It should be noted that the *designs within a family are not necessarily close to each other*. Indeed, they may tend to congregate in different areas (corresponding to different local optima of

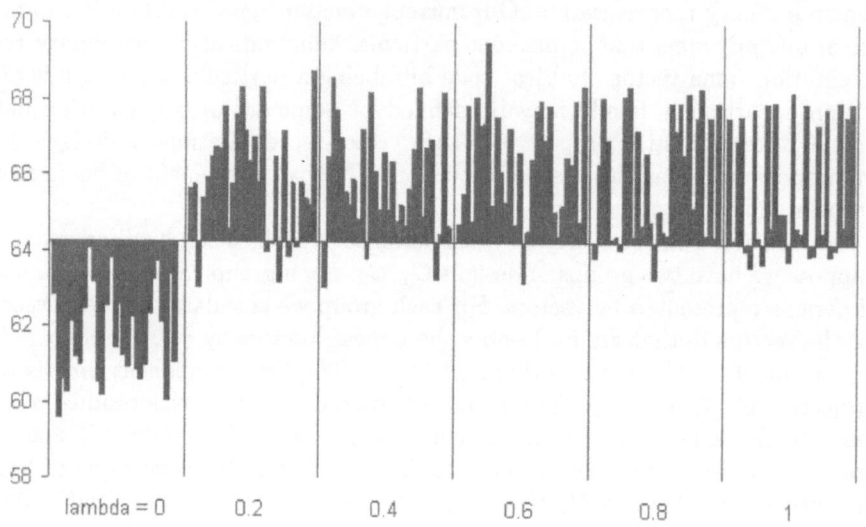

Fig. 4. Squares of projections, for different values of λ. The axis is drawn at the maximum value of the $\lambda = 0$ statistic, to highlight the contrasting behaviour for other values of λ.

cost function used). A malicious designer might take advantage of this weakness or similar weaknesses of other discrimination techniques used. This we now show.

4.3 Games and Meta-games

If the discrimination technique is known then it is possible to recast the malicious design problem with a cost function

$$cost(d) = (1 - \lambda)h(d) + \lambda t(d) + easeDetect(d) \qquad (6)$$

Here *easeDetect* represents how easily the solution may be identified as a maliciously derived one. Given the scheme above a suitable candidate would be to punish solutions whose summary statistic was too large. If a variety of known discriminants are used each of these can similarly be taken into account. Of course, it becomes more difficult to find a good solution, but the malicious designer is really just looking to gain an edge.

Malicious designers can attempt to evade detection in this way only if they know the discrimination tests to be applied. Analysts are free to choose fairly arbitrarily which tests to apply. This makes the task of malicious designers more difficult. They can appeal to their intuition as to what sorts of tests might be sensible (as indeed must the analyst), but they will be second guessing the analyst at large.

4.4 Meta-meta Games

The work described has concentrated on how trapdoors can be inserted and about how claims to have used particular cost functions can be checked. We stated in the introduction that a designer could simply publish the cost function, the search algorithm, and its initialisation data; the doubting analyst could simply replay the search to obtain the same result. This is undoubtedly so, but the implication that this gives us *honest design* must be questioned. We must question in fact what it means to be 'honest'.

Our "honest" cost function of Equation 4 uses an exponent of 3. Why 3? Would 2.9 be dishonest? It really depends on what our *intention* is for choosing a particular value. As it happens we have found that 3.0 gives good results for finding the desired properties, but other values may perform similarly well. Our honest cost function is indeed significantly different from those used by other researchers, whose cost functions are actually more intuitively clear, but seem to perform less well.

It would not be surprising if functions derived using different honest cost functions gave rise to results which have different properties as far as illicit trapdoor functionality is concerned. Large scale experimentation could lead to the discovery of cost functions that give results with excellent public properties but which *happen* to have a good element of specific trapdoor functionality. Such higher level games could be played most easily by those with massive computational capability. Perhaps this could be countered by cost function cryptanalysis!

As a final game, consider the following. Suppose you discover a search technique that works better than existing ones. Suppose, for example, it could find with ease a balanced Boolean function of 8 variables with non-linearity of 118. (This would answer an open research question, since the best least upper bound is 118, but the best actual non-linearity achieved is 116.) If you so desired you could use you enhanced search capability to find functions that had, say, non-linearity 116 but good trapdoor properties too. *You* know that you have sacrificed a little non-linearity to achieve your malicious goal, but the public analyst will compare your results against the best *published* ones. Enhanced search can also be provided by increased computational power. Of course, the malicious designer will have to do without the kudos that would be attached to publishing results to open problems (an academic would of course never do this!).

5 Conclusions

This paper builds on extant optimisation-based cryptographic design synthesis work and extends it to show the optimisation techniques offer the potential for planting trapdoors. We have explored some of the consequences. We know of no comparable work.

Different cost functions solve different problems. Someone might be solving a different problem from the one you assumed; someone may have added a trapdoor *to their alleged cost function*. However, you can use optimisation to evolve a design against a criterion many times. The resulting *sample* of solutions may provide a yardstick against which to judge proffered designs. It provides therefore a potential means of resolving whether a trapdoor was present in the fitness function actually used. Currently, it is hard to generate a large sample of good designs by other techniques (without the benefit of automatic search it may be difficult to demonstrate a single good design).

That designs evolved against different criteria should exhibit different functionality is clearly no surprise — it means that optimisation based approach works! This paper has observed that different designs have different bit level representations and that different groups of designs can be detected by using statistical properties of the representations.

The following consequences have emerged from the work:

An optimisation-based design process may be open and reproducible. A designer can publish the criteria (that is, the cost function) and the optimisation search algorithm (together with any initialistion data such as random number seed).The search can be repeated by interested parties.

Optimisation can be abused. If optimisation 'works' then 'use' means one approves of the cost criteria and 'abuse' means one doesn't!

Optimisation allows a family of representative designs to be obtained. The search process may be repeated as often as desired to give a set of designs from which statistical distributions of designs evolved to possess particular properties may be extracted.

Designs developed against different criteria just look different! If designs have different functionality then they are different in some discernible way. In certain cases this difference may be detected by *examination of the design representations alone*.

The games just do not stop. Cryptographic design and analysis go hand in hand and evolve together. When the smart cryptanalyst makes an important discovery, the designers must counter.

These ideas are at an early stage at the moment, but there would appear to be several natural extensions. The simple projection method used for illustration in this paper loses a very large amount of information. A study of available discrimination techniques should bring great benefit. We aim to evolve larger artifacts with trapdoor functionality, and are currently attempting to evolve DES-family cryptosystems with built-in trapdoors (by allowing a search to range over S-Box configurations). We acknowledge that we may have been beaten to it! How could we tell?

6 Coda: The Authors Are $\lambda = 0$ People

Our aim in this paper has been to point out what is *possible* or indeed what might already have been attempted (secretly). We do not wish to suggest that the techniques *should* be used to plant trapdoors. Indeed, this is the reason why we have begun to investigate the abuse of optimisation via discriminatory tests. The paper is, we hope, an interesting, original and rather playful exploration of an intriguing issue.

References

1. John A Clark and Jeremy L Jacob. Searching for a Solution: Engineering Trade-offs and the Evolution of Provably Secure Protocols. In *Proceedings 2000 IEEE Symposium on Research in Security and Privacy*, pages 82–95. IEEE Computer Society, May 2000.
2. John A Clark and Jeremy L Jacob. Two Stage Optimisation in the Design of Boolean Functions. In Ed Dawson, Andrew Clark, and Colin Boyd, editors, *5th Australasian Conference on Information Security and Privacy, ACISP 2000*, pages 242–254. Springer Verlag LNCS 1841, july 2000.
3. John A Clark and Jeremy L Jacob. Fault Injection and a Timing Channel on an Analysis Technique. In *Advances in Cryptology Eurocrypt 2002*. Springer Verlag LNCS 1592, 2002.
4. John A Clark, Jeremy L Jacob, Susan Stepney, Subhamoy Maitra, and William Millan. Evolving Boolean Functions Satisfying Multiple Criteria. In Alfred Menezes and Palash Sarkar, editors, *INDOCRYPT*, volume 2551 of *Lecture Notes in Computer Science*. Springer, 2002.
5. C. Ding, G. Xiao, and W. Shan. *The Stability of Stream Ciphers*, volume 561 of *Lecture Notes in Computer Science*. Springer-Verlag, 1991.
6. Giddy J.P. and Safavi-Naini R. Automated Cryptanalysis of Transposition Ciphers. *The Computer Journal*, XVII(4), 1994.
7. S. Kirkpatrick, Jr. C. D. Gelatt, and M. P. Vecchi. Optimization by Simulated Annealing. *Science*, 220(4598):671–680, May 1983.
8. Robert A J Mathews. The Use of Genetic Algorithms in Cryptanalysis. *Cryptologia*, XVII(2):187–201, April 1993.
9. W. Millan, A. Clark, and E. Dawson. An Effective Genetic Algorithm for Finding Highly Non-linear Boolean Functions. pages 149–158, 1997. Lecture Notes in Computer Science Volume 1334.
10. W. Millan, A. Clark, and E. Dawson. Boolean Function Design Using Hill Climbing Methods. In Bruce Schneier, editor, *4th Australian Conference on Information Security and Privacy*. Springer-Verlag, april 1999. Lecture Notes in Computer Science Volume 1978.
11. National Bureau of Standards. Data Encryption Standard. *NBS FIPS PUB 46*, 1976.
12. Bruce Schneier. *Applied Cryptography*. Wiley, 1996.
13. Richard Spillman, Mark Janssen, Bob Nelson, and Martin Kepner. Use of A Genetic Algorithm in the Cryptanalysis of Simple Substitution Ciphers. *Cryptologia*, XVII(1):187–201, April 1993.
14. Forsyth W.S. and Safavi-Naini R. Automated Cryptanalysis of Substitution Ciphers. *Cryptologia*, XVII(4):407–418, 1993.

GenTree: An Interactive Genetic Algorithms System for Designing 3D Polygonal Tree Models

Clare Bates Congdon[1] and Raymond H. Mazza[2]

[1] Department of Computer Science, Colby College
5846 Mayflower Hill Drive, Waterville, ME 04901 USA
ccongdon@colby.edu
[2] Entertainment Technology Center, Carnegie Mellon University
5000 Forbes Avenue, Doherty Hall 4301, Pittsburgh, PA 15213 USA
rmazza@andrew.cmu.edu

Abstract. The creation of individual 3D models to include within a virtual world can be a time-consuming process. The standard approach to streamline this is to use procedural modeling tools, where the user adjusts a set of parameters that defines the tree. We have designed GenTree, an interactive system that uses a genetic algorithms (GA) approach to evolve procedural 3D tree models. GenTree is a hybrid system, combining the standard parameter adjustment and an interactive GA. The parameters may be changed by the user directly or via the GA process, which combines parameters from pairs of parent trees into those that describe novel trees. The GA component enables the system to be used by someone who is ignorant of the parameters that define the trees. Furthermore, combining the standard interactive design process with GA design decreases the time and patience required to design realistic 3D polygonal trees by either method alone.

1 Introduction

This paper describes GenTree, an interactive system that models 3D polygonal trees as a set of parameters for a procedural tree description and uses genetic algorithms (GA's) to evolve these parameters.

Realistic-looking organic objects for 3D virtual worlds are time consuming to model. Typically, designers must become skilled in the use of a commercial modeling program such as 3D Studio Max or Maya in order to be able to construct realistic tree models to be used in simulations. Creation of a single tree by even a skilled modeler would typically take several hours using this approach.

Procedural modeling systems, such as described in [11] (or commercial systems SpeedTree and Tree Storm) are intended to streamline the tree model creation process. They allow users to adjust a number of parameters that control the appearance of a tree without having to focus on low-level components of the model.

In procedural systems, many of the possible parameters interact with each other; for example, the length of branches, the base width of branches relative to the parent branch, and the rate at which they taper will combine to determine the width of the tip of each branch. Thus, it can be time consuming and difficult to design a tree with a desired look, and it is difficult for the user to gain an understanding of the effects and interactions of the parameters. Since the trees are constructed one at a time, exploring variations (and creating multiple models) is further time consuming. Furthermore, more realism and variation in the space of possible trees is achieved in procedural tree programs by increasing the parameter space. Increasing the size of this search space increases the burden of using and learning to use such a system.

GenTree seeks to extend the capabilities of the procedural tree-building approach by adding a GA component to the search process. GenTree enables users, even those with no experience in using 3D modeling tools, to quickly design trees for use in virtual worlds. The system is interactive, allowing (but not requiring) the user to adjust the parameters that define each tree. The system also allows (but does not require) the user to assign a numeric rating to each tree; this information is then used by the GA to favor the more appealing trees as parents in the evolutionary process.

This work does not focus on rendering the trees smoothly, and is instead focused on specifying the parameters that define the general shapes of the trees. Consequently, there is much room for improvement in the component of the system that renders the trees, depending on the intended use for these trees.

2 Background

There are several examples of evolutionary approaches to design in recent literature, including [1][2]. These collections in particular include several examples of aesthetic evolutionary projects, but differ in focus from the project presented here.

Sims evolves 3D creatures [1] (and [10][9]), but his focus was on the mobility of the creatures (for example, joints) and not a natural appearance of the creatures. Soddu [2] evolves cityscapes, composed of individual buildings composed in turn of architectural elements. These do not have the added constraint of intended realism in modeling natural objects. Hancock and Frowd's work in the generation of faces is also relevant[2]; however, this project is aimed at altering photographs, rather than constructing 3D models of the faces.

Bentley's system [1] evolves solutions to a wide variety of 3D modeling problems, but focuses on problems that can be evaluated independently of aesthetic considerations. From these collections, the work here is most similar to an extension to Bentley, reported in [2], which allowed the GADES system to interact with a user. The system is able to evolve a wide variety of shapes in response to the user-provided fitness feedback.

L-systems an an established approach to modeling the development of plants, for example [5][7], and has been used directly or inspired previous work on the evolution of plant forms, for example [6]. Finding a set of rules to achieve a desired appearance is a search problem, not unlike the search for parameters

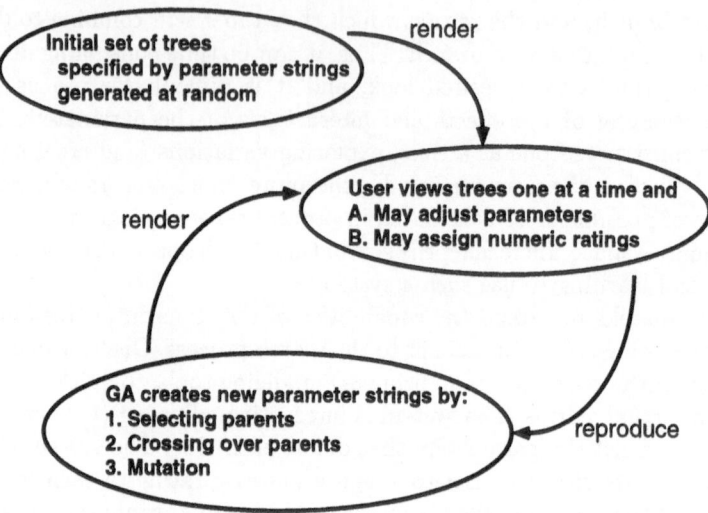

Fig. 1. GenTree iterates between user input and the genetic algorithm in the creation of new trees.

in a procedural tree-modeling program. However, since these systems are evolving rule sets or hierarchical structures, the concerns are different from those in this paper. Also, since the underlying representations are more complex, these approaches are much harder for users to attempt without a genetic algorithms approach (or another form of help with the search process).

Sims [8] evolved plant shapes using parameter strings, but the system is strictly evolutionary and does not allow for interactive parameter tuning. In [4], we describe a GA approach to designing 3D polygonal trees that led to the work reported here. However, this work was an initial exploration of the concept and uses a limited parameter set, resulting in fanciful, rather than realistic, trees. Also, like [8], the system is strictly evolutionary, so the trade-offs between a strictly procedural approach and an evolutionary approach are not explored, nor is the potential for a hybrid approach explored.

3 System Details

As demonstrated in [8] and [4], a GA approach is able to help navigate the search space of a procedural tree-building program. The parameters that define the trees are easily represented as strings, and the user-provided rating of the quality of the solutions serves as a fitness function. Finally, new (offspring) parameter sets can be derived from existing (parent) parameter sets using typical GA processes of crossover and mutation.

This paper extends these ideas to combine the interactive approach of a standard procedural program (where the user adjusts parameters) with an interactive GA.

```
main
  create (random) initial population
  getFitness
  while (maxTrials not reached)
    generateNew
    getFitness

generateNew
  select parents
  mutate
  crossover
  save best (elitism)
```

Fig. 2. High-level pseudocode for the GA system in GenTree.

```
getFitness
  (all trees start with default fitness of 0.0)
  while (not DoneWithGeneration)
    renderTree -- display each tree for user
      (each tree rotates to show all angles)
    user may increment or decrement fitness for each tree
    user may advance to next tree or backup to previous tree
    user may change any of the parameters that define the tree
    user may signal DoneWithGeneration
```

Fig. 3. High-level pseudocode for the interactive component.

3.1 System Overview

GenTree starts with an initial population of trees whose parameters are generated at random. The user may look at each tree in turn, may change any of the parameters that define any of the trees, and may assign each tree a numeric rating that indicates its quality. When signaled by the user, a new generation of trees is produced from the old, using the GA process. Most notably, the parameter sets from two different trees "cross over", producing two new parameter sets. The new set of trees is viewed by the user, and may again be altered and/or rated, and a new set of trees produced from the old via the GA process. A high-level view of the system is illustrated in Figure 1; pseudocode is shown in Figures 2 and 3.

3.2 GA Component

To hasten the development of our system, we used Genesis [3] for the genetic algorithms component of GenTree; the Genesis code has been combined with a modest OpenGL system (developed by the authors) to display the trees for evaluation by the user. We named the system GenTree to reflect a genetic algorithms approach to creating tree models.

Table 1. A description of the values represented in the strings evolved by GenTree and rendered as 3D polygonal trees.

Gene #	Name	Description	Range of Values
0	Number of Branches	The number of sub-branches that protrude from any one branch (modified by noise)	[1..8]
1	Length Ratio	The ratio of the length of a branch to that of the parent branch	[0.2 .. 0.9]
2	Width Ratio	The ratio of the width of the base of a sub-branch to that of the parent branch	[0.2 .. 0.9]
3	Branch Taper	The ratio of the end of a branch to the base of a branch	[0.2 .. 0.9]
4	Branch Proximity	From the tip of the parent branch, the percent of the parent to be used for distributing the sub-branches along the parent.	[0.1 .. 0.8]
5	Tree Depth	The number of branches that can be traversed from the trunk to the tips (excluding the trunk)	[2 .. 9]
6	Branch Angle Delta	The change in x and z direction in the vector that determines the angle of a sub-branch, relative to the parent branch	[0.0 .. 0.1]
7	Vertical change	Added to direction inherited from parent branch.	[0.0 .. 1.0]
8	Parent influence	Strength of influence of parent vector on offspring branches.	[0.0 .. 1.0]
9	Branch Direction Noise	Noise in the x, y, and z direction of the growth of a branch section (off the defined normal vector for the branch section)	[0.0 .. 1.0]
10	Branch Number Noise	Noise in determining the number of branches at each level of growth (higher noise tends to add branches)	[0.0 .. 1.0]
11	Subtree Depth Noise	Noise in determining how deep a subtree is (higher noise tends to truncate subtrees)	[0.0 .. 1.0]
12	Random Seed	Stored with tree parameters so that tree will render consistently	[0.0 .. 1.0]

The mechanics of the genetic algorithm (GA) component of the system are largely unchanged from the original Genesis code, although the interactive nature of the system changes the flow of control, since mouse clicks and keystrokes from the user affect the GA. This will be described later in this section.

3.3 Procedural Tree Component

In designing the parameters used to define the trees, we looked to existing procedural tree programs, plus a bit of trial and error, to determine what would make good looking trees. The parameters evolved by the system and their ranges are provided in Table 3.3.

```
renderTree
  drawSpan(trunk)

drawSpan
  setDirection
  draw conic section
  flip biased coin to possibly add a branch (bias determined by gene 10)
  if (subBranches)
    for each branch
      flip biased coin to possibly make a smaller subtree
        (bias determined by gene 11)
      drawSpan(branch)

setDirection
  if trunk, straight up
  else, combine normal vectors for:
    parent direction, weighted by gene 8
    random compass direction, weighted by gene 6
    random y direction, weighted by gene 7
```

Fig. 4. High-level pseudocode for the rendering component.

3.4 Rendering a GA String into OpenGL Trees

The trees are rendered recursively from the trunk through the branches. High-level pseudocode for the rendering process is provided in Figure 4. Each GA string uniquely translates into a specific rendering, including a seed used for random numbers (gene 12), so the process is deterministic for a given string.

The trunk and each branch are rendered as primitive conic sections[1]. The trees are generated without leaves, but green triangles can be generated at the tips of the branches to assess what a tree would look like if it had leaves on it.

4 Results

One is first tempted to evaluate the GenTree system by judging how good the resulting trees look, and based on this criteria, the system is able to generate good looking trees in a short period of runtime.

---

[1] In OpenGL, conic sections are drawn using a specified number of faces. Because the rendering component is not the focus of the work reported here, the number of faces used here is four to keep the polygon counts low and the rendering fast. Thus, the sections are rectangular.

Fig. 5. A variety of trees from the final population.

4.1 Resulting Trees

Subjective results show that with an initial population of 20 trees[2], several interesting and realistic looking trees can be produced within 5-10 generations of the GA and with less than an hour of the user's time. Because the GA is working with several models simultaneously, the end result of a run is a population of trees, several of which are likely to be attractive and useful. Furthermore, similar looking but distinct trees can be produced from a favorite tree by altering its parameters. Therefore, it is possible to easily construct a variety of trees as the result of one run of the system.

Figure 5 shows a variety of trees from the final population from a run of 7 generations of 20 trees, with both interactive fitness and interactive adjustment used.

Figure 6 shows another example tree produced at the end of the same run, and then shows this same tree with leaves added.

Recall that since the trees (and leaves) are rendered using simple cylinders (and triangles), it is the structure of the trees that should be considered, and not the surfaces of the trees.

---

[2] A population size of more than 20 starts to become tedious from the usability perspective, as the user will typically view and interact with each tree before advancing to the next generation.

GenTree: An Interactive Genetic Algorithms System   2041

Fig. 6. One of the best trees from the final population, shown with and without leaves.

Fig. 7. Five example trees from the initial population.

4.2 Initial Population

As a reference point, a sampling of trees from the initial population are shown in Figure 7. This sample (every fourth tree) includes the two from the initial population that most resemble trees (the third and the fifth shown), as well as one with branches growing into the ground and two with branches growing inside each other. The three clearly errant trees shown in this figure illustrate typical features of the randomly generated trees.

4.3 Comparison of Interaction Styles

The primary question to be asked in the context of the work reported here is whether the GA component assists with the development of the 3D tree models. To assess this, the system was run four different ways:

1. Without providing fitnesses for trees and without altering them in any way.
2. Altering tree parameters for trees that can be improved upon each generation, but not rating the trees.
3. Rating the trees, but not altering them beyond the GA process.
4. Altering tree parameters for trees that can be improved upon each generation and also rating these trees before they are used in the GA process.

In the first instance, the system starts with randomly generated trees and recombines them, so it is not surprising that it does not produce anything interesting. It has been mentioned here merely for completeness.

The second instance is not typical of the GA process in that no information is provided to the GA about solutions that are "better" or "worse" than others, and this information is typically used for preferential parent selection. However, in this case, the alternative mechanism of "eugenics" (helping there be more good solutions to be used as parents) achieves a similar role of recombining primarily good solutions.

The third instance corresponds to the standard use of Genetic Algorithms, and the fourth instance corresponds to the full capacity of the GenTree system to take input from the user in two forms.

The system was run by the authors in all four variants with the general goal of finding good-looking trees with minimal effort. Although this remains a subjective test and evaluation, the authors explicitly did not fine tune any trees in the adjustment process. A population size of 20 trees was used, with a crossover rate of 60 percent and a mutation rate of 0.1 percent (each bit has a .1 percent chance of being set to a random 0 or 1 value). With each approach save the first, for each variation of altering trees and supplying ratings, GA generations were run until approximately an hour had elapsed. This resulted in 5 generations for variation 2, 20 generations for variation 3, and 7 generations for variation 4.

An explanation for the varying numbers of generations is that with the second variation, the most time was spent adjusting the parameters to each tree, since all trees are equally likely to be "parents" to the next generation. In variation 4, when the ratings were used, bad trees (those that could not easily be adjusted) were simply given low ratings. This happens sometimes, for example, when a very angular tree pops up. (Such a tree can be useful "grist for the mill" or might be useful when constructing deliberately fanciful trees, but can sometimes be difficult to smooth out without considering how parameters interact and perhaps changing the parameter set severely. Thus, it can be expedient to simply give a difficult tree a low rating.) With the third variation, no alterations were allowed, so each tree was viewed and assigned a rating, allowing more generations in the time frame.

All three approaches are capable of producing comparable trees, and the assessment of "best" may have to do with the user's background. As users familiar with the available parameters, it is frustrating to do the third run and restrain oneself from "touching up" a promising tree. However, the ease with which a naive user can design trees using only ratings and without knowing the parameters has been witnessed several times in curious colleagues trying the software. For more knowledgeable users, there seems to be little reason to run the system with the second variation because the ability to weed out bad trees or favor interesting ones can be expedient, even if used only occasionally.

The differences observed between the second and fourth variations are that when both ratings and alterations are used, the trees in the final population tend to show more variation than when just alterations (and no ratings) are used. This is a somewhat counterintuitive result in that in many GA applications, the population tends to converge, and it is impossible to discern at this point whether this is a quirk of the author's use of the system or a more meaningful effect. However a possible explanation can be constructed in that when adjusting most of the trees each generation (and not being allowed to simply indicate a tree is undesirable and leave it behind), it appears likely that one gets into the habit of adjusting the same parameters into similar values. Whereas when the ratings are used, less parameter adjusting happens and therefore less bias in adjusting particular values is introduced. (This is just one possible explanation of observed differences in the final populations.)

The *question must also be asked* about the results of altering parameters only (without advancing any generations in the GA process) for an hour, during which time a knowledgeable user would surely be able to produce one or more attractive trees. Using the system this way, more skill would be required in assessing the relative contributions of the different parameters and the interactions of the various parameter settings with each other. The addition of the GA process (with or without ratings) considerably lessens the need for understanding the effects of individual parameters, or even knowing what the parameters are.

Further Observations

Overall, we are quite pleased with the wide variety of shapes that are possible in the system and its ability to yield some trees that are approaching realism. Not surprisingly, the noise parameters appear to be instrumental in making the trees look more realistic and providing more variations of "similar" trees.

5 Conclusions and Future Work

The GenTree system is able to produce a variety of trees with realistic shapes. Although it is difficult to effectively evaluate a system whose output is judged on aesthetic criteria, the addition of the GA mechanisms to a procedural tree modeling system simplifies the development of 3D tree models in that users do not need to understand the effects of individual parameters or even know what the parameters are in order to develop useful models. Rather than having to identify why a tree is good or bad (or the ways in which a tree is good or bad), the user can assign a subjective rating.

Future Work

The introduction of more parameters into the procedural definition of the tree would be useful in that it would allow for a greater variety of trees, including pine trees and palm trees. However, this would also make the space of possible trees more difficult for users to navigate. Thus, as the parameter space increases, the GA mechanisms are expected to be even more useful for helping the user discover appealing trees.

The trees would be significantly more realistic by improving the rendering engine, even if the GA and interactive facets of the system were left unchanged. This would include adding textures to the surfaces and smoother transitions from branches to sub-branches. Furthermore, it may be desirable to parameterize the rendering component, for example, to allow favoring low-polygon-count trees (for some applications) or favoring realism.

Additional benefits could be seen from parameterizing leaf characteristics and improving the rendering of the leaves. Root structure would also make trees more realistic. In addition, different colors and textures could be specified for bark and for leaves, and the colors and textures could be subject to evolution. The ability to add flowers and fruit, and variations on these, could also be added as parameters.

In a few years, we will no longer be constrained by polygonal limits in virtual worlds; some applications will call for trees of high intricacy, which will be too time consuming even to attempt to model through regular methods. There is an obvious advantage to using GenTree in this situation.

References

1. P. J. Bentley. *Evolutionary Design By Computers*. Morgan Kaufmann, San Francisco, CA, 1999.
2. P. J. Bentley and D. W. Corne. *Creative Evolutionary Systems*. Morgan Kaufmann, San Francisco, CA, 2002.
3. J. J. Grefenstette. A user's guide to GENESIS. Technical report, Navy Center for Applied Research in AI, Washington, DC, 1987. Source code updated 1990; available at http://www.cs.cmu.edu/afs/cs/project/ai-repository/ai/areas/genetic/ga/systems/genesis/.
4. R. H. Mazza III and C. B. Congdon. Towards a genetic algorithms approach to designing 3D polygonal tree models. In *Proceedings of the Congress on Evolutionary Computation (CEC-2002)*. IEEE Computer Society Press, 2002.
5. C. Jacob. Genetic L-system programming. In *Parallel Problem Solving from Nature Proceedings (PPSN III), Lecture Notes in Computer Science*, volume 866. Springer, 1994.
6. G. Ochoa. On genetic algorithms and Lindenmayer systems. In *Parallel Problem Solving from Nature Proceedings (PPSN V), Lecture Notes in Computer Science*, volume 1498. Springer, 1998.
7. P. Prusinkiewicz, M. Hammel, R. Mech, and J. Hanan. The artificial life of plants. In *SIGGRAPH course notes, Artificial Life for Graphics, Animation, and Virtual Reality*. ACM Press, 1995.
8. K. Sims. Artificial evolution for computer graphics. In *Proceedings of the 18th annual conference on Computer graphics and interactive techniques (SIGGRAPH91)*, pages 319–328. ACM Press, 1991.
9. K. Sims. Evolving 3D morphology and behavior by competition. In R. Brooks and P. Maes, editors, *Artificial Life IV Proceedings*, pages 28–39. MIT Press, 1994.
10. K. Sims. Evolving virtual creatures. In *Proceedings of the 21st annual conference on Computer graphics and interactive techniques (SIGGRAPH94)*, pages 15–22. ACM Press, 1994.
11. J. Weber and J. Penn. Creation and rendering of realistic trees. In *Proceedings of the 22nd annual conference on Computer graphics and interactive techniques (SIGGRAPH95)*, pages 119–128. ACM Press, 1995.

Optimisation of Reaction Mechanisms for Aviation Fuels Using a Multi-objective Genetic Algorithm

Lionel Elliott[1], Derek B. Ingham[1], Adrian G. Kyne[2], Nicolae S. Mera[2], Mohamed Pourkashanian[3], and Chritopher W. Wilson[4]

[1] Department of Applied Mathematics, University of Leeds, Leeds, LS2 9JT, UK
{lionel,amt6dbi}@amsta.leeds.ac.uk
[2] Centre for Computational Fluid Dynamics, Energy and Resources Research Institute, University of Leeds, Leeds, LS2 9JT, UK
{fueagk,fuensm}@sun.leeds.ac.uk
[3] Energy and Resources Research Institute, University of Leeds, Leeds, LS2 9JT, UK, fue6lib@sun.leeds.ac.uk
[4] Centre for Aerospace Technology, QinetiQ group plc, Cody Technology Park Ively Road, Pyestock, Farnborough, Hants, GU14 0LS, UK
cwwilson1@qinetiq.com

Abstract. In this study a multi-objective genetic algorithm approach is developed for determining new reaction rate parameters for the combustion of kerosene/air mixtures. The multi-objective structure of the genetic algorithm employed allows for the incorporation of both perfectly stirred reactor and laminar premixed flame data into the inversion process, thus producing more efficient reaction mechanisms.

1 Introduction

Reduction of engine development costs, improved predictions of component life or the environmental impact of combustion processes, as well as the improved assessment of any industrial process making use of chemical kinetics, is only possible with an accurate chemical kinetics model. However, the lack of accurate combustion schemes for high-order hydrocarbon fuels (e.g. kerosene) precludes such techniques from being utilised in many important design processes. To facilitate further development, a procedure for optimising chemical kinetic schemes must be devised in which a reduced number of reaction steps are constructed from only the most sensitive reactions with tuned reaction rate coefficients. A novel method for automating the optimisation of reaction mechanisms of complex aviation fuels is developed in this paper.

Complex hydrocarbon fuels, such as kerosene, require more than 1000 reaction steps with over 200 species, see [1]. Hence, as all the reaction rate data is not well known, there is a high degree of uncertainty in the results obtained using these large detailed reaction mechanisms. A reduced reaction mechanism for kerosene combustion was developed in [2] and their mechanism is used in this paper to test the ability of GAs to retrieve the reaction rate coefficients.

We note that various methods have been proposed in order to find a set of reaction rate parameters that gives the best fit to a given set of experimental data. However, in the case of complex hydrocarbon fuels, the objective function is usually highly structured, having multiple ridges and valleys and exhibiting multiple local optima. For objective functions with such a complex structure, traditional gradient based algorithms are likely to fail. Optimisation methods based upon linearization of the objective function, see [3] and [4], fall into the same category. On the other hand, genetic algorithms are particularly suitable for optimising objective functions with complex, highly structured landscapes.

A powerful inversion technique, based on a single objective genetic algorithm, was developed in [5] in order to determine the reaction rate parameters for the combustion of a hydrogen/air mixture in a perfectly stirred reactor (PSR). However, many practical combustors, such as internal combustion engines, rely on premixed flame propagation. Moreover, burner-stabilized laminar premixed flames are very often used to study chemical kinetics in a combustion environment. Such flames are effectively one-dimensional and can be made very steady, thus facilitating detailed experimental measurements of temperature and species profiles. Therefore, for laminar premixed flames it is much easier to obtain accurate experimental data to be used in the genetic algorithm inversion procedure. Therefore, in order to obtain a reaction mechanism which can be used for a wide range of practical problems it is required to use an inversion procedure which incorporates multiple objectives. Another reason to use this approach is the fact that sometimes in practical situations the amount of experimental data available is limited and it is important to use all the available experimental data, which may come from different measurements for perfectly stirred reactors or laminar premixed flames.

Therefore the single objective optimisation technique proposed in [5] is further extended to a multi objective genetic algorithm in order to include into the optimisation process data from premixed laminar flames. The algorithm was found to provide good results for small scale test problem, i.e. for the optimisation of reaction rate coefficients for a hydrogen/air mixture. It is the purpose of this paper to extend this technique to the case of complex hydrocarbon fuels.

An inversion procedure based on a reduced set of available measurements is also considered. In many experimental simulations, some of the species concentrations are subject to large experimental errors both in perfectly stirred rectors and laminar premixed flames. In order to facilitate the incorporation of various types of experimental data in the GA based inversion procedures, this study also investigates such GA calculations based on such incomplete or reduced sets of data.

2 Description of the Problem

2.1 Reaction Rate Parameters

In a combustion process the net chemical production or destruction rate of each species results from a competition between all the chemical reactions involving that species. In this study it is assumed that each reaction proceeds according to the law of mass action and the forward rate coefficients are of three parameter functional Arrhenius form, namely

$$k_{f_i} = A_i T^{b_i} \exp\left(-\frac{E_{a_i}}{RT}\right) \quad (1)$$

for $i = 1, \ldots, N_R$, where R is the universal gas constant and there are N_R competing reactions occurring simultaneously. The rate equations (1) contains the three parameters A_i, b_i and E_{a_i} for the ith reaction. These parameters are of paramount importance in modelling the combustion processes since small changes in the reaction rates may produce large deviations in the output species concentrations. In both perfectly stirred reactors as well as in laminar premixed flames, if the physical parameters required are specified, and the reaction rates (1) are given for every reaction involved in the combustion process, then the species concentrations of the combustion products may be calculated. It is the possibility of the determination of these parameters for each reaction, based upon outlet species concentrations, which is investigated in this paper.

2.2 The Perfectly Stirred Reactor (PSR) Calculations and the Laminar Premixed Flames (PREMIX) Calculations

Various software packages may be used for the direct calculations to determine the output species concentrations if the reaction rates are known. In this study the perfectly stirred rector calculations are performed using the PSR FORTRAN computer program that predicts the steady-state temperature and composition of the species in a perfectly stirred reactor, see [6] while the laminar premixed flame structure calculations were performed using the PREMIX code, see [7].

We use X_k to denote the mole fraction of the kth species, where $k = 1, \ldots, K$ and K represents the total number of species. The temperature and composition which exit the reactor are assumed to be the same as those in the reactor since the mixing in the reactor chamber is intense.

If PSR calculations are undertaken for N_S^{PSR} different sets of reactor conditions then the output data which is used in the GA search procedure will consist of a set of mole concentrations, $(X_{jk}, j = 1, \ldots, N_S^{PSR}, k = 1, \ldots, K)$, where X_{jk} represents the mole concentration of the kth species in the jth set of reactor conditions.

The laminar premixed flame structure calculations were performed using the PREMIX code for burner stabilised flames with a known mass flow rate. The code is capable of solving the energy equation in determining a temperature profile. However, throughout this investigation a fixed temperature profile was preferred and the reason for fixing the temperature was two fold. Firstly, in many flames there can be significant heat losses to the external environment, which are of unknown and questionable origin and thus troublesome to model. Secondly, the CPU time involved in arriving at the final solution was decreased significantly.

If all the physical operating conditions are specified then for a given set of reaction rates (1) the profiles of the species concentration and the burning velocity as functions of the distance from the burner's surface are calculated. If PREMIX calculations are performed for N_S^{PREMIX} different sets of operating conditions then the output data of the code which is used by the GA in the matching process consists of a set of species concentration profiles ($Y_{jk}(x)$, $j = 1, ..., N_S^{PREMIX}$, $k = 1, ..., K$) and a set of burning velocity profiles ($V_j(x)$, $j = 1, ..., N_S^{PREMIX}$), where Y_{jk} is the profile of the mole concentration along the burner for the kth species in the jth set of operating conditions and V_j is the burning velocity profile for the jth set of operating conditions.

It is the purpose of this paper to determine the reaction rate coefficients A's, b's and E_a's in equation (1) given species concentrations X_{jk} in a perfectly stirred reactor and/or profiles of species concentration Y_{jk} and the burning velocity profiles V_j in laminar premixed flames.

3 Reformulation of the Problem as an Optimisation Problem

An inverse solution procedure attempts, by calculating new reaction rate parameters that lie between predefined boundaries, to recover the profiles of the species (to within any preassigned experimental uncertainty) resulting from numerous sets of operating conditions. The inversion process aims to determine the unknown reaction rate parameters $((A_i, b_i, E_{a_i}), i = 1, ..., N_R)$ that provide the best fit to a set of given data. Thus, first a set of output concentration measurements of species is simulated (or measured). If numerical simulations are performed for N_S^{PSR} different sets of PSR operating conditions and/or N_S^{PREMIX} different sets of PREMIX operating conditions then the data will consist of a set of KN_S^{PSR} concentration measurements of species for PSR calculations and $(K + 1)N_S^{PREMIX}$ profiles of species concentrations and burning velocity for PREMIX calculations.

Genetic algorithms based inversion procedures seek for the set of reaction rate parameters that gives the best fit to these measurements. In order to do so we consider two objective functions which compare predicted and measured species concentrations for PSR and PREMIX simulations

$$f_{PSR}((A_i, b_i, E_{a_i})_{i=1, N_R}) = \left\{ 10^{-8} + \sum_{j=1}^{N_S} \sum_{k=1}^{K} W_k \frac{\left|X_{jk}^{calc} - X_{jk}^{orig}\right|}{X_{jk}^{orig}} \right\}^{-1} \quad (2)$$

$$f_{PREMIX}((A_i, b_i, E_{a_i})_{i=1, N_R}) =$$
$$\left\{ 10^{-8} + \sum_{j=1}^{N_S} \left(\frac{\|V_j^{calc} - V_j^{orig}\|_{L^2}}{\|V_j^{orig}\|_{L^2}} \right) + \sum_{k=1}^{K} W_k \frac{\|Y_{jk}^{calc} - Y_{jk}^{orig}\|_{L^2}}{\|Y_{jk}^{orig}\|_{L^2}} \right\}^{-1} \quad (3)$$

where

- X_{jk}^{calc}, Y_{jk}^{calc} and V_j^{calc} represent the mole concentrations of the kth species in the jth set of operating conditions and the burning velocity in the jth set of

operating conditions using the set of reaction rate parameters $((A_i, b_i, E_{a_i}), i = 1, ..., N_R)$,
- X_{jk}^{orig}, Y_{jk}^{orig} and V_j^{orig} are the corresponding original values which were measured or simulated using the exact values of the reaction rate parameters,
- $\|\cdot\|_{L^2}$ represents the L^2 norm of a function which is numerically calculated using a trapezoidal rule,
- W_k are different weights that can be applied to each species depending on the importance of the species.

It should be noted that the fitness function (2) is a measure of the accuracy of species concentrations predictions obtained by a given reaction mechanism for PSR simulations. The second fitness function (3) is a measure of the accuracy in predicting species concentrations and burning velocity profiles in the case of laminar premixed flames.

4 The Multi-objective Genetic Algorithm Inversion Technique

There is a vast variety of genetic algorithms approaches to multi-objective optimisation problems. GAs are in general a very efficient optimization technique for problems with multiple, often conflicting, object functions. However, most GA approaches require that all the object functions are evaluated for every individual generated during the optimisation process. For the problem considered in this paper, the object function f_{PREMIX} requires a long computational run time. For example, in the kerosene case a PREMIX run for one operating condition, $N_S^{PREMIX} = 1$, was found to take an average of 94 seconds CPU time on a Pentium IV processor running at 1.7GHz. This leads to a running time of more than 65 days, for evaluating say 1000 GA generations with 60 individuals generated at every iteration. On the other hand, PSR calculations are much faster, requiring for example an average of 14 seconds CPU time for evaluating one individual at $N_S^{PSR} = 18$ different operating conditions. Therefore, in this paper we propose a multi-objective GA that avoids PREMIX evaluations at every generation. The algorithm is based on alternating the fitness functions that are used for selecting the survivors of the populations generated during the GA procedure. Under the assumptions that the optimisation problem considered has n objective functions the multi-objective GA employed in this paper consists of the following steps

1. Construct n fitness functions $f_1, f_2, ..., f_n$ and assign a probability for each of these functions $p_1, p_2, ..., p_n$.
2. Construct an initial population.
3. Select a fitness function f_i with the probability p_i using a roulette wheel selection
4. Using the fitness function selected at step 3 construct a new population
    - Use the fitness function f_i to select a mating pool.

- Apply the genetic operators (crossover and mutation) to create population of offsprings.
- Use the fitness function f_i to construct a new population by merging the old population with the offsprings population.
5. Repeat steps 3-4 until a stopping condition is satisfied, i.e. convergence is achieved.

For the problem considered in this paper we have $n = 2$, $f_1 = f_{PSR}$, $f_2 = f_{PREMIX}$. Different values can be used for the fitness selection probabilities p_{PSR} and p_{PREMIX}. It is worth noting that the fitness selection probabilities p_{PSR} and p_{PREMIX} can be used to bias the convergence of the algorithm towards accurate predictions in PSR or PREMIX, respectively. However they do not act as weights since at every generation only one of the objective functions (2) and (3) is evaluated and not a linear combination of both of them. Instead these objective functions are alternated during the evolution process. This method maintains the multi-objective character of the optimisation process while avoiding the evaluation of the expensive fitness function (3) during every generation.

The CPU time required by the algorithm depends on the values set for the probabilities p_{PSR} and p_{PREMIX}. In this paper we have considered $p_{PSR} = 0.9$ and $p_{PREMIX} = 0.1$. For these values the genetic algorithm was found to require an average of 22 minutes/generation on a Pentium IV processor running at 1.7GHz. Numeircal experiments have been performed for various values for the selection probabilities p_{PSR} and p_{PREMIX} and it was found that the quality of the solution found in the end is not affected for a large range of values. However the CPU time can be controlled by choosing suitably these probabilities according to the number of PSR and PREMIx runs required.

We note that the GA employed is based on a floating point encoding of the variables. The genetic operators and the GA parameters were taken to be as follows: population size $n_{pop} = 50$, number of offspring $n_{child} = 60$, uniform arithmetic crossover, crossover probability $p_c = 0.65$, tournament selection, tournament size $k = 2$, tournament probability $p_t = 0.8$, non-uniform mutation, mutation probability $p_m = 0.5$, elitism parameter $n_e = 2$.

The new sets of rate constants generated every iteration are constrained to lie between predefined boundaries that represent the uncertainty associated with the experimental findings listed in the National Institute of Standards and Technology (NIST) database. The constraints are imposed by associating a very low fitness function to the individuals that violate these constraints.

It is worth noting that when working with accurate experimental data or numerically simulated data the two fitness functions (2) and (3) have a common global optimum given by the real values of the reaction rate coefficients or the values that were used to generate the data, respectively. Therefore for the problem considered the two objective functions are not conflicting and the method we have proposed can be used although it searches for a single common optima, rather than a set of non-dominated solutions. It is worth noting that if the experimental data is corrupted with a large level of noise then the two objective functions (2) and (3) may be conflicting and then Pareto oriented techniques have to be used in order to construct the front of non-dominated solutions, i.e. a set of possible optima. In this situation other considerations have to be used to make a decision on what is the best location of the optimum.

5 Numerical Results

In order to test the multi-objective genetic algorithm inversion procedure, the test problem considered is the recovery of the species concentration profiles predicted by a previously validated kerosene/air reaction mechanism for a wide range of operating conditions which was presented in [2]. The reaction scheme used to generate the data involves $K = 63$ species across $N_R = 168$ reactions. Ideally, experimental data would have been preferred but at this developmental stage in order to avoid the uncertainty associated with experimental data it was decided to undertake numerical computations on a set of numerically simulated data. Since the purpose of this study is to develop and to assess the technical feasibility of the multi-objective GA techniques for the development of new reaction mechanisms for complex high order fuels in order to employ the most efficient schemes for future investigations concerning aviation fuels, for this study numerically simulated data is as efficient as experimental data.

The test problem considered is the combustion of a kerosene/air mixture at constant atmospheric pressures, $p = 10 atm$ and $p = 40 atm$. A set of $N_S^{PSR} = 18$ different operating conditions have been considered for PSR runs corresponding to various changes to the inlet temperature from $780K$ to $1150K$. The inlet composition is defined by a specified fuel/air equivalence ratio, $\Phi = 1.0$ or $\Phi = 1.5$. Rather than solving the energy equation we assume that there is no heat loss from the reactor, i.e. $Q = 0$ and the temperature profile is specified.

For PREMIX calculations $N_S^{PREMIX} = 1$ for a steady state burner stabilized premixed laminar flames with specified temperature, pressure $p = 1 atm$ and fuel/air equivalence ratio, $\Phi = 1.0$.

In many practical situations measurements are available only for the most important species. Therefore it is the purpose of this section to test the GA inversion procedure proposed also for the case of incomplete sets of data. Two reaction mechanisms are generated using two sets of data simulating complete and incomplete sets of measurements as follows:

– COMPLETE - is generated using multi-objective optimisation based on PSR and PREMIX data for all the output species
– REDUCED - is generated using multi-objective optimisation based on PSR and PREMIX data only for the following species: kerosene (C10H22+C6H5CH3), O2, C6H6, CO, CH2O, CO2, H2, C2H4, CH4, C3H6, C2H6, C2H2, C4H8

These two reaction mechanisms are used to produce estimations of the output species concentrations and these estimations are compared with the predictions given by the original mechanism (ORIGINAL), i.e. the mechanism which was used to simulate the data numerically, see [2].

5.1 Output Species Predictions for PSR and PREMIX Calculations

Figures 1 presents the unburned fuel concentration calculated with the PSR program, using the reaction mechanisms generated in the GA inversion procedure and the original mechanism for various temperatures and the air/fuel ratio given by $\Phi = 1.0$ at constant atmospheric pressures, $p = 10 atm$. It can be seen that both the COMPLETE and REDUCED mechanism which were generated by the multi-objective GA are accurate in reproducing the output species concentrations as predicted by the original mechanism. Although not presented here, it is reported that also for the other species similar results were obtained. If the pro-

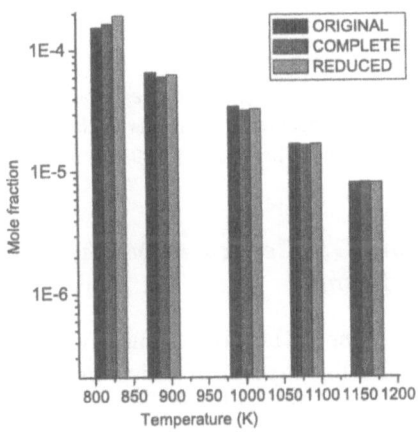

Fig. 1. The mole fraction of unburned fuel for various temperatures and air/fuel ratio given by $\Phi = 1.0$ at constant atmospheric pressures, $p = 10 atm$ as predicted by various reaction mechanisms in PSR calculations.

files of the mole fractions in the PREMIX calculations are investigated, then the GA generated reaction mechanisms are again found to be efficient in predicting the species profiles, see Figure 2 which presents the output species profiles of four major species for a pressure p = 1atm and fuel/air equivalence ratio, $\Phi = 1.0$. It can be seen that both mechanisms generated by our inversion procedure are efficient in predicting the mole fractions even for very low concentrations. Many other species have been investigated for PSR and PREMIX calculations for operating conditions within the range used for the optimisation process and overall it was found that the reaction mechanisms generated by the GA are accurately reconstructing the output species predicted by the original mechanism.

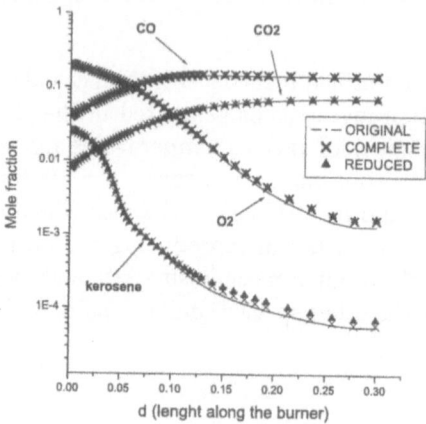

Fig. 2. The species profiles of various output species for premixed laminar flames as predicted by various reaction mechanisms, namely the original mechanism (-), the COMPLETE mechanism (×) and the REDUCED mechanism (▲).

5.2 Validation of the GA Generated Mechanisms for Various Operating Conditions

The previous section presented the predictions given by the constructed reaction mechanisms for operating conditions which were included in the optimisation process. However, the aim of the inversion procedure is to develop reaction mechanisms that are valid over wide ranges of operating conditions. Therefore, next the GA generated reaction mechanisms were tested at operating conditions outside the range included in the optimisation procedure. Figure 3 presents the species concentrations obtained by PSR calculations for the two GA generated reaction mechanisms, in comparison with those generated by the original mechanism for kerosene concentration at temperatures outside the range used for optimisation, i.e. $T = 1000K - 2000K$. Again it can be seen that the mechanisms COMPLETE and REDUCED give accurate predictions in comparison with the original mechanism. Figure 4 presents the predicted concentrations profiles of four of the main species obtained by PREMIX calculations at a constant atmospheric pressure $p = 2atm$ which is outside the range of operating conditions used for optimisation. It is worth noting that both COMPLETE and REDUCED reaction mechanisms are able to accurately predict the species concentration profiles for PREMIX operating conditions outside the range used for optimisation. It is worth noting that for some species, see Figure 4 it was found that the REDUCED mechanism is more efficient than the COMPLETE mechanism in particular in the part of the species profiles where the concentration is becoming very low. This can be explained by the fact that the REDUCED mechanism is biased towards some of the species and it predicts them very accurately while the COMPLETE mechanism is forced to predict with reasonable accu-

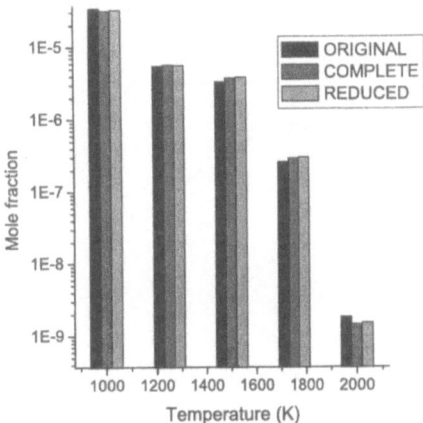

Fig. 3. The mole fraction of unburned fuel for various temperatures $T=1000K$-$2000K$ as predicted by various reaction mechanisms in PSR calculations.

racy all the species, which is undertaken at the expense of slightly less accurate predictions for the main species. The REDUCED mechanism outperformed the COMPLETE mechanism only for those species which were included in the REDUCED inversion procedure while the COMPLETE mechanism outperformed the REDUCED mechanism if all the species are taken into consideration. It is worth mentioning that when working with numerically simulated data both mechanisms can be generated but in practice, it is impossible to obtain measurements for all the output species and therefore a reaction mechanism of the type REDUCED is generated when working with experimental data. Similar conclusions have been obtained for a wide range of operating conditions. Overall, the reaction mechanisms generated by the multi-objective GA proposed in this study were found to be very efficient in predicting the output species concentrations and burning velocity for both PSR and PREMIX simulations. Thus we may conclude that the inversion procedure proposed efficiently generates a general reaction mechanism that can be used to predict the combustion products for various combustion problems at a variety of operating conditions.

5.3 Ignition Delay Time Validations

To provide additional validation of the results obtained by the inversion procedures presented in this paper, the predicted ignition delay times are computed using all the reaction mechanisms generated in this study. In testing the ignition delay time, the SENKIN code of Sandia National Laboratories [8] was used to predict the time dependent chemical kinetic behaviour of a homogeneous gas mixture in a closed system.

Figure 5 compares the ignition delay times predicted by the reaction mechanisms generated in this study with the ignition delay times generated by the

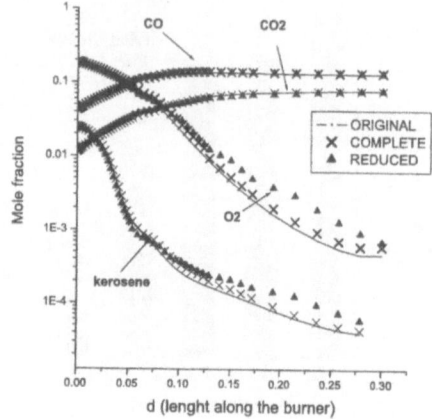

Fig. 4. The species profiles of various output species for premixed laminar flames at pressure $p = 2atm$ and fuel/air equivalence ratio, $\Phi = 1.0$ as predicted by various reaction mechanisms, namely the original mechanism (-), the COMPLETE mechanism (×) and the REDUCED mechanism (▲).

original mechanism at four different initial temperatures and the air/fuel ratio $\Phi = 1.0$. We note that both the COMPLETE and the REDUCED mechanisms based on complete or incomplete sets of data, respectively, reproduce the same ignition delay times as the original mechanism and this indicates that the GA does not just generate a mechanism whose validity only extends as far as those operating conditions included in the optimisation process.

6 Conclusions

In this study we have developed a multi-objective GA inversion procedure that can be used to generate new reaction rate coefficients that successfully predict the profiles of the species produced by a previously validated kerosene/air reaction scheme. The new reaction rate coefficients generated by GAs were found to successfully match mole fraction values in PSR simulations and burning velocity and species profiles in PREMIX simulations, as well as ignition delay time predicted by the original starting mechanism over a wide range of operating conditions, despite their origin being based solely on measurements for a restricted range of operating conditions.

Acknowledgments. The authors would like to thank the EPSRC and the corporate research program for the MOD and DTI Program for the funding and permission to publish this work.

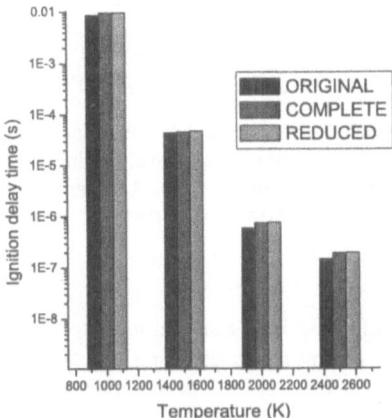

Fig. 5. The ignition delay time at various temperatures T=1000K-2500K as predicted by various reaction mechanisms.

References

1. Dagaut, P., Reuillon, M., Boetner, J.C. and Cathonnet, M.: Kerosene combustion at pressures up to 40atm: experimental study and detailed chemical kinetic modelling, Proceedings of the Combustion Institute, 25, (1994), 919–926.
2. Kyne, A.G., Patterson, P.M., Pourkashanian, M., Williams, A. and Wilson, C.J. Prediction of Premixed Laminar Flame Structure and Burning Velocity of Aviation Fuel-Air Mixtures, Proceedings of Turbo Expo 2001: ASME TURBO EXPO2001: LAND, SEA AND AIR, June 4-7, New Orleans, USA, (2001)
3. Milstein, J., The inverse problem: estimation of kinetic parameters, in: K.H. Ezbert, P. Deuflhard and W. Jager (Eds), Modelling of Chemical Reaction Systems, , Springer, Berlin (1981).
4. Bock, H.G., Numerical treatment of inverse problems in chemical reaction kinetics, in Ezbert,K.H., Deuflhard,P. and Jager, W. (Eds), Modelling of Chemical Reaction Systems , Springer, Berlin , (1981).
5. Harris, S.D., Elliott,L., Ingham, D.B. Pourkashanian, M. Wilson, C.W. The Optimisation of Reaction Rate Parameters for Chemical Kinetic Modelling of Combustion using Genetic Algorithms, Computer methods in applied mechanics and engineering, 190, (2000), 1065–1083.
6. Glarborg, P., Kee, R.J., Grcar, J.F., Miller, J.A.: PSR: A FORTRAN program for modelling well-stirred reactors, Sandia National Laboratories Report SAND86-8209, (1988).
7. Kee, R.J., Grcar, J.F., Smooke, M.D., Miller, J.A.: A FORTRAN program for modelling steady laminar one-dimensional premixed flames, Sandia National Laboratories Report SAND85-8240, (1985).
8. Lutz, A.E., Kee, R.J. and Miller, J.A., SENKIN: A FORTRAN program for predicting homogeneous gas phase chemical kinetics with sensitivity analysis, Sandia National Laboratories Report SAND87-8248, (1987).

System-Level Synthesis of MEMS via Genetic Programming and Bond Graphs

Zhun Fan[1], Kisung Seo[1], Jianjun Hu[1],
Ronald C. Rosenberg[2], and Erik D. Goodman[1]

[1]Genetic Algorithms Research and Applications Group (GARAGe)
[2]Department of Mechanical Engineering
Michigan State University, East Lansing, MI, 48824
{hujianju,ksseo,fanzhun,rosenber,goodman}@egr.msu.edu

Abstract. Initial results have been achieved for automatic synthesis of MEMS system-level lumped parameter models using genetic programming and bond graphs. This paper first discusses the necessity of narrowing the problem of MEMS synthesis into a certain specific application domain, e.g., RF MEM devices. Then the paper briefly introduces the flow of a structured MEMS design process and points out that system-level lumped-parameter model synthesis is the first step of the MEMS synthesis process. Bond graphs can be used to represent a system-level model of a MEM system. As an example, building blocks of RF MEM devices are selected carefully and their bond graph representations are obtained. After a proper and realizable function set to operate on that category of building blocks is defined, genetic programming can evolve both the topologies and parameters of corresponding RF MEM devices to meet predefined design specifications. Adaptive fitness definition is used to better direct the search process of genetic programming. Experimental results demonstrate the feasibility of the approach as a first step of an automated MEMS synthesis process. Some methods to extend the approach are also discussed.

1 Introduction

Mechanical systems are known to be much more difficult to address with either systematic design or clean separation of design and fabrication. Composed of parts involving multiple energy domains, lacking a small set of primitive building blocks such as the NOR and NAND gates in used VLSI, and lacking a clear separation of form and function, mechanical systems are so diverse in their design and manufacturing procedures that they present more challenges to a systematic approach and have basically defied an automated synthesis attempt.

Despite the numerous difficulties presented in automated synthesis of macro-mechanical systems, MEMS holds the promise of being amenable to structured automated design due to its similarities with VLSI, provided that the synthesis is carried out in a properly constrained design domain.

Due to their multi-domain and intrinsically three-dimensional nature of MEMS, their design and analysis is very complicated and requires access to simulation tools with finite element analysis capability. Computation cost is typically very high. A common representation that encompasses multiple energy domains is thus needed for modeling of the whole system. We need a system-level model that reduces the number of degrees of freedom from the hundreds and thousands of degrees of freedom characterizing the meshed 3-D model to as few as possible. The bond graph, based on power flow, provides a unified model representation across multiple energy domain system and is also compatible with 3-D numerical simulation and experimental results in describing the macro behavior of the system, so long as suitable lumping of components can be done to obtain lumped-parameter models. It can be used to represent the behavior of a subsystem within one energy domain, or the interaction of multiple domains. Therefore, the first important step in our method of MEMS synthesis is to develop a strategy to automatically generate bond graph models to meet particular design specifications on system level behaviors.

For system-level design, hand calculation is still the most popular method in current design practice. This is for two reasons: 1) The MEMS systems we are considering, or designing are relatively simple in dynamic behavior -- especially the mechanical parts -- largely due to limitation in fabrication capability. 2) There is no powerful and widely accepted synthesis approach to automated design of multi-domain systems.

The BG/GP approach, which combines the capability of genetic programming to search in an open-ended design space and the merits of bond graphs for representing and modeling multi-domain systems elegantly and effectively, proves to be a promising method to do system-level synthesis of multi-domain dynamical systems [1][2]. In the first or higher level of system synthesis of MEMS, the BG/GP approach can help to obtain a high-level description of a system that assembles the system from a library of existing components in an automated manner to meet a predefined design specification. Then in the second or lower level, other numerical optimization approaches [3], as well as evolutionary computation, may be used to synthesize custom components from a functionality specification. It is worthwhile to point out that for the system designer, the goal of synthesis is not necessarily to design the optimum device, but to take advantage of rapid prototyping and "design reuse" through component libraries; while for the custom component designer, the goal may be maximum performance. These two goals may lead to different synthesis pathways. Figure 1 shows a typical structured MEMS synthesis procedure, and the BG/GP approach aims to solve the problem of system-level synthesis in an automated manner in the first level.

However, in trying to establish an automated synthesis approach for MEMS, we should take cautious steps. Due to the limitations of fabrication technology, there are many constraints in design of MEMS. Unlike in VLSI, which can draw on extensive sets of design rules and programs that automatically test for design-rule violations, the MEMS field lacks design verification tools at this time. This means that no design automation tools are available at this stage capable of designing and verifying any kind of geometrical shapes of MEMS devices. Thus, automated MEMS synthesis tools must solve sub-problems of MEMS design in particular application domains for which a small set of predefined and widely used basic electromechanical elements are available, to cover a moderately large functional design space.

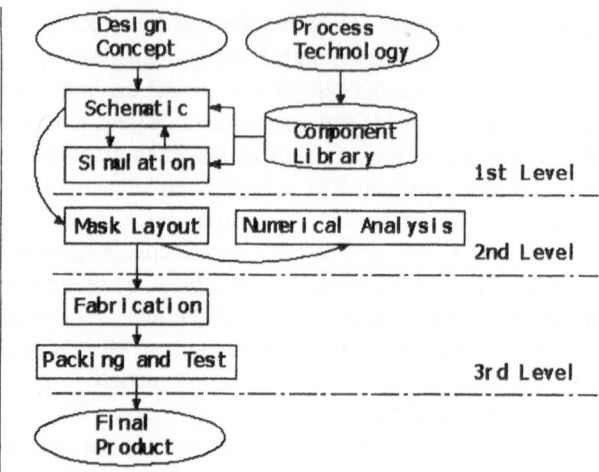

Fig. 1. Structured MEMS design flow

Automated synthesis of an RF MEM device, namely, a micro-mechanical band pass filter, is taken as an example in this paper. As designing and micromachining of more complex structures is a definite trend, and research into micro-assembly is already on its way, the BG/GP approach is believed to have many potential applications. More work to extend this approach to an integrated evolutionary synthesis environment for MEMS across a variety of design layers is also discussed at the end.

2 Design Methodology

Over the past two decades, computational design algorithms based on Charles Darwin's principles of evolution have developed from academic curiosities into practical and effective tools for scientists and engineers. Gero, for example, investigates evolutionary systems as computational models of creative design and studies the relationships among genetic engineering, style emergence, and complex evolution [13]. Goodman et al. [14] studied evolution of engineering artifacts parallel using genetic algorithms.

Shah [15] identifies the importance of bond graphs for unifying multi-level design of multi-domain systems. Tay et al. [12] use bond graphs and GA to generate and analyze dynamic system designs automatically. This approach adopts a variational design method, which means they make a complete bond graph model first, and then change the bond graph topologically using a GA, yielding new design alternatives. However, the efficiency of his approach is hampered by the weak ability of GA to search in both topology and parameter spaces simultaneously. Other researchers have begun to explore combination of bond graphs and evolutionary computation [16]. Campell [18] also uses the idea of both bond graphs and genetic algorithm in his A-

Design framework. In this research, we use an approach combining genetic programming and bond graphs to automate the process of design of dynamic systems to a significant degree.

2.1 Bond Graphs

The bond graph is a modeling tool that provides a unified approach to the modeling and analysis of dynamic systems, especially hybrid multi-domain systems including mechanical, electrical, pneumatic, hydraulic components, etc. It is the explicit representation of model topology that makes the bond graphs a good candidate for use in open-ended design search. For notation details and methods of system analysis related to the bond graph representation, see [4].

Bond graphs have four embedded strengths for design applications, namely, the wide scope of systems that can be created because of the multi- and inter-domain nature of bond graphs, the efficiency of evaluation of design alternatives, the natural combinatorial features of bond and node components for generation of design alternatives, and the ease of mapping to the engineering design process. Those attributes make bond graphs an excellent candidate for modeling and design of a multi-domain system.

2.2 Combining Bond Graphs and Genetic Programming

The most common form of genetic programming [5] uses trees to represent the entities to be evolved. Defining of a proper function set is one of the most significant steps in using genetic programming. It may affect both the search efficiency and validity of evolved results and is closely related to the selection of building blocks for the system being designed. In this research, a basic function set and a modular function set are presented and listed in Tables 1 and 2. Operators in the basic function set basically aim to construct primitive building blocks for the system, while operators in the modular function set purport to utilize relatively modular and predefined building blocks composed of primitive building blocks. Notice that numeric functions are included in both function sets, as they are needed in both cases. In other research, we hypothesize that usage of modular operators in genetic programming has some promise for improving its search efficiency. However, in this paper, we concentrate on another issue, proposing the concept of a realizable function set. By using only operators in a realizable function set, we seek to guarantee that the evolved design is physically realizable and has the potential to be manufactured. This concept of realizability may include stringent fabrication constraints to be fulfilled in some specific application domains. This idea is to be illustrated in the design example of an RF MEM device, namely, a micro-mechanical band pass filter.

Examples of modular operators, namely insert_BU and insert_CU operators, are illustrated in Figures 2 and 3. Examples of basic operators are available in our earlier work [6]. Figure 2 explains how the insert_BU function works. A Bridging Unit (BU) is a subsystem that is composed of three capacitors with the same parameters, attached together with a 0-junction in the center and 1-junctions at the left and right ends. After execution of the insert_BU function, an additional modifiable site (2) appears at the rightmost newly created bond. As illustrated in Figure 3, a resonant unit (RU), composed of one I, R, and C component all attached to a 1-junction, is inserted in an original bond with a modifiable site through the insert_RU function. After the insert_RU function is executed, a new RU is created and one additional modifiable site, namely bond (3), appears in the resulting phenotype bond graph, along with the original modifiable site bond (1). The new added 1-junction also has an additional modifiable site (2). As components C, I, and R all have parameters to be evolved, the insert_RU function has three corresponding ERC-typed sites, (4), (5), and (6), for numerical evolution of parameters. The reason these representations are chosen for the RU and BU components is discussed in the next, case study, section.

Table 1. Operators in Basic Function Set

Basic Function Set	
add_C	Add a C element to a junction
add_I	Add a I element to a junction
add_R	Add a R element to a junction
insert_J0	Insert a 0-junction in a bond
insert_J1	Insert a 1-junction in a bond
replace_C	Replace the current element
replace_I	Replace the current element
replace_R	Replace the current element
+	Sum two ERCs
-	Subtract two ERCs
enda	End terminal for add functions
endi	End terminal for insert func-
endr	End terminal for replace func-
erc	Ephemeral Random Constant

Table 2. Operators in Modular Function Set

Modular Function Set	
insert_RU	Insert a Resonant Unit
insert_CU	Insert a Coupling Unit
insert_BU	Insert a Bridging Unit
add_RU	Add a Resonant Unit
insert_J01	Insert a 0-1-junction com-
insert_CIR	Insert a special CIR com-
insert_CR	Insert a special CR compound
Add_J	Add a junction compound

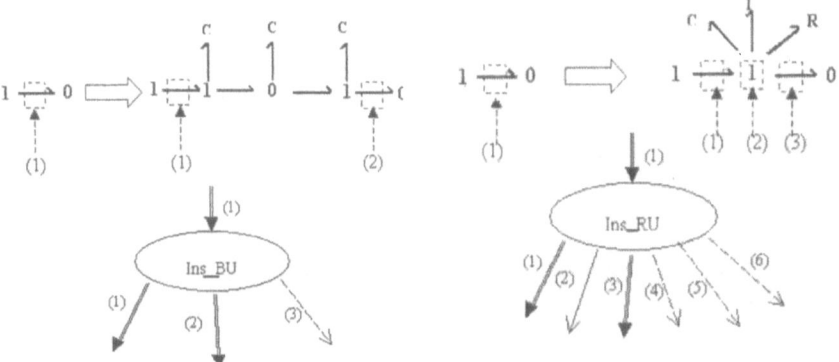

Fig. 2. Operator to Insert Bridging Unit   **Fig. 3.** Operator to Insert Resonant Unit

3 MEM Filter Design

3.1 Filter Topology

Automated synthesis of a RF MEM device, micro-mechanical band pass filters is used as an example in this paper [7]. Through analyzing two popular topologies used in surface micromachining of micro-mechanical filters, we found that they are topologically composed of a series of RUs and Bridging Units (BUs) or RUs and Coupling Units (CUs) concatenated together. Figure 4, 5, 6 illustrates the layouts and bond graph representations of filter topology I and II.

3.2 Design Embryo

All individual genetic programming trees create bond graphs from an embryo. Selection of the embryo is also an important topic in system design, especially for multi-port systems. In our filter design problems, we use the bond graph shown in Figure 7 as our embryo.

3.3 Realizable Function Set

BG/GP is a quite general approach to automate synthesis of multidisciplinary systems. Using a basic set of building blocks, BG/GP can perform topologically open composition of an unconstrained design. However, engineering systems in the real world are often limited by various constraints. So if BG/GP is to be used to synthesize real-world engineering systems, it must enforce those constraints.

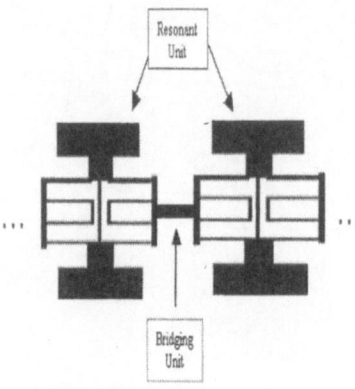

Fig. 4. Layout of Filter Topology I: Filter is composed of a series of Resonator Units (RUs) connected by Bridging Units (BUs).

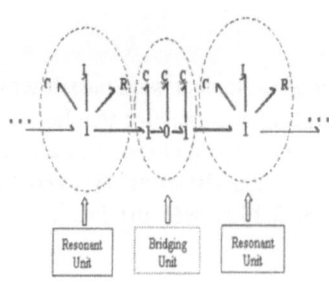

Fig. 5. Bond Graph Representation of Filter Topology I

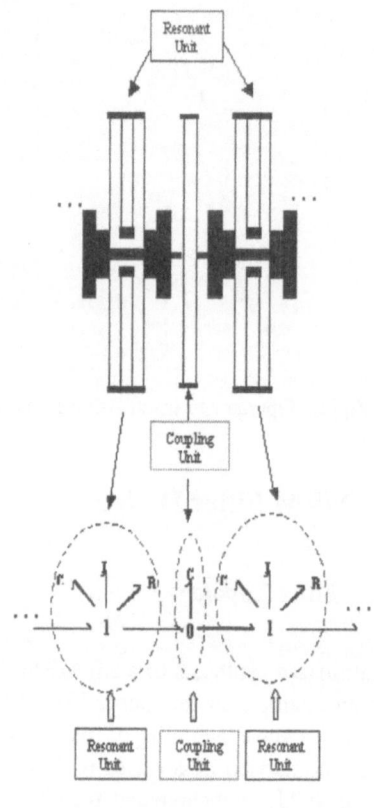

Fig. 6. Layout of Filter Topology II: Filter is composed of a series of Resonator Units coupled by Coupling Units. Its corresponding bond graph representation is also shown.

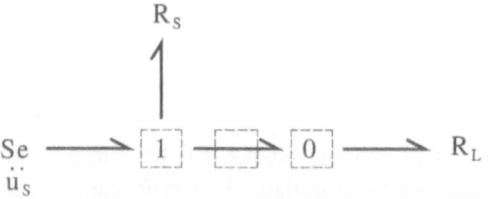

Fig. 7. Embryo of Design

Unlike our previous designs with basic function sets, which impose fewer topological constraints on design, MEMS design features relatively few devices in the component library. These devices are typically more complex in structure than those primitive building blocks used in the basic function set. Only evolved designs represented by bond graphs matching the dynamic behavior of those devices belonging to the component library are expected to be manufacturable under current or anticipated technology. Thus, an important and special step in MEMS synthesis with the BG/GP approach is to define a *realizable* function set that, throughout execution, will produce only phenotypes that can be built using existing or expected technology.

Analyzing the system of MEM filters of [7] from a bond graph viewpoint, the filters are basically composed of Resonator Units (RUs) and Coupling Units (CUs). Another popular MEM filter topology includes Resonator Units and Bridging Units (BUs). A realizable function set for these design topologies often includes functions from both the basic set and modular set. In many cases, multiple realizable function sets, rather than only one, can be used to evolve realizable structures of MEMS. In this research, we used the following function set, along with traditional numeric functions and end operators, for creating filter topologies with coupling units and resonant units.

$$\Re 1 = \{f\_tree, f\_insert\_J1, f\_insert\_RU,$$
$$f\_insert\_CU, f\_add\_C, f\_add\_R, f\_add\_I\}$$
$$\Re 2 = \{f\_tree, f\_insert\_J1, f\_insert\_RU,$$
$$f\_insert\_BU, f\_add\_C, f\_add\_R, f\_add\_I\}$$

3.4 Adaptive Fitness Function

Within the frequency range of interest, $f_{range} = [f_{min}, f_{max}]$, uniformly sample 100 points. Here, $f_{range} = [0.1, 1000K]$ Hz. Compare the magnitudes of the frequency response at the sample points with target magnitudes, which are 1.0 within the pass frequency range of [316, 1000] Hz, and 0.0 otherwise, between 0.1 and 1000KHz.

Compute their differences and get a sum of squared differences as raw fitness, defined as $Fitness_{raw}$.

If $Fitness_{raw} <$ Threshold, change f_{range} to $f_{range}^* = [f_{min}^*, f_{max}^*]$. Usually $f_{range}^* \subset f_{range}$. Repeat the above steps and obtain a new $Fitness_{raw}$.

Then normalized fitness is calculated according to:

$$Fitness_{norm} = 0.5 + Norm / (Norm + Fitness_{raw})$$

The reason to use adaptive fitness evaluation is that after a GP population has reached a fairly high fitness value as a group, the differences of frequency responses of individuals need to be centered on a more constrained frequency range. In this circumstance, if there is not sufficient sampling within this much smaller frequency range, the GP may lack sufficient search pressure to push the search forward. The normalized fitness is calculated from the sampling differences between the frequency response magnitudes of the synthesized systems and the target responses. Therefore, we adaptively change and narrow the frequency range to be heavily sampled. The effect is analogous to narrowing the search window on a smaller yet most significant area, magnifying it, and continuing to search this area with closer scrutiny.

3.5 Experimental Setup

We used a strongly-typed version of lilgp to generate bond graph models. The major GP parameters were as shown below:

Population size: 500 in each of thirteen subpopulations
Initial population: half_and_half
Initial depth: 4-6
Max depth: 50   Max_nodes 5000
Selection: Tournament (size=7)
Crossover: 0.9   Mutation: 0.3

Three major code modules were created in this work. The algorithm kernel of HFC-GP was a strongly typed version [17] of an open software package developed in our research group -- lilgp. A bond graph class was implemented in C++. The fitness evaluation package is C++ code converted from Matlab code, with hand-coded functions used to interface with the other modules of the project. The commercial software package 20Sim was used to verify the dynamic characteristics of the evolved design. The GP program obtains satisfactory results on a Pentium-IV 1GHz in 1000~1250 minutes.

3.6 Experimental Results

Experimental results show the strong topological search capability of genetic programming and feasibility of our BG/GP approach for finding realizable designs for micro-mechanical filters. Although significant fabrication difficulty is currently presented when fabricating a micro-mechanical filter with more than 3 resonators, it does not invalidate our research and the topological search capability of the BG/GP approach BG/BP shows potential for exploring more complicated topologies of future MEMS design and the ever-progressing technology frontiers of MEMS fabrication.

In Figure 8, K is the number of resonant units appearing in the best design of the generation on the horizontal axis. The use of hierarchical fair competition [8] is facilitating continual improvement of the fitness. As fitness improves, the number of resonant units, K, grows – unsurprising because a higher-order system with more resonator units has the potential of better system performance than its low-order counterpart. The plot of corresponding system frequency responses at generations 27, 52, 117 and 183 are shown in Figure 9.

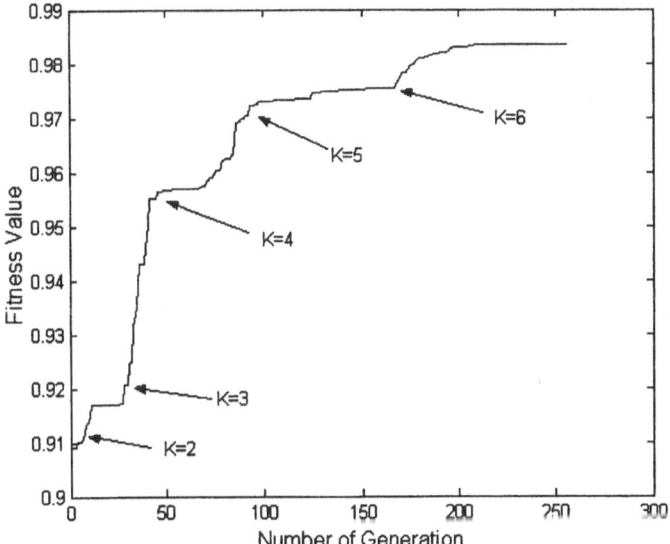

Fig. 8. Fitness Improvement Curve

A layout of a design candidate with three resonators and two bridging units as well as its bond graph representation is shown below in Figure 10. Notice that the geometry of resonators may not show the real sizes and shapes of a physical resonator and the layout figure only serves as a topological illustration. The parameters are listed in Table 3.

Using the BG/GP approach, it is also possible to explore novel topologies of MEM filter design. In this case, we may not necessary use a strictly realizable function set. Instead, a semi-realizable function set is used to relax the topological constraints with the purpose of finding new topologies not realized before but still realizable after careful design. Figure 11 gives an example of a novel topology for a MEM filter design. An attempt to fabricate this kind of topology is being carried out in a university research setting.

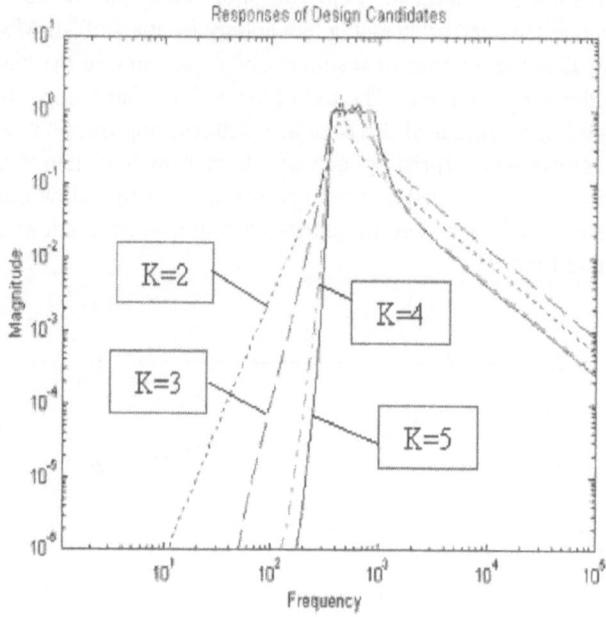

Fig. 9. Frequency responses of a sampling of design candidates, which evolved topologies with larger numbers, K, of resonators as the evolution progressed. All results are from one genetic programming run of the BG/GP approach.

4 Extensions

In MEMS, there are two or three levels of designs that need to be synthesized. Usually the design process starts with basic capture of the schematic of the overall system, and then goes on through layout and construction of a 3-D solid model. So the first design level is the system level, which includes selection and configuration of a repertoire of planar devices or subsystems. The second level is 2-D layout of basic structures like beams to form the elementary planar devices. In some cases, if the MEMS is basically a result of a surface-micro machining process and no significant 3-D features are present, design of this level will end one cycle of design. More generally, modeling and analysis of a 3-D solid model for MEMS is necessary.

For the second level -- two-dimensional layout designs of cell elements -- layout synthesis usually takes into consideration a large variety of design variables and design constraints. The most popular synthesis method seems to be based on conventional numerical optimization methods. The design problem is often first formulated as a nonlinear constrained optimization problem and then solved using an optimization software package [3]. Geometric programming, one special type of convex optimization method, is reported to synthesize a CMOS op-amp. The method is

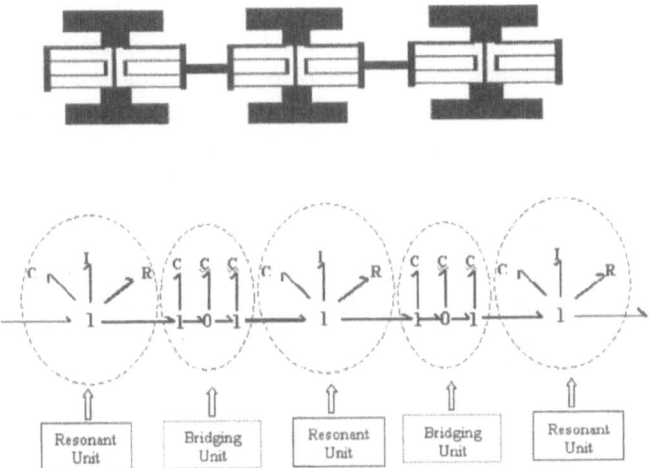

Fig. 10. Layout and bond graph representation of a design candidate from the experiment with three resonator units coupled with two bridging units.

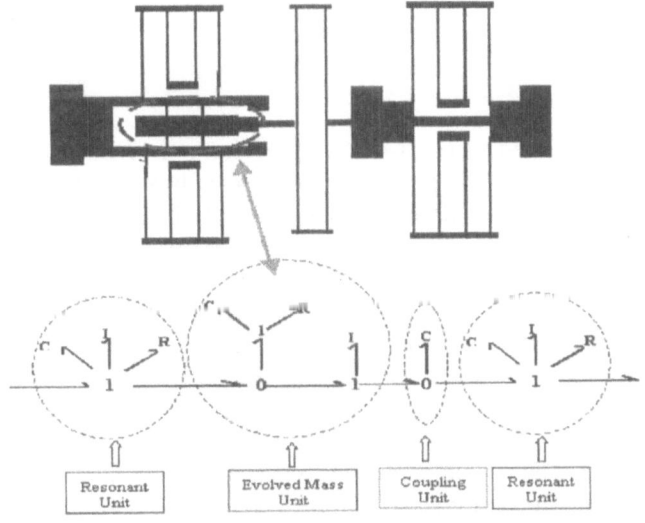

Fig. 11. A novel topology of MEM filter and its bond graph representation

claimed to be both globally optimal and extremely fast. The only disadvantage and limitation is that the design problem has to be carefully formatted first to make it suitable for the treatment of the geometric programming algorithm. However, all the above approaches are based on the assumption that the structures of the cell elements are relatively fixed and subject to no radical topology changes [9]. A multi-objective evolutionary algorithm approach is reported for automatic synthesis of topology and sizing of a MEMS 2-D meandering spring structure with desired stiffnesses in certain directions [10].

The third level design calls for FEA (Finite Element Analysis). FEA is a computational method used for analyzing mechanical, thermal, electrical behavior of complex structures. The underlying idea of FEA is to split structures into small pieces and determine behaviors of each piece. It is used for verifying results of hand calculations for simple model, but more importantly, for predicting behavior of complex models where 1st order hand calculations are not available or insufficient. It is especially well suited for iterative design. As a result, it is quite possible that we can use an evolutionary computation approach to evolve a design using evaluation by means of FEA to assign fitness. Much work in this area has already been reported and it should also be an ideal analysis tool for use in the synthesis loop for final 3-D structures of MEMS. However, even if we have obtained an optimized 3-D device shape, it is still very difficult to produce a proper mask layout and correct fabricate procedures. Automated mask layout and process synthesis tools will be very helpful to relieve the designers from considering the fabrication details and focus on the functional design of the device and system instead [11]

Our long time task of research is to include computational synthesis for different design levels, and to provide support for design engineers in the whole MEMS design process.

5 Conclusions

This paper has suggested a design methodology for automatically synthesizing system-level designs for MEMS. For design of systems like the MEM filter problem, with strong topology constraints and fewer topology variations allowed, the challenge is to define a realizable function set that assures the evolved design is physically realizable and can be built using existing or anticipated technologies. Experiments show that a mixture of functions from both a modular function set and a basic function set form a realizable function set, and that the BG/GP algorithm evolves a variety of designs with different levels of topological complexity that satisfy design specifications.

Acknowledgment. The authors gratefully acknowledge the support of the National Science Foundation through grant DMII 0084934. The authors are grateful to Prof. Steven Shaw, Michigan State University, for an introduction to various aspects of MEMS design.

References

1. Fan Z., Hu J., Seo K., Goodman E., Rosenberg R., and Zhang B.: Bond Graph Representation and GP for Automated Analog Filter Design. GECCO-2001 Late-Breaking Papers, San Francisco. (2001) 81–86

2. Fan Z., Seo K., Rosenberg R. C., Hu J., Goodman E. D.: Exploring Multiple Design Topologies using Genetic Programming and Bond Graphs. Proceedings of the Genetic and Evolutionary Computation Conference, GECCO-2002, New York. (2002) 1073–1080.
3. Zhou Y.: Layout Synthesis of Accelerometers. Thesis for Master of Science. Department of Electrical and Computer Engineering, Carnegie Mellon University. (1998)
4. Rosenberg R. C.: Reflections on Engineering Systems and Bond Graphs, Trans. ASME J. Dynamic Systems, Measurements and Control, (1993) 115: 242–251
5. Koza J. R.: Genetic Programming II: Automatic Discovery of Reusable Programs, The MIT Press (1994)
6. Rosenberg C., Goodman E. D. and Seo K.: Some Key Issues in using Bond Graphs and Genetic Programming for Mechatronic System Design, Proc. ASME International Mechanical Engineering Congress and Exposition, New York (2001)
7. Wang K. and Nguyen C. T. C.: High-Order Medium Frequency Micromechanical Electronic Filters, Journal of Microelectromechanical Systems. (1999) 534–556
8. Hu J., Goodman E. D.: Hierarchical Fair Competition Model for Parallel Evolutionary Algorithms. CEC 2002, Honolulu, Hawaii, May, (2002)
9. Hershenson M. M, Boyd, S. P., and Lee T.H.: Optimal Design of a CMOS Op-Amp via Geometric Programming. Computer-Aided Design of Integrated Circuits and Systems Vol. 20. No. 1 (2001) 1–21
10. Zhou N., Zhu B., Agogino A., Pister K.: Evolutionary Synthesis of MEMS design. ANNIE 2001, IEEE Neural Networks Council and Smart Engineering System Design conference, St. Louis, MO, Nov 4–7, (2001).
11. Ma L. and Antonsson E.K.: Automated Mask-Layout and Process Synthesis for MEMS. Technical Proceedings of the 2000 International Conference on Modeling and Simulation of Microsystems (2000) 20–23
12. Tay E. H., Flowers W. and Barrus J.: Automated Generation and Analysis of Dynamic System Designs. Research in Engineering Design (1998) 10: 15–29.
13. Gero J. S.: Computers and Creative Design, in M. Tan and R. Teh (eds), The Global Design Studio, National University of Singarpo (1996) 11–19
14. Eby, D., Averill, R., Gelfand, B., Punch, W., Mathews, O., Goodman, E.: An Injection Island GA for Flywheel Design Optimization. 5th European Congress on Intelligent Techniques and Soft Computing EUFIT'97 Vol. 1. (1997) 687–691
15. Vargas-Hernandez N., Shah J., Lacroix Z.: Development of a Computer Aided Conceptual Design Tool for Complex Electromechanical Systems. Computational Synthesis: From Basic Building Blocks to High Level Functionality, Papers from the 2003 AAAI Symposium Technical Report SS-03-02 (2003) 255–261
16. Wang, J. and Terpenny, J.: Interactive Evolutionary Solution Synthesis in Fuzzy Set-based Preliminary Engineering Design, Special Issue on Soft Computing in Manufacturing, Journal of Intelligent Manufacturing, Vol. 14. (2003) 153–167
17. Luke S.: Strongly-Typed, Multithreaded C Genetic Programming Kernel, http://www.cs.umd.edu/users/-seanl/gp/patched-gp/ (1997)
18. Campbell, M., Cagan J. and Kotovsky K.: Agent-based Synthesis of Electro-Mechanical Design Configurations, Journal of Mechanical Design, Vol. 122. No. 1, (2000) 61–69

Congressional Districting Using a TSP-Based Genetic Algorithm

Sean L. Forman* and Yading Yue

Mathematics and Computer Science Department
Saint Joseph's University
Philadelphia, PA 19131, USA**

Abstract. The drawing of congressional districts by legislative bodies in the United States creates a great deal of controversy each decade as political parties and special interest groups attempt to divide states into districts beneficial to their candidates. The genetic algorithm presented in this paper attempts to find a set of compact and contiguous congressional districts of approximately equal population. This genetic algorithm utilizes a technique based on an encoding and genetic operators used to solve Traveling Salesman Problems (TSP). This encoding forces near equality of district population and uses the fitness function to promote district contiguity and compactness. A post-processing step further refines district population equality. Results are provided for three states (North Carolina, South Carolina, and Iowa) using 2000 census data.

1 Problem History

The United States Congress consists of two houses, the Senate (containing two members from each of the fifty states) and the House of Representatives. The House of Representatives has 435 members, and each state is apportioned a congressional delegation in proportion to its population as determined by a national, decennial census.

Each state (usually the state's legislative body) is responsible for partitioning its state into a number of districts (a *districting plan*) equal to its apportionment. Through years of case law, the courts have outlined several requirements for the drawing of districts [1].

- The districts must be contiguous.
- The districts must be of equal population following the "one-man one-vote" principle.[1]
- The districts should be of a pleasing shape.[2]

---

* Corresponding author: sean.forman@sju.edu, http://www.sju.edu/~sforman/
** Special thanks to Alberto Segre for providing support and a sounding board for this work, Saint Joseph's University for support of this work, the U.S. Census Bureau Geography Division for their guidance, and the reviewers for their comments.

[1] Despite a census error of 2-3%, a federal court recently threw out a Pennsylvania plan in which the smallest and largest districts varied by 19 people in a state with 12.3 million people [2].

[2] The 1990 North Carolina plan was thrown out due to a district where one candidate quipped he could hit everyone in the district by driving down Interstate-75 with both doors open [3].

- The districts should strive not to divide regions with a common interest.
- The districts should not be a dramatic departure from previous plans.

For the purposes of this paper, a districting plan will be considered *valid* if each district created is (1) contiguous and (2) its population is within 1% of the state's total population divided by the number of districts (the population for perfect district population equality). Within these constraints, the algorithm described will attempt to find districts with a compact (pleasing) shape.

2 Current Practices

The redistricting process is largely a partisan process with the political party in power drawing districts likely to keep them in power. This can lead to elongated, poorly shaped districts, or gerrymanders–a contraction of long ago Massachusetts Governor Eldridge Gerry and a salamander. Due to this politicization of the redistricting process, automated redistricting is an appealing alternative.

Automated redistricting typically attempts to find a partition of the state into contiguous districts that minimizes some objective function promoting compactness and an equal distribution of the population. The number of entities to partition can vary from counties (10-150 per state) to census tracts (100-10,000). However, one can quickly see that the number of potential solutions grows very quickly as smaller and smaller entities are used.

In the 1960's, there was a flurry of efforts to apply new computational technology to the redistricting problem beginning with Vickrey's initial work in 1961 [4]. In 1963, Weaver and Hess [5] used an operations research approach similar to the methods used to locate warehouses. In 1965, Kaiser and Nagel developed methods that take current districting plans and improve them by swapping units between adjoining districts [6,7]. In 1973, Liittschwager applied Vickrey's technique to redistricting in Iowa [8].

More recently, several groups have attempted to use local search techniques to find optimal or near-optimal solutions to the redistricting problem [9]. Altman sketches a solution based on node partitioning [10] where the fitness function promotes equality of population and compactness. Mehrotra, et al [11], propose a constrained graph partitioning solution with pre and post-processing steps. Their solution of a transshipment problem to completely balance district populations is mimicked in this paper.

di Cortona [1] and Altman [12] provide excellent backgrounds on redistricting techniques. Altman also provides a proof that the redistricting problem is, in fact, NP-complete by relating it to the class of set partitioning problems.

3 A Redistricting Genetic Algorithm

The genetic algorithm described here (*ConRed*, for *Con*gressional *Red*istricting) takes as an input the number of districts to be drawn, a set of *tracts*, which could mean any partition of a state (counties, townships, census tracts, census blocks, or some combination), each with a population and an area, a list of all inter-tract borders and their lengths, and a list of all borders between a tract and a neighboring state or shoreline and their respective lengths. It then outputs the best districting plan it finds as a list of tracts assigned to districts.

3.1 Fitness Function

The encoding used (Section 3.2) forces every solution to have approximately equal (at worst, 5%-8% error) populations among the created districts.[3] Therefore, the fitness function focuses primarily on contiguity and compactness. Some consideration is given to population equality to encourage solutions within 1% of perfectly balanced.

Methods used to determine the compactness of individual districts and districting plans fall into three primary categories [13,14]:

- Dispersion measures – district length versus width; ratio of district area to that of the smallest circumscribed circle; moment of inertia.
- Perimeter-based measures – sum of all perimeters; ratio of perimeter to area.
- Population measures – population's moment of inertia; ratio of population in district to the population within the smallest circumscribed circle.

Each of these techniques has its own biases and pathological examples. The Schwartzberg Index [15], a perimeter-based measure, was chosen for ConRed because it can be computed quickly and incrementally as tracts are added and subtracted from districts. The Schwartzberg Index measures the compactness of a district as the perimeter of the district squared divided by its area. For this measurement, lower is better, and it is minimized by a circular shape.

The fitness function for ConRed consists of a measurement of the plan's shape and discontiguity (Eq. 1) and a measurement of the variance from equal district populations (Eq. 2) with the final fitness a linear combination of these two parts (Eq. 3).

The shape fitness function (Eq. 3) modifies the Schwarzberg Index to reward contiguity by multiplying this value by one plus the number of excess discontiguous pieces ($pieces_i$) found in the district (preferably, zero) weighted by a parameter, ϕ. The more discontiguous pieces a proposed district has the more severe the penalty will be. The products for individual districts are then averaged across all n districts to give the overall districting plan's shape fitness.[4] Unlike some possible techniques, the encoding does not force a possible plan to be contiguous, but a lack of contiguity is penalized severely in the fitness function by the multiplication of the district's compactness with the number of pieces.

$$Fitness_{shape} = \frac{\sum_{i=1}^{n}(1 + \phi\,(pieces_i - 1))\frac{(perimeter_i)^2}{area_i}}{n} \qquad (1)$$

The plan's population variance function depends on the *ideal district population* ($idealDistPop$ = state population divided by the number of districts) and the maximum error allowed in a valid solution ($Var_k = idealDistPop * k$, in our case $k = 1\%$). The first term of Eq. 2 serves as a penalty function (γ determines just how severe) for each district within a districting plan that violates the population constraint. The second term of Eq. 2 drives the population differences towards zero once a valid plan has been found.

---

[3] Recall that according to the courts, the primary determinants of a hypothetical districting plan's fitness are the equality of its population distribution and the compactness and contiguity of its districts.

[4] One could also choose to find the minimax compactness or any similar measurement.

$$Fitness_{pop} = \frac{\gamma \sum_{i=1}^{n} MAX(|pop_i - idealDistPop| - Var_{1\%}, 0)}{n \times idealDistPop} \qquad (2)$$
$$+ \frac{\sum_{i=1}^{n} |pop_i - idealDistPop|}{n \times idealDistPop}$$

Then adding the two together gives the overall fitness (Eq. 3). The parameter θ can be tuned to balance population and shape considerations.

$$Fitness = Fitness_{shape} + \theta \, Fitness_{pop} \qquad (3)$$

3.2 Encoding

The encoding chosen for this genetic algorithm is the path representation used for Traveling Salesman Problems (TSP) [16,17,18]. As in the TSP, a single chromosome travels through each tract, and as the tracts are traversed, districts are formed by the sequence of tracts.

This differs significantly from GA techniques used to solve node partitioning problems (NPP) or clustering problems [19]. ConRed does not store the partition of the state in its chromosome per se, but extracts it from the order of the entries. Though details may vary, NPP solutions typically encode the actual partitions of the graph vertices as chromosomes [20]. Clustering problems may be solved similarly, or they may encode the chromosomes as coordinates for the location of the cluster center with data points assigned to the nearest cluster center [19]. The clustering approach would guarantee compactness and contiguity, but the population equality constraint will be difficult to satisfy since it may not always be possible to draw compact districts for a given problem. An NPP encoding that would maintain contiguity and population equality in every chromosome and optimize on compactness would require a significant amount of post-mutation and post-crossover processing to maintain chromosome validity.

Encoding the redistricting problem in a manner similar to the TSP causes each chromosome to have districts with approximately equal populations and relies upon the fitness function to create compactness and contiguity. The trade-off is that the search space is dramatically enlarged with redundant solutions as will be shown below.

To demonstrate how the encoding is translated from a permutation of tracts to a set of districts, take a fictitious state named Botna, with three districts and nine tracts arranged in a 3×3 array (tract population is given in the subscripts), and a chromosome, which is just a permutations of length nine.

1_{30}	2_{20}	3_{10}
4_{10}	5_{20}	6_{30}
7_{10}	8_{20}	9_{30}

$A = 1 - 4 - 5 - 2 - 3 - 6 - 9 - 8 - 7$

Now to convert this permutation into a districting plan, one travels along the chromosome summing the populations until some threshold population is met. The threshold chosen is typically the $idealDistPop \pm \delta$, where δ is typically 5% of the $idealDistPop$. In cases where the number of tracts is small, there is no guarantee that the population will be split so neatly as they are in this example, but for a large numbers of tracts the approximate equality of district populations is virtually guaranteed.

Table 1. Converting chromosome **A** to a districting plan. State population is 180, so the $idealDistPop = 60$.

Tract	Pop.	$\sum$Pop.	District
1	30	30	1
4	10	40	1
5	20	60	1
2	20	20	2
3	10	30	2
6	30	60	2
9	30	30	3
8	20	50	3
7	10	60	3

```
1 2 3
4 5 6
7 8 9
```

In the districting plan, **A**, shown in Table 1, every tract has an area of one and each inter-tract boundary is of length one. The $fitness_{shape}$ of this chromosome, **A**, can then be evaluated using Equation (1) with $\phi = 2$.

District 1 (1-4-5) Perim. = 8, Area = 3, Fitness = 21.33
District 2 (2-3-6) Perim. = 8, Area = 3, Fitness = 21.33
District 3 (9-8-7) Perim. = 8, Area = 3, Fitness = 21.33

Taking the average of the three districts gives $fitness_{shape} = 21.33$. Here is another example, which illustrates a problem with this encoding (again $\phi = 2$).

B = 1 − 6 − 3 − 2 − 5 − 7 − 9 − 8 − 4

Table 2. Example displaying encoding shortcomings.

Tract	Pop.	$\sum$Pop.	District
1	30	30	1
6	30	60	1
3	10	10	2
2	20	30	2
5	20	50	2
7	10	60	2
9	30	30	3
8	20	50	3
4	10	60	3

District 1 (1-6) Perim. = 8, Area = 2, and Pieces = 2, Fitness = 96.
District 2 (3-2-5-7) Perim. = 12, Area = 4, and Pieces = 2, Fitness = 108.
District 3 (9-8-4) Perim. = 10, Area = 3, and Pieces = 2, Fitness = 100.
Taking the average of the three districts gives $fitness_{shape} = 101.33$.

So while this encoding does not force the districts to be contiguous, the fitness function does penalize those districting plans which are not contiguous.

An additional problem with this encoding is that it allows a large number of redundant solutions to be considered. For instance, $(1-6)-(3-2-5-7)-(9-8-4)$ and $(1-6)-(2-3-5-7)-(9-8-4)$ would produce the exact same districting plans in the example above.

3.3 Initial Population & Selection

Several heuristics are used to find an initial population. The algorithm begins each new chromosome at a randomly selected border tract.[5] Adjoining, unvisited tracts are added randomly to the chromosome. The direction of the next selected tract is random, but biased toward the selection of tracts adjoining previously visited elements. This gives some preliminary structure to the permutations. If the chromosome finds no adjoining, unvisited tracts to add, the chromosome jumps randomly to a non-adjoining tract elsewhere in the state and continues adding tracts to the chromosome.[6]

After the fitnesses are calculated for each member of the population, they are ranked and are selected proportionally by their ranks [21]. For example, in a population of ten chromosomes the best chromosome is ten times more likely to be selected than the worst chromosome. Additionally, a copy of the best chromosome is passed to the next generation.

3.4 Genetic Operators

Of the many different operators used on the traveling salesman problem, a crossover, mutation and heuristic operator have been chosen.

Maximal Preservative Crossover. This operator uses a donor and a receiver chromosome [22]. From the donor, a random substring is chosen, call it Ω. All of the elements in Ω are then deleted from the receiver chromosome, and Ω is inserted where the first element of Ω occurs in the receiver. This implementation is slightly altered from Mühlenbein, et al.'s original implementation as they placed Ω at the beginning of the receiver chromosome.

Suppose the following chromosomes are chosen,
$A = 1-4-5-8-7-2-3-6-9$
$B = 1-6-3-2-5-4-7-8-9$

---
[5] While testing was inconclusive, this appeared to be a somewhat better strategy than using the largest tract (by area or population) or a random tract.
[6] This method does provide some structure, but none of the millions of initial chromosomes created in the tests performed (Section 4) represented valid or even contiguous districting plans.

where **A** is the donor and **B** is the receiver. Suppose $\Omega = 5 - 8 - 7 - 2$. In **B**, Ω will be inserted after the 3 since 2 is in Ω.
$$\mathbf{B} = 1 - 6 - 3 - - - 4 - - - - 9 \Rightarrow \mathbf{B} = 1 - 6 - 3 - 5 - 8 - 7 - 2 - 4 - 9$$

Exchange Mutation. Exchange mutation [23] cuts a piece out of chromosome and then switches it with a piece of the same size somewhere else in the chromosome. The size of this cut is a parameter swapSize that may be entered by the user. Suppose swapSize = 2, cutPoint = 2 and pastePoint = 6.
$$\mathbf{B} = 1 - (6-3) - 2 - 5 - (4-7) - 8 - 9 \Rightarrow \mathbf{B} = 1 - (4-7) - 2 - 5 - (6-3) - 8 - 9$$

Discontiguity Patch. In the course of building chromosomes, one often finds islands of tracts from a district trapped within another district (see Table 2). A heuristic clean-up of these islands can significantly improve a chromosome's fitness while not altering its underlying character. The clean-up process involves searching for discontiguous districts in a districting plan, removing one of the smaller pieces from the string of tracts and then re-inserting it after a tract that borders the first tract in the island piece. For example with Table 2, tracts 4, 8, and 9 are part of the same district, but tract 4 is discontiguous from tracts 8 and 9. The heuristic snips tract 4 from the string, and then randomly inserts it after one of the tracts that it borders (1, 5, or 7). The resulting plan removes several of the discontiguities and has an improved fitness of 75.33.
$$1 - 6 - 3 - 2 - 5 - 7 - 9 - 8 - (4) \Rightarrow 1 - (4) - 6 - 3 - 2 - 5 - 7 - 9 - 8$$

3.5 Pre-processing

Population equality tends to improve and compactness and contiguity tends to worsen as the number of tracts increases. The coarsest tract structure available is the county level. But since most states have at least some counties with very large populations, it is often necessary to partition a large county into a set of subtracts, ideally each with an equal population. For instance, a large county can be divided into a number of smaller tracts[7] and then used with the other unbroken counties to give a data set with the population more evenly divided among the input tracts. To create these subtracts (itself essentially a districting problem), ConRed was applied on individual counties with populations above a prescribed threshold. In Table 3, the number of tracts increases as this threshold is lowered. A low population threshold for subdividing a county will lead to more tracts in the input set and possibly less contiguity, but it will increase the probability that a solution will have a low population variance as Table 3 shows.

3.6 Post-processing

The encoding used provides approximate equality of district populations, but often only within 5% of the ideal district population rather than the 1% that is considered valid in this context. To bring the district populations into this desired range, a post processing step has been added. This step is applied following the fitness evaluation and only to

---
[7] For these experiments, this number was capped at nine.

chromosomes that already generate contiguous plans. A desired population reapportionment is calculated using a transshipment algorithm. Guided by this reapportionment, ConRed then proceeds to swap boundary tracts between districts until no further population balancing is possible. The transshipment algorithm [24] is a common operations research technique used to determine the optimal shipments between a set of known supply and demand points (commonly factories, warehouses and stores). Each possible branch has a cost associated with it, so the objective is to minimize the overall cost while satisfying the given demand using the given supply.

As applied to this problem, the districts with larger-than-ideal populations are suppliers for demanders which are districts with smaller-than-ideal populations. The costs are structured to guarantee that swapping occurs only between districts bordering each other, therefore some districts may act as go-betweens (both suppliers and demanders) for districts needing to adjust their populations. This problem is solved using the transportation simplex method. This technique for balancing district population was suggested by Mehrotra, et al [11].

3.7 Parameters

The primary parameters for tuning are the probabilities for the three operators, the maximum and minimum lengths of the segments removed and inserted by the operators, and the fitness function's parameters: γ, θ, and ϕ. Some work has been done on tuning parameters for this GA, but it is an area where considerable improvement may still be possible. For the experimental results that follow, $\phi = 2 * districts$, $\gamma = 10$, $\theta = 500$, $p_{mpx} = .40$, $p_{exc} = .10$, and $p_{disc} = .40$. The length of any exchange or patch is at most ten tracts and the crossover can be of any length.

4 Experimental Results

ConRed[8] performed a total of 80 runs for each of three states: North Carolina (13 districts), South Carolina (6 districts), and Iowa (5 districts).[9] For each state, 20 runs were made on 4 different tract layouts created by decrementing the pre-processing population threshold from 100,000 to 25,000 (Section 3.5). For these 20 runs, GA population sizes were incremented from 500 to 2000 chromosomes (by 500) for 5 runs each. The number of generations was chosen such that there were 2,000,000 plan evaluations for each run.

Results are presented in Table 3. *Contig* gives the number of runs whose best, final solutions are contiguous and *N valid* is the number of those plans that are both contiguous and have a population variation within 1.0% of ideal. *PopVar* is a plan's largest district population variance from the ideal. Time is in seconds. The results in the **Best** column

---

[8] ConRed is written in ANSI C and was tested on a 867MHz Pentium III using RedHat Linux.
[9] The U.S. Census Bureau provides a pair of files that are needed to run tests on actual state data. *Summary File 1* (SF1) [25] provides information on population, area, and latitude and longitude for a variety of entity sizes (county, county subdistrict, census tract, census block) for each state. *TIGER/Line files* [26] contain all boundary information between various entities found in the SF1 files. A series of Perl scripts were written to build the necessary input files and generate the postscript files used to visualize final solutions.

are the best fitness across all runs and the minimax population variance across all of the runs, so the two numbers may be from different runs. The best, valid solutions are shown in Figure 1.

The results show that for small states with six or fewer representatives (26 of 50 states), ConRed will draw compact districts of essentially equal populations. For a larger state like North Carolina (42 of the 50 states have 13 representatives or less), ConRed can produce contiguous districts, but is not able to consistently bring the populations close enough together to be a valid solution. This may be a matter of parameter tuning. A higher setting for θ may produce better results. Still these results could be used as rough plans and then manipulated along the borders to produce an equal population distribution.

Unfortunately, it has been difficult to locate other existing solvers with which to compare ConRed. Comparisons to the existing districting plans are interesting, but legislatures are not attempting to form compact districting plans, so they are of little informational value. Almost any computational technique will produce a solution more compact than those drawn by a legislature. In the 1960's, Kaiser-Nagel [6] and Weaver [5] produced plans with compactness, but population variances greater than 1%. In 1996, Hayes [3] produced a twelve-district plan for North Carolina with population variation of less that 1%, but no concern for compactness. In 1998, Mehrotra [11] produced a plan for South Carolina beginning with 48 tracts. They initially found a solution with a population variation of 4.4%. A transshipment step determined the amount of population that needed shifting to achieve district population equality, and then by hand (it appears) tracts were sliced away on district boundaries until an equal population distribution was created. A visual inspection suggests that the valid solutions produced by ConRed are as compact as their plan.

Table 3. ConRed experiments. See text (Section 4) for an explanation.

	Tracts	Runs	Contig	N valid	Average Fitness	PopVar%	Time (sec)	Best Fitness	PopVar%
IA	108	20	20	20	27.8	0.66	568	25.1	0.19
5 dist.	115	20	20	20	28.1	0.60	612	26.1	0.20
	124	20	20	18	29.2	0.65	624	25.4	0.25
	164	20	20	20	28.5	0.49	722	25.2	0.12
SC	69	20	20	15	37.1	0.89	571	32.4	0.48
6 dist.	83	20	20	13	35.0	0.90	607	30.6[8]	0.58
	115	20	20	18	39.9	0.84	725	33.3	0.55
	183	20	20	19	38.7	0.69	883	34.1	0.43
NC	147	20	17	0[9]	71.5	2.33	1669	43.3	1.36
13 dist.	166	20	16	0	79.6	2.58	1797	38.7	1.06
	212	20	17	0	73.9	1.65	1792	43.0	1.08
	352	20	8	2	173.3	1.95	1715[10]	41.9	0.83

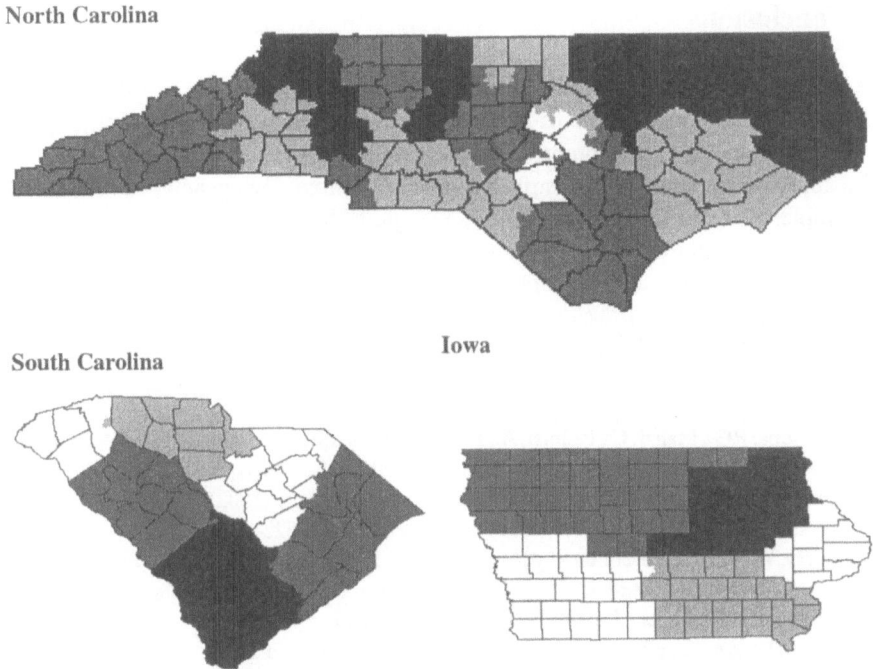

Fig. 1. ConRed's best, valid, final solutions for North Carolina, South Carolina and Iowa. North Carolina's solution divides 352 tracts into 13 districts and has a fitness of 45.1 and a maximum population variation of 0.86%. There are better shaped solutions than this, but their population variation fell just outside the 1% cutoff. South Carolina's solution divides 83 tracts into 6 districts and has a fitness of 30.6 and a maximum population variation of 0.81%. Iowa's solution divides 108 tracts into 5 districts and has a fitness of 25.1 and a maximum population variation of 0.44%. All interior borders shown are county borders.

5 Future Work

Additional heuristics, improved genetic operators, and additional parameter tuning may provide further improvement in the quality of the solutions produced. Thus far, most of the work has focused on the heuristics and the visualization aspects of the project, so further attention to parameter tuning and the choice of operators will be the next area of study. In addition, politicians may also be interested in the distribution of political affiliations within districts, and applications to other location or partitioning problems have yet to be considered.

---

[8] This particular solution was found by ConRed on four separate runs.

[9] Of the 58 contiguous solutions for NC, maximum population variation ranged from 0%-1% for two solutions, 1.0%-1.5% for 23, 1.5%-2.0% for 18, and above two for 15.

[10] Due to the lower number of contiguous solutions, the post-processing step was performed less often than in other problems, hence the lower time required despite a larger number of tracts.

6 Conclusions

The genetic algorithm presented in this paper, ConRed, uses an encoding adapted from traveling salesman problems to produce possible congressional districting plans. ConRed's fitness function promotes compactness and contiguity and its encoding provides approximate equality of district population for free. Two genetic operators have been implemented in ConRed along with a single heuristic operator. Experimental results on three small to medium states show significant success at producing compact, contiguous districts of nearly equal population.

References

1. di Cortona, P.G., Manzi, C., Pennisi, A., Ricca, F., Simeone, B.: Evaluation and Optimization of Electoral Systems. Monographs on Discrete Mathematics and Applications. SIAM (1999)
2. Milewski, M.: Court overturns redistricting. The Intelligencer (2002)
3. Hayes, B.: Machine politics. American Scientist **84** (1996) 522–526
4. Vickrey, W.: On the prevention of gerrymandering. Political Science Quarterly **76** (1961) 105
5. Hess, S., Weaver, J., Siegfeldt, H., Whelan, J., Zitlau, P.: Nonpartisan political redistricting by computer. Operations Research **13** (1965) 998–1006
6. Kaiser, H.: An objective method for establishing legislative districts. Midwest Journal of Political Science (1966)
7. Nagel, S.: Simplified bipartisan computer redistricting. Stanford Law Review **17** (1965) 863–899
8. Liittschwager, J.: The Iowa redistricting system. Annals of the New York Academy of Science **219** (1973) 221–235
9. Ricca, F., Simeone, B.: Political redistricting: Traps, criteria, algorithms, and trade-offs. Ricera Operativa **27** (1997) 81–119
10. Altman, M.: Is automation the answer? the computational complexity of automated redistricting. Rutgers Computer and Technology Law Journal **23** (2001) 81–142
11. Mehrotra, A., Johnson, E.L., Nemhauser, G.L.: An optimization based heuristic for political districting. Management Science **44** (1998) 1100–1114
12. Altman, M.: Modeling the effect of manadatory district compactness on partisan gerrymanders. Political Geography **17** (1998) 989–1012
13. Young, H.: Measuring compactness of legislative districts. Legislative Studies Quarterly **13** (1988) 105–111
14. Niemi, R., Grofman, B., Carlucci, C., Hofeller, T.: Measuring compactness and the role of a compactness standard in a test for partisan and racial gerrymandering. Journal of Politics **52** (1990) 1152–1182
15. Schwartzberg, J.: Reapportionment, gerrymanders, and the notion of compactness. Minnesota Law Review **50** (1966) 443–452
16. Jog, P., Suh, J.Y., van Gucht, D.: Parallel genetic algorithms applied to the traveling salesman problem. SIAM Journal of Optimization **1** (1991) 515–529
17. Grefenstette, J., Gopal, R., Rosimaita, B., van Gucht, D.: Genetic algorithms for the traveling salesman problem. In: International Conference on Genetic Algorithms and their Applications. (1985) 160–168
18. Larra naga, P., Kuijpers, C., Murga, R., Inza, I., Dizdarevic, S.: Genetic algorithms for the traveling salesman problem: A review of representations and operators. Artificial Intelligence Review **13** (1999) 129–170

19. Sarkar, M., Yegnanarayana, B., Khemani, D.: A clustering algorithm using an evolutionary programming-based approach. Pattern Recognition Letters **18** (1997) 975–986
20. Chandrasekharam, R., Subramanian, S., Chaudhury, S.: Genetic algorithm for node partitioning problem and applications in VLSI design. IEEE Proceedings: Comput. Digit. Tech. **140** (1993) 255–260
21. Baker, J.: Adaptive selection methods for genetic algorithms. In Grefenstette, J., ed.: Proc. of the 1st International Conference on Genetic Algorithms and their Applications, Hillsdale, NJ, Lawrence Erlbaum Associates (1985) 101–111
22. Mühlenbein, H., Gorges-Schleuter, M., Krämer, O.: Evolution algorithms in combinatorial optimization. Parallel Computing **7** (1988) 65–85
23. Banzhaf, W.: The "molecular" traveling salesman. Biological Cybernetics **64** (1990) 7–14
24. Winston, W.L.: Operations Research: Applications and Algorithms. Volume 3. Duxbury Press (1994)
25. U.S. Census Bureau Washington, DC: Census 2000 Summary File 1. (2001)
26. U.S. Census Bureau Washington, DC: Census 2000 TIGER/Line Files. (2000)
27. Goldberg, D.: Genetic Algorithms in Search, Optimization and Machine Learning. Addison-Wesley (1989)

Active Guidance for a Finless Rocket Using Neuroevolution

Faustino J. Gomez and Risto Miikkulainen

Department of Computer Sciences
University of Texas
Austin, TX, 78712 USA

Abstract. Finless rockets are more efficient than finned designs, but are too unstable to fly unassisted. These rockets require an active guidance system to control their orientation during flight and maintain stability. Because rocket dynamics are highly non-linear, developing such a guidance system can be prohibitively costly, especially for relatively small-scale rockets such as sounding rockets. In this paper, we propose a method for evolving a neural network guidance system using the Enforced SubPopulations (ESP) algorithm. Based on a detailed simulation model, a controller is evolved for a finless version of the Interorbital Systems RSX-2 sounding rocket. The resulting performance is compared to that of an unguided standard full-finned version. Our results show that the evolved active guidance controller can greatly increase the final altitude of the rocket, and that ESP can be an effective method for solving real-world, non-linear control tasks.

1 Introduction

Sounding rockets carry a payload for making scientific measurements to the Earth's upper atmosphere, and then return the payload to the ground by parachute. These rockets serve an invaluable role in many areas of scientific research including high-G-force testing, meteorology, radio-astronomy, environmental sampling, and micro-gravity experimentation [1,2]. They have been used for more than 40 years; they were instrumental e.g., in discovering the first evidence of X-ray sources outside the solar system in 1962 [3]. Today, sounding rockets are much in demand as the most cost-effective platform for conducting experiments in the upper atmosphere.

Like most rockets, sounding rockets are usually equipped with fins to keep them on a relatively straight path and maintain stability. While fins are an effective "passive" guidance system, they increase both mass and drag on the rocket which lowers the final altitude or *apogee* that can be reached with a given amount of fuel. A rocket with smaller fins or no fins at all can potentially fly much higher than a full-finned version. Unfortunately, such a design is unstable, requiring some kind of *active* attitude control or guidance to keep the rocket from tumbling.

Finless rockets have been used for decades in expensive, large-scale launch vehicles such as the USAF Titan family, the Russian Proton, and the Japanese H-IIA. The guidance systems on these rockets are based on classical feedback control such as Proportional-Integral-Differential (PID) methods to adjust the thrust angle (i.e. thrust vectoring) of the engines. Because rocket flight dynamics are highly non-linear, engineers must make simplifying assumptions in order to apply these linear methods, and take great care to ensure that these assumptions are not violated during operation. Such an undertaking requires detailed knowledge of the rocket's dynamics that can be very costly to acquire. Recently, non-linear approaches such as neural networks have been explored primarily for use in guided missiles (see [4] for an overview of neural network control architectures). Neural networks can implement arbitrary non-linear mappings that can make control greatly more accurate and robust, but, unfortunately, still require significant domain knowledge to train.

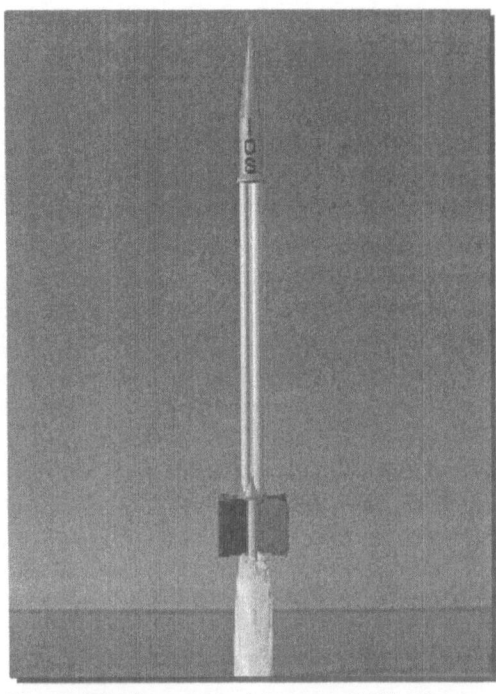

Fig. 1. The Interorbital Systems RSX-2 Rocket. The RSX-2 is capable of lifting a 5 pound payload into the upper atmosphere using a cluster of four liquid-fueled thrusters. It is currently the only liquid-fueled sounding rocket in production. Such rockets are desirable because of their low acceleration rates and non-corrosive exhaust products.

In this paper, we propose a method for making the development of finless sounding rockets more economical by using Enforced SubPopulations (ESP; [5,6]) to evolve a neural network guidance system. As a test case, we will focus on a finless version of the Interorbital Systems RSX-2 rocket (figure 1). The RSX-2 is a liquid-fueled sounding rocket that uses the differential thrust of its four engines to control attitude. By evolving a neural network controller that maps the state of the rocket to thrust commands, the guidance problem can be solved without the need for analytical modeling of the rocket's dynamics or prior knowledge of the appropriate kind of control strategy to employ. Using ESP, all that is required is a sufficiently accurate simulator and a quantitative measure of the guidance system's performance, i.e. a fitness function.

In the next three sections we describe the Enforced SubPopulations method used to evolve the guidance system, the controller evolution simulations, and experimental results on controlling the RSX-2 rocket.

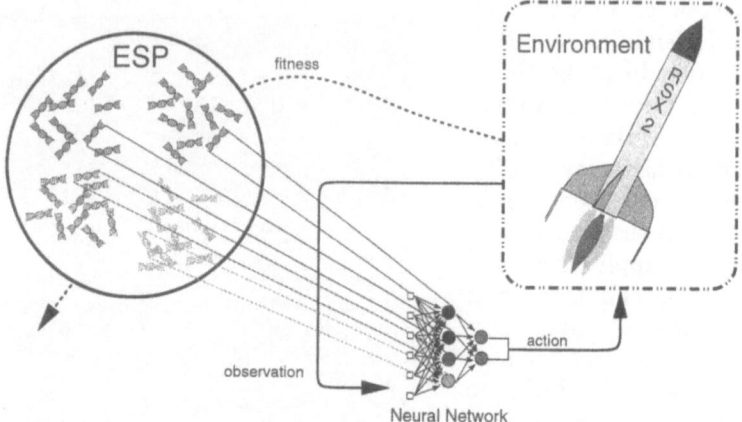

Fig. 2. The Enforced Subpopulations Method (ESP; color figure). The population of neurons is segregated into subpopulations shown here in different colors. Networks are formed by randomly selecting one neuron from each subpopulation. A neuron accumulates a fitness score by adding the fitness of each network in which it participated. This score is then normalized and the best neurons within each subpopulation are mated to form new neurons. By evolving neurons in separate subpopulations, the specialized sub-functions needed for good networks are evolved more efficiently.

2 Enforced Subpopulations (ESP)

Enforced Subpopulations[1] [5,6] is a neuroevolution method that extends the Symbiotic, Adaptive Neuroevolution algorithm (SANE; [7]). ESP and SANE differ from other NE methods in that they evolve partial solutions or *neurons* instead of complete networks, and a subset of these neurons are put together to form a complete network. In SANE, neurons are selected from a single population to form networks. In contrast, ESP makes use of explicit subtasks; a separate subpopulation is allocated for each of the h units in the network, and a neuron can only be recombined with members of its own subpopulation (figure 1). This way the neurons in each subpopulation can evolve independently and specialize rapidly into good network sub-functions.

Evolution in ESP proceeds as follows:

1. Initialization. The number of hidden units h in the networks that will be formed is specified and a subpopulation of neuron chromosomes is created. Each chromosome encodes the input and output connection weights of a neuron with a random string of real numbers.
2. Evaluation. A set of h neurons is selected randomly, one neuron from each subpopulation, to form the hidden layer of a complete network. The network is submitted to a *trial* in which it is evaluated on the task and awarded a fitness score. The score is added to the *cumulative fitness* of each neuron

---

[1] The ESP software package is available at:
http://www.cs.utexas.edu/users/nn/pages/software/abstracts.html#esp-cpp

that participated in the network. This process is repeated until each neuron has participated in an average of e.g. 10 trials.
3. Recombination. The average fitness of each neuron is calculated by dividing its cumulative fitness by the number of trials in which it participated. Neurons are then ranked by average fitness within each subpopulation. Each neuron in the top quartile is recombined with a higher-ranking neuron using 1-point crossover and mutation at low levels to create the offspring to replace the lowest-ranking half of the subpopulation.
4. The Evaluation–Recombination cycle is repeated until a network that performs sufficiently well in the task is found.

Evolving networks at the neuron level has proven to be a very efficient method for solving reinforcement learning tasks such as pole-balancing [6], robot arm control [8], and game playing [7]. ESP is more efficient that SANE because the subpopulation architecture makes the evaluations more consistent in two ways: first, the subpopulations that gradually form in SANE are already present by design in ESP. The species do not have to organize themselves out of a single large population, and their progressive specialization is not hindered by recombination across specializations that usually fulfill relatively orthogonal roles in the network. Second, because the networks formed by ESP always consist of a representative from each evolving specialization, a neuron is always evaluated on how well it performs its role in the context of all the other players.

The accelerated specialization in ESP makes it more efficient than SANE, but it also causes diversity decline over the course of evolution like that of a normal GA. This can be a problem because a converged population cannot easily adapt to a new task. To deal with premature convergence, ESP is combined with *burst mutation*. The idea is to search for optimal modifications of the current best solution. When performance has stagnated for a predetermined number of generations, new subpopulations are created by adding noise to each of the neurons in the best solution. Each new subpopulation contains neurons that represent differences from the best solution. Evolution then resumes, but now searching the space in a "neighborhood" around the best previous solution. Burst mutation can be applied multiple times, with successive invocations representing differences to the previous best solution. Assuming the best solution already has some competence in the task, most of its weights will not need to be changed radically. To ensure that most changes are small while allowing for larger changes to some weights, ESP uses the Cauchy distribution to generate noise:

$$f(x) = \frac{\alpha}{\pi(\alpha^2 + x^2)} \quad (1)$$

With this distribution 50% of the values will fall within the interval $\pm\alpha$ and 99.9% within the interval $\pm 318.3\alpha$. This technique of "recharging" the subpopulations keeps diversity in the population so that ESP can continue to make progress toward a solution even in prolonged evolution.

Burst mutation is similar to the Delta-Coding technique of [9] which was developed to improve the precision of genetic algorithms for numerical optimization problems. Because our goal is to maintain diversity, we do not reduce the range of the noise on successive applications of burst mutation and we use Cauchy rather that uniformly distributed noise.

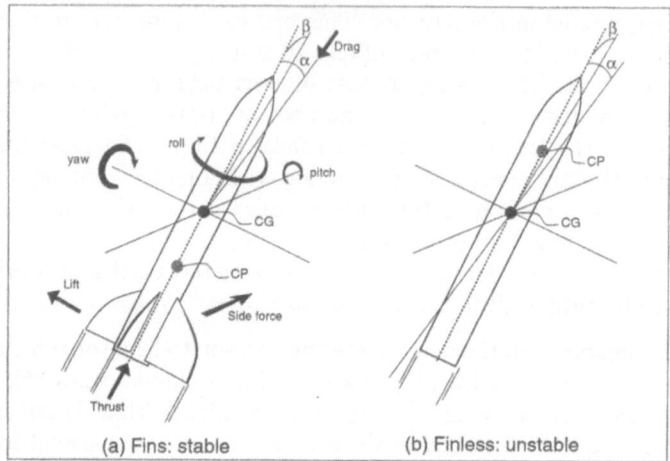

(a) Fins: stable   (b) Finless: unstable

Fig. 3. Rocket Dynamics. The rocket (a) is stable because the fins increase drag in the rear of the rocket moving the center of pressure (CP) behind the center of gravity (CG). As a result, any small angles α and β are automatically corrected. In contrast, the finless rocket (b) is unstable because the CP stays well ahead of the CG. To keep α and β from increasing, i.e to keep the rocket from tumbling, active guidance is needed to counteract the destabilizing torque produced by drag.

3 The Finless Rocket Guidance Task

In previous work, ESP was shown to outperform a wide range of reinforcement learning algorithms including Q-learning, SARSA(λ), Evolutionary Programming, and SANE on several difficult versions of the pole-balancing or inverted pendulum task [6]. The rocket guidance domain is similar to pole-balancing in that both involve stabilizing an inherently unstable system. Figure 1 gives a basic overview of rocket dynamics. The motion of a rocket is defined by the translation of its center of gravity (CG), and the rotation of the body about the CG in the pitch, yaw, and roll axes. Four forces act upon a rocket in flight: (1) the thrust of the engines which propel the rocket, (2) the drag of the atmosphere exerted at the *center of pressure* (CP) in roughly the opposite direction to the thrust, (3) the *lift force* generated by the fins along the yaw axis, and (4) the *side force* generated by the fins along the pitch axis. The angle between the direction the rocket is flying and the longitudinal axis of the rocket in the yaw-roll plane is known as the *angle of attack* or α, the corresponding angle in the pitch-roll plane is known as the *sideslip angle* or β. When either α or β is greater than 0 degrees the drag exerts a torque on the rocket that can cause the rocket to tumble if it is not stable. The arm through which this torque acts is the distance between the CP and the CG.

In figure 1a, the finned rocket is stable because the CP is behind the rocket's CG. When α (β) is non-zero, a torque is generated by the lift (side) force of the fins that counteracts the drag torque, and tends to minimize α (β). This situation corresponds to a pendulum hanging down from its hinge; the pendulum

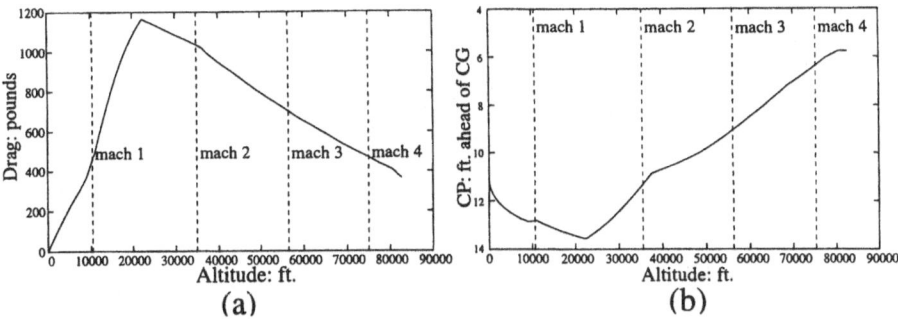

Fig. 4. The time-varying difficulty of the guidance task. Plot (a) shows the amount of drag force acting on the finless rocket as it ascends through the atmosphere. More drag means that it takes more differential thrust to control the rocket. Plot (b) shows the position of the center of pressure in terms of how many feet ahead it is of the center of gravity. From 0 to about 22,000ft the control task becomes more difficult due to the rapid increase in drag and the movement of the CG away from the nose of the rocket. At approximately 22,000ft, drag peaks and begins a decline as the air gets thinner, and the CP starts a steady migration towards the CG. As it ascends further, the rocket becomes progressively easier to control as the density of the atmosphere decreases.

will return to this stable equilibrium point if it is disturbed. When the rocket does not have fins, as in figure 1b, the CP is ahead of the CG causing the rocket to be unstable. A non-zero α or β will tend to grow causing the rocket to eventually tumble. This situation corresponds to a pendulum at its unstable upright equilibrium point where any disturbance will cause it to diverge away from this state.

Although the rocket domain is similar to the inverted pendulum, the rocket guidance problem is significantly more difficult for two reasons: the interactions between the rocket and the atmosphere are highly non-linear and complex, and the rocket's behavior continuously changes throughout the course of a flight due to system variables that are either not under control or not directly observable (e.g. air density, fuel load, drag, etc.).

Figure 3 shows how the difficulty of stabilization varies over the course of a successful flight for the finless rocket. In figure 3a, drag is plotted against altitude. From 0ft to about 22,000ft, the rocket approaches the sound barrier (Mach 1) and drag rises sharply. This drag increases the torque exerted on the rocket in the yaw and pitch axes for a given α and β, making it more difficult to control its attitude. In figure 3b, we see that also during this period the distance between the CG and CP increases because the consumption of fuel causes the CG to move back, making the rocket increasingly unstable. After 22,000ft, drag starts to decrease as the air becomes less dense, and the CP steadily migrates back towards the CG, so that the rocket becomes easier to control.

For ESP, this means that the fitness function automatically scales the difficulty of the task in response to the performance of the population. At the beginning of evolution the task is relatively easy. As the population improves and individuals are able to control the rocket to higher altitudes, the task becomes progressively harder. Although figure 3 indicates that above 22,000ft the task again becomes easier, progress in evolution continues to be difficult because the controller is constantly entering an unfamiliar part of the state space. A fitness function that gradually increases in difficulty is usually desirable because it allows for sufficient selective pressure

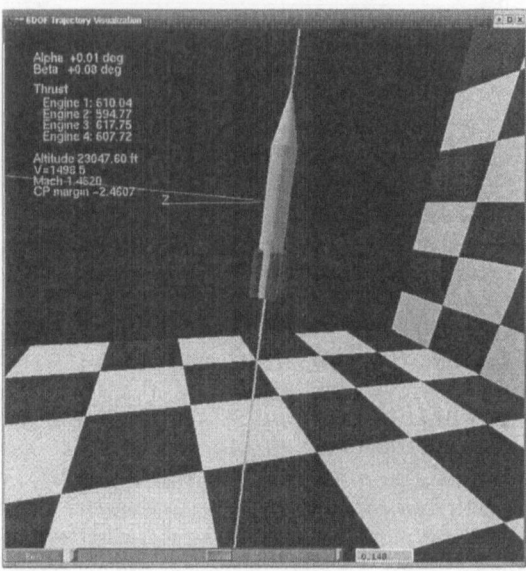

Fig. 5. RSX-2 Rocket Simulator. The picture shows a 3D visualization of the JSBSim rocket simulation used to evolve the RSX-2 guidance controllers. The simulator provides a realistic environment for designing and verifying aircraft dynamics and guidance systems.

at the beginning of evolution to direct the search into a favorable region of the solution space. However, the rocket control task is already too hard in the beginning—all members of the initial population perform so poorly that the evolution stalls and converges to a local maxima. In other words, direct evolution does not even get started on this very challenging task. One way to overcome this problem is to make the initial task easier and then increase the difficulty as the performance of the population improves. In the experiments below we employ such an *incremental evolution* approach [5]. Instead of trying to evolve a controller for the finless rocket directly, a more stable version of the rocket is used initially to make the ultimate task more accessible.

The following section describes the simulation environment used to evolve the controller, the details of how a guidance controller interacts with the simulator, and the experimental setup for evolving a neural network controller for this task.

4 Rocket Control Experiments

4.1 The RSX2 Rocket Simulator

As an evolution environment we used the JSBSim Flight Dynamics Model[2] adapted for the RSX-2 rocket by Eric Gullichsen of Interorbital Systems. JSB-

---
[2] More information about the free JSBSim software package is available at: http://jsbsim.sourceforge.net/

Sim is an open source, object-oriented flight dynamics simulator with the ability to specify a flight control system of any complexity. JSBSim provides a realistic simulation of the complex dynamic interaction between the airframe, propulsion system, fuel tanks, atmosphere, and flight controls. The aerodynamic forces and moments on the rocket were calculated using a detailed geometric model of the RSX-2. Four versions of the rocket with different fin configurations were used: full fins, half fins (smaller fins), quarter fins (smaller still), and no fins, i.e. the actual finless rocket. This set of rockets allowed us to observe the behavior of the RSX2 at different levels of instability, and provided a sequence of increasingly difficult tasks with which to evolve incrementally. All simulations used Adams-Bashforth 4th-order integration with a time step of 0.0025 seconds.

4.2 Neural Guidance Control Architecture

The rocket controller is represented by a feedforward neural network with one hidden layer (figure 6). Every 0.05 seconds (i.e. the control time-step) the controller receives a vector of readings from the rocket's onboard sensors that provide information about the current orientation (pitch, yaw, roll), rate of orientation change, angle of attack α, sideslip angle β, the current throttle position of the four thrusters, altitude, and velocity in the direction of flight. This input vector is propagated through the sigmoidal hidden and output units of the network to produce a new throttle position for each engine determined by:

$$u_i = 1.0 - o_i/\delta, \; i = 1..4 \quad (2)$$

where u_i is the throttle position of thruster i, o_i is the value of network output unit i, $0 \leq u_i, o_i \leq 1$, and $\delta \geq 1.0$. A value of ω for u_i means that the controller wants thruster i to generate $\omega \times 100\%$ of maximum thrust. The parameter δ controls how far the controller is permitted to "throttle back" an engine from 100% thrust.

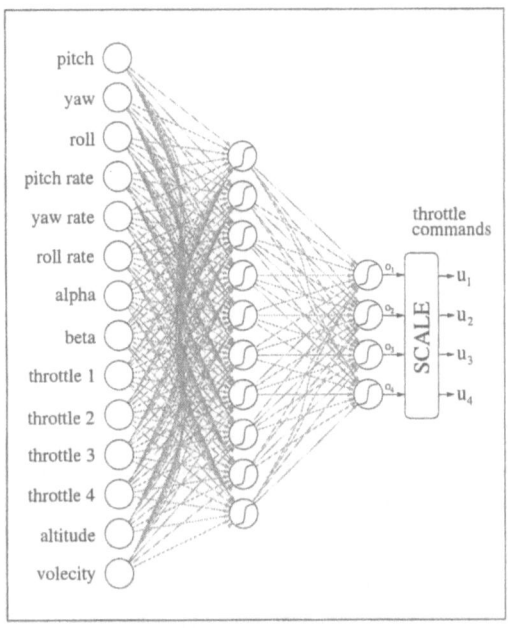

Fig. 6. Neural Network Guidance. The control network receives the state of the rocket every time step through its input layer. The input consists of the rocket's orientation, the rate of change in orientation, α, β, the current throttle position of each engine, the altitude, and the velocity of the rocket in the direction of flight. These values are propagated through the network to produce a new throttle command (the amount of thrust) for each engine.

4.3 Experimental Setup

The objective of the experiments was to determine whether ESP could evolve a controller to stabilize the finless version of the RSX-2 rocket. To do so, the neural guidance system has to control the thrust of each of the four engines in order to maintain the rocket's angle of attack α and sideslip angle β within ±5 degrees from the time of ignition to burnout when all of the fuel has been expended. Exceeding the ±5 degree boundary indicates that the rocket is about to tumble and is therefore considered a catastrophic failure. Each ESP network was evaluated in a single trial that consisted of the following four phases:

1. At time t_0, the rocket is attached to a launch rail that will guide it on a straight path for the first 50 feet of flight. The fuel tanks are full and the engines are ignited.
2. At time $t_1 > t_0$, the rocket begins its ascent as the engines are powered to full thrust.
3. At time $t_2 > t_1$, the rocket leaves the launch rail and the controller begins to modulate the thrust as described in section 4.2 according to equation 2.
4. While controlling the rocket one of two events occurs at time $t_f > t_2$:

    a) α or β exceeds ±5 degrees, in which case the controller has failed.

    b) the rocket reaches burnout, in which case the controller has succeeded.

    In either case, the trial is over and the altitude of the rocket at t_f becomes the fitness score for the network.

In a real launch, the rocket would continue after burnout and coast to apogee. Since we are only concerned with the control phase, for efficiency the trials were limited to reaching burnout. This fitness measure is all that is needed to encourage evolutionary progress. However, there is a large locally maximal region in the network weight space corresponding to the policy $o_i = 1.0$, $i = 1..4$; the policy of keeping all four engines at full throttle. Since it is very easy to randomly generate networks that saturate their outputs, this policy will be present in the first generations. Such a policy clearly does not solve the task, but because the rocket is so unstable, no better policy is likely to be present in the initial population. Therefore, it will quickly dominate the population and halt progress toward a solution. To avoid this problem, all controllers that exhibited this policy were penalized by setting their fitness to zero. This procedure ensured that the controller was not rewarded for doing nothing.

The simulations used 10 subpopulations of 200 neurons (i.e. the networks were composed of 10 hidden units), and δ was set to 10 so the network could only control the thrust in the range between 90% and 100% for each engine. It was determined in early testing that this range produced sufficient differential thrust to counteract side forces, and solve the task.

As was discussed in section 3, evolving a controller directly for the finless rocket was too difficult and an incremental evolution method was used instead. We first evolved a controller for the quarter-finned rocket. Once a solution to this easier task was found, the evolution was transitioned to the more difficult finless rocket.

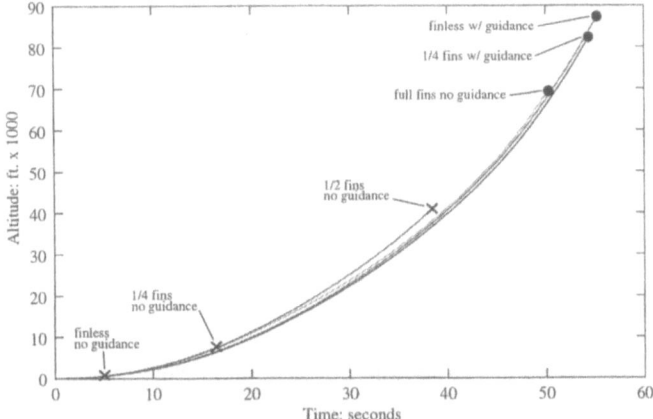

Fig. 7. Comparison of burnout altitudes for different fin-size rockets with and without guidance. The crosses indicate the altitude at which a particular rocket becomes unstable (i.e. either α or $\beta > \pm 5$ degrees). The circles indicate the altitude of a successful rocket that remained stable all the way to burnout. The guided quarter-finned and finless rockets fly significantly higher than the unguided full-finned rocket.

5 Results

ESP solved the task of controlling the quarter-finned rocket in approximately 600,000 evaluations. Another 50,000 evaluations were required to successfully transition to the finless rocket. Figure 4.3 compares the altitudes that the various rockets reach in simulation. Without guidance, the full-finned rocket reaches burnout at approximately 70,000ft, whereas the finless, quarter-finned, and half-finned rockets all fail before reaching burnout. However, with neural network guidance the quarter-finned and finless rockets do reach burnout and exceed the full-finned rocket's altitude by 10,000ft and 15,000ft, respectively. After burnout, the rocket will begin to coast at a higher velocity in a less dense part of the atmosphere; the higher burnout altitude for the finless rocket translates into an apogee that is about 20 miles higher than that of the finned rocket.

Figure 5a shows the behavior of the four engines during a guided flight for the finless rocket. The controller makes smooth changes to the thrust of the engines throughout the flight. This very fine control is required because any abrupt changes in thrust at speeds of up to Mach 4 can quickly cause failure. Figure 5b shows α and β for the various rockets with and without guidance. Without guidance, the quarter-finned and even the half-finned rocket start to tumble as soon as α or β start to diverge from 0 degrees. Using guidance, both the quarter-finned and finless rockets keep α and β at very small values up to burnout. Note that although the finless controller was produced by further evolving the quarter-finned controller, the finless controller not only solves a more difficult task, but does so with more optimal performance.

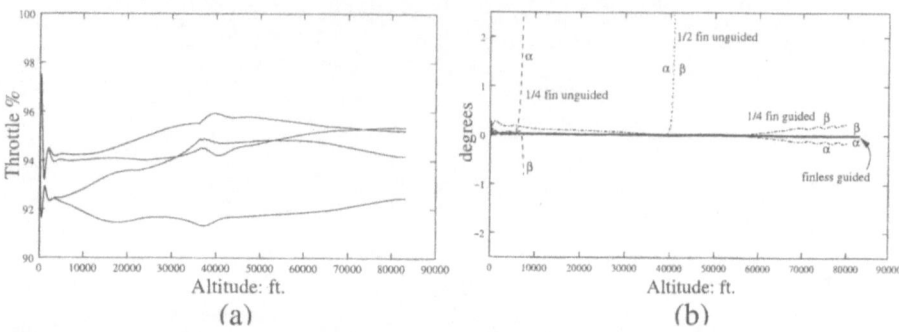

Fig. 8. Controller performance for the finless rocket. Plot (a) shows the policy implemented by the controller. Each curve corresponds to the percent thrust of one of the four rocket engines over the course of a successful flight. After some initial oscillation the control becomes very fine, changing less than 2% of total thrust for any given engine. Plot (b) compares the values α and β for various rocket configurations and illustrates how well the neural guidance system is able to minimize α and β. The unguided quarter-finned and half-finned rockets maintain low α and β for a while, but as soon as either starts to grow the rocket tumbles. In contrast, the guidance systems for the quarter-finned and finless rockets are able to contain α and β all the way up to burnout. The finless controller, although evolved from the quarter-finned controller, is more optimal.

6 Discussion and Future Work

The rocket control task is representative of many real world problems such as manufacturing, financial prediction, and robotics that are characterized by a complex non-linear interaction between system components. The critical advantage of using ESP over traditional engineering approaches is that it can produce a controller for these systems without requiring formal knowledge of system behavior or prior knowledge of correct control behavior. The experiments in this paper demonstrate this ability by solving a high-precision, high-dimensional, non-linear control task.

Of equal importance is the result that the differential thrust approach for the finless version of the current RSX-2 rocket is feasible. Also, having a controller that can complete the sounding rocket mission in simulation has provided valuable information about the behavior of the rocket that would not otherwise be available.

In the future, we plan on making the task more realistic in two ways: (1) the controller will no longer receive α and β values as input, and (2) instead of a generating continuous control signal, the network will output a binary vector indicating whether or not each engine should "throttle back" to a preset low throttle position. This scheme will greatly simplify the control hardware on the real rocket. Once a controller for this new task is evolved, work will focus on varying environmental parameters and incorporating noise and wind so that the controller will be forced to utilize a robust and general strategy to stabilize the rocket. This phase will be critical to our ultimate goal of transferring the controller to the RSX-2 and testing it in an actual rocket launch.

7 Conclusion

In this paper, we propose a method for learning an active guidance system for a dynamically unstable, finless rocket by using the Enforced Subpopulation neuroevolution algorithm. Our experiments show that the evolved guidance system is able to stabilize the finless rocket and greatly improve its final altitude compared to the full-finned, stable version of the rocket. The results suggest that neuroevolution is a promising approach for difficult nonlinear control tasks in the real world.

Acknowledgments. This research was supported in part by the National Science Foundation under grant IIS-0083776, and the Texas Higher Education Coordinating Board under grant ARP-003658-476-2001. The authors would like to give special thanks to Eric Gullichsen of Interorbital Systems (egullich@colo.to) for his collaboration and invaluable assistance in this project.

References

1. Corliss, W.R.: NASA sounding rockets, 1958–1968: A historical summary. Technical Report NASA SP-4401, National Aeronautics and Space Administration, Washington, D.C. (1971)
2. Seibert, G.: A world without gravity. Technical Report SP-1251, European Space Agency (2001)
3. Giacconi, R., Gursky, H., Paolini, F., Rossi, B.: Evidence for X-rays from sources outside the solar system. Physical Review Letters 9 (1962) 439–444
4. Liang Lin, C., Wen Su, H.: Intelligent control theory in guidance and control system design: an overview. Proc. Natl. Sci, Counc. ROC(A) 24 (2000) 15–30
5. Gomez, F., Miikkulainen, R.: Incremental evolution of complex general behavior. Adaptive Behavior 5 (1997) 317–342
6. Gomez, F., Miikkulainen, R.: Robust non-linear control through neuroevolution. Technical Report AI02-292, Department of Computer Sciences, The University of Texas at Austin (2002)
7. Moriarty, D.E.: Symbiotic Evolution of Neural Networks in Sequential Decision Tasks. PhD thesis, Department of Computer Sciences, The University of Texas at Austin (1997) Technical Report UT-AI97-257.
8. Moriarty, D.E., Miikkulainen, R.: Evolving obstacle avoidance behavior in a robot arm. Technical Report AI96-243, Department of Computer Sciences, The University of Texas at Austin (1996)
9. Whitley, D., Mathias, K., Fitzhorn, P.: Delta-Coding: An iterative search strategy for genetic algorithms. In Belew, R.K., Booker, L.B., eds.: Proceedings of the Fourth International Conference on Genetic Algorithms, San Francisco, CA: Morgan Kaufmann (1991) 77–84

Simultaneous Assembly Planning and Assembly System Design Using Multi-objective Genetic Algorithms

Karim Hamza[1], Juan F. Reyes-Luna[1], and Kazuhiro Saitou[2] *

[1] Graduate Student, [2]Assistant Professor, Mechanical Engineering Department
University of Michigan, Ann Arbor, MI 48109-2102, USA
{khamza,juanfr,kazu}@.umich.edu

Abstract. This paper aims to demonstrate the application of multi-objective evolutionary optimization, namely an adaptation of NSGA-II, to simultaneously optimize the assembly sequence plan as well as selection of the type and number of assembly stations for a production shop that produces three different models of wind propelled ventilators. The decision variables, which are the assembly sequences of each product and the machine selection at each assembly station, are encoded in a manner that allows efficient implementation of a repair operator to maintain the feasibility of the offspring. Test runs are conducted for the sample assembly system using a crossover operator tailored for the proposed encoding and some conventional crossover schemes. The results show overall good performance for all schemes with the best performance achieved by the tailored crossover, which illustrates the applicability of multi-objective GA's. The presented framework proposed is generic to be applicable to other products and assembly systems.

1 Introduction

The optimization of product assembly processes is a key issue for the efficiency of manufacturing system, which involves several different types of decisions such as selecting the assembly sequences of products, assigning tasks to the assembly stations, and selecting the number and type of machines at each assembly station.

Research on assembly sequence planning was originated by two pioneer works in late eighties: De Fazio and Whitney [1] and de Mello *et al.* [2] independently presented graph-based models of assemblies and algorithms for enumerating all feasible assembly sequences. Since then, numerous work has been conducted on assembly sequence planning[1]. However, a few attentions have been paid to the integration of assembly sequence planning and assembly system design.

Assembly system design is a complex problem that may have several objectives such as minimizing overall cost, meeting production demand, increasing productivity, reliability and/or product quality. Assembly sequence planning is a precedence-

---

* Corresponding Author
[1] A comprehensive bibliography of the area is found at www.cs.albany.edu/~amit/bib/ASSPLAN.html

constrained scheduling problem, which is known to be NP-complete [3]. Furthermore, allocating machines to assembly tasks is resource constrained scheduling which is also known to be NP-complete [3]. In real-life workshops, there is a need to consider assembly sequence and machine allocation simultaneously which results in a doubled difficulty that makes such problems beyond the feasibility of full enumeration. Thus, assembly systems design provides rich opportunities for heuristics approaches such multi-objective GA's. For instance, process planning using GA's was considered in the late eighties [4]. More recently, Awdah et al. [5] proposed a computer-aided process-planning model based on GA's. Kaeschel et al. [6] applied an evolutionary algorithm to shop floor scheduling and multi-objective evolutionary optimization of flexible manufacturing systems was considered by Chen and Ho [7]. Saitou et al.[8] applied GA for robust optimization of multi-product manufacturing systems subject to production plan variations.

This paper presents the application of multi-objective GA's to simultaneously optimize the assembly sequence, assembly stations' type and number selection, based on data extracted from a real assembly shop that assembles specially designed wind-propelled ventilators (Fig. 1). The reminder of the paper first describes the problem formulation, a special encoding and crossover schemes, the results of simulation runs with the proposed crossover scheme as well as arithmetic projection, multipoint and uniform crossovers. Finally, discussion and future extensions are provided.

2 Wind Propelled Ventilators

The family of products considered is the three models of wind propelled ventilators. Shown in Fig. 1, is a photo of model A, the basic model used in ventilating industrial or storage hangars in dry regions. The basic idea of operation is that the ventilators are placed atop ventilating ducts in the ceiling of the hangars. When the lateral wind blows across the hangar, it spins the spherically shaped blades, which in turn perform a sucking action that draws air from the top of the hangar. Model B has the same exoskeleton and blades as model A, but has different internal shaft as well as an additional rotor that improves the air suction out of the ventilation duct. Model C is the same as Model B, except that its outer blades have improved shape design. The three models share several components, and are assembled in the same assembly shop and may use the same assembly stations. Table 1 provides a listing of all components in the three models and in which models they are being used.

The problem of designing an assembly system that can assemble the three types of ventilators, models A, B and C, is formulated as a multi-objective optimization problem with two objective functions to me minimized: assembly cost f_1 and production shortage within a given production period f_2. These are the functions for the three decision variable categories: the assembly sequences of each products, the type of machines at each assembly station, and the number of machines for each type.

The model of the assembly system used in this study is simple one, which computes the production cost f_1 by summing over the startup and hourly operation rate of the assembly stations. Production volume is estimated according to average cycle times, for the numbers and types of machines assigned to each assembly task:

Minimize $\quad f_1 = \sum_{\text{Machines}} \text{Investment Cost} + \sum_{\text{Machines}} \text{Processing Time} \times \text{Hourly Rate} \quad$ (1)

$\quad\quad\quad\quad\quad f_2 = \sum_{\text{Products}} \max(0, \text{Target Production} - \text{Production Volume}) \quad$ (2)

Decision Var. Assembly Sequence, Type of assembly stations, Number of assembly stations of every used type

Fig. 1. A wind-propelled ventilator – Model A

While the simulations provided in this paper are based on the real assembly shop data in a company, the actual production volumes and assembly station rates are not revealed due to proprietary nature of the information.

Table 1. Listing of components in three models of the wind-propelled ventilator.

Component Label	Description	Used in Model		
		A	B	C
1	Duct and Main Body	x	x	x
2	Lower Bracket – Type #1	x		
3	Upper Bracket	x	x	x
4	Axle	x		
5	Lower Ball Bearing – Type #1	x		
6	Upper Ball Bearing	x	x	x
7	Lower Hub – Type #1	x		
8	Upper Hub – Type #1	x		
9	Blades – Type #1	x	x	
10	Lower Bracket – Type #2		x	x
11	Shaft		x	x
12	Lower Ball Bearing – Type #2		x	
13	Lower Hub – Type #2		x	x
14	Upper Hub – Type #2		x	x
15	Inner Rotor		x	x
16	Blades Type #2			x

3 Problem Formulation for Multi-objective GA

A software system, ASMGA (ASsembly using Multi-objective GA) was implemented, whose overall structure is illustrated in Fig. 2. It is capable of communicating with user-provided models of assembly systems involving multi-products.

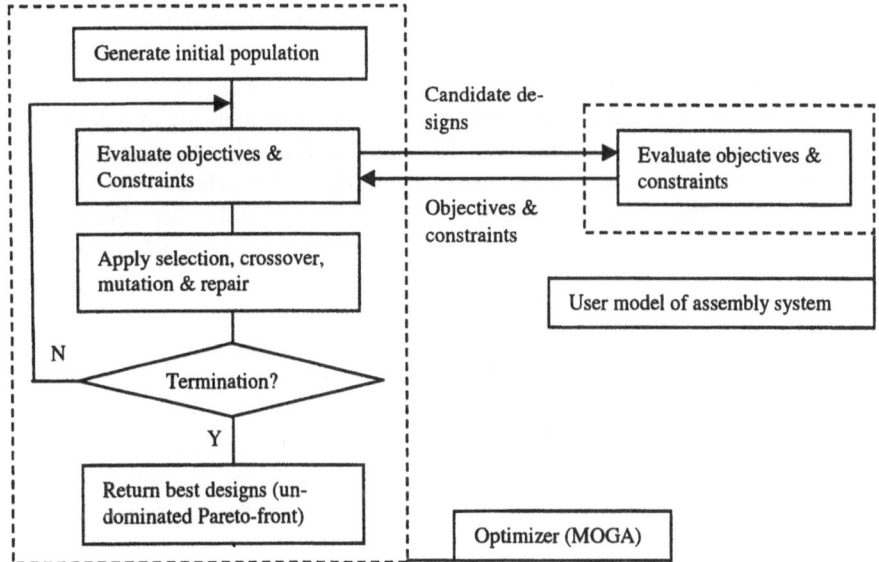

Fig. 2. Overall structure of the ASMGA (ASsembly using Multi-objective GA).

3.1 Encoding Scheme of Chromosomes

The following three decision variable *sets* are considered:

- Assembly sequence of every product;
- Choice of type of assembly station for every assembly operation;
- Number of assembly stations of every type.

For every product consisting of N components, there are $2(N - 1)$ variables that control the assembly sequence, $(N - 1)$ variables controlling the type of assembly stations and $(N - 1)$ variables to specify the number of each type of assembly stations in the platform. In the rest of the paper, a set of values for the decision variables is sometimes referred to as a *candidate design* of the assembly system.

The above decision variables can be efficiently encoded in a fixed-length string of integers in a manner that allows efficient implementation of a repair operator to maintain the feasibility of the offspring for assembly precedence constraints. The layout of the integer string chromosome is shown in Fig. 3. The basic building block of the chromosome is a set of *four integer numbers* that define *one assembly operation*. Every product is assembled through a set of assembly operations. Performing assembly operations to obtain an assembled product out of its individual components may be mentally visualized by thinking of a hypothetical bin, which initially contains all separate components of the product. In every assembly operation, one reaches into the bin, grabs two items assembles them together (provided they can be assembled together) and puts them back into the hypothetical bin. It can be shown that one gets a product consisting of N components, in this manner, in exactly N-1 assembly operations. These *N-1* assembly operations are laid on a linear chromosome.

The four-integer assembly operation information translates as follows: The first number refers to one of the components (or sub-assemblies) inside the bin of the product. The second number refers to the component in the bin, which will be joined to the one selected by the first number. The third number refers to the assembly station chosen to perform the operation and the fourth number refers to the number of those assembly stations that should be present in the assembly shop. For example, chromosome c=11421021 for the simple product shown in Fig. 4 would be translated as follows:

- In the first assembly operation (first four numbers) the component, which has the order of 1 (*i.e,*. the component labeled 2, because ordering starts at zero) is selected.
- The component labeled 2 is assembled to the second component that can be coupled to it (*i.e.*, the component labeled 3) to produce a subassembly (2,3), using the fifth type of assembly station that can perform the task and there are two such stations in the assembly plant
- In the second assembly operation, according to the new order of the components and subassemblies in the bin, the subassembly (2,3) is selected.
- The subassembly (2,3) is assembled the first (and only) component that can be coupled to it (*i.e.*, the component labeled 1) to produce the full product (1,2,3), using the second type of assembly station that can perform the task and there is one such station in the assembly plan

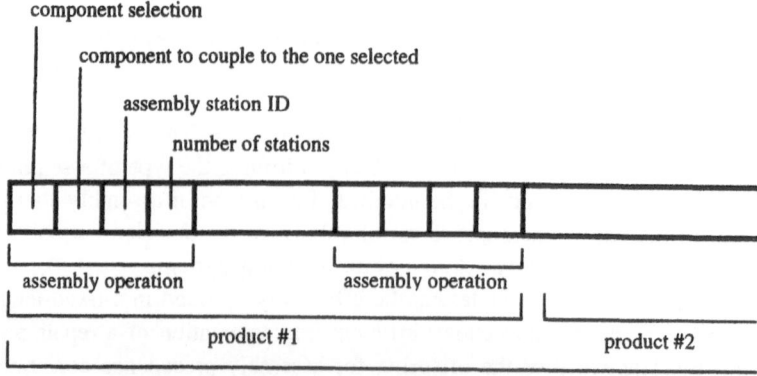

Fig. 3. Layout of chromosomes.

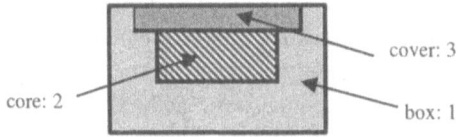

Fig. 4. A simple product.

3.2 Assembly Constraints and Repair Operator

Due to the geometrical and other quality requirements, assembly precedence relationships exist as constraints upon the assembly sequence in almost every real product. Thus it makes sense to define all the assembly constraints during a *pre-processing* stage, and then make (or attempt to make) all candidate designs of the assembly system conform to those constraints throughout the optimization. The ASMGA software includes a pre-processor that guides the user to define all feasible assembly operations and automatically generates a set of all sub-assemblies that may be encountered.

Chromosome reproduction through crossover and mutation may result in an assembly sequences that violate such precedence relationships. One way of enforcing the feasibility of the assembly sequences is through penalizing the objective functions of infeasible candidate designs. However, such penalizing means relying on the selection according to fitness, while allowing the integer numbers on the chromosome to take on any value within the maximum and minimum ranges, which would lead to an impossibly large search space that is probably beyond the capabilities of GA or any other optimization technique. Instead, after crossover and mutation, a repair operator is invoked that systematically increments the integer number on the chromosome so that the assembly sequence remains feasible. Such integer incrementing is performed for every four-integer assembly operation till it represents one of the assembly operations defined on the set of feasible assembly operations, that can be invoked to join two of the components or sub-assemblies that are in the hypothetical bin of the product.

3.3 Crossover

In the following example, four crossover schemes are tested: Arithmetic Projection, Multi-Point, Uniform, and a special crossover operator especially tailored for the proposed encoding scheme.

Arithmetic Projection Crossover

This is a blend of arithmetic crossover [9] and heuristic crossover [9]. Both arithmetic and linear projection crossover schemes are mainly applied in real coded GA's and they are tested in this paper although the chromosome is integer coded, because some portion of the chromosome has underlying continuity (the number of assembly stations). When viewing the chromosome in multi-dimensional vector space, such crossover schemes set the gene values of the offspring chromosome to some point along the line joining two parent points. In arithmetic crossover, the offspring point is in between the parents, while in projection crossover the point is outside the parents, projected in the direction of the better (more fit or higher rank) parent. Thus in the combined version of arithmetic-projection, the gene values of the offspring are given by:

$$g_c = g_{p1} + r(g_{p2} - g_{p1}) \qquad (3)$$

Where g_c is the offspring gene value, g_{p1} is the gene value of the less fit (or lower rank) parent, g_{p2} is the gene value of the more fit parent, and r is a uniformly distributed random number within a certain range. If the range of r is between zero and 1.0, Eq. 1 would represent arithmetic crossover, while if the range is between 1.0 and some number bigger than 1.0, Eq. 1 would represent heuristic crossover. In this paper, r is uniformly distributed between zero and 2.5, so it represents a blend of both arithmetic and heuristic crossover.

Multipoint Crossover

Similar to single-point crossover but exchange of chromosome segments between parents to produce the offspring chromosome occurs at several points [4]. It is believed that multipoint crossover works better for long chromosomes.

Uniform Crossover

Can be viewed as an extreme case of multipoint crossover, where every building block can be exchanged between the parent chromosomes, with a specified probability.

Special Crossover Scheme

A Special crossover operator is designed based on understanding of the nature of the encoding scheme of the chromosomes. The basic idea is that the *meaning* of the numbers of a four-integer building block is dependent on the *state* of the components and subassemblies present in the hypothetical bin of the product. Thus if crossover changes some building block on a chromosome, it means crossover has subsequently changed the meaning of *all* the building blocks that follow the changed building block for that product. The special crossover operator is designed as follows:

- One single-point crossover is used for every product on the chromosome. This is slightly different from multipoint crossover as one crossover occurs in every product, while in multipoint crossover, one product can have several crossover points, while another has none.
- Crossover can occur only at intervals in between the building blocks (which are of length four, each defining an assembly operation).

- In order to grow good schemata in the population of chromosomes, the crossover location in the initial population has higher probability of occurring nearer to the start of the string, but as the search progresses, the crossover location probability *centeroid* is shifted gradually towards the end of the string.

It is worth mentioning that within one product, single-point crossover is chosen and not two-point or multi-point crossover because of the heavy backward dependency in the chromosome introduced by the encoding scheme. This backward dependency makes a second crossover point within the same product no more than a random mutation for the remaining part of the chromosome.

3.4 Mutation

When arithmetic projection, multipoint or uniform crossover are used, mutation of the building blocks occurs in a classical fashion, where according to a user specified probability, random changing of the numbers on the chromosome to any number within their allowed range with uniform probability is performed [9]. However, when the special crossover is employed, a corresponding special mutation is used.

In the special mutation, separate probabilities are assigned to each of the four numbers of the building block. Typically, the first number has the least mutation probability, and then following numbers the next have higher and higher mutation probabilities. The intuition behind this is that the *meaning* of the second number on the building block is dependent on the first number, so changing the first subsequently changes the second. Also the first two numbers on the building block play the role of defining the assembly *sequence*, so changing any of them subsequently changes the *meaning* of the whole chromosome portion that follows the location of the mutation.

It is noted that the special mutation scheme could also be applied to the classical forms of crossover considered. However, the authors preferred keeping the study of specialized operators (crossover and mutation) separate from classical ones in order to highlight the benefit (if any) that is gained by introducing prior knowledge of the problem structure into the GA operators.

3.5 Multi-objective GA

The implemented multi-objective GA is similar to NSGA-II [10, 11] with a few modifications in enforcing elitism, where the un-dominated members are copied into a separate elite population that is preserved across generations unless it grows too big. General highlights of NSGA-II that differ from single objective GA include: i) selection is based on ranking through Pareto-based dominance and ii) use of a Nitching function in selecting from with members that have the same rank. The Nitching function serves to give higher probability of selection for members that have no others near to them and thereby enhances the spread of the population over the Pareto-front. The pseudo-code for the implemented multi-objective GA is given as:

```
1.  Generate Initial Population
2.  Loop Until Termination: (Max. Number of Generations or Function
    Evaluations)
3.  Perform Pareto-Ranking of Current Population
4.  Add Un-dominated Members of Current Population Members into Elite
    Population
5.  Re-Rank Elite Population and Kill any members that become dominated
6.  If Elite Population Grows beyond an allowed size, select a number of
    members equal to the allowed elite population size according to a
    Nitching function and kill the rest of the members
7.  Loop until New Population is full
8.  Select Parents for reproduction
        a.  Perform a binary tournament to select a rank
        b.  Select a parent member from within a chosen rank according
            to a Nitching function
9.  Perform Crossover, mutation and Repair then place new offspring in
    New Population
10. When New Population is full, replace Current Population with the new
    one and repeat at Step #2.
```

4 Results and Discussion

The feasible assembly operations satisfying the assembly precedence relations for each of the ventilator models are shown in a tree representation in Fig. 5. A list of assembly stations is provided in Table 2. Ten multi-objective GA runs are performed for each of the crossover schemes, using a population size of 150, for 100 generations, with a limit on the elite population size of 30 individuals. Crossover and mutation follow the general recommendation [4] of high crossover probability (about 0.9), low mutation rate (about 2%).

Figure 6 shows typical populations at the beginning and end of the search and their corresponding un-dominated elite front. Since the objective is to minimize both assembly cost f_1 and production shortage f_2, the closer a candidate design gets to the lower left corner of the plot, the better the design is. The difference between the initial and final populations is quite apparent. The initial population is scattered over a wide range on the $f_1 - f_2$ space, while final population has its un-dominated front closer to the lower left corner of the plot and the whole population is gathered pretty close to the un-dominated front. As such, the plot implies that convergence is achieved and that the choice of number of generations is sufficient.

Of particular interest, is the extreme point along the Pareto-front, which has zero production volume shortage (*i.e.* $f_2=0$), meaning that it has minimal production cost while meeting the required production volume during a given production period. This point on the Pareto-front is referred to as the *best point*. In the best-known solution of the problem, the best point has the assembly sequences shown using thicker lines in

Fig. 5. The best-known solution also utilizes types and number of assembly stations as shown in the last column of Table 2. It is observed, that whenever possible, the best solution avoids employing assembly stations that incur extra cost to do an operation.

The history of the average value of f_1 for the best point during the search is displayed in Fig. 7. The percentage success in attaining the best-known solution is given in Table 3. It is seen in Fig. 7 that given enough generations, all the considered crossover schemes succeeded in bringing down the objective function to the near-optimal region. Arithmetic-Projection crossover seems to have the fastest early descent because of its embedded gradient following heuristic, but has the least probability of success in attaining the best-known solution, which is probably due to premature convergence. Multi-point crossover has the slowest convergence rate but has a good chance at reaching the best-known solution. The proposed crossover scheme, which is tailored according to the structure of the encoded chromosome, has a moderate convergence rate, but superior chance at reaching the best-known solution. A possible reason is that the employed encoding scheme introduces heavy backward dependency to the chromosome genes, which presents difficulty to traditional crossover schemes, but the special crossover scheme is better geared to operate with such encoding.

Overall, all the considered crossover schemes performed well at getting close to the best-known solution, which implies the effectiveness of using multi-objective GA's, and in particular the adapted version of NSGA-II, for similar problems.

Table 2. List of assembly stations.

ID	Operation	Inv. Cost	Hr. Rate	# in Opt.	ID	Operation	Inv. Cost	Hr. Rate	# in Opt.
1	Frame to L. Bracket	1000	5.0	1	11	Blades	4000	10.0	2
2	Frame to U. Bracket	3000	10.0	1	12	L. Brckt 2 to L. Brg. 2	2000	8.0	1
3	L. Brckt to Axle 1	2000	8.0	1	13	U. Brckt to U. Brg.	2000	8.0	1
4	L. Brckt to Axle 2	3000	8.0	NA	14	L. Brg. 2 to Shaft	4000	10.0	1
5	U. Brckt to Axle	4000	8.0	1	15	U. Brg. to Shaft	2000	8.0	1
6	Axle to L. Brg.	1000	8.0	1	16	Inner Rotor 1	3000	8.0	2
7	Axle to U. Brg. 1	1000	8.0	1	17	Inner Rotor 2	5000	10.0	NA
8	Axle to U. Brg. 2	4000	8.0	NA	18	Shaft to L. Hub	3000	8.0	1
9	L. Brg. to L. Hub	1000	8.0	1	19	Shaft to U. Hub	3000	8.0	1
10	U. Brg. to U. Hub	1000	8.0	1					

5 Conclusion

This paper presented the application of an adapted version of NSGA-II for multi-objective optimization of a real multi-product assembly shop. Although using real data, the model of an assembly system is rather simple. Further extension of this study, would test the same approach on more sophisticated assembly models through linking to random discrete time event simulation software such as ARENA [12]. Also presented in this paper, was a special encoding scheme for the decision parameters and associated special crossover, mutation and repair operators. The obtained results for the simplified assembly model are encouraging and motivate further exploration.

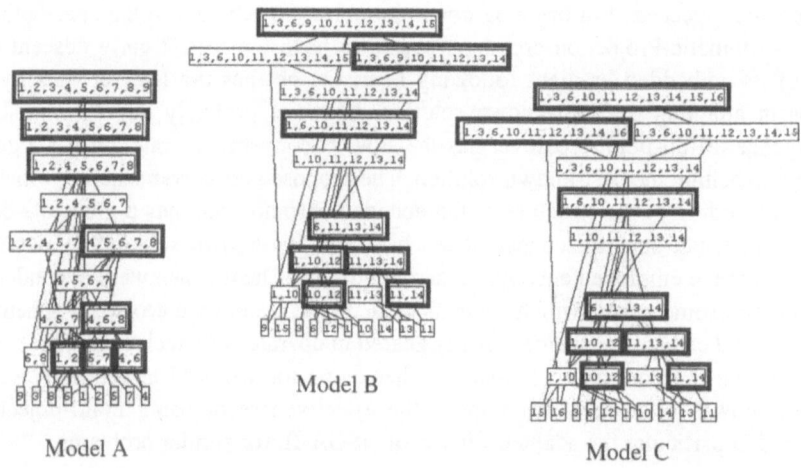

Fig. 5. Feasible assembly operations and sequences for best-known solution (thick lines).

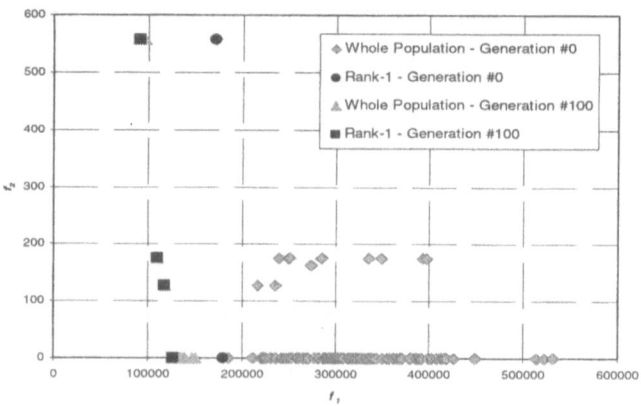

Fig. 6. Typical display of initial and final populations and their un-dominated Pareto-fronts.

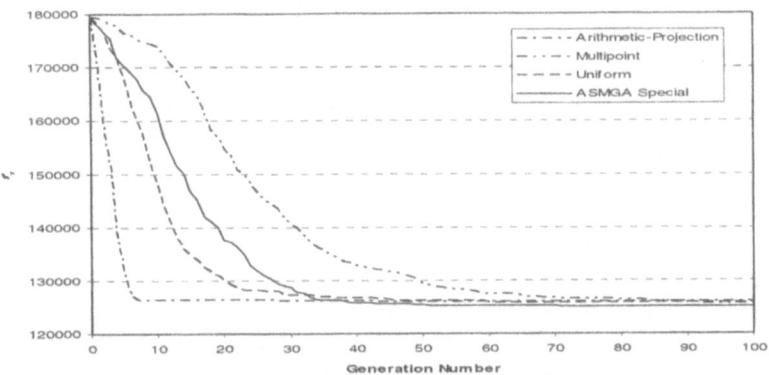

Fig. 7. History of search average progress for the best point.

Table 3. Percentage success in attaining the best-known solution.

Crossover Scheme	Percentage Success
Arithmetic Projection Crossover	30%
Multipoint Crossover	50%
Uniform Crossover	40%
ASMGA Special Crossover	90%

Acknowledgements. This work is an extension of a course project for ME588: Assembly Modeling conducted during the Fall 2002 semester at the University of Michigan, Ann Arbor. MECO: Modern Egyptian Contracting provided the data for the wind-propelled ventilators.

References

1. De Fazio, T., Whitney, D.: Simplified Generation of All Mechanical Assembly Sequences. IEEE J. of Robotics and Automation, Vol. 3, No. 6, (1987) 640–658.
2. De Mello, H., Luiz S., Sanderson, A.: A correct and complete algorithm for the generation of mechanical assembly sequences. IEEE Transactions on Robotics and Autonomous Systems, Vol. 7, No. 2 (1991) 228–240.
3. Garey, M., Johnson, D.,: Computers and Intractability: A Guide to the Theory of NP-Completeness. W.H. Freeman and Company, NY, (1979).
4. Goldberg, D.: Genetic Algorithms in Search Optimization and Machine Learning. Addison – Wesley Publishing Company (1989).
5. Awdah, N., Sepehri, N., Hwalwshka, O.: A Computer-Aided Process Planning Model Based o Genetic Algorithms. Computers and Operations Research, Vol. 22, No. 8,. (1995) 841–856.

6. Kaeschel, J., Meier, B., Fisher, M., Teich, T.: Evolutionary Real-World Shop Floor Scheduling using Parallelization and Parameter Coevolution. Proceedings of the Genetic and Evolutionary Computation Conference, Las Vegas, NV (2000) 697–701.
7. Chen, J., Ho, S.: Multi-Objective Evolutionary Optimization of Flexible Manufacturing Systems. Proceedings of the Genetic and Evolutionary Computation Conference, San Francisco, CA (2001) 1260–1267.
8. Saitou, K., Malpathak, S., Qvam, H.: Robust Design of Flexible Manufacturing Systems using Colored Petri Net and Genetic Algorithm. Journal of Intelligent Manufacturing, Vol. 13, No. 5, (2002) 339–351.
9. Michalewicz, Z.: Genetic Algorithms + Data Structures = Evolution Programs. 3rd edn. Springer-Verlag, Berlin Heidelberg New York (1996).
10. Deb, K., Argawal, S., Pratab, A., Meyarivan, T.: A Fast Elitist Non-Dominated Sorting Genetic Algorithm for Multi-Objective Optimization: NSGA-II. Proceedings of the Parallel Problem Solving from Nature VI Conference, Paris, France (2000) 849–858.
11. Coello, C., Van Veldhuizen, D., Lamont, G.: Evolutionary Algorithms for Solving Multi-Objective Problems. Kluwer Academic Publishers (2002).
12. Kelton, W., Sadowski, R., Sadowski, D.: Simulation with ARENA. 2nd edn, McGraw Hill (2002).

Multi-FPGA Systems Synthesis by Means of Evolutionary Computation

J.I. Hidalgo[1], F. Fernández[2], J. Lanchares[1], J.M. Sánchez[2], R. Hermida[1],
M. Tomassini[3], R. Baraglia[4], R. Perego[4], and O. Garnica[1]

[1] Computer Architecture Department (DACYA)
Universidad Complutense de Madrid, Spain
{hidalgo,julandan,rhermida,ogarnica}@dacya.ucm.es
[2] Departamento de Informática. Universidad de Extremadura, Spain
{fcofdez,jmsanchez}@unex.es
[3] Computer Science Institute. University of Laussane, Switzerland
Marco.Tomassini@iismail.unil.ch
[4] Istituto di Scienza e tecnologie dell'informazione "Alessandro Faedo", CNR, Italy
{Ranieri.Baraglia,Raffaele.Perego}@cnuce.cnr.it

Abstract. Multi-FPGA systems (MFS) are used for a great variety of applications, for instance, dynamically re-configurable hardware applications, digital circuit emulation, and numerical computation. There are a great variety of boards for MFS implementation. In this paper a methodology for MFS design is presented. The techniques used are evolutionary programs and they solve all of the design tasks (partitioning placement and routing). Firstly a hybrid compact genetic algorithm solves the partitioning problem and then genetic programming is used to obtain a solution for the two other tasks.

1 Introduction

An FPGA (Field Programmable Gate Array) is an integrated circuit capable of implementing any digital circuit by means of a configuration process. FPGAs are made up of three principal components: configurable logic blocks, input-output blocks and connection blocks. Configurable logic blocks (CLBs) are used to implement all the logic circuitry of a given design. They are distributed in a matrix way in the device. On the other hand, the input-output blocks (IOBs) are the responsible for connecting the circuit implemented in the FPGA with the external world. This "world" may be the application environment for which it has been designed, or a set of different FPGAs. Finally, the connection blocks (switch-boxes) and interconnection lines are the elements available to the designer for making the routing of the circuit. In most occasions we need to use some of the CLBs to accomplish the routing. The internal structure of an FPGA is shown in Figure 1. Logic blocks, IOBs and the interconnection resources are indicated on it [1].

Sometimes, the size of an FPGA is not enough to implement large circuits and it is necessary the use of Multi-FPGA system (MFS). There is a great number of MFSs, each of them suited for different applications [2]. These systems include not only integrated circuits but, also memories and connection circuitry. Nowadays MFS are used for a great variety of applications, for instance, dynamically re-configurable hardware applications [3], digital circuit emulation [4], and numerical computation [5]. There are a lot of different boards and topologies for MFS implementation. The two most common topologies are the mesh and crossbar types. In a mesh, the FPGAs

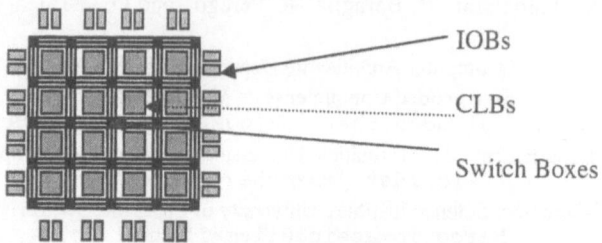

Fig. 1. General structure of an FPGA

are connected in the nearest-neighbor pattern. This kind of board has a simple routing methodology as well as an easy expandability and all programmable devices within the board have identical functionality. Figure 2a shows and FPGA with mesh-topology. On the other hand crossbar topologies, Figure 2b, separate the system components into logic and routing chips. This topology could be suitable for some specific problems, but crossbar topologies usually waste not only logic resources but also routing resources. For these reasons we have focused on mesh topologies

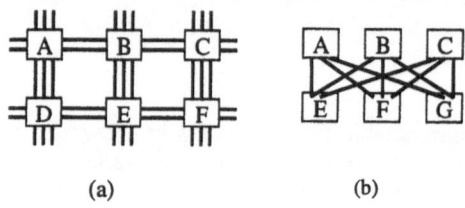

Fig. 2. Multi-FPGA Mesh (a) and crossbar (b) topologies

As in any integrated circuit device, MFSs design flow has three major tasks: partitioning, placement and routing. Frequently two of these tasks are tackled together, because when accomplishing the partitioning, the placement must be considered or vice versa. The solution that we are proposing on this paper covers the whole process. Firstly, we obtain the partitions of the circuit; each of them will be later implemented into a different FPGA. Second, we need to place and route the circuit using the FPGA resources. As we will see in the following sections, we use two different evolutionary algorithms: a hybrid compact genetic algorithm (cGA) for the partitioning step and a genetic programming technique for the routing and placement step. The rest of the paper is organized as follows. Section 2 describes the partitioning methodology. Section 3 shows how genetic programming can finish the design process within the FPGAs, while section 4 contains the experimental results and finally section 5 drafts our conclusions and the future work.

2   MFS Partitioning and Placement

Methodology: Partitioning deals with the problem of dividing a given circuit into several parts, in order to be implemented on a MFS. When using a specific board we must bear in mind several constraints related to the board topology. Some of these constraints are the number of available I/O pins and logic capacity. Although the logic capacity is usually a difficulty, the number of available pins is the hardest problem, because FPGA devices have a reduced number of them comparing with their logic capacity. In addition we must reserve some of the pins to interconnect the parts of the circuit placed on non-adjacent FPGAs. Most of the research related to the problem of partitioning on FPGAs has been adapted from other VLSI areas, and hence, they disregard the specific features of these systems. In this paper a new method for solving the partitioning and placement problem in MFSs is presented. We apply the graph theory to describe a given circuit, and then a compact genetic algorithm (cGA) with a local search improvement [17] is applied with a problem-specific encoding. This algorithm not only preserves the original structure of the circuit but also evaluates the I/O-pins consumption due to direct and indirect connections among FPGAs. It is done by means of a fuzzy technique. We have used the Partitioning93 benchmarks [6], described in the XNF hardware description language (Xilinx Netlist Format) [7].

Circuit Description: Some authors use hypergraphs for representing a circuit netlist, although there are some approximations, which use graphs. We have used an undirected graph representation to describe the circuit. This selection has been motivated because it can be adapted to the compact genetic algorithm code. We will describe later how the edges of his spanning tree can be used to represent a k-way partitioning. A spanning tree of a graph is a tree, which has been obtained selecting edges from this graph [8]. Then we use a hybrid compact genetic algorithm to search the optimal partitioning which works basically as follows. First we obtain a graph from the netlist description of the circuit, and then a spanning tree of that graph is obtained. From this tree, we select k-1 edges and they are eliminated in order to obtain a k-way partition. The partitions are represented by those deleted edges.

Genetic Representation: The compact genetic algorithm (cGA) uses the encoding presented in [9], which is directly connected with the solution of the problem. The code for our problem is based on the edges, which belong to the spanning tree. We have seen above how the partition is obtained by the elimination of some edges. We assign a number to every edge of the tree. Consequently a chromosome will have k-1 genes for a k-way partitioning, and the value of these genes can be any of the order values of the edges. For example, chromosome (3 4 6) for a 4-way partitioning, represents a solution obtained after the suppression of edge numbers 3, 4, and 6 from its spanning tree. So the alphabet of the algorithm is {0, 1... n-1} where n is the number of vertices of the target graph (circuit), because the spanning tree has n-1 edges.

Hybrid Compact Genetic Algorithm: The *cGA* does not manage a population of solutions but only mimics its existence [10]. It represents the population by means of a vector of values, $p_i \in [0,1]$, $\forall_i = 1,...,l$, where l is the number of alleles needed to represent the solutions. In order to design a cGA for MFS partitioning we adopted the edge representation explained below and we consider the frequencies of the edges occurring in the simulated population. A vector V with the same dimension as the number of nodes minus one was used to store these frequencies. Each element v_i of V represents the proportion of individuals whose partition use the edge e_i. The vector elements v_i were initialised to 0.5 to represent a randomly generated population in which each edge has equal probability to belong to a solution [11]. Sometimes it is necessary to increase the selection pressure (Ps)rate to reach good results with a Compact Genetic Algorithm. A value for Ps near to 4 is usually a good value for MFS partitioning. It is not recommendable to increase this value very much because the computation time grows. Additionally, for some problems we need a complement for the cGA. We can combine heuristics techniques with local search algorithms to obtain this additional tool called hybrid algorithms. We have implemented a cGA with local search after a certain number of iterations in order to improve the solutions obtained by the only use of cGA. In [12] a compact genetic algorithm for MFSs partitioning was presented, and in [13] a Hybrid cGA was explained. Authors combine a cGA with the Lin-Kernighan (LK) local search algorithm, to solve TSP problems. The cGA part explores the most interesting areas of the search space and LK task is the fine-tuning of those solutions obtained by cGA. Following this structure we have implemented the hybrid cGA for MFS partitioning presented in [17]. In this paper we have used other heuristic different from LK, which is more feasible to the problem are solving.

Most of the local search algorithms try to perform search as exhaustive as possible. But, this can implies an unacceptable amount of computation time. In MFS problem, the ideal implementation of local search is to explore all the neighbour solutions to the current best solutions after a certain number of iterations. Unfortunately, the most computational expensive step of our cGA is the evaluation of the individuals. We have employed a local search heuristic every *n* iterations and as in parallel genetic algorithms we need to fix the value of *n* to keep the algorithm search in good working order. After an empirical study for different values the local search frequency, we obtain that n must be located between 20 and 60 with an optimal value (that depends on the benchmark) near to 50. So for our experiments we fixed the local search frequency *n* to 50 iterations, i.e. we develop a local search process every 50 iterations of the compact GA.

Remember that a chromosome has k-1 genes for a k-way partitioning, and the value of these genes are the edges eliminated to obtain a partitioning solution. In order to explain the algorithm we must define what a neighbour solution is. We say that solution A is a neighbour solution of B (and B is a neighbour solution of A) if the difference between their chromosomes is just one gene. For example solution represented by chromosome (1 43 56 78 120 **345** 789) is a neighbour solution of the partition represented by (1 43 56 78 120 **289** 789), in an 8-way partitioning problem. Our local search heuristic explores only one neighbour solution for each gene, that is k-1 neighbour solutions of the best solution every n iterations. For the sake of clarity we reproduce the explanation of the local search process presented in [17].

The local search process works as follows. Every n iterations, we obtain the best solution up to that time, which is denoted by BS. To obtain BS, first we explore the compact GA probability vector and select the k-1 most used genes (edges) to form MBS (vector best individual). We also have the best individual generated up to now (GBS) (similar to elitism). The best individual between MBS and GBS (i.e. which of them has the best fitness value) will be BS. After BS has been deduced at iteration n, the first random neighbour solution (TS1) to BS is generated substituting the first gene (edge) of the chromosome by a random one not used in BS. Then we evaluate the fitness value of BS (FV_{BS}) and the fitness value of TS1 (FV_{TS1}). If FV_{TS1} is better than FV_{BS}, TS1 is dropped to BS and the initial BS is eliminated, otherwise TS1 is eliminated. Then we repeat the same process using the new BS but with the second gene, to generate TS2. If the fitness value of TS2 (FV_{TS2}) is better than the present FV_{BS} then TS2 will be our new BS or, if FV_{TS2} is worst than FV_{BS}, there will be no change in BS. The process is repeated for all genes until the end of the chromosome, that is, k-1 times for a k-way partitioning. Although only a very small part of the solution neighbourhood space is explored, we improve the performance of the algorithm significantly (in terms of quality of our solutions) without drastically degrading its total computation time.

In order to clarify the explanation about the proposed local search method we can see an example. Let us suppose a graph with 12 nodes and its spanning tree, for a 5-way partitioning problem (i.e. is we want to divide the circuit into five parts). As we have explained, we will use individuals with 4 genes. Let us also suppose a local search frequency (n) of 50 and that after 50 iterations we have reached to a best solution represented by:

$$BS = 3, 4, 6, 7 \qquad (2)$$

The circuit graph has 12 nodes, so its spanning tree is formed by 11 edges. The whole set of possible edges to obtain a partitioning solution is called E:

$$E = \{0, 1, 2, 3, 4, 5, 6, 7, 8, 9, 10\} \qquad (3)$$

In order to generate TS1 we need to know the available edges A_{LS} for random selection, as we have said we eliminate the edges within BS from E to obtain A_{LS}:

$$A_{LS} = \{0, 1, 2, 5, 8, 9, 10\} \qquad (4)$$

Now we randomly select an edge (suppose 0) to build TS1 substituying it by the first gene (3) in BS:

$$TS1 = 0, 4, 6, 7 \qquad (5)$$

The third step is the evaluation of TS1 (suppose FV_{TS1}=12) and comparing (suppose a minimization problem) with FV_{BS} (suppose FV_{BS} = 25). As FV_{TS1} is better than FV_{BS}, TS1 will be our new BS and the original BS is eliminated. Those changes also affect to A_{LS} because our new A_{LS} is:

$$A_{LS} = \{1, 2, 3, 5, 8, 9, 10\} \qquad (6)$$

Table 1 represents the rest of the local search process for this example.

Table 1. Local search example

i	A_{LS}	BS	FV	Random gene	TS	FV	New BS
1	0,1,2,5,8,9,10	3,4,6,7	25	0	0,4,6,7	12	0,4,6,7
2	1,2,3,5,8,9,10	0,4,6,7	12	1	0,1,6,7	37	*0,1,6,7*
3	1,2,3,5,8,9,10	0,4,6,7	12	9	0,4,9,7	10	*0,4,9,7*
4	1,2,3,5,6,8,10	0,4,9,7	10	8	0,4,8,9	11	0,4,9,7
Pre-Local Search Best Solution: 3, 4, 6, 7							
Post-Local Search Best Solution: 0, 4, 9, 7							

3 Placement and Routing on FPGAs

Once different partitions have been obtained and assigned to different FPGAs, we must place components into FPGAs' CLBs and connect CLBs within each of the FPGAs. To do so, we use Genetic Programming (GP). A wide description of this technique can be found in [14].

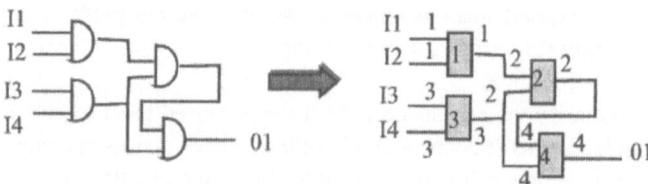

Fig. 3. Representing a circuit wit black boxes and labeling connections

Partitions Representation Using Trees: The main goal for this step is to implement a partition (circuit) -that has been obtained in the previous step- into an FPGA. We have thus to place each of the circuit component into a CLB and then to connect all the CLBs according to the circuit's topology. We have used Genetic Programming (GP) based in tree structures in this task. Therefore, circuits will be encoded here as trees. A given circuit is made up of components and connections. If we forget the name and function of each of the simple components (considering each of them as a black box), and instead we use only one symbol for representing any of them, a circuit could be represented in a similar way as the example depicted in figure 3. Given that components compute very easy logic function, any of them can be implemented by using any of the CLBs available within each FPGA. This means that we only have to connect CLBs from the FPGA according to the interconnection model that a given circuit implements, and then we can configure each of the CLB with the function that each component performs in the circuit. After this couple of simple steps we have got the circuit in the FPGA. Given that we employ Genetic Programming we have to encode the circuit in a tree. We can firstly number each component from the circuit, and then assign the number of those components to the ends of wires connected to them (see figure 3). Wires could now be disconnected without loosing any information. We could even rebuild the circuit by using labels as a guide.

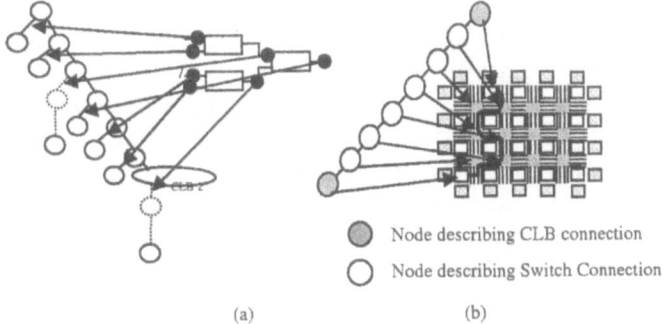

Fig. 4. Encoding circuits by means of binary trees. a) Each branch of the tree describe a connection from the circuit. Dotted lines indicates a number internal nodes in the branch. b) Making connections in the FPGA according to nodes.

We may now describe all the wires by means of a tree by connecting each of the wires as a branch of the tree and keeping them all together in the same tree. By labeling both extremes of branches, we will have all the information required to reconstructing the circuits. This way of representing circuits allows us to go back and construct the real graph. Moreover, any given tree, randomly generated, will always correspond to a particular graph, regardless of the usefulness of the associated circuit. In this proposal, each node from the tree is representing a connection, and each branch is representing a wire. The next stage is to encode the path of wires into an FPGA. Each branch of the tree will encode a wire from the circuit. We have now to decide how each of the tree's branches can encode a set of connections. As seen in previous sections, mesh FPGAs contains CLBs, switch blocks and wire segments. Each wire segment can connect adjacent blocks, both CLBs and switch blocks. Several wire segments must be connected through switch blocks when joining two CLBs' pins according to a given circuit description. A connection in a circuit can be placed into an FPGA in many different ways. For example, there are as many choices in the selection of each CLB as the number of rows multiplied by the number of columns available in the FPGA (see figure 1, section 1). Moreover, which of the pins of the CLB will be used and how the wire traverses the FPGA has to be decided from among the incredibly high number of possibilities. Of course, the same connection can be made in many different ways, with more or fewer switch blocks being crossed by the wire.

Every wire in an FPGA is made up of two ends - these can connect to a CLB or to an IOB. On the other hand, as said above, a given number of switch connections may conform the path of the wire. In the representation we have used a branch from tree for codifying wires, CLB and IOB connections are described as each of the two end nodes which make up a branch. In order to describe switch connections, we add as many new internal nodes to the branch as switch blocks are traversed by wires (see figure 4b). Each internal node requires some extra information: if the node corresponds to a CLB we need to know information about the position of the CLB in the FPGA, *the number of pin to which one of the ends of the wire is connected, and* which of the wires of the wire block we are using; if the node represents a switch connection, we need information about that connection (Figure 4 graphically depicts how a tree describes a circuit).

It may well happen that when placing a wire into an FPGA, some of the required connections specified in the branch can not be made, because, for instance, a switch block connection has been previously used for routing another wire segment. In this case the circuit is not valid, in the sense that not all the connections can be placed into a physical circuit. In order for the whole circuit to be represented by means of a tree, we will use a binary tree, whose left most branch will correspond to one of its connections, and the left branch will consist of another subtree constructed recursively in the same way (left-branch is a connection and right-branch a subtree). The last and deepest right branch will be the last circuit connection. Given that all internal nodes are binary ones we can use only a kind of function with two descendants.

GP Sets: When solving a problem by means of GP one of the first things to do once the problem has been analyzed is to build both the function and terminal sets. The function set for our problem contains only one element: F={SW}, Similarly, the terminal set contains only one element T={CLB}. But SW and CLB may be interpreted differently depending on the position of the node within a tree. Sometimes a terminal node corresponds to an IOB connection, while sometimes it corresponds to a CLB connection in the FPGA (see figure 4,a). Similarly, a SW node sometimes corresponds to a CLB connection, while others affects switch connections in the FPGA. Each of the nodes in the tree will thus contain different information:

- If we are dealing with a terminal node, it will have information about the position of CLBs, the number of pins selected, the number of wires to which it is connected, and the direction we are taking when placing the wire.
- If we are instead in a function node, it will have information about the direction we are taking. This information enables us to establish the switch connection, or in the case of the first node of the branch, the number of the pin where the connection ends.

If we look at figure 4, we can observe that wires with IOBs at one of their ends are shorter –only needs a couple of nodes- than those that have CLBs at both ends –they require internal nodes for expressing switch connections-. Wires expressed in the latest position of trees have less space to grow, and so we decided to place IOB wires in that position, thus leaving the first parts of the trees for long wires joining CLBs.

Evaluating Individuals: In order for GP to work, individuals from the population have to be evaluated and reproduced employing the GP algorithm. For evaluating an individual we must convert the genotype (tree structure) to the phenotype (circuit in the FPGA), and then compare it to the circuit provided by the partitioning algorithm. We developed an FPGA simulator for this task. This software allows us to simulate any circuit and checks its resemblance to other circuit. Therefore, this software tool is in charge of taking an individual from the population and evaluating every branch from the tree, in a sequential way, establishing the connections that each branch specifies. Circuits are thus mapped by visiting each of the useful nodes of the trees and making connections on the virtual FPGA, thus obtaining phenotype. Each time a connection is made, the position into the FPGA must be brought up to date, in order to be capable of making new connections when evaluating the following nodes. If we evaluate each branch, beginning with the terminal node, thus establishing the first end of the wire, we could continue evaluating nodes of the branch from the bottom to the top. Nevertheless, we must be aware that there are several terminals related to each branch, because each function node has two different descendants. We must decide which of the terminals will be taken as the beginning of the wire, and then drive the evaluation to the top of the branch. We have decided to use the terminal that is reached when going down through the branch using always the left descendant (see figure 5).

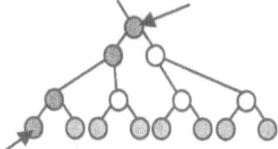

Fig. 5. Evaluating trees

In one sense there is a waste of resources when having so many unused nodes. Nevertheless they represent new possibilities that can show up after a crossover operation (in nature, there always exist recessive genes, which from time to time appear in descendants). These nodes are hidden, in the sense that they don't take part in the construction of the circuit and may appear in new individuals after some generations. If they are useful in solving the problem, they will remain in descendants in the form of nodes that express connections. The fitness function is computed as the difference between the circuit provided and the circuit described by the individual.

4 Experimental Results

- *Partitioning and Placement onto the FPGAs*

The algorithm has been implemented in C and it has been run on a Pentium II 450 MHz. We have used the partitioning 93 benchmarks in XNF format. As the number and characteristics of CLBs depend on the device used for the implementation, we have supposed that each block of the circuits uses one CLB. We use the Xilinx's 4000 series. Table 2 contains some experimental results. It has seven columns which express: the name of the test circuit (Circuit), its number of CLBs (CLB), the number of connections between CLBs (Edges), the distribution of CLBs among the FPGAs (Distribution), the number of I/O pins lacking (p), the device proposed for the implementation (FPGA type), and the CPU time in seconds necessary to obtain a solution (T(sec)). From the results we can observe that there are some unbalanced distributions. This is a logic result because we need some circuits to pass the nets from one device to another. In addition our fitness function has been developed to achieve two objectives, so that the cGA works. In summary, the algorithm succeeds in solving the partitioning problem with board constraints. We have implemented a board and we have checked some basic circuits so we can conclude that the algorithms works.

Table 2. Partitioning and Placement Results for different benchmarks

Circuit	CLBs	Edges	Distribution	p	FPGA type	T (sec)
S208	127	200	16,15,17,21,11,11,19,17	0	4003	20.70
S298	158	306	20,23,29,20,20,20,25,11	0	4003	5.92
S400	217	412	28,47,23,37,16,20,33,13	0	4005	52.96
S444	234	442	27,38,36,29,41,37,16,27	0	4005	52.99
S510	251	455	34,41,38,42,32,19,24,2	0	4005	96.74
S832	336	808	230,11,25,14,17,18,16,5	0	4008	96.74
S820	338	796	237,17,24,14,21,8,7,10	0	4008	138.93
S953	494	882	168,60,82,31,101,289,15	0	4008	194.65
S838	495	800	100,92,37,77,67,60,43,17	0	4008	293.45
S1238	574	1127	91,11,293,56,50,25,17,31	0	4008	320.14
C1423	829	1465	525,52,114,14,37,51,28,8	0	4020	273.65
C3540	1778	2115	614,135,88,89,56,26,28,2	0	4020	844.16

- **Inter-FPGA Placement and Routing**

Several experiments with different sizes and complexities have been performed for testing the placement and routing process. [10]. One of them is shown in figure 6. We worked on a SUN workstation 167 Mhz. The main parameters employed were the following: Population size = 200, Number of generations = 500, Population size: 200, Maximum depth: 30, Steady State Tournament size: 10. Crossover probability=98%, Mutation probability=2%, Creation type: Ramp Half/Half. Add best to New Population. The GP tool we used is described in [16]. Figures 6 and 7 show one of the proposed circuits and one of the solutions found, respectively. More solutions found for this circuit are described in [15].

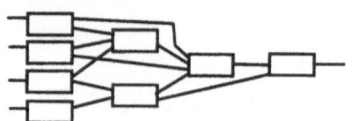

Fig. 6. Circuit to be tested.

5 Conclusions and Future Work

In this paper a methodology for circuit design using MFSs has been presented. We have used evolutionary computation for all steps in the design process. First, a compact genetic algorithm with a local search heuristic was used on achieving partitioning and placement for intra-FPGA systems and, for the Inter-FPGA tasks Genetic programming was used. This method can be applied for different boards and solves the whole design flow process. As future work, we are working now on the parallelization of all of the steps and studying Multi-Objective Genetic Algorithms (MOGA) techniques.

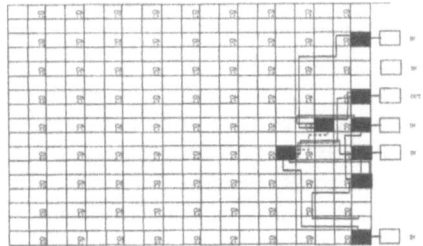

Fig. 7. A solution found for example on Figure 6

Acknowledgement. Part of this research has been possible thanks to the Spanish Government research projects number TIC2002-04498-C05-01 and TIC2002/750.

References

1. S. Trimberger. "Field Programmable Gate Array Technology". Kluwer 1994.
2. S .Hauck,: Multi-FPGA systems. Ph. D. dissertation. University of Washington. 1994
3. M. Baxter. "Icarus: A dinamically reconfigurable computer architecture" IEEE Symposium on FPGAs for Custom Computing machines, 1999, 278–279.
4. R. Macketanz, W. Karl. "JVX: a rapid prototyping system based on Java and FPGAs". In Field Programmable Logic: From FPGAs to Computing Paradigm, pages 99–108. Spinger Verlag, 1998
5. M.I. Heywood and A.N. Zincir-Heywood. "Register based genetic programming on FPGA computing platforms". Euro GP 2000, 44–59.
6. CAD Benmarching Laboratory, http://vlsicad.cs.ud.edu/
7. *XNF: Xilinx Netlist Format*", http://www.xilinx.com
8. F. Harary. "Graph Theory". Addison-Wesley 1968
9. J.I. Hidalgo, J. Lanchares, R. Hermida. "Graph Partitioning methods for Multi-FPGA systems and Reconfigurable Hardware based on Genetic algorithms", Proceedings of the 1999 Genetic and Evolutionary Computation Conference Workshop Program, Orlando (USA), 1999, 357–358.
10. G.R. Harik, F.G. Lobo, D. E. Goldberg "The Compact Genetic Algorithm". Illigal Report N° 97006, August 1997. University of Illinois at Urbana-Champaign
11. G.R. Harik, F.G. Lobo, D. E. Goldberg "The Compact Genetic Algorithm". IEEE Transactions on Evolutionary Computation. Vol. 3, No. 4, pp. 287–297, 1999.
12. J.I. Hidalgo. R.Baraglia, R. Perego, J. Lanchares, F. Tirado. "A Parallel compact genetic algorithm for Multi-FPGA partitioning" Euromicro PDP 2001, 113–120. IEEE Press.
13. R, Baraglia, J.I.Hidalgo, and R. Perego. "A Hybrid Heuristic for the Travelling Salesman Problem ". IEEE Transactions on Evolutionary Computation. Vol. 5, No. 6, pp. 613–622, December 2001.
14. J.R. Koza: Genetic Programming. On the programming of computers by mens of natural selection. Cambridge MA: The MIT Press

15. F. Fernández, J.M. Sánchez, M. Tomassini, "Placing and routing circuits on FPGAs by means of Parallel and Distributed Genetic Programming ". Proceedings 4th international conference on Evolvable systems ICES 2001.
16. M. Tomassini, F. Fernández, L. Vannexhi, L. Bucher, "An MPI-Based Tool for Distributed Genetic Programming" In Proceedings of IEEE International Conference on Cluster Computing CLUSTER2000, IEEE Computer Society. Pp. 209–216
17. J.I. Hidalgo, J. Lanchares, A.ibarra,R. Hermida. A Hybrid Evolutionary Algorithm for Multi-FPGA Systems Design. Proceedings of Euromicro Symposium on Digital System Design, DSD 2002. Dortmund, Germany, September 2002. IEEE Press, pp. 60–68.

Genetic Algorithm Optimized Feature Transformation – A Comparison with Different Classifiers

Zhijian Huang[1], Min Pei[1], Erik Goodman[1], Yong Huang[2], and Gaoping Li[3]

[1] Genetic Algorithms Research and Application Group (GARAGe)
Michigan State University, East Lansing, MI
{huangzh1,pei,goodman}@egr.msu.edu
[2] Computer Center, East China Normal University, Shanghai, China
siewl@online.sh.cn
[3] Electrocardiograph Research Lab, Medical College
Fudan University, Shanghai, China
gpli@shmu.edu.cn

Abstract. When using a Genetic Algorithm (GA) to optimize the feature space of pattern classification problems, the performance improvement is not only determined by the data set used, but also depends on the classifier. This work compares the improvements achieved by GA-optimized feature transformations on several simple classifiers. Some traditional feature transformation techniques, such as Principal Components Analysis (PCA) and Linear Discriminant Analysis (LDA) are also tested to see their effects on the GA optimization. The results based on some real-world data and five benchmark data sets from the UCI repository show that the improvements after GA-optimized feature transformation are in reverse ratio with the original classification rate if the classifier is used alone. It is also shown that performing the PCA and LDA transformations on the feature space prior to the GA optimization improved the final result.

1 Introduction

The genetic algorithm (GA) has been tested as an effective search method for high-dimensional complex problems, taking advantage of its capability for sometimes escaping local optima to find optimal or near optimal solutions. In pattern classification, GA is widely used for parameter tuning, feature weighting [1] and prototype selection [2].

Feature extraction and selection is a very important phase for a classification system, because the selection of a feature subset will greatly affect the classification result. GA has recently also been used in the area of feature extraction. The optimization of feature space using GA can be linear [3], or non-linear [4], where in both cases, the GA stochastically, but efficiently, searches in a very high-dimensional data space that makes traditional deterministic search methods run out of time. The GA approach can also be combined with other traditional feature transformation methods.

Prakash presented the combination of GA with Principal Components Analysis (PCA), where instead of the few largest Principal Components (PCs), a subset of PCs from the whole spectrum was chosen by the GA to get the best performance [5].

In this work, three classifiers – a kNN classifier, a Bayes classifier and a Linear Regression classifier – are tested, together with the PCA and LDA transformations. One new, real-world dataset, the Electrocardiogram (ECG) data, and five benchmark datasets from the UCI Machine Learning Repository [6] are used to test the approach.

The paper starts with an introduction to GA approaches in the area of pattern classification in Section 2, followed by our solution designed in Section 3. Section 4 presents the results on ECG data with detailed comparison with regard to both classifier choice and the use of PCA/LDA transformations. Section 5 extends the tests to five benchmark pattern classification test sets, by using the best solution on PCA/LDA combination found in Section 4. Section 6 concludes the paper and Section 7 proposes some possible future work.

2 GA in Pattern Classification

Generally, the GA-based approaches to pattern classification can be divided into two groups:
- Those directly applying GA as part of the classifier.
- Those optimizing parameters in pattern classification.

2.1 Direct Application of GA as Part of the Classifier

When the GA is directly applied as part of the classifier, the main difficulty is how to represent the classifier using the GA chromosome. Bandyopadhyay and Murthy proposed an idea using a GA to perform a direct search on the partitions of an N-dimensional feature space, where each partition represents a possible classification rule [7]. In this approach, the decision boundary of the N-dimensional feature space is represented by H lines. The genetic algorithm is used to find those lines that minimize the misclassification rate of the decision boundary. The number of lines, H, turns out to be a parameter similar to the k in the kNN classifier. More lines (higher H) do not necessarily improve the classification rate, due to the effect of over-fitting.

In addition to using lines as space separators, Srikanth et al [8] also gave a novel method clustering and classifying the feature space by ellipsoids.

2.2 Optimizing Parameters in Pattern Classification by GA

However, most of the approaches using GA in pattern classification do not design the classifier using GA. Instead, GA is used to estimate the parameters of the pattern classification system, which can be categorized into the following four classes:

GA-Optimized Feature Selection and Extraction. Feature selection and extraction are the most widely used applications of GA in pattern classification. The classification rate is affected indirectly when different weights are applied to the features. A genetic algorithm is used to find a set of optimal feature weights that can improve the classification performance on training samples. Before GA-optimized feature extraction and selection, traditional feature extraction techniques such as the Principal Components Analysis (PCA) can be applied [5], while after that, a classifier should be used to calculate the fitness function for the GA. The most commonly used classifier is the k-Nearest Neighbor classifier [9], [1].

GA-Optimized Prototype Selection. In supervised pattern classification, the reference set or training samples are critical for the classification of testing samples. A genetic algorithm can be also used in the selection of prototypes in case-based classification [2], [10]. In this approach, a subset of the most typical samples is chosen to form a prototype, on which the classification for testing and validation samples is based.

GA-Optimized Classifier. GA can be used to optimize the input weight or topology of a Neural Network (NN) [4]. It is intuitive to give weights for each connection in a NN. By evolving the weights using GA, it is possible to throw away some connections of the neural network if their weights are too small, thus improving the topology of the NN, too.

GA-Optimized Classifier Combination. The combination of classifiers, sometimes called Bagging and Boosting [11], may also be optimized by Genetic Algorithm. Kuncheva and Jain [12], in their design of the Classifier Fusion system, not only selected the features, but also selected the types of the individual classifiers using a genetic algorithm.

3 GA-Optimized Feature Transformation Algorithm

This section first reviews the two models for feature extraction and feature selection in pattern classification. Then a GA-optimized feature weighting and selection algorithm based on the wrapper model [13] is proposed, outlining the structure of the experiment in this paper.

3.1 The Filter Model and the Wrapper Model

For the problem of feature extraction and selection in pattern classification, two models play important roles. The *filter model* chooses features by heuristically determined "goodness", or knowledge, while the *wrapper model* does this by the feedback of the classifier evaluation, or experience. Fig. 1 illustrates the differences between these two models.

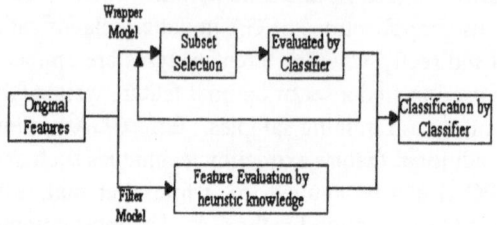

Fig. 1. Comparison of filter model and wrapper model

Research has shown that the wrapper model performs better than the filter model, comparing the predictive power on unseen data [14]. Some widely used feature extraction approaches, such as Principal Components Analysis (PCA), belong to the Filter model because they rank the features by their intrinsic properties: the eigenvalues of the covariance matrix. Most recently developed feature selection or extraction techniques are categorized to be Wrapper models, taking into consideration the classification results of a particular classifier. For example, the GFNC (Genetically Found, Neurally Computed) approach by Firpi [4] uses a GA and a Neural Network to perform non-linear feature extraction with the feedback from a kNN classifier as the fitness function for the Genetic Algorithm. A modified PCA approach proposed by Prakash also uses genetic algorithms to search for the optimal feature subset with the feedback result of the classifier [5].

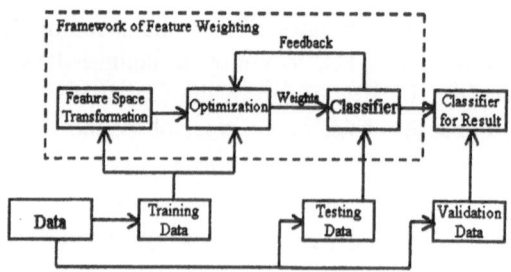

Fig. 2. Classification system with feature weighting

3.2 GA-Optimized Pattern Classification with Feature Weighting

Consider the wrapper model introduced above; in any of the pattern classification systems with weighted features, there are five components that need to be determined as illustrated in Fig. 2.
- The dataset used.
- The feature space transformation to be applied on the original feature space, known as the feature extraction phase in traditional pattern classification systems.

- The optimization algorithm which searches for the best weighting for the features;
- The classifier used to get the feedback, or fitness, for the GA, of the feature weighting. (The Induction Algorithm in the Wrapper Model [13])
- The classifier to calculate the final result for the classification problem based on the newly weighted features (the classifier for the result).

From Fig. 2, we can see that among these components, the *feature space transformation*, the *optimization* and the *classifier* are the plug-in procedures for feature weighting optimization. The feature-weighting framework is the plug-in procedure in traditional classification systems that transforms the feature space, weights each feature, evaluates and optimizes the weight attached to each feature.

In this paper, we can replace each of the components listed above by the specific choices we made, as follows:

- *Data*: ECG data and five other data sets from UCI repository.
- *Feature Space Transformation*: PCA, LDA and their combinations.
- *Optimization Algorithm*: Genetic algorithm or none. When implemented with different classifiers: feature weighting for kNN classifier, and feature selection for other classifiers. The reason for using feature weighting for the kNN classifier is because of its distance metric, that will affect the classification result by changing its weight [1], while for the Bayes classifier and the linear regression classifier, feature weighting has no further effect on the training error when compared with feature selection [15].
- *Classifier* (induction algorithm): A kNN classifier, a Bayes classifier and a linear regression classifier.
- *Classifier for Result*: Same classifier as used for the induction algorithm.

4 Test Results for ECG Data

The test results for ECG data, with various settings regarding the feature space transformation, using the GA or not, and using various classifiers, are presented and compared in this section. We will first discuss the experimental settings, and then move on to the results.

4.1 Experimental Settings

The ECG data used in this paper is directly extracted from the 12-lead, 10-second digital signal acquired from Shanghai Zhongshan Hospital, containing 618 normal ECG cases and 618 abnormal cases. The abnormal cases contain three different kinds of diseases with roughly the same number of cases of each. Altogether, 23 features, including 21 morphological features and 2 frequency domain features, are extracted from the original signal.

For a non-GA-optimized classifier, the data is partitioned into training samples and validation samples; for a GA-optimized algorithm, the data is partitioned into training data, testing data and validation data, in an n-fold cross-validation manner. If not specifically indicated, here the n is set to be 10. Table 1 lists the details of the data partitioning.

A simple genetic algorithm (SGA) is used here to do the optimization. The crossover rate is 30% and the mutation rate is set to 0.03 per bit. The program runs for 200 generations with a population size of 50 individuals (80 individuals for kNN feature weighting). When there has been no improvement within the last 100 generations, the evolution is halted.

Classifiers: A 5-nearest neighbor classifier is used. A Bayes Plug-In classifier with its parameters estimated by Maximum Likelihood Estimation (MLE) is implemented, and a linear regression classifier uses the simple multivariate regression combined with a threshold decision to predict the class labels.

With the kNN classifier, the feature weighting is allowed to range among 1024 different values between 0.0 and 10.0, with minimum changes of about 0.01, as determined by the GA, by setting the chromosome for each feature to be 10 binary digits. With the Bayesian classifier and linear regression classifier, only feature selection was tested.

Table 1. Summary of Data Partitioning

Experiments	Training	Testing	Valid
Non-GA	40%	N/A	60%
GA (n-fold)	$40\% \times \frac{n-1}{n}$	$60\% \times \frac{n-1}{n}$	$\frac{1}{n}$

Table 2. Result of kNN classifier ($k = 5$)

Settings	kNN ($k = 5$) Results of Classification Rate in %				
Fea. Trans	Non-GA	GA	Trn	Improve	P-Value
None	73.09	74.54	78.29	1.45±3.23	0.3008
PCA	74.10	77.16	79.39	3.06±1.59	**0.0043**
LDA	73.09	77.00	78.72	3.91±2.24	**0.0065**
Both	72.41	75.38	79.82	2.97±2.78	**0.0404**
Overall P value:					**0.0000**

4.2 Results and Conclusions

Tables 2-4 list the results of GA-optimized feature extraction using a kNN classifier (k=5), Bayes Plug-In classifier and linear regression classifier. The improvement after GA optimization is represented by the average improvement, with a two-tailed t-test with a 95% confidence interval. A P value indicating the probability of the Null Hypothesis (the improvement is 0) is also given, among which results having a 95% significant improvement are in bold font.

In addition to the row-wise statistics, an overall improvement significance level based on the improvement percentage is calculated for each classifier, which is a P value on all the improvement ratio values of that classifier. This significance indicator can be considered as a final summary of the GA improvement based on a particular classifier and is listed at the bottom of each table.

Table 3. Result of Bayes classifier

Settings	Bayes Results of Classification Rate in %				
Fea. Trans	Non-GA	GA	Trn	Improve	P-Val
None	71.92	74.19	76.51	2.27±3.69	0.1912
PCA	72.53	74.27	77.23	1.75±1.95	0.0730
LDA	72.95	75.47	77.23	2.52±3.11	0.0968
Both	71.85	76.99	78.33	5.14±2.54	**0.0014**
Overall P value:					**0.0000**

Table 4. Result of Linear Regression classifier

Settings	Linear Regression Results, Rate in %				
Fea. Trans	Non-GA	GA	Trn	Improve	P-Value
None	77.68	76.49	79.08	-1.19±3.23	NA
PCA	76.48	78.02	79.87	1.53±3.09	0.25
LDA	78.44	77.42	77.36	-1.02±3.15	NA
Both	77.59	77.41	77.84	-0.18±2.79	NA
Overall P value:					NA

Conclusions:

1. For the ECG data, the utility of GA-optimized feature weighting and selection depends on the classifier used. The GA-kNN feature weighting and GA-Bayes feature selection yield significant improvement, with three rows showing a significance level of more than 90%, and the fourth showing less improvement, but in the same direction. As a result, the overall significance based on the improvement ratio is 99.99%. In contrast, the GA-optimized linear regression classifier does not show improved performance, and the inconsistency of change direction makes it likely that any systematic improvement due to the GA-optimized weights, if it exists, is very small.

2. The PCA and/or LDA transformation is useful, in combination with GA optimization. As we can see from the kNN and Bayes classifiers, although applying the PCA and/or LDA transformation on the non-GA optimized classifiers yields no major progress, their combination with GA yields significant improvement as well as better final classification rates. In some sense, the PCA and LDA transformations can help the GA to break the "barrier" of the optimum classification rate.

3. The more accurate the original classifier is, the less improvement GA optimization yields. From the data, we can see that the linear regression classifier is the most powerful classifier if used alone, and also the least improved classifier after GA optimization. Fig. 3 illustrates this conclusion.

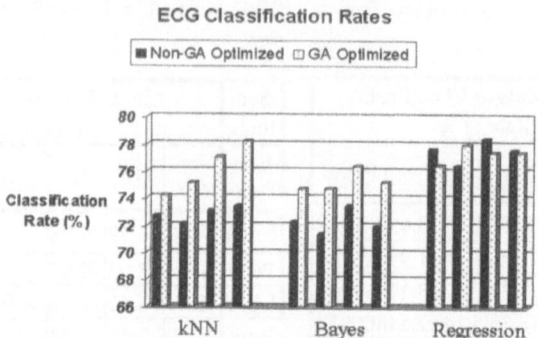

Fig 3. Summary of the ECG classification rate

4.3 Results of GA Search

For small numbers of features (here we set the standard: ≤ 15) in feature selection, it is possible to apply an exhaustive search in the whole feature space, thus providing the possibility to determine whether the GA can find the best solution or not. At the same time, some information about the usefulness of features can be traced from the terrain graph of the whole feature space.

In some cases, the global optimum was found. However, in Fig. 4, although the GA found quite a good result, it was not the global optimum. But since the classification rate for the validation samples is related to but not linearly dependent on the training rate obtained by the GA, such a near-optimal result seems to produce good performance for a pattern classification problem.

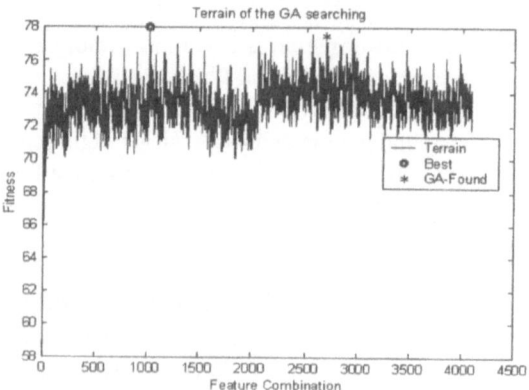

Fig. 4. The GA search space (Best not found)

5 Test Results for Other Data

Five datasets from UCI repository were tested to further validate our result from the previous section. A brief introduction to the data sets is given first, followed by the results and discussion.

5.1 The Testing Datasets

- WDBC: The Wisconsin Diagnostic Breast Cancer [16] data contains 30 features expanded from the original 10 features by taking mean, standard error and extreme value of the originally measured features. The dataset has 357 benign and 212 malignant samples, with highest reported classification accuracies around 97%.
- LIVER: The BUPA Liver Disorders data has 6 numerical features on 345 instances, with 2 classes.
- PIMA: The Pima Indians Diabetes Database has 8 features on female diabetes patients, classified to be positive and negative. The total of 768 samples has 500 negative and 268 positive samples.
- SONAR: The Sonar data compares mines versus rocks based on their reflected sonar signals. With the 111 metal patterns and 97 rock patterns, each of them has 60 feature values [17].
- IONO: The Ionosphere database from Johns Hopkins University [18] contains 351 instances of radar signals returned from the ionosphere, with 34 features. It contrasts "Good" and "Bad" radar returns that show or not show evidence of some types of structure in the ionosphere.

5.2 Results for Benchmark Datasets

The results presented here are all based on both PCA and LDA transformation, which were shown to be useful in GA-optimized pattern classification in Section 4.

Table 5. kNN classifier (Benchmark Datasets)

DATA	Non-GA	GA	Train	Improve	P-Value
WDBC	91.38	94.19	92.24	2.81±2.93	0.0571
LIVER	66.36	66.64	72.89	0.28±6.28	0.9126
PIMA	70.44	72.65	76.97	2.21±2.90	0.1076
SONAR	69.28	74.97	64.06	5.69±5.91	0.0563
IONO	79.19	80.32	69.98	1.13±3.13	0.3986
Overall P value:					**0.0001**

Table 6. Bayes classifier (Benchmark Datasets)

DATA	Non-GA	GA	Train	Improve	P-Value
WDBC	94.78	94.71	96.66	-0.07±2.44	NA
LIVER	56.96	64.00	67.40	7.04±6.57	**0.0334**
PIMA	72.91	74.74	76.30	1.84±3.73	0.2864
SONAR	46.83	68.27	76.37	21.45±10.77	**0.0015**
IONO	64.88	90.64	94.17	25.76±11.52	**0.0000**
Overall P value:					**0.0000**

From the results shown in Table 5—7, we can see that:
1. More than half of the row-wise results show an improvement with significance above 90%. For some settings, such as Bayes classifier for SONAR data, the original classification rate is very low, but GA can make up for this and yield a decent result. The effect of GA optimization here is to reach a fairly good result, if not the best, when the original settings of the classifier are not very good.

Table 7. Regression classifier (Benchmark Datasets)

DATA	Non-GA	GA	Train	Improve	P-Value
WDBC	94.43	95.26	95.49	0.83±2.50	0.4737
LIVER	66.04	67.23	66.33	1.19±6.21	0.6752
PIMA	76.57	76.69	77.55	0.11±3.14	0.9364
SONAR	63.65	72.44	74.78	8.79±9.91	0.0757
IONO	85.12	82.61	84.98	-2.51±5.78	NA
Overall P value:					0.1364

2. The linear regression classifier has the highest classification rate among the three. After GA-optimized feature weighting and selection, the gaps in performance among the various classifiers became smaller, as shown in Fig. 5.

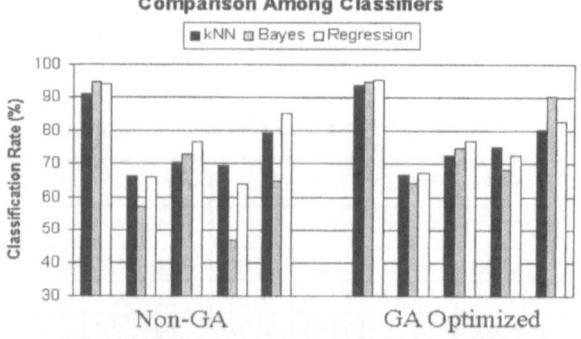

Fig. 5. Comparison, before and after GA optimization

6 Conclusions

The utility of GA-optimized feature weighting and selection depends on both the classifier and the data set. Especially in those cases when a particular classifier has a better classification rate than some other classifiers, the potential improvement from GA optimization of the better classifier seems to be quite limited in comparison with its performance improvement of the poorer classifiers.

Clearly, to evaluate a new approach involving optimization on feature space, it is necessary to test on different classifiers, and the improvement in the best classifier will be the most convincing evidence of the utility of that method.

In this work, the Genetic Algorithm shows powerful searching ability in high-dimensionality feature spaces. By comparing it with an exhaustive search algorithm on small-scale problems, it was determined that the GA found the optimal or a nearly optimal solution with a computational complexity of $O(n)$.

The results from Sections 4 and 5 indicate that over-fitting exists in various approaches. While the training performances can be significantly improved, the improvements on the validation samples lag behind in every case.

The tests run in Section 4 show that the PCA and LDA transformations are very useful in pattern classification. The significance levels of GA optimization are greatly improved for the kNN and Bayes classifiers, though the absolute values after PCA and LDA transformation without GA optimization do not differ much from the non-PCA and non-LDA cases. The point is, however, that GA-optimized feature extraction and selection extend the utilities of those traditional feature transformation techniques.

7 Future Works

To solve the problem of over-fitting, one possible approach is to evaluate the solution not only by the classification rate on training data, but also to consider the margin between classes and the boundary, because a larger margin means a more general and more robust classification boundary. Support Vector Machines (SVM) are classification systems that separate the training patterns by maximizing the margins between support vectors (those nearest patterns) and the decision boundary in a high-dimensional space. Work by Gartner and Flach [19] that used SVM rather than a GA to optimize the feature space yielded a statistically significant result on Bayes classifiers.

Another possible improvement may be non-linear feature construction using GA [4]. Non-linear feature construction can generate more artificial features so that the GA can search for more hidden patterns. However, the problem of over-fitting still theoretically exists.

This work was supported in part by the National Science Foundation of China (NSFC) under Grant 39970210.

References

1. M. Pei, E. D. Goodman, W. F. Punch. "Feature Extraction Using Genetic Algorithms", *Proceeding of International Symposium on Intelligent Data Engineering and Learning'98* (IDEAL'98) Hong Kong, Oct. 1998.
2. David B. Skalak. "Using a Genetic Algorithm to Learn Prototypes for Case Retrieval and Classification", *Proceedings of the AAAI-93 Case-Based Reasoning Workshop*, pages 64–69, Washington, D.C., American Association for Artificial Intelligence, Menlo Park, CA, 1994.
3. W.F. Punch, E.D. Goodman, Min Pei, Lai Chia-Shun, P. Hovland and R. Enbody. "Further Research on Feature Selection and Classification Using Genetic Algorithms", In *5th International Conference on Genetic Algorithms*, Champaign IL, pages 557–564, 1993
4. Hiram A. Firpi Cruz. "Genetically Found, Neurally Computed Artificial Features With Applications to Epileptic Seizure Detection and Prediction", Master's Thesis, University of Puerto Rico, Mayaguez, 2001.
5. M. Prakash, M. Narasimha Murty. "A Genetic Approach for Selection of (Near-) Optimal Subsets of Principal Components for Discrimination", *Pattern Recognition Letters*, Vol. 16, pages 781–787, 1995.
6. Blake, C.L. & Merz, C.J. UCI Repository of machine learning databases [http://www.ics.uci.edu /~mlearn/MLRepository.html]. Irvine, CA: University of California, Department of Information and Computer Science (1998).
7. S. Bandyopadhyay, C.A. Murthy. "Pattern Classification Using Genetic Algorithms", *Pattern Recognition Letters*, Vol. 16, pages 801–808, 1995.
8. R. Srikanth, R. George, N. Warsi, D. Prabhu, F.E.Petry, B.P.Buckles. "A Variable-Length Genetic Algorithm for Clustering and Classification", *Pattern Recognition Letters*, Vol. 16, pages 789–800, 1995.
9. W. Siedlecki, J. Sklansky. "A Note on Genetic Algorithms for Large-Scale Feature Selection", *Pattern Recognition Letters*, Vol. 10, pages 335–347, 1989.
10. Ludmila I. Kuncheva. "Editing for the k-Nearest Neighbors Rule by a Genetic Algorithm", *Pattern Recognition Letters*, Vol. 16, pages 809–814, 1995.
11. Richard O. Duda, Peter E. Hart, David G. Stock. "Pattern Classification", Second Edition, Wiley 2001.
12. Ludmila I. Kuncheva and Lakhmi C. Jain. "Designing Classifier Fusion Systems by Genetic Algorithms", *IEEE Transactions on Evolutionary Computation*, Vol. 4, No. 4, September 2000.
13. Ron Kohavi, George John. "The Wrapper Approach", *Feature Extraction, Construction and Selection: A Data Mining Perspective*. Edited by Hiroshi Motoda, Huan Liu, Kluwer Academic Publishers, July 1998.
14. George H. John, Ron Kohavi, Karl Pfleger. "Irrelevant Features and the Subset Selection Problem", *Proceedings of the Eleventh International Conference of Machine Learning*, pages 121–129, Morgan Kaufmann Publishers, San Francisco, CA, 1994.
15. Zhijian Huang. "Genetic Algorithm Optimized Feature Extraction and Selection for ECG Pattern Classification", Master's Thesis, Michigan State University, 2002.
16. O.L. Mangasarian, W.N. Street and W.H. Wolberg. "Breast Cancer Diagnosis and Prognosis via Linear Programming", *Operations Research*, 43(4), pages 570–577, July-August 1995.
17. Gorman and T. J. Sejnowski. "Learned Classification of Sonar Targets Using Massively Parallel Network", *IEEE Transactions on Acoustic, Speech and Signal Processing*, 36 (7), pages 1135–1140, 1988.

18. V.G. Sigilito, S.P. Wing, L.V. Hutton and K.B. Baker. "Classification of Radar Returns from the Ionosphere Using Neural Networks", *Johns Hopkins APL Technical Digest*, Vol. 10, pages 262–266, 1989.
19. Thomas Gartner, Peter A. Flach. "WBC_{SVM}: Weighted Bayesian Classification based on Support Vector Machine", *Eighteenth International Conference on Machine Learning (ICML-2001)*, pages 156–161. Morgan Kaufmann, 2001.

Web-Page Color Modification for Barrier-Free Color Vision with Genetic Algorithm

Manabu Ichikawa[1], Kiyoshi Tanaka[1], Shoji Kondo[2],
Koji Hiroshima[3], Kazuo Ichikawa[3], Shoko Tanabe[4], and Kiichiro Fukami[4]

[1] Faculty of Engineering, Shinshu Unversity
4-17-1 Wakasato, Nagano-shi, Nagano, 380-8553 Japan
{manabu, ktanaka}@iplab.shinshu-u.ac.jp
[2] Kanto Electronics & Research, Inc.
5134 Yaho, Kunitachi-shi, Tokyo, 186-0914 Japan
[3] Social Insurance Chukyo Hospital
1-1-10 Sanjyo, Minami-ku, Nagoya-shi, Aichi, 457-8510 Japan
[4] Research Institute of Color Vision
12-26 Sanbonmatsu-cho, Atta-ku, Nagoya-shi, Aichi, 456-0032 Japan

Abstract. In this paper, we propose a color modification scheme for web-pages described by HTML markup language in order to realize barrier-free color vision on the internet. First, we present an abstracted image model, which describes a color image as a combination of several regions divided with color information, and define some mutual color relations between regions. Next, based on fundamental research on the anomalous color vision, we design some fitness functions to modify colors in a web-page properly and effectively. Then we solve the color modification problem, which contains complex mutual color relations, by using Genetic Algorithm. Experimental results verify that the proposed scheme can make the colors in a web-page more recognizable for anomalous vision users through not only computer simulation but also psychological experiments with them.

1 Introduction

Due to the rapid development of computers, internet, and display and printing techniques we can take advantage of color to describe and deliver digital information. The colors themselves used in the description of a message often contain very important information. For example, various colors of text, graphics, and images have been used in the web pages on the internet, in which color difference often classifies the importance of information, such as a linkage to another page, and so on. However, from a medical point of view, colorful information description does not always provide real convenience and easy understanding of the information to all users. It is well-known that 5-8% of men, with so-called anomalous vision, have difficulties recognizing certain colors and color differences rather than normal people. Those who have a different color vision from normal people (anomalous vision people) may miss the information that can be recognized by normal vision people, which causes the disparity of capability getting information between them. In this paper, we focus on this problem and try to modify the colors used on the web pages to colors that could be more recognizable for anomalous vision people in order to realize barrier-free color vision in the IT society.

The anomalous vision has been mainly verified by lots of psychological experiments in medical field so far [1,2]. From an engineering point of view, Kondo proposed a color vision model which can describe the anomalous color vision to simulate (display) the colors that the anomalous vision people may observe [3]. In this model, the anomalous vision can be considered as if a certain unit or a channel between units in the model is broken or their functions are deteriorated. In this paper, we follow this fundamental research and propose a novel web page color modification method to realize a barrier-free color vision. In this paper, we first propose an abstracted image model for general color images and apply it for web pages. Next, we propose an optimization method of color arrangement in the abstracted image model for web pages. When we increase the number of colors used in a web page, the complexity of color arrangement increases rapidly. In this paper we use Genetic Algorithm (GA) for color arrangement optimization of web pages, which is widely known as a robust optimization method [4]. We show some experimental results and give a discussion about those.

2 Color Vision Model

First of all, we show a normal vision model based on the stage theory in Fig. 1, which is an equivalent circuit to explain the phenomena on color vision. In this figure, R, G and B denote cones that produce output signals depending on input stimuli. The output signals from cones produce a Luminance signal L through a unit denoted as V, and two opponent color signals, C_{rg} and C_{yb} through opponent processes denoted by r-g and y-b via intermediate units r and y. Here the function proposed by Smith and Pokorny [5] is used as the cone sensitivity function in Eq.(1), where $\bar{x}(\lambda)$, $\bar{y}(\lambda)$ and $\bar{z}(\lambda)$, are the color matching functions.

$$\begin{pmatrix} S_r(\lambda) \\ S_g(\lambda) \\ S_b(\lambda) \end{pmatrix} = \begin{pmatrix} 0.15514 & 0.54312 & -0.03286 \\ -0.15514 & 0.45684 & 0.03286 \\ 0 & 0 & 1 \end{pmatrix} \begin{pmatrix} \bar{x}(\lambda) \\ \bar{y}(\lambda) \\ \bar{z}(\lambda) \end{pmatrix} \quad (1)$$

L, C_{rg} and C_{yb} in Fig. 1 are obtained by Eq.(2), which is derived from a linear combination of cone sensitivity functions given by Eq.(1) with a suitable normalization and rounding off small fractions for simplicity.

$$\begin{pmatrix} C_{rg}(\lambda) \\ C_{yb}(\lambda) \\ L(\lambda) \end{pmatrix} = \begin{pmatrix} 2.00 & -2.00 & 0.00 \\ 0.00 & 1.00 & -1.00 \\ 0.00 & 1.00 & 0.00 \end{pmatrix} \begin{pmatrix} \bar{x}(\lambda) \\ \bar{y}(\lambda) \\ \bar{z}(\lambda) \end{pmatrix} \quad (2)$$

Next, we show an anomalous vision model [3], in which some units and channels in the normal vision model are partially changed. First, in the model of "protan", the normal cone R changes to an abnormal one R', which changes the

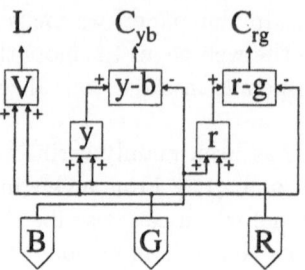

Fig. 1. A normal vision model

channel from R and the output from R' goes only to r as shown in Fig. 2(a). Because the output signal from R' is very weak and fragile, we can explain the phenomena of "protan" without contradiction to medical knowledge. In Fig. 2(a), the parameter $K_P (0 \leq K_P < 1)$ denotes the degree of color weakness. The case $K_P = 0$ corresponds to a complete "protanopia", $K_P > 0$ does to a "protanomalia". (The case $K_P = 1$ means the hypothetical "protan" who has an anomalous cone but a perfect hue discrimination capability like normal ones.) The opponent signals C_{rg} and C_{yb} and the luminance one L are obtained by

$$\begin{pmatrix} C_{rg}(\lambda) \\ C_{yb}(\lambda) \\ L(\lambda) \end{pmatrix} = \begin{pmatrix} 0.252 K_P & -0.203 K_P & -0.011 K_P \\ -0.155 & 0.457 & -0.251 \\ -0.155 & 0.457 & 0.033 \end{pmatrix} \begin{pmatrix} \bar{x}(\lambda) \\ \bar{y}(\lambda) \\ \bar{z}(\lambda) \end{pmatrix}, \qquad (3)$$

where the coefficients are derived from numerical experiments. Similarly, in the model of "deutan", the normal cone G changes to an abnormal one G', which changes the channel from G and the output from G' goes only to r-g as shown in Fig. 2(b). In this figure, the parameter $K_D (0 \leq K_D < 1)$ denotes the degree of color weakness. The case $K_D = 0$ corresponds to a complete "deutan", $K_D > 0$ does to a "deuter anomalopsia". (The case $K_D = 1$ means the hypothetical "deuteranopia".) The output signals C_{rg}, C_{yb} and L are obtained by

$$\begin{pmatrix} C_{rg}(\lambda) \\ C_{yb}(\lambda) \\ L(\lambda) \end{pmatrix} = \begin{pmatrix} 0.105 K_D & -0.132 K_D & 0.020 K_D \\ 0.155 & 0.543 & -0.969 \\ 0.155 & 0.543 & -0.033 \end{pmatrix} \begin{pmatrix} \bar{x}(\lambda) \\ \bar{y}(\lambda) \\ \bar{z}(\lambda) \end{pmatrix}. \qquad (4)$$

We can simulate the anomalous color vision by using Eq.(3) and (4) and the inverse transformation of Eq.(2). That is, first we input RGB signals into the anomalous color vision model and output the corresponding C_{rg}, C_{yb} and L signals. Then we input them from the output-side of Fig. 1, and transform them to RGB signals by the inverse transformation of Eq.(2).

Through this calculation, we can obtain the colors that the anomalous vision people may observe. If two colors obtained through this process are similar (close to each other), we can judge that it might be a hard combination of colors for anomalous vision people. When we select a number of color pairs, which are hard to recognize the difference for anomalous vision people, and plot them on

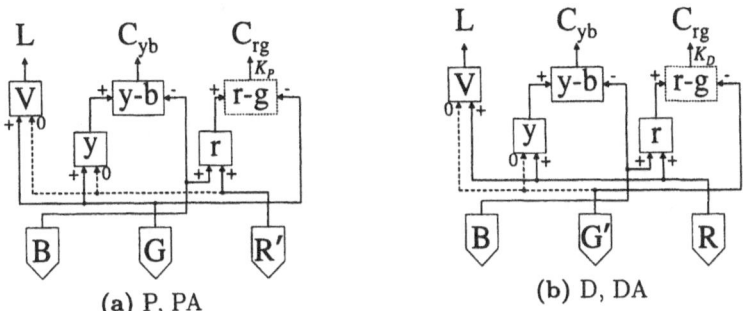

Fig. 2. Models of anomalous vision

Table 1. Chromaticity coordinates for protanopic and deuteranopic convergence points

	x	y
Protanopic	0.7465	0.2535
Deuteranopic	1.4000	-0.4000

the chromaticity diagram, they tend to be on some straight lines. Such lines obtained are called "confusion lines" and they converge to a point called "convergence point" of confusion lines. This unique point is experimentally obtained and several values have been so far proposed. In this paper, we adopt the results proposed by Smith and Pokorny [5].

3 Abstracted Image Model and Color Modification

3.1 Abstracted Image Model

We consider a color image as a combination of several color regions, each of which contains similar colors in it. In this paper, we call a color image, which is divided into several regions with colors, as the abstracted image model as shown in Fig. 3. In this model we can define some relations between regions as follows. We define "included" as the relation that a region completely contains another region (regions). The region containing a region is called "parent region" and the region included by the "parent region" is called "child region". In Fig. 3, since region D contains F, we describe this relation as

$$D \supset F. \tag{5}$$

On the other hand, we define "even" as the relation in which two regions have a same "parent" region. Note that the outside of an image can be considered as the "parent" region. In Fig. 3, since regions D and E are "even" with each other, we describe this relation as

$$\{D, E\}. \tag{6}$$

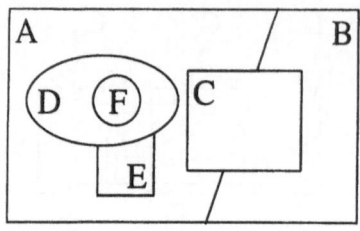 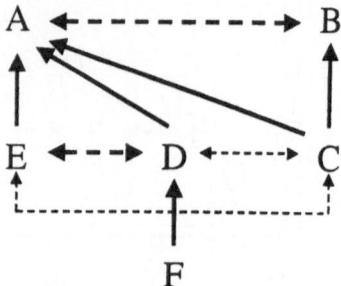

Fig. 3. An example of the abstracted image model

Fig. 4. Mutual color relations to be considered

These "included" and "even" relations are often satisfied inter-regionally. For example, since the region A contains two regions D and E, which are "even" with each other, we can describe $A \supset \{D, E\}$. As another example, since both regions A and B contain region C, we can describe $\{A, B\} \supset C$. Therefore, we can describe the entire relations in Fig. 3 as

$$\{\{A \supset \{\{D \supset F\}, E\}\}, B\} \supset C. \tag{7}$$

3.2 Color Modification for Divided Regions

When we human beings recognize regions by using colors, we make use of color difference between regions. Because in case of "included" and "parent" region and its "child" region are neighbors with each other, it could be easier for us to recognize the boundary if we enhance the color difference between regions. In case of "even", we can make the boundary more recognizable by enhancing color difference between regions but we should additionally consider the distance between regions. That is, the necessity to enhance the color difference between "even" regions varies depending on the distance between regions. We depict the graph which shows mutual color relations to improve the recognizability of Eq.(7) in Fig. 4. The solid and broken lines denote "included" and "even", respectively. And bold line shows the necessity of color difference enhancement between regions. In this paper, we modify all colors of regions by considering such mutual color relations in the image in order to obtain more recognizable color combinations for anomalous vision users.

3.3 Applying the Abstracted Image Model to Web-Pages

Let us consider the application of the abstracted image model to web-pages described by the HTML markup language. In the HTML description we can specify colors for a background, characters, part of characters, figures, tables, and so on. The background and characters on it have an "included" relation between them. Also, two different colors of characters have an "even" relation. Since the background should be "parent" region and the characters on the background should be "child" regions of it, the "child" regions have a single "parent" region and might not be contained in multiple "parent" regions. In this case, since the "child" region (a color of characters) is being neighbors with the "parent" region (a color of background), we enhance the color difference between them in order to make a given web-page more recognizable. In case of "even" we should consider mutual color relation between characters, a portion of characters has a different color from the color of other characters, for example, for specifying a linkage function to users. In this case, we enhance the color difference between two colors in characters. We should consider balancing the weight for color modification between "included" and "even" cases.

When we enhance the color difference in HTML description, we should consider the controllable range of it. Because we specify the colors to be used with RGB color space and the dynamic range of (R, G, B) signals are limited to 0-255, we can not describe all colors in the chromaticity diagram [6]. Also, since (R, G, B) signals must be positive, the colors which can be displayed are limited to a narrow range when the lightness of the color is low or high. In this paper, we distinguish the color into lightness and spectral colors, and separately evaluate them for optimization.

The number of colors used in the entire web-page could be an important parameter when we optimize to modify all colors in it. In general, the mutual color relations to be considered increase as the number of colors n increases. Thus in this paper we use Genetic Algorithm (GA) [4] to solve the complex color modification problem.

4 Color Modification with Genetic Algorithm

4.1 Individual Representation

We try to generate the optimum color vector $(C_1^*, C_2^*, \cdots, C_n^*)$ for the original color vector $(C_1, C_2, \cdots, C_n)$ in terms of making the web-page more recognizable for anomalous vision users. To accomplish this purpose we design the individual which has all n kinds of color information in it. However, the amount of information an individual has becomes large, which enlarge the solution space and thus the search speed might be deteriorated remarkably. We can specify mutual color relations in the HTML description by analyzing the scanned web-page. Thus we divided the entire solution space into n sub-spaces for each color $C_i (i = 1, 2, \cdots, n)$. Therefore, each individual is described with 24 bits binary representation (8 bits for each (R, G, B) component). (The way of evaluation by considering mutual color relations will be mentioned in 4.3.)

4.2 Design of Fitness Functions

We design the function that can evaluate a fitness value between the color represented by an individual and a basis color C_B. We achieve the evaluation in the Luv perceptually uniform color space [6]. Here we denote the color represented by an individual as $C = (L, u, v)$ and its filtered one by the computer simulation of anomalous vision as $C' = (L', u', v')$. Similarly, we denote the basis color and its filtered one by the computer simulation of anomalous vision as $C_B = (L_B, u_B, v_B)$ and $C'_B = (L'_B, u'_B, v'_B)$ respectively. Also, we denote the original color and its filtered one as $C_O = (L_O, u_O, v_O)$ and $C'_O = (L'_O, u'_O, v'_O)$ respectively. Also, we define a normalized spectral color difference between two colors C_x and C_y in Luv color space as

$$d_{(C_x, C_y)} = \frac{\sqrt{(u_x - u_y)^2 + (v_x - v_y)^2}}{d_{max}}, \tag{8}$$

where d_{max} is the maximum color distance in Luv space. Furthermore, we define a normalized lightness difference between two colors as

$$b_{(C_x, C_y)} = \frac{|L_x - L_y|}{b_{max}}, \tag{9}$$

where b_{max} is the maximum brightness in Luv space.

(1) Weighting Coefficient between Spectral Colors and Brightness. As described in 3.3 it is difficult to control the color difference with spectral colors when the lightness of the color is low or high. In this case, we stress on the lightness enhancement. On the other hand, when the lightness of the color is medium, we stress on color difference enhancement with spectral colors. In this paper, we define the weighting coefficient $\alpha(0 \leq \alpha \leq 1)$ between spectral colors and brightness as a function of L'_B. We designed $\alpha(L'_B)$ such that $\alpha(L'_B) \approx 0$ as L'_B is low or high and $\alpha(L'_B) \approx 1$ as L'_B is intermediate.

(2) Evaluation for Spectral Colors. We positively evaluate the color difference $d_{(C', C'_B)}$ between C' and C'_B obtained through the simulation of anomalous vision, but negatively does the difference $d_{(C, C_O)}$ between C and C_O in order to keep the original colors as possible as we can. Thus in this paper we use the following function to evaluate the color C represented by an individual

$$f_c = d_{(C', C'_B)} \left[1 - \{d_{(C, C_O)}\}^{g_c\left(d_{(C'_O, C'_B)}\right)} \right], \tag{10}$$

where $g_c\left(d_{(C'_O, C'_B)}\right)$ is a constraint function on the original spectral colors. The constraint gradually increases as g_c approaches to 0.

(3) Evaluation for Brightness. We evaluate the brightness of the color properly to control it depending on the brightness of C'_B and the brightness difference between C and C_B by

$$f_b = b_{(C', C'_B)} \left[1 - \{b_{(C, C_O)}\}^{g_b\left(b_{(C'_O, C'_B)}\right)} \right], \tag{11}$$

where $g_b\left(d_{(C'_O, C'_B)}\right)$ is a constraint function on the original brightness. The constraint gradually increases as g_b approaches to 0.

(4) Combined Fitness function. We combine two evaluation functions f_c(Eq.(10)) and f_b(Eq.(11)) with the weighting coefficient α as

$$f = \alpha f_c + (1-\alpha)f_b. \tag{12}$$

With this equation, we finally evaluate the color represented by an individual C and the basis color C_B.

4.3 Evaluation Method by Considering Mutual Color Relations

Each color $C_i (i = 1, 2, \cdots, n)$ has some mutual color relations with other colors in $(C_1, C_2, \cdots, C_n)$ obtained by scanning a HTML web-page data such as "included" and "even" as described in 3.1. Thus we should construct an evaluation method such that we can consider such mutual relations in colors. Fig. 5 shows the way of mutual evaluation method between two colors C_1 and C_2. Although C_1 and C_2 should be determined by considering their partner colors, both colors will be changing (evolving) along with the alternation of generations by GA. In the proposed scheme, the individuals in the population P_1 for the determination of C_1 are evaluated with the best individual $C_2^{*(t-1)}$ by $(t-1)$-th generation in the population P_2 for the determination of C_2. Note that the best individual in P_1 at $(t-1)$-th generation, $C_1^{*(t-1)}$, is re-evaluated with $C_2^{*(t-1)}$ and we determine the best individual $C_1^{*(t)}$ among the individuals in $P_1^{(t)}$ and $C_2^{*(t-1)}$. Similarly, the individuals in the population P_2 for the determination of C_2 are evaluated with the best individual $C_1^{*(t-1)}$. When a color has more than two relations, the fitness values for multiple basis colors are averaged. In this way the proposed scheme synchronizes to evolve all the evolution of the populations $P_i (i = 1, 2, \cdots, n)$ corresponding to the colors $C_i (i = 1, 2, \cdots, n)$ generation by generation. Finally, the best individuals $(C_1^*, C_2^*, \cdots, C_n^*)$ collected from fully matured populations are output and displayed as the modified colors.

4.4 Genetic Operators in GA

In the proposed scheme we employ an improved GA (GA-SRM) [7,8] to achieve efficient and reliable optimization. GA-SRM uses two kinds of cooperative and competitive genetic operators and combines them with an extinctive selection method. It has been verified that this framework is quite effective for 0/1 multiple-knapsack problems [7,8], NK-Landscape problems [9,10], and image halftoning problems [11,12]. In this paper, we follow the basic framework of GA-SRM. That is, we create offspring with CM (Crossover and serial background Mutation) and SRM (Self-Reproduction with Mutation) operators in parallel. Their offspring compete for survival through (μ, λ) proportional selection [13]. We adopt one-point crossover and background mutation with the mutation probability $P_m^{(CM)}$ for CM. On the other hand, for SRM we adopt ADP (Adaptive Dynamic Probability) mutation strategy [7] which adaptively reduces the mutation probability $P_m^{(SRM)}$ depending on a mutants survival ratio.

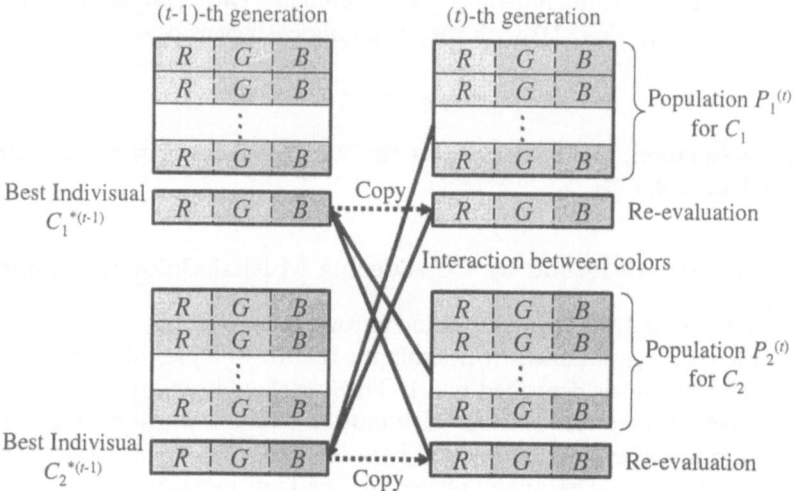

Fig. 5. Mutual evaluation method between two colors

5 Experimental Results and Discussion

In this paper, we use the following functions as $\alpha(L'_B)$, g_c and g_b. The parameters in these equations are experimentally determined.

$$\alpha(L'_B) = 0.1 + 0.65 \left\{ 0.5 \cos \left(\frac{|L'_B/b_{max} - 0.5|}{0.5} \pi \right) + 0.5 \right\} \quad (13)$$

$$g_c\left(d_{(C'_O, C'_B)}\right) = 0.6 \left\{ 0.7 \left(1 - d_{(C'_O, C'_B)}\right) + 0.3 \right\} \quad (14)$$

$$g_b\left(b_{(C'_O, C'_B)}\right) = 2 \left\{ 0.7 b_{(C'_O, C'_B)} + 0.3 \right\} \quad (15)$$

First, we prepared two kinds of test images (image A and B) described with HTML markup language as shown in Fig. 6(a) and (b). In these examples, we can describe the color relations as $C_1 \supset \{C_2, C_3\}$ with a background color C_1, and two kinds of colors for characters C_2 and C_3. First we show the results through the computer simulation of anomalous color vision [3] in Fig. 6(c)~(f), where the degree of color weakness, K_P and K_D, are set to 0. We can consider these images as the color appearance observed by anomalous vision people. We can see that the colors in all images are deteriorated and it becomes hard to recognize the characters from the background and between set of characters.

Next we show the results after applying the proposed scheme to the test images in Fig. 7. Here we set the weight between "included" and "even" relations to 2:1 and use the genetic parameters in Table 2. In Fig. 7, we show the images with modified colors and their appearance through the computer simulation of anomalous color vision. Compared with the ones without modification in Fig. 6, we can see that color differences are enhanced, and becomes more recognizable

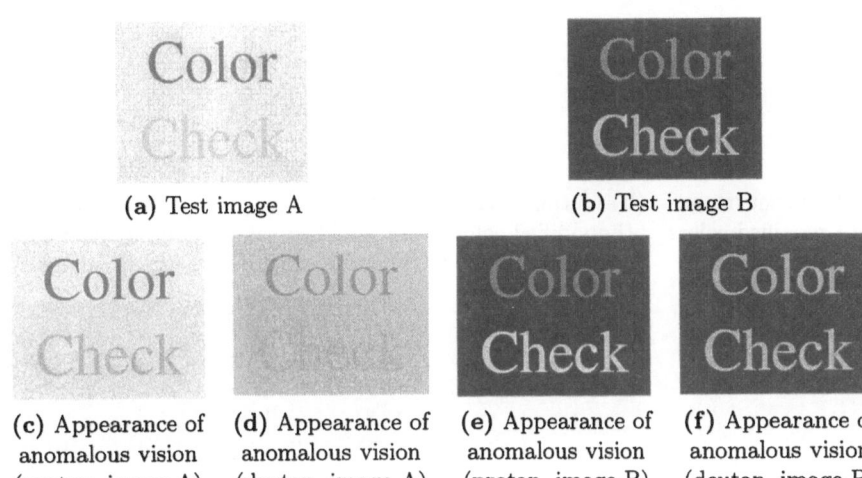

Fig. 6. Test images and their appearance through computer simulation of anomalous color vision

Table 2. Genetic parameters of GAs

	cGA	GA-SRM
Crossover ratio P_c	0.6	1.0
Mutation ratio P_m	$\frac{1}{24}$	$P_m^{(CM)} = \frac{1}{24}$ $P_m^{(SRM)} = [\frac{1}{8}, \frac{1}{24}]$ [7,8]
Selection method	Proportional selection (μ, λ)Proportional selection	(μ, λ) Proportional selection
Population size N	64	64
Evaluation numbers T	128,000	128,000

(distinguishable) the characters from the background and between set of characters by our scheme. It could be an evidence that the fitness functions designed in 4.2 are working effectively.

Next we show the evolution of solutions achieved by GA-SRM in Fig. 8(a). The figure also shows results by cGA (canonical GA) and GA(μ, λ) (cGA with (μ, λ) proportional selection but no parallel varying mutation). These plots are averaged over 100 random runs. From this figure we can see that GA-SRM can generate a high-fitness solution with less evaluation times compared with other configulations. Also we depict the standard deviation σ around the average of 100 runs corresponding to Fig. 8(a) in Fig. 8(b). We can see that the standard deviation attained by GA-SRM is very small and we can achieve reliable solution search with GA-SRM.

Furthermore, we conducted psychological experiments with the cooperation of actual anomalous vision people (a protan, a protanomalia, a dutan and a duteranopia, totally 4 examinees). We prepared 15 sample pages including two colors (a background color and a character color) and 16 samples with three col-

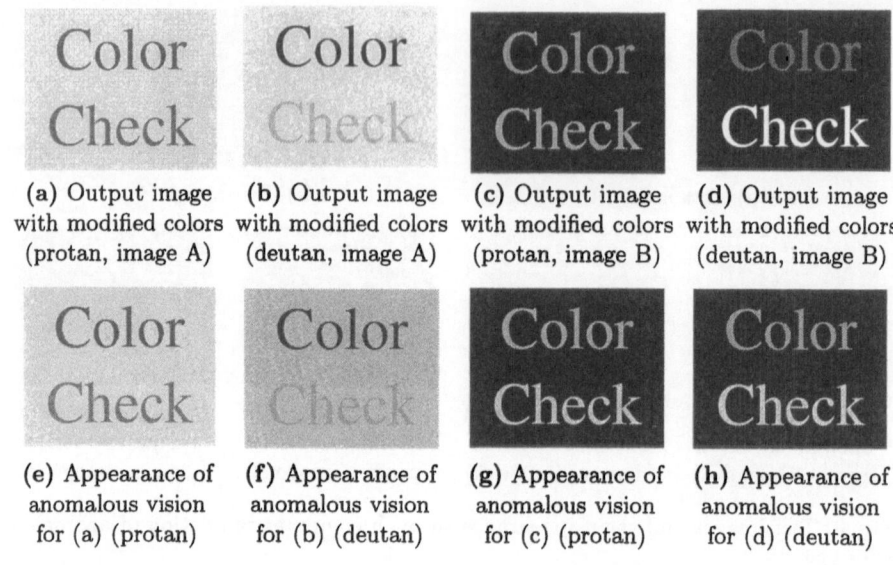

Fig. 7. Output images with modified colors and their appearance through computer simulation of anomalous color vision

ors (a background color and two character colors). All pages have different color combinations. We showed two pages for each sample to the examinee, one was the original page and the other was its modified one by the proposed scheme. The original and its modified page were randomly displayed on the right-hand side or left-hand side on the screen. To avoid that the examinees expect either one to be surely recognizable from another one, we randomly included some dummy pages, in which both pages were the same (no change). We asked examinees which page is recognizable or even, and how recognizable (somehow/remarkable) in case the examinee chose either one. We keep score as follow: 0 point to even, +1 point to somehow recognizable for the modified page, +2 points to remarkably recognizable for modified page, -1 point to somehow recognizable to the original page, and -2 points to remarkably recognizable to the original page. Finally, we sum up all points for all sample pages to measure the color recognizability improvement. The obtained results are shown in Table 3. It can be seen that around +1.0 points are obtained for achromatopsia examinees ($K_P = K_D = 0$) in average and a positive points (around +0.7) obtained for dyschromatopsia ones. The reason why lower points were obtained for the latter case is that some original pages without modification can be recognizable for dyschromatopsia examinees. In general, the effect of color modification increases as the font size decreases. This is becasue it is sometimes very hard even for normal color vision people to distinguish the color of small font characters on a background color. Also, we can see the improvement is slightly deteriorated in case of three colors, which is mainly caused by a compromise between mutual color relations. Nevertheless, we have never scored negative point for the modified page for all samples and achieved a positive +0.85 points as the entire average. Therefore, we can

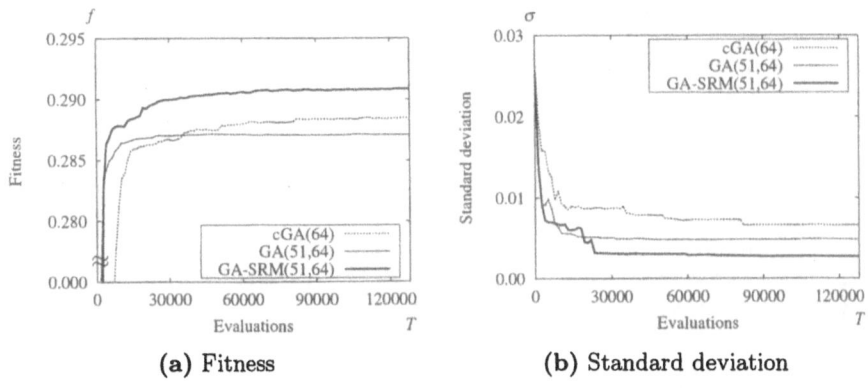

Fig. 8. Performance achieved by cGA and GA-SRM

Table 3. Results by psychological experiments with actual anomalous vision people

type	protan	protanomalia	deutan	deuteranomalia
Two colors (10pt)	+1.23	+0.83	+1.10	+0.83
Two colors (12pt)	+1.20	+0.87	+1.13	+0.77
Two colors (18pt)	+1.10	+0.37	+1.03	+0.77
Two colors (Average)	+1.18	+0.69	+1.09	+0.79
Three colors (10pt)	+0.88	+0.94	+0.84	+0.72
Three colors (12pt)	+0.94	+0.84	+0.78	+0.63
Three colors (18pt)	+0.88	+0.22	+0.91	+0.59
Three colors (Average)	+0.90	+0.67	+0.84	+0.65
Total average	+1.03	+0.68	+0.96	+0.72

say that the proposed scheme successfully modifies colors on web pages to more recognizable ones for anomalous vision users.

6 Conclusions

In this paper, we have proposed a color modification scheme for web-pages described by HTML markup language in order to realize barrier-free color vision on the internet. We solved the color modification problem, which contains complex mutual color relations, by using an improved Genetic Algorithm (GA-SRM). Through computer simulation and psychological experiments we verified that the proposed scheme can make the colors in a web-page more recognizable for anomalous vision users. The proposed scheme can be implemented in web servers or PCs on the internet. With these options, because all web pages that transit our system will be displayed after color modification, anomalous vision users can always view recognizable pages.

As future works, we should further investigate on the implementation aspects of this scheme. Also, we are planning to extend this scheme which can modify colors in any kind of color images, such as natural images, to further realize barrier-free color vision in the internet society.

References

1. Joel Pokorny, Vivianne C. Smith, Guy Verriest, A.J.L.G. Pinckers, *Congenital and Acquired Color Vision Defects*, Grune and Stranton, 1979.
2. Committee on Vision Assembly of Behavioral and Social Science National Research Council, *Procedures for Testing Color Vision Report of Working Group 41*, National Academy Press, 1981.
3. S. Kondo, "A Computer Simulation of Anomalous Color Vision", *Color Vision Deficiencies*, Kugler & Ghedini Pub., pp.145–159, 1990.
4. D.E. Goldberg, *Genetic Algorithms in Search, Optimization and Machine Learning*, Addision-Wesley, 1989.
5. Smith, V. C and Pokorny, J., Spectral sensitivity of the foveal cone photopigments between 400 and 500 nm, *Vis. Res.* 15, pp.161–171, 1975.
6. Andrew S. Glassner, *Principles of Digital Image Synthesis*, Morgan Kaufmann Publishers, Inc., 1995.
7. H. Aguirre, K. Tanaka and T. Sugimura, "Cooperative Model for Genetic Operators to Improve GAs", *Proc. IEEE ICIIS*, pp.98–109, 1999.
8. H. Aguirre, K. Tanaka, T. Sugimura and S. Oshita, "Cooperative-Competitive Model for Genetic Operators: Contributions of Extinctive Selection and Parallel Genetic Operators", *Proc. Late Breaking Papers at GECCO*, pp.6–14, 2000.
9. M. Shinkai, H. Aguirre and K. Tanaka, "Mutation Strategy Improves GA's Performance on Epistatic Problems", *Proc. IEEE CEC*, pp.968–973, 2002.
10. Hernan Aguirre and Kiyoshi Tanaka, "Genetic algorithms on NK-landscapes: effects of selection, drift, mutation, and recombination", *Proc. 3rd European Workshop on Evolutionary Computation in Combinatorial Optimization*, LNCS(Springer), Vol.2611, to appear, 2003.
11. H. Aguirre, K. Tanaka and T. Sugimura, "Accelerated Halftoning Technique Using Improved Genetic Algorithm with Tiny Populations", *Proc. IEEE SMC*, pp.905–910, 1999.
12. H. Aguirre, K. Tanaka, T. Sugimura and S. Oshita, "Halftone Image Generation with Improved Multiobjective Genetic Algorithm", *Proc. EMO'01*, LNCS(Springer), Vol.1993, pp.501–515, 2001.
13. T.Bäck, *Evolutionary Algorithms in Theory and Practice*, Oxford Univ. Press, 1966.

Quantum-Inspired Evolutionary Algorithm-Based Face Verification

Jun-Su Jang, Kuk-Hyun Han, and Jong-Hwan Kim

Dept. of Electrical Engineering and Computer Science
Korea Advanced Institute of Science and Technology (KAIST)
Guseong-dong, Yuseong-gu, Daejeon, 305-701, Republic of Korea
{jsjang,khhan,johkim}@rit.kaist.ac.kr

Abstract. Face verification is considered to be the main part of the face detection system. To detect human faces in images, face candidates are extracted and face verification is performed. This paper proposes a new face verification algorithm using Quantum-inspired Evolutionary Algorithm (QEA). The proposed verification system is based on Principal Components Analysis (PCA). Although PCA related algorithms have shown outstanding performance, the problem lies in the selection of eigenvectors. They may not be the optimal ones for representing the face features. Moreover, a threshold value should be selected properly considering the verification rate and false alarm rate. To solve these problems, QEA is employed to find out the optimal distance measure under the predetermined threshold value which distinguishes between face images and non-face images. The proposed verification system is tested on the AR face database and the results are compared with the previous works to show the improvement in performance.

1 Introduction

Most approaches to face verification fall into one of two categories. They are either based on local features or on holistic templates. In the former category, facial features such as eyes, mouth and some other constraints are used to verify face patterns. In the latter category, 2-D images are directly classified into face groups using pattern recognition algorithms.

We focus on the face verification under the holistic approach. The basic approach in verifying face patterns is a training procedure which classifies examples into face and non-face prototype categories. The simplest holistic approaches rely on template matching [1], but these approaches have poor performance compared to more complex techniques like neural networks.

The first neural network approach to face verification was based on multi-layer perceptrons [2], and advanced algorithms were studied by Rowely [3]. The neural network was designed to look through a window of 20×20 pixels and was trained by face and non-face data. Based on window scanning technique, the face detection task was performed. It means that face verification network was applied to input image for possible face locations at all scales.

One of the most famous methods among holistic approaches is Principal Components Analysis (PCA), which is well known as eigenfaces [4]. Given an ensemble of different face images, the technique first finds the principal components of the face training data set, expressed in terms of eigenvectors of the covariance matrix of the face vector distribution. Each individual face in the face set can then be approximated by a linear combination of the eigenvectors. Since the face reconstruction by its principal components is an approximation, a residual reconstruction error is defined in the algorithm as a measure of faceness. The residual reconstruction error which they termed as "distance-from-face space"(DFFS) gives a good indication of the existence of a face [5]. Moghaddam and Pentland have further developed this technique within a probabilistic framework [6].

PCA is an appropriate way of constructing a subspace for representing an object class in many cases, but it is not necessarily optimal for distinguishing between the face class from the non-face class. Face space might be better represented by dividing it into subclasses, and several methods have been proposed for doing this. Sung and Poggio proposed the mixture of multidimensional Gaussian model and they used an adaptively changing normalized Mahalanobis distance metric [7]. Afterward, many face space analysis algorithms have been investigated and some of them have outstanding performance. The problem of the PCA related approaches lies in the selection of eigenvectors. They may not be the optimal ones for representing the face features. Moreover, a threshold value should be selected properly considering the verification rate and false alarm rate. By employing QEA, the performance of the face verification is improved enough to distinguish between face images and non-face images.

In this paper, eigenfaces are constructed based on PCA and a set of weight factors is selected by using Quantum-inspired Evolutionary Algorithm (QEA) [8]. QEA has lately become a subject of special interest in evolutionary computation. It is based on the concept and principles of quantum computing such as a quantum bit and superposition of states. Instead of binary, numeric or symbolic representation, it uses a Q-bit as a probabilistic representation. Its performance was tested on the knapsack problem, which produced on outstanding result [8].

This paper is organized as follows. Section 2 describes QEA briefly. Section 3 presents the PCA and density estimation. Section 4 presents how the QEA is applied to optimize the decision boundary between face images and non-face images. Section 5 presents the experimental results and discussions. Finally, conclusion and further works follow in Section 6.

2 Quantum-Inspired Evolutionary Algorithm (QEA)

QEA [8] can treat the balance between exploration and exploitation more easily when compared to conventional GAs (CGAs). Also, QEA can explore the search space with a small number of individuals (even with only one individual for real-time application) and exploit the global solution in the search space within a short span of time. QEA is based on the concept and principles of quantum computing, such as a quantum bit and superposition of states. However, QEA is not a quantum algorithm, but a novel evolutionary algorithm. Like other

Procedure QEA
begin
   $t \leftarrow 0$
i) initialize $Q(t)$
ii) make $P(t)$ by observing the states of $Q(t)$
iii) evaluate $P(t)$
iv) store the best solutions among $P(t)$ into $B(t)$
  while (**not** termination condition) **do**
  begin
    $t \leftarrow t + 1$
v) make $P(t)$ by observing the states of $Q(t-1)$
vi) evaluate $P(t)$
vii) update $Q(t)$ using Q-gates
viii) store the best solutions among $B(t-1)$ and $P(t)$ into $B(t)$
ix) store the best solution **b** among $B(t)$
x) **if** (global migration condition)
   then migrate **b** to $B(t)$ globally
xi) **else if** (local migration condition)
   then migrate $\mathbf{b}_j^t$ in $B(t)$ to $B(t)$ locally
  end
end

Fig. 1. Procedure of QEA.

evolutionary algorithms, QEA is also characterized by the representation of the individual, the evaluation function, and the population dynamics.

QEA is designed with a novel Q-bit representation, a Q-gate as a variation operator, an observation process, a global migration process, and a local migration process. QEA uses a new representation, called Q-bit, for the probabilistic representation that is based on the concept of qubits, and a Q-bit individual as a string of Q-bits. A Q-bit is defined as the smallest unit of information in QEA, which is defined with a pair of numbers, (α, β), where $|\alpha|^2 + |\beta|^2 = 1$. $|\alpha|^2$ gives the probability that the Q-bit will be found in the '0' state and $|\beta|^2$ gives the probability that the Q-bit will be found in the '1' state. A Q-bit may be in the '1' state, in the '0' state, or in a linear superposition of the two. A Q-bit individual is defined as a string of m Q-bits. QEA maintains a population of Q-bit individuals, $Q(t) = \{\mathbf{q}_1^t, \mathbf{q}_2^t, \cdots, \mathbf{q}_n^t\}$ at generation t, where n is the size of population, and $\mathbf{q}_j^t$, $j = 1, 2, \cdots, n$, is a Q-bit individual.

Fig. 1 shows the standard procedure of QEA. The procedure of QEA is explained as follows:

i) In the step of 'initialize $Q(t)$,' α_i^0 and β_i^0, $i = 1, 2, \cdots, m$, of all $\mathbf{q}_j^0$, are initialized to $\frac{1}{\sqrt{2}}$. It means that one Q-bit individual, $\mathbf{q}_j^0$ represents the linear superposition of all possible states with the same probability.

ii) This step generates binary solutions in $P(0)$ by observing the states of $Q(0)$, where $P(0) = \{\mathbf{x}_1^0, \mathbf{x}_2^0, \cdots, \mathbf{x}_n^0\}$ at generation $t = 0$. One binary solution,

$\mathbf{x}_j^0$, is a binary string of length m, which is formed by selecting either 0 or 1 for each bit by using the probability, either $|\alpha_i^0|^2$ or $|\beta_i^0|^2$ of $\mathbf{q}_j^0$, respectively.

iii) Each binary solution $\mathbf{x}_j^0$ is evaluated to give a level of its fitness.

iv) The initial best solutions are then selected among the binary solutions, $P(0)$, and stored into $B(0)$, where $B(0) = \{\mathbf{b}_1^0, \mathbf{b}_2^0, \cdots, \mathbf{b}_n^0\}$, and $\mathbf{b}_j^0$ is the same as $\mathbf{x}_j^0$ at the initial generation.

v, vi) In the **while** loop, binary solutions in $P(t)$ are formed by observing the states of $Q(t-1)$ as in step ii), and each binary solution is evaluated for the fitness value. It should be noted that $\mathbf{x}_j^t$ in $P(t)$ can be formed by multiple observations of $\mathbf{q}_j^{t-1}$ in $Q(t-1)$.

vii) In this step, Q-bit individuals in $Q(t)$ are updated by applying Q-gates defined as a variation operator of QEA. The following rotation gate is used as a basic Q-gate in QEA:

$$U(\Delta\theta_i) = \begin{bmatrix} \cos(\Delta\theta_i) & -\sin(\Delta\theta_i) \\ \sin(\Delta\theta_i) & \cos(\Delta\theta_i) \end{bmatrix}, \quad (1)$$

where $\Delta\theta_i$, $i = 1, 2, \cdots, m$, is a rotation angle of each Q-bit. $\Delta\theta_i$ should be designed in compliance with the application problem.

viii, ix) The best solutions among $B(t-1)$ and $P(t)$ are selected and stored into $B(t)$, and if the best solution stored in $B(t)$ is a better solution fitting than the stored best solution **b**, the stored solution **b** is replaced by the new one.

x, xi) If a global migration condition is satisfied, the best solution **b** is migrated to $B(t)$ globally. If a local migration condition is satisfied, the best one among some of the solutions in $B(t)$ is migrated to them. The migration condition is a design parameter, and the migration process can induce a variation of the probabilities of a Q-bit individual. A local-group in QEA is defined to be the subpopulation affected mutually by a local migration, and a local-group size is the number of the individuals in a local-group. Until the termination condition is satisfied, QEA is running in the **while** loop.

3 PCA and Density Estimation

In this section, we present the PCA concept and density estimation using Gaussian densities. It should be noted that this method is a basic technique in pattern recognition and it lays the background of this study.

3.1 PCA Concept

A technique commonly used for dimensionality reduction is PCA. In the late 1980's, Sirovich and Kirby [9] efficiently represented human faces using PCA and it is currently a popular technique.

Given a set of m×n pixels images $\{I_1, I_2, \ldots, I_K\}$, we can form a set of 1-D vectors $X = \{\mathbf{x}_1, \mathbf{x}_2, \ldots, \mathbf{x}_K\}$, where $\mathbf{x}_i \in \Re^{N=mn}, i = 1, 2, \ldots, K$. The basis functions for the Karhunen-Loeve Transform(KLT) [10] are obtained by solving the eigenvalue problem

$$\Lambda = \Phi^T \Sigma \Phi \quad (2)$$

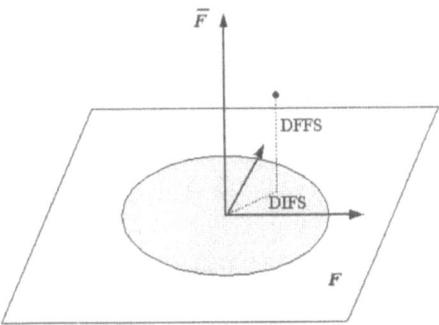

Fig. 2. Two subspaces

where Σ is the covariance matrix of X, Φ is the eigenvector matrix of Σ, and Λ is the corresponding diagonal matrix of eigenvalues. We can obtain M largest eigenvalues of the covariance matrix and their corresponding eigenvectors. Then feature vector is given as follows:

$$\mathbf{y} = \Phi_M^T \tilde{\mathbf{x}} \qquad (3)$$

where $\tilde{\mathbf{x}} = \mathbf{x} - \bar{\mathbf{x}}$ is the difference between the image vector and the mean image vector, and Φ_M is a submatrix of Φ containing the M largest eigenvectors. These principal components preserve the major linear correlations in the given set of image vectors. By projecting to Φ_M^T, original image vector $\mathbf{x}$ is transformed to feature vector $\mathbf{y}$. It is a linear transformation which reduces N dimensions to M dimensions as follows:

$$\mathbf{y} = T(\mathbf{x}) : \Re^N \longrightarrow \Re^M. \qquad (4)$$

By selecting M largest eigenvectors, we can obtain two subspaces. One is the principal subspace (or feature space) F containing the principal components, and another is the orthogonal space $\bar{F}$. These two spaces are described in Fig. 2, where DFFS stands for "distance-from-feature-space" and DIFS "distance-in-feature-space".

In a partial KL expansion, the residual reconstruction error is defined as

$$\epsilon^2(\mathbf{x}) = \sum_{i=M+1}^{N} y_i^2 = \|\tilde{\mathbf{x}}\|^2 - \sum_{i=1}^{M} y_i^2 \qquad (5)$$

and this is the DFFS as stated before which is basically the Euclidean distance. The component of $\mathbf{x}$ which lies in the feature space F is referred to as the DIFS.

3.2 Density Estimation

In the previous subsection, we obtained DFFS and DIFS. DFFS is an Euclidean distance, but DIFS is generally not a distance norm. However, it can be interpreted in terms of the probability distribution of $\mathbf{y}$ in F. Moghaddam estimated

DIFS as the high-dimensional Gaussian densities [6]. This is the likelihood of an input image vector **x** formulated as follows:

$$P(\mathbf{x}|\Omega) = \frac{\exp[-\frac{1}{2}(\mathbf{x}-\bar{\mathbf{x}})^T \Sigma^{-1}(\mathbf{x}-\bar{\mathbf{x}})]}{(2\pi)^{N/2}|\Sigma|^{1/2}} \qquad (6)$$

where Ω is a class of the image vector **x**. This likelihood is characterized by the Mahalanobis distance

$$d(\mathbf{x}) = (\mathbf{x}-\bar{\mathbf{x}})^T \Sigma^{-1}(\mathbf{x}-\bar{\mathbf{x}}) \qquad (7)$$

and it can be also calculated efficiently as follows:

$$\begin{aligned} d(\mathbf{x}) &= \tilde{\mathbf{x}}^T \Sigma^{-1} \tilde{\mathbf{x}} \\ &= \tilde{\mathbf{x}}^T [\Phi \Lambda^{-1} \Phi^T] \tilde{\mathbf{x}} \\ &= \mathbf{y}^T \Lambda^{-1} \mathbf{y} \\ &= \sum_{i=1}^{N} \frac{y_i^2}{\lambda_i} \end{aligned} \qquad (8)$$

where λ is the eigenvalue of the covariance matrix. Now, we can divide this distance into two subspaces. It is determined as

$$d(\mathbf{x}) = \sum_{i=1}^{M} \frac{y_i^2}{\lambda_i} + \sum_{i=1+M}^{N} \frac{y_i^2}{\lambda_i}. \qquad (9)$$

It should be noted that the first term can be computed by projecting **x** onto the M-dimensional principal subspace F. However, the second term cannot be computed explicitly in practice because of the high-dimensionality. So, we use the residual reconstruction error to estimate the distance as follows:

$$\begin{aligned} \hat{d}(\mathbf{x}) &= \sum_{i=1}^{M} \frac{y_i^2}{\lambda_i} + \frac{1}{\rho} \sum_{i=M+1}^{N} y_i^2 \\ &= \sum_{i=1}^{M} \frac{y_i^2}{\lambda_i} + \frac{\epsilon^2(\mathbf{x})}{\rho}. \end{aligned} \qquad (10)$$

The optimal value of ρ can be determined by minimizing a cost function, but $\rho = \frac{1}{2}\lambda_{M+1}$ may be used as a thumb rule [11].

Finally, we can extract the estimated probability distribution using (6) and (10). The estimated form is determined by

$$\begin{aligned} \hat{P}(\mathbf{x}|\Omega) &= \frac{\exp\left(-\frac{1}{2}\sum_{i=1}^{M}\frac{y_i^2}{\lambda_i}\right)}{(2\pi)^{M/2}\prod_{i=1}^{M}\lambda_i^{1/2}} \cdot \frac{\exp\left(-\frac{\epsilon^2(\mathbf{x})}{2\rho}\right)}{(2\pi\rho)^{(N-M)/2}} \\ &= P_F(\mathbf{x}|\Omega) \cdot \hat{P}_{\bar{F}}(\mathbf{x}|\Omega). \end{aligned} \qquad (11)$$

Using (11), we can distinguish the face class from the non-face class by setting a threshold value for $\hat{P}(\mathbf{x}|\Omega)$, which is the Maximum Likelihood (ML) estimation method. In this case, the threshold value becomes the deciding factor between the verification rate and false alarm rate. If the threshold value is too low, the verification rate would be quite good but the false alarm rate would also increase. For this reason, the threshold value has to be carefully selected.

In the following section, we propose an optimization procedure for selecting a set of eigenvalues to determine the decision boundary between the face and non-face classes. By optimizing the Mahalanobis distance term in (11) under the given threshold value, we can find a better distance measure and need not refer to the threshold value.

4 Optimization of Decision Boundary

In this section, we describe a framework for decision boundary optimization using QEA. In the case of ML estimation, the decision boundary is determined by a probability equation and an appropriate threshold value. To improve the performance of the verification rate and reduce the false alarm rate, we attempt to find a better decision boundary.

The Mahalanobis distance-based probability guarantees quite good performance, but it is not optimal. In (11), an eigenvalue is used as the weight factor of the corresponding feature value. These weight factors can be optimized on a training data set. To perform the optimization, we construct the training data set. It consists of two classes: face class (positive data) and non-face class (negative data). Fig. 3 shows an example of a face training data set. A non-face training data set consists of arbitrarily chosen images, including randomly generated images.

To search for the weight factors, QEA is used. The number of weight factors to be optimized is M, which is the same as the number of principal components. Using the weight factors obtained by QEA, we can compute the probability distribution as follows:

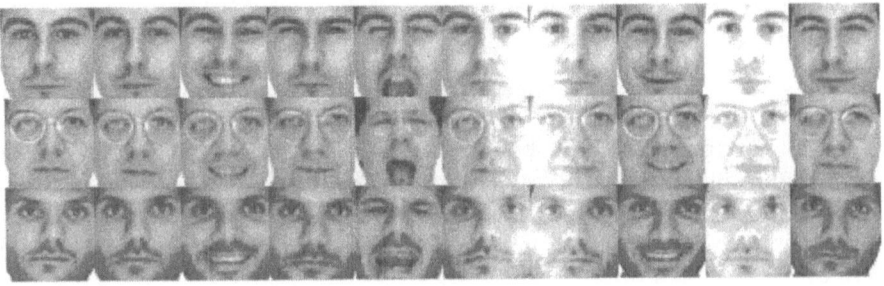

Fig. 3. Example of the face training data set

$$P_{opt}(\mathbf{x}|\Omega) = \frac{\exp\left(-\frac{1}{2}\sum_{i=1}^{M}\frac{y_i^2}{\omega_i}\right)}{(2\pi)^{M/2}\prod_{i=1}^{M}\lambda_i^{1/2}} \cdot \frac{\exp\left(-\frac{\epsilon^2(\mathbf{x})}{2\rho}\right)}{(2\pi\rho)^{(N-M)/2}}. \tag{12}$$

It is the same as (11) except for the weight factors $\omega_i, i = 1, 2, \ldots, M$. To apply (12) to face verification, the threshold value should be assigned. But, since QEA yields optimized weight factors to the predetermined threshold value, we need not assign the threshold value.

To evaluate the fitness value, we calculate the score. The score is added by $+1$ for every correct verification. The score is used as a fitness measure considering both the verification rate (P_score) and the false alarm rate (N_score) because the training data set consists of both face and non-face data. Then the fitness is evaluated as

$$Fitness = P\_score + N\_score \tag{13}$$

where P_score is for the face class (positive data) and N_score is for the non-face class (negative data). Using this fitness function, we can find the optimal weight factors for training data set under the predetermined threshold value.

5 Experimental Results and Discussions

We constructed 3 types of database for the experiment. First, 70 face images were used for extracting principal components. Second, 560 images (280 images for face and 280 images for non-face) were used for training weight factors. Third, 1568 images (784 images for face and 784 images for non-face) were used for the generalization test.

All images are 50×50 pixels with 256 gray levels. We chose 50 principal components from the 70 face images. For pre-processing, histogram equalization was performed to normalize the lighting condition.

Positive data were produced from the face region of the AR face database [12]. An example of a face training data set is shown in Fig. 3. Variations of the facial expression and illumination were allowed. Negative data consisted of both randomly generated images and natural images excluding the face images. Position-shifted face images and different-scale face images were also included as negative data.

The following boundary of each weight factor was considered as a domain constraint:

$$0.1\lambda_i < \omega_i < 10\lambda_i, \quad (1 \leq i \leq 50). \tag{14}$$

By setting the constraint of the boundary using the eigenvalue, it becomes a constraint optimization problem. We performed QEA for 560 training images using the parameters in Table 1. In (1), rotation angles should be selected properly. For each Q-bit, $\theta_1 = 0, \theta_2 = 0, \theta_3 = 0.01\pi, \theta_4 = 0, \theta_5 = -0.01\pi, \theta_6 = 0, \theta_7 = 0, \theta_8 = 0$ were used.

The termination condition was given by the maximum generation. The perfect score was 560 points. If the score does not reach 560 points before maximum

Table 1. Parameters for QEA

Parameters	No.
Population size	15
No. of variables	50
No. of Q-bits per variable	10
No. of observations	2
Global migration period	100
Local migration period	1
No. of individuals per group	3
Max. generation	2000

Table 2. Results for generalization test

	P_score (784)	Verification rate(%)	N_score (784)	False Alarm rate(%)	$Fitness$ (1568)
DFFS classifier	726	92.60	716	0.087	1442
ML classifier	728	92.86	726	0.074	1454
QEA-based classifier	741	94.52	740	0.056	1481

generation, the evolution process stops at maximum generation. After the searching procedure, we performed a generalization test to 1568 images using the weight factors obtained by QEA. We also compared the results with the DFFS and the ML classifier. For the DFFS and the ML classifier, we selected the threshold value that provoked the best score. For QEA-based classifier, we used the same threshold value set for the ML classifier. It should be noted that there is no need to choose a threshold value for better performance in our classifier because the weight factors have been already optimized under the predetermined threshold value.

Table 2 shows the results for the generalization test. The results show that the proposed method performs better than the DFFS or the ML classifier.
The results described above suggest that the QEA-based classifier works well not only in terms of the verification rate (P_score), but also in terms of the false alarm rate (N_score). The verification rate of the QEA-based classifier was 1.66% higher than that of the ML classifier. The false alarm rate was 0.018% lower than that of the ML classifier. The advantage of the proposed classifier is that more training data can improve its performance.

6 Conclusion and Further Works

In this paper, we have proposed a decision boundary optimization method for face verification using QEA. The approach is basically related to eigenspace density estimation technique. To improve the previous Mahalanobis distance-based probability, we have used a new distance which consists of the weight factors optimized at the training set. The proposed face verification system has been tested by face and non-face images extracted from AR database, and very good results have been achieved both in terms of the face verification rate and false alarm rate.

The advantage of our system can be summarized in two aspects. First, our system does not need an exact threshold value to perform optimally. We only need to choose an appropriate threshold value and QEA will find the optimal decision boundary based on the threshold value. Second, our system can be adapted to various negative data. A fixed structured classifier such as the ML classifier can not change its character in frequently failure situation. But our system can be adapted to this case by reconstructing the training data and following the optimization procedure.

As a future research, we will construct face detection system using this verification method. In face detection, it is clear that verification performance is very important. Most of the image-based face detection approaches apply a window scanning technique. It is an exhaustive search of the input image for possible face locations at all scales. In this case, overlapping detection arises easily. A powerful verification method is therefore needed to find exact face locations. We expect that our verification method will also work well for the face detection task.

References

1. Holst, G.: Face detection by facets : Combined bottom-up and top-down search using compound templates. Proc. of the International Conference on Image Processing (2000) TA07-08
2. Propp, M., Samal, A.: Artificial neural network architecture for human face detection. Intell. Eng. Systems Artificial Neural Networks 1 (1992) 535-540
3. Rowley, H.A., Baluja, S., Kanade, T.: Neural network-based face detection. IEEE Trans. Pattern Anal. Mach. Intell. 20 (1998) 23-38
4. Turk, M., Pentland, A.: Eigenfaces for recognition. J. Cog. Neurosci. 3 (1991) 71-86
5. Pentland, A., Moghaddam, B., Strarner, T.: View-based and modular eigenspaces for face recognition. IEEE Proc. of Int. Conf. on Computer Vision and Pattern Recognition (1994) 84-91
6. Moghaddam, B., Pentland, A.: Probabilistic visual learning for object representation, IEEE Trans. Pattern Anal. Mach. Intell. 19 (1997) 696-710
7. Sung, K.-K., Poggio, T.: Example-based learning for view-based human face detection, IEEE Trans. Pattern Anal. Mach. Intell. 20 (1998) 39-51
8. Han, K.-H., Kim, J.-H.: Quantum-inspired evolutionary algorithm for a class of combinatorial optimization. IEEE Trans. Evolutionary Computation 6 (2002) 580-593
9. Sirovich, L., Kirby, M.: Low-dimensional procedure for the characterization of human faces. J. Opt. Soc. Amer. 4 (1987) 519-524
10. Loeve, M.: Probability Theory. Princeton, N.J., Van Nostrand (1955)
11. Cootes, T., Hill, A. Taylor, C., Haslam, J.: Use of active shape models for locating structures in medical images. Image and Vision Computing 12 (1994) 355-365
12. AR face database: http://rvl1.ecn.purdue.edu/~aleix/aleix_face_DB.html

Minimization of Sonic Boom on Supersonic Aircraft Using an Evolutionary Algorithm

Charles L. Karr[1], Rodney Bowersox[2], and Vishnu Singh[3]

[1] Associate Professor and Head, Aerospace Engineering and Mechanics, The University of Alabama, Box 870280, Tuscaloosa, AL, 35487-0280,
ckarr@coe.eng.ua.edu
[2] Associate Professor, Department of Aerospace Engineering, Texas A&M University, 701 H.R. Bright Bldg., TAMU-3141, College Station, TX 77843-3141
[3] Graduate Student, Aerospace Engineering and Mechanics, The University of Alabama, Box 870280, Tuscaloosa, AL, 35487-0280

Abstract. The aerospace community has an increasing interest in developing super sonic transport class vehicles for civil aviation. One of the concerns in such a project is to minimize the sonic boom produced by the aircraft as demonstrated by its ground signature. One approach being considered is to attach a spike/keel on the front of the aircraft to attenuate the magnitude of an aircraft's ground signature. This paper describes an effort to develop an automatic method for designing the spike/keel area distribution that satisfies constraints on the ground signature of a specified aircraft. In this work a genetic algorithm is used to perform the design optimization. A modified version of Whitham's theory is used to generate the near field pressure signature. The ground signature is computed with the NFBOOM atmospheric propagation code. Results indicate that genetic algorithms are effective tools for solving the design problem presented.

1 Introduction

The minimum achievable sonic boom of any aircraft can be computed – it is proportional to the weight of the aircraft divided by its length raised to the 1.5 power [1,2]. Given that the minimum sonic boom can be computed, the natural extension is to attempt to design an aircraft so that its sonic boom is minimized. McLean [3] was the first to consider trying to achieve this objective by eliminating shocks in the ground signature of a supersonic aircraft.

The motivation for minimizing a sonic boom is many-fold, but the driving force behind the current effort is simply human comfort (from the perspective of an observer on the ground). The acceptability of a finite rise time overpressure as apposed to the N-wave shock structure was demonstrated in human subject tests [4,5]. This data indicated that a rise time of about 10 msec rendered realistic (0.6 psf) overpressures acceptable to all the subjects tested. Increasing rise time reduces the acoustic power in the frequency range to which the ear is most sensitive. Unfortunately, the aircraft length required to achieve a 10 msec rise time has consistently been proven to be impractical. McLean considered a Super Sonic Transport (SST) class vehicle (600K lb, flying at 60K ft and Mach 2.7) and concluded that a 570 ft long vehicle would be required (atmospheric effects resulting in midfield signature freezing were included).

Miller and Carlson [6,7] considered the projection of both heat and a force field upstream of a SST class vehicle. Their work focused on projecting a "phantom body" in front of the SST, with the goal of a finite (10 msec) ground rise time. Linear theory indicates that the optimum shape for a finite rise time is a 5/2-power area distribution (i.e., an isentropic spike). They concluded the power required to project the phantom body was the same as that needed to propel the vehicle. Marconi [8], using two-dimensional computational fluid dynamics, confirmed this power requirement. However, a significant wave drag reduction associated with replacing the vehicle nose shock with an isentropic pressure rise was also found.

There are a number of challenging problems associated with projecting anything far upstream of a supersonic vehicle. In particular, a significant reduction in efficiency is incurred when trying to project energy. These projection problems are avoided with the off-axis volume control introduced by Batdorf [9]. Figure 1 depicts his concept as implemented by Marconi et al [10]. Linear theory predicts that the asymptotic far field flow is independent of the details of the vehicle and is influenced only by the cross sectional area (and equivalent area due to lift) distribution of the configuration. In particular, in a supersonic flow the cross sectional area distribution in planes inclined at the Mach angle (μ in Fig.1) governs the far field flow downstream of those Mach planes. The theory predicts that whether the volume being cut is centered along the axis of the vehicle (i.e., a nose spike) or shifted off-axis (i.e., the keel of Fig.1) should not influence the resulting far field flow.

Batdorf's concept was tested by Swigart [11] at Mach 2 and shown to be feasible. Swigart tested an equivalent body of revolution (typical of an SST) with an all-solid keel in addition to a keel with a portion of its volume replaced by a heated region. Swigart also tested a lifting wing body combination with an all-solid keel. The keels (both solid and thermal) were design to match the Mach plane area of a 5/2 power isentropic spike and achieved a nearly linear near-field pressure distribution as expected. The solid keel tested by Swigart was very large, extending the entire length of the configuration. Only the thermal keel that was tested seemed practical.

Marconi et al. [10] extended the Batdorf concept. In that study, the initial spike/keel sizing was accomplished using a modified version of the classical linear Whitham Theory [13], and detailed 3-D flowfields were analyzed using a nonlinear Euler solver. That study demonstrated that single keels produced significant 3-D mitigation over entire ground signature. In addition, they examined multiple approaches to minimizing the impact of the keel on the vehicle performance. Lastly, Marconi et al demonstrated that the spike/keel area distribution could be tailored to produce a desired ground signature.

The purpose of the present work is to develop an optimization procedure for the design of the spike/keel area distribution. A modified version of Whitham's

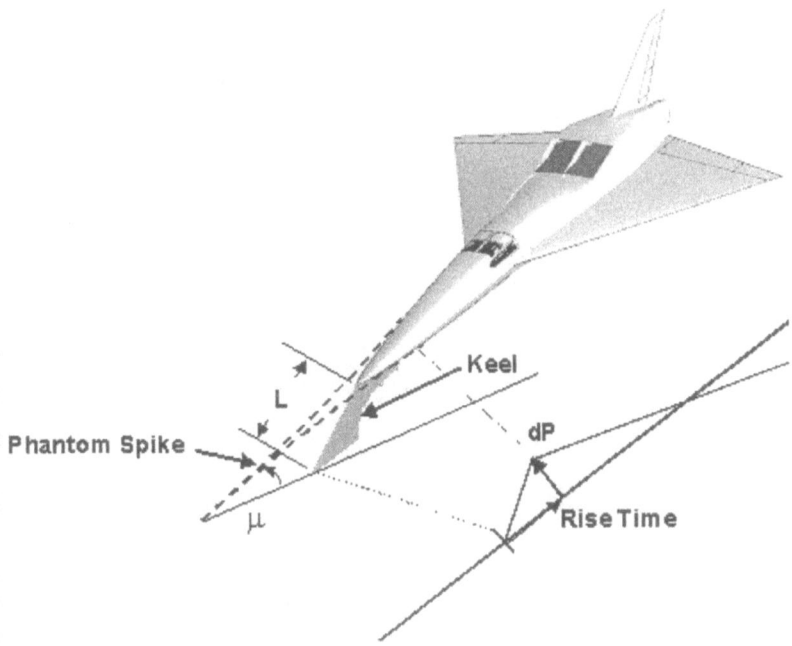

Fig. 1. Design Optimization - Genetic Algorithms (Forward Swept Keel → Length Amplification Factor = M_∞)

theory [12] is used to generate the near field pressure signature. The ground signature is computed with the NFBOOM [13] atmospheric propagation code. Genetic algorithms (GAs) are used for the optimization.

2 Flowfield Analysis – Genetic Algorithnm Fitness Function

Naturally, as in any GA application, determining a fitness function for the problem at hand is a central issue. In the current problem an attempt is being made to design a keel for a SST aircraft that will minimize the sonic boom felt on the ground. Thus, the fitness function must ultimately be able to take a keel design (the coding issue is addressed later) and provide a maximum decibel level in the associated ground signature.

Seebass [2] presents an algorithm for using linear theory to predict sonic boom ground signatures. In summary, the midfield pressure signature is computed from the aircraft equivalent body of revolution using the quasi-linear theory of Whitham [11] making far field approximations; i.e., computation of the "F-function". The midfield signature is then propagated to the ground using acoustic theory. A modified version of this theory (described next) was used for initial sizing of the phantom isentropic spike (dashed-lined-body extending the airplane nose in Fig.1).

In the present work, the far field approximations in the classical Whitham theory were relaxed; this was facilitated by the use of modern computational resources. Specifically, the small-disturbance perturbation velocity field solution, with the Whitham [12] modifications, to the linear form of the velocity potential equation,

$$\phi_{rr} + \phi_r/r - \beta^2 \phi_{xx} = 0 \tag{1}$$

is given by

$$\frac{u}{U_\infty} = -\int_0^y \frac{f'(\eta)d\eta}{\sqrt{(y-\eta)(y-\eta+2\beta r)}}$$
$$\frac{v}{U_\infty} = \frac{1}{r}\int_0^y \frac{(y-\eta+\beta r)f'(\eta)d\eta}{\sqrt{(y-\eta)(y-\eta+2\beta r)}} \tag{2}$$

In Eqn. (2), $y(x,r) = constant$ is the nonlinear characteristic curve along which $dx/dr = \cot(\mu + \theta) \approx \beta + (\gamma+1)M^4 u/2\beta - M^2(v + \beta u) + o(u^2 + v^2)$ [11]. Neglecting the second order terms, substituting for the velocity components [Eqn. (2)] and integrating $dx \approx [\beta + (\gamma+1)M^4 u/2\beta - M^2(v + \beta u)]dr$ from the body surface results in

$$x - \beta r = \begin{cases} -\frac{(\gamma+1)M_\infty^4}{2\beta^2} \int_0^y \frac{\sqrt{(y-\eta)+2\beta r} - \sqrt{y-\eta+2\beta R_B}}{\sqrt{y-\eta}} f'(\eta)d\eta \\ -M_\infty^2 \left\{ \int_0^y \log\left[\frac{\sqrt{(y-\eta)+2\beta r}-\sqrt{y-\eta}}{\sqrt{(y-\eta)+2\beta r}+\sqrt{y-\eta}}\right] f'(\eta)d\eta \\ -\int_0^y \log\left[\frac{\sqrt{(y-\eta)+2\beta R_B}-\sqrt{y-\eta}}{\sqrt{(y-\eta)+2\beta R_B}+\sqrt{y-\eta}}\right] f'(\eta)d\eta \right\} + y \end{cases} \tag{3}$$

$y(x,r)$ is the value of $x - \beta r$ where the characteristic meets the body surface; $f(x)$ represents the source distribution, and when the tangency boundary condition is applied, $f(x) = A'(x)/2\pi$. Consistent with the small disturbance theory, the pressure field was computed from the linearized form of the compressible Bernoulli's equation $(p + \rho U^2 = constant)$; i.e.,

$$\frac{p - p_\infty}{p_\infty} = -\gamma M_\infty^2 \frac{u}{U_\infty} \tag{4}$$

Using the above nonlinear characteristic theory improves the accuracy [12] over traditional linear theory, however multi-valued solutions due to wave crossing (break points) are possible. Physically, the overlap regions correspond to coalescence of the Mach waves into shock wave, followed by a discontinuous change in flow properties across the shock. Because near field solutions ($r/L \sim 0.2$) were computed here, break points were rare, and if they did occur, the spatial extent was very small. Hence, a simple shock-fitting algorithm, similar to that of Whitham was incorporated. The main simplification was that the shock was placed at the midpoint of the break region.

In summary, Equations (2-4), with 200 characteristics, were used to predict the velocity and pressure fields for a given airplane area distribution (with and without the isentropic spikes) including the additional equivalent area due to lift [2]. The integrals in Equation (2) were evaluated using a fourth order accurate numerical integration scheme; 5000 increments in η were used to ensure numerical convergence to within 0.1%. A computer program was written to perform this near field analysis. The program was validated with the 5/2-power frontal spike experiment given Swigart [11]; a comparison of the present algorithm to the experimental data is shown in Fig.2. The agreement was considered sufficient for the present sonic boom mitigation analysis.

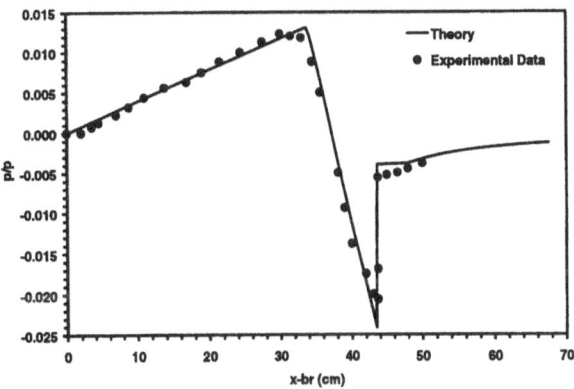

Fig. 2. Comparison of Quasi-Linear Theory with the Frontal Spike (r = 50.8 cm) Data of Swigart [11].

The near field solutions were extrapolated to the ground using the NASA Ames sonic boom FORTRAN routines in NFBOOM [13]. Specifically, the ANET subroutine, which extrapolates the signal through the non-uniform atmosphere, was incorporated into the Whitham theory program. A reflection factor of 1.9 was taken and no wind effects were assumed. In addition, the perceived ground pressure levels [4,5] were estimated with the PLdB subroutine also present in NFBOOM.

3 Design Optimization – Genetic Algorithm Coding Scheme

For sonic boom mitigation, the GA is given the task of minimizing the loudness of the sonic boom created by the body shape of the aircraft. This body shape is represented by graphing the effective area of the aircraft versus the position along the length of the aircraft. To optimize this body shape, an assumption is made that the body shape may be represented by continuous fourth-order polynomials.

The GA's task is then to modify the coefficients of this polynomial to minimize the loudness while also satisfying certain constraints that are placed on the body shape. One of the constraints placed on the GA is that the derivatives of the body shape must not fluctuate beyond a certain pre-set level. This is done to insure that the body shape will not be oscillatory in nature. Second, the GA must develop a body shape that visually lies between a "minimum" and "maximum" body shape. Thus, the plot of the polynomial generated by the GA must fit between two pre-defined curves. This focuses the scope of the GA's search and allows it to be more efficient.

In order to quantitatively evaluate the quality of each body shape that the GA proposes, each body shape is analyzed using Whitham theory. This results in a set of data that can then be read into NFBOOM program code to calculate the effective loudness of the sonic boom. This loudness serves as the driving factor of the GA's search. Those solutions with lower decibel levels are recombined with other quality solutions to form new solutions to test, while those solutions with higher decibel levels are discarded in favor of better solutions. This allows the GA to bias the population of potential solutions towards better solutions over time.

This approach of allowing a GA to propose coefficients of several fourth-order polynomials (constrained such that the first and second derivatives of the curve are continuous) allows for reasonable, if not effective, area distributions. Figure 3 shows a sample area distribution proposed by a GA.

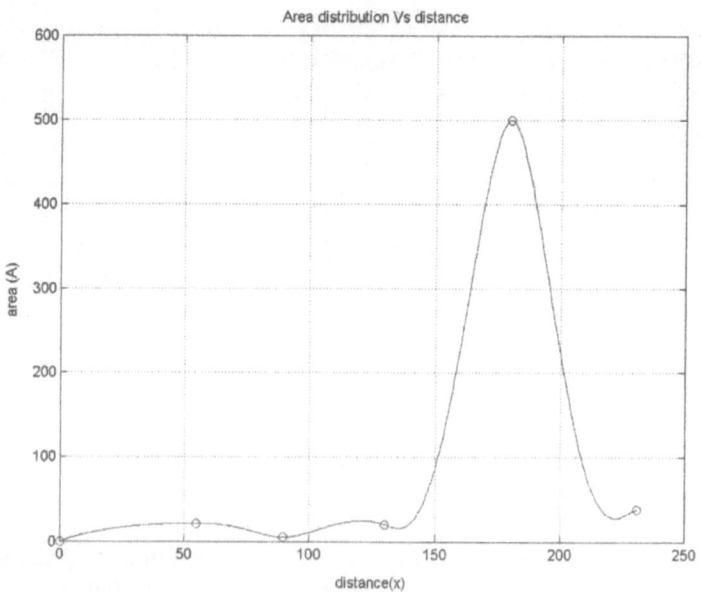

Fig. 3. Sample area distribution as proposed by a GA.

4 Results

With the definition of a fitness function and a coding scheme, a floating point GA can be used to effectively design keels that can be added to SST aircraft that minimize the sonic boom for given flight conditions. For this study, the following GA parameters are used:

- *population size* = 50
- *maximum generations* = 250
- *mutation probability* = 0.3
- *mutation operator* = Gaussian mutation
- *crossover operator* = standard two-point crossover

The remainder of this section describes the effectiveness of the GA in determining keel designs.

4.1 Initial Isentropic Spike Design

Phantom isentropic spikes (extending the nose of the airplane as shown in Fig.1 and 2) were designed to lower the initial shock pressure rise to 0.3 psf (Marconi et al [10] goal). The length and strength (K) of the 5/2-spike ($A = Kx^{5/2}$) were the design parameters. The spike was blended to the body, and the body geometry was fit with five fourth-order polynomial curves. The coefficients were selected to ensure that the body fit was continuous through the third derivative. Figure 4 shows an example aircraft area distribution (including the equivalent area due to lift) with the isentropic (forward) spike and the body fits. This parameterization proved sufficient for all configurations tested. Because the 5/2-spike removes the shock wave altogether, the perceived sound level (PLdB) [4] was used to convert the linear ground-signature waveform sound level into an equivalent-shock-wave signature, where the perceived sound levels were matched. Figure 5 shows an example result, where a spike was designed to reduce the initial shock pressure rise from just over 0.6 psf for the baseline aircraft down to 0.3 psf (the design goal). The length of the corresponding swept forward keel [10] would be 8.1% of the baseline aircraft length.

4.2 Initial Genetic Algorithm Designs – Feasibility Study

The first step in determining the effectiveness of using a GA to perform keel design for sonic boom mitigation is to match published data. As described in Marconi et al. [10], the upstream spike area distribution can be used as a design variable to produce a desired ground signature. Hence, it the present work, the 5/2-spike requirement was relaxed, and genetic algorithms were used to produce an "optimum" design. In the present work, the optimum was determined to be the shortest spike/keel that produced a ground signature shock overpressures < 0.3 psf; signatures with multiple shocks that met this requirement were acceptable.

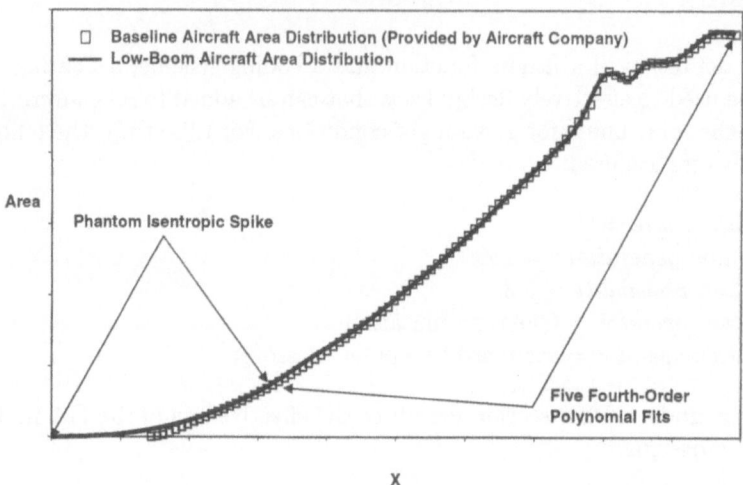

Fig. 4. Example Equivalent Mach Plane Area Distribution (Scales intentionally left blank).

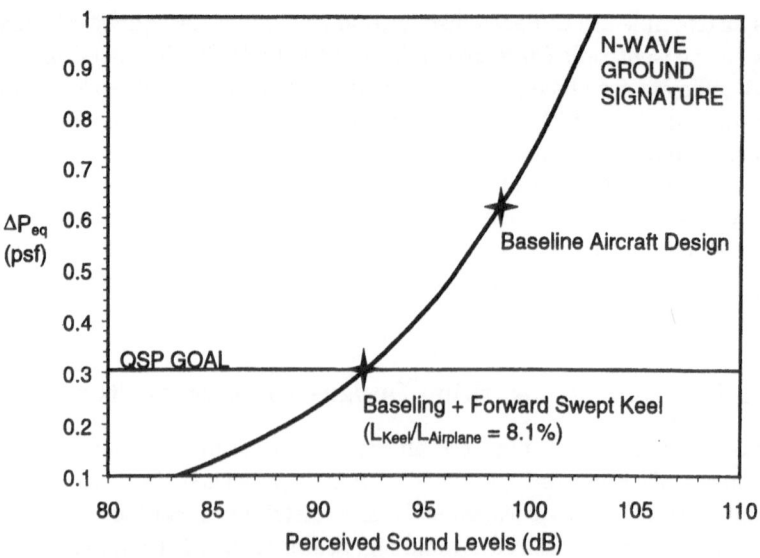

Fig. 5. Example Boom Reduction Results.

Minimization of Sonic Boom on Supersonic Aircraft   2165

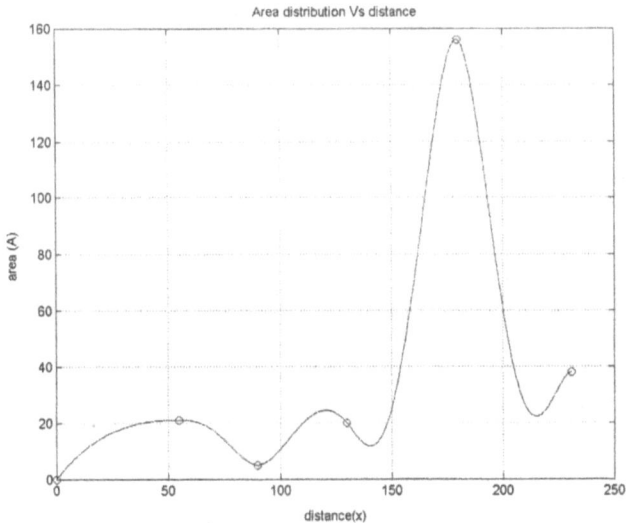

Fig. 6. The area distribution above represents the best solution determined using a GA.

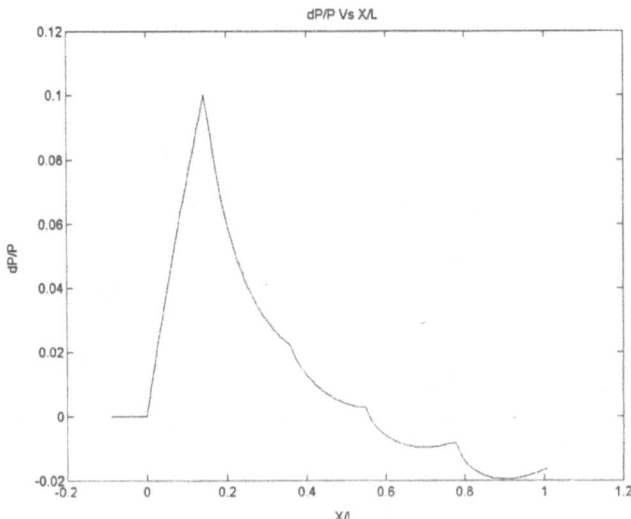

Fig. 7. The ground signature above results in a sonic boom of magnitude 104.0 dB.

Given the GA's ability to solve this simple design problem, the more complex problem of determining a keel shape that minimizes the decibel level in a ground signature was undertaken.

4.3 Initial Genetic Algorithm Designs – Feasibility Study

Based on the results presented in Fig.4 and 5, the GA was considered to be an appropriate tool for determining the keel shape (as depicted in an area distribution) that minimizes the sonic boom of a SST aircraft. Figures 6 and 7 show the optimal design and associated ground signature determined in this manner. It is important to note that the GA found a design that had a sonic boom of magnitude 104.0 dB; the best solution to date determined by a human designer had a sonic boom of magnitude 106.3 dB.

References

1. Seebass, R. and Argrow, B., "Sonic Boom Minimization Revisited," AIAA Paper 98-2956, 1998.
2. Seebass, R., "Sonic Boom Theory", *AIAA Journal of Aircraft*, Vol. 6, No. 3, 1969.
3. McLean, E., "Configuration Design for Specific Pressure Signature Characteristics," Sonic Boom Research, edited by I.R. Schwartz, NASA SP-180, 1968.
4. Leatherwood, J.D. and Sullivan, B.M., "Laboratory Study of Sonic Boom Shaping on Subjective Loudness and Acceptability," NASA TP-3269, 1992.
5. McCurdy, D.A., "Subjective Response to Sonic Booms Having Different Shapes, Rise Times, And Duration," NASA TM 109090, 1994
6. Miller, D. S., and Carlson, H. W., "A Study of the Application of Heat or Force Field to the Sonic-Boom-Minimization Problem," NASA TND-5582, 1969.
7. Miller, D. S., and Carlson, H. W., "On the Application of Heat or Force Field to the Sonic-Boom-Minimization Problem," AIAA Paper 70-903, 1970.
8. Marconi, F., "An Investigation of Tailored Upstream Heating for Sonic Boom and Drag Reduction", AIAA Paper 98-0333, 1998.
9. Batdorf, S. B., "On Alleviation of the Sonic Boom by Thermal Means," AIAA Paper 70-1323, 1970.
10. Marconi, F., Bowersox, R. and Schetz, J., "Sonic Boom Alleviation Using Keel Configurations," AIAA-2002-0149, 40th Aerospace Science Meeting, Reno NV, Jan. 2002.
11. Swigart, R.J. "Verification of the Heat-Field Concept for Sonic-Boom Alleviation", *AIAA Journal of Aircraft*, Vol. 12, No. 2, 1975.
12. Whitham, G., "The Flow Pattern of a Supersonic Projectile," *Communications on Pure and Applied Mathematics*, Vol. V, 1952, pp. 301-348.
13. Durston, D.A., "NFBOOM User's Guide, Sonic Boom Extrapolation and Sound-Level Prediction", NASA Ames Research Center, Unpublished Document (Last Updated Nov. 2000)
14. Marconi, F., Bowersox, R. and Schetz, J., "Sonic Boom Alleviation Using Keel Configurations," AIAA-2002-0149, 40th Aerospace Science Meeting, Reno NV, Jan. 2002.

Nomenclature
A Airplane Area Distribution
K 5/2-Area Distribution Strength (i.e., $A = Kx^{5/2}$)
L Length
M Mach Number
p Pressure
R_b Body Radius
r Radial Coordinate
u Axial velocity perturbation
U Axial Velocity
v Radial velocity perturbation
x Axial coordinate
y Curved Characteristic Line
β $\sqrt{M^2 - 1}$
γ Ratio of Specific Heats
μ Mach Wave Angle
ρ Density

Subscripts
∞ Flight Condition at Altitude

Optimizing the Order of Taxon Addition in Phylogenetic Tree Construction Using Genetic Algorithm

Yong-Hyuk Kim[1], Seung-Kyu Lee[2], and Byung-Ro Moon[1]

[1] School of Computer Science & Engineering, Seoul National University
Shilim-dong, Kwanak-gu, Seoul, 151-742 Korea
{yhdfly, moon}@soar.snu.ac.kr
[2] NHN Corp., 7th floor, Startower,
737 Yoksam-dong, Kangnam-gu, Seoul, Korea
spin30@soar.snu.ac.kr

Abstract. Phylogenetics has gained in public favor for the analysis of DNA sequence data as molecular biology has advanced. Among a number of algorithms for phylogenetics, the fastDNAml is considered to have reasonable computational cost and performance. However, it has a defect that its performance is likely to be significantly affected by the order of taxon addition. In this paper, we propose a genetic algorithm for optimizing the order of taxon addition in the fastDNAml. Experimental results show that the fastDNAml with the optimized order of taxon addition constructs more probable evolutionary trees in terms of the maximum likelihood.

1 Introduction

As the revolutions in molecular biology have produced a huge amount of DNA sequence data, extracting useful information from them has been considered to be of paramount importance. One of the most important issues includes *phylogenetics*.

Phylogenetics [27] [18] [24] is to infer the most probable evolutionary relationships among species from DNA sequence data. The inferred relationships among species are typically represented by a tree, also called *phylogeny*, which consists of nodes and branches connecting nodes; each node represents a species and each branch represents the amount of genetic variation between two species. It is known that constructing the most probable phylogenetic tree is NP-complete [5] [12]. We are usually interested in the most probable tree in terms of both tree topology and branch lengths.

A number of algorithms for constructing evolutionary trees have been proposed. Parsimony [6] [7] [1] is one of the most popular methods. However, it has a severe problem in that it constructs an inconsistent evolutionary tree when the amounts of genetic changes in different lineages are sufficiently unequal [8].

In contrast to the parsimony which make full use of the information available in the DNA sequence, there have been simpler approaches that exploit only the pairwise similarity between DNA sequences. The least-squares [3] is a popular method among them. While the least-squares has explicit statistical justification, it also constructs an inconsistent tree if the rates of evolution are sufficiently unequal in different lineages [4] [8], as in the parsimony.

To estimate more consistent and probable trees, statistical methods using a probabilistic model of evolution are proposed. One of the most robust method is considered to be the maximum likelihood [8], motivated from the earlier probabilistic models of evolution [25]. The approaches using the maximum likelihood can be classified into two categories: constructive approach and non-constructive one.

The constructive approach, which is more popular, builds an evolutionary tree by adding one taxon at a time, starting at an empty tree, with some heuristic information. DNAml [9] and its improved variant, fastDNAml [26], are the representative of them. Although fastDNAml is one of the most widely used method in the phylogenetics literature, its performance is limited due to its incremental nature in constructing trees. In particular, the performance of fastDNAml is notably affected by the order of taxon addition.

As an alternative, non-constructive approaches have also been applied for phylogeny reconstruction. They include all the algorithms without the explicit taxon addition. Recently, evolutionary algorithms such as genetic algorithms [15] [11] [23] have been proposed for constructing evolutionary trees [20] [17] [16] [21]. However, most of them conducted experiments with limited data sets and required considerably high computational cost compared with the constructive approach. They need to be more elaborate to be useful as practical algorithms with reasonable performance.

In this paper, we propose a genetic algorithm for optimizing the order of taxon addition in fastDNAml. The rest of the paper is organized as follows. In Section 2, we describe the maximum likelihood and the fastDNAml. In Section 3, we explain our genetic algorithm in detail and present our experimental results in Section 4. Finally, we make our conclusions in Section 5.

2 Preliminaries

2.1 Maximum Likelihood

Maximum likelihood method [8] is a method for reconstructing phylogenetic trees, or evolutionary trees. Its distinctive feature is that it requires a model of sequence evolution which designates how the sequence evolves. The maximum likelihood method consists of three elements: an evolutionary model, a tree, and the observed sequence.

The maximum likelihood method computes the likelihood of obtaining the observed sequence with a given tree topology, assigned branch lengths, and a given evolutionary model. Since the likelihood is mostly very small, we usually work with log-likelihood rather than the likelihood itself. The log likelihood of obtaining the observed sequence is defined by:

$$\ln L = \sum_{i=1}^{k} \ln L_i$$

where k is the number of sites and L_i is the likelihood of obtaining the nucleotide, one of $\{A, C, G, T\}$, at site i. Based on the maximum likelihood, trees with higher log-likelihoods are considered better.

2.2 fastDNAml

The fastDNAml [26] is one of the most popular programs with reasonable performance and running time. It is an improved version of its predecessor, DNAml [9], in terms of both performance and running time.

The main motivation for the fastDNAml was to reduce the computational cost of DNAml. The DNAml was effective in reconstructing phylogenetic trees with high likelihoods but it required considerably long time to find the trees. To alleviate the cost, fastDNAml uses Newton-Raphson method for finding optimal branch lengths and limits the effort concerning to the branch length optimization. With the two alterations, fastDNAml considerably outperformed the DNAml in terms of both performance and running time.

Figure 1 shows the outline of tree construction in fastDNAml. Note that the phylogenetic tree with three taxa has only one topology. Details for partial tree check and full tree check are described in [26].

The fastDNAml is the representative of the constructive approaches, which build an evolutionary tree by adding one taxon at a time, starting at an empty tree. The performance of fastDNAml is greatly affected by the order of taxon addition. Figure 2 shows two example phylogenetic trees with different orders of taxon addition for an instance with eight taxa (instance *algae*). It suggests that the order of taxon addition can greatly affect the qualities of the resultant trees.

3 A Genetic Algorithm

We propose a genetic algorithm (GA) for finding an optimal order of taxa addition. It conducts a search using an evaluation function related with distance between taxa. The order can be found by enumerating and testing all possibilities. The search space with n taxa has $n!$ elements if all possibilities are considered. Our GA provides an alternative search method to find a good order of taxa addition.

A genetic algorithm hybridized with local optimizations is called a hybrid GA. A considerable number of studies about hybridization of GAs [30] [29] [19] have been proposed. Figure 3 shows a typical steady-state hybrid genetic algorithm. In the next subsection, we describe each part of the hybrid GA that we used for this work.

```
fastDNAml()
// n : the final number of taxa
// i : the number of taxa in the current tree
// A : the next taxon to be inserted
// T_i : the current estimate of the best tree size i
// T_p : the tree after partial tree check (minor changes)
// T_f : the tree after full tree check (greater changes)
// P_i : the set of all the possible tree topologies by adding A to T_i

Compute the optimal tree T_3;
i ← 3;
do {
    Pick the next taxon A;
    Construct the set P_i;
    for each tree in P_i
        { Compute the optimal branch lengths and corresponding likelihood; }
    Set T_{i+1} to be the best tree in P_i;
    i ← i + 1;
    do
        Generate a modified tree T_p from T_i (partial tree check);
    until (none of T_p's is better than T_i);
    T_i ← the best among T_p's;
} until (i = n);
do
    Generate a modified tree T_f from T_n (full tree check);
until (none of T_f's is better than T_n);
T_n ← the best among T_f's;
return T_n;
```

Fig. 1. The outline of tree construction in fastDNAml

3.1 Genetic Operators

- *Encoding*: A chromosome corresponds to an order of taxon addition. The number of genes in the chromosome is equal to the number of taxa. Each gene corresponds to a taxon.
- *Initialization*: All the chromosomes are created at random. Any valid permutation of order is allowed. We set the population size to be 50 in our algorithm.
- *Selection*: The roulette-wheel-based *proportional selection* is used. The fitness value F_i of a chromosome i is calculated as follows:

$$F_i = (O_w - O_i) + (O_w - O_b)/3$$

where O_w is object value of the worst chromosome in the population, O_b is object value of the best chromosome in the population, O_i is object value of chromosome i. Each chromosome is selected as a parent with a probability proportional to its fitness value.

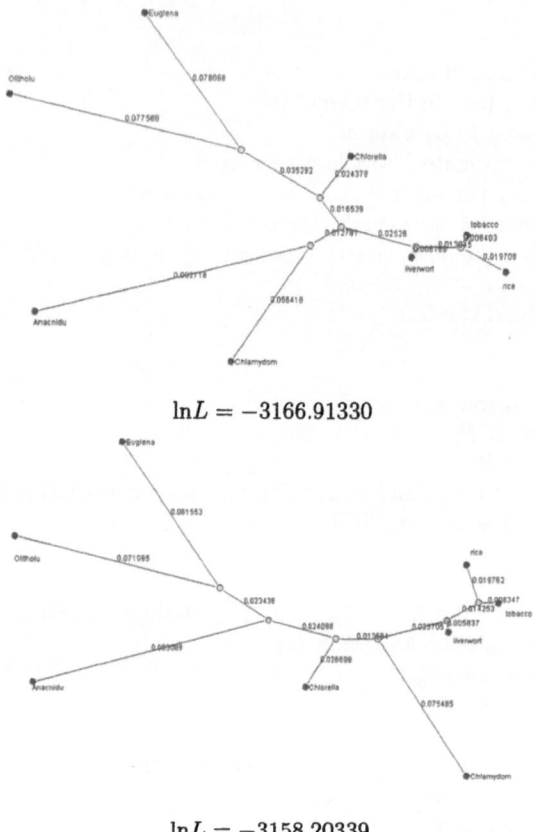

$\ln L = -3166.91330$

$\ln L = -3158.20339$

Fig. 2. Two example phylogenetic trees with different orders of taxon addition for *algae*

```
Create initial population of fixed size;
do {
    Choose parent1 and parent2 from population;
    offspring ← crossover(parent1, parent2);
    mutation(offspring);
    local-optimization(offspring);
    if suited(offspring) then replace(population, offspring);
}until (stopping condition);
return the best answer;
```

Fig. 3. A typical steady-state hybrid genetic algorithm

```
iterative-improvement()
// c_i : i^th gene of chromosome C
// f_C : fitness of chromosome C

prev ← f_C;
do {
    flag ← false;
    for all i, j pairs (i < j)
    {
        Swap c_i and c_j;
        current ← f_C;
        gain ← prev − current;
        if (gain < 0) then Swap c_i and c_j; // undo swapping
        else {
            flag ← true;
            prev ← current;
        }
    }
} until (flag = false);
```

Fig. 4. An iterative improvement heuristic

- *Crossover:* Since a chromosome designates an order, an order-based crossover is a natural choice. We use the PMX (Partially Matched Crossover) [10], one of the most popular order-based crossovers. PMX proceeds as follows. 1) Two chromosomes are aligned. 2) Two crossing points are selected at random along the chromosomes, defining a matching section. 3) The genes in the matching section are exchanged. 4) Repair for a valid permutation is performed.
- *Mutation:* Two genes are randomly chosen and swapped. The swaps are repeated for a predetermined times.
- *Local Optimization:* Hybrid genetic algorithms have been considered natural in solving a difficult problem to get desirable performance since genetic algorithms are not so good at fine tuning near local optima. In this study, we use an iterative improvement heuristic for local minimization and it is applied to the offspring after mutation. Figure 4 shows the iterative heuristic.
- *Replacement:* The preselection [2] is used. The offspring replaces the worse parent. The preselection is advantageous in maintaining the diversity of the population.
- *Stopping Criterion:* Our GA stops when one of the two conditions is satisfied: i) the number of generations reaches 5,000, ii) when the fitness of the worst chromosome is equal to the fitness of the best one.

Table 1. Comparison of Two Addition Orders

	Max-relation order	Min-relation order
HIVenvSweden	−1159.93528	−1159.68256
algae	−3159.07438	−3158.20339
hasegawa5	−2682.76961	−2682.75376
exampleTipDate	−3869.25646	−3869.25645

3.2 Evaluation Function

It is ideal to use fastDNAml itself for the fitness evaluation of the GA. However, because of the serious time requirement of fastDNAml, we use a heuristic method for the fitness evaluation of a taxon order.

We performed some experiments to get insights on good orders of taxon addition. Firstly, we tried to iteratively add a taxon that highly relates with previously added taxa. We call this order "Max-relation order." On the other hand, we also tried the opposite. In this heuristic, we prefer a taxon most *unrelated* with previously added taxa, the order is called "Min-relation order." Table 1 shows the ln$L$ scores for some instances by the two addition orders. The results of "Min-relation order" were better than those of "Max-relation order." This result is contrary to our expectation. The performance of "Max-relation order" seems to be limited in that it is likely to form too strong a shape in the early stage of tree construction.

We attempt to find a Min-relation order. We suspect that such an order first makes a global sketch of the tree topology and then adjusts the details. Our GA minimizes the following formula:

$$object\ function = \sum_{i>j}(D_{ij} - w \cdot ((n-1) - (i-j)))^2$$

where D_{ij} is the gene distance between taxa i and j, n is the number of taxa, and the balancing factor $w = \sum_{i>j} D_{ij} / \sum_{i>j}(i-j)$.

4 Experimental Results

4.1 Data Sets

Nine instances were tested. Table 2 shows the number of taxa and the number of sites for each instance. The number of taxa ranges from 7 to 55. The number of sites ranges from 232 up through 1,485. Brief descriptions about the instances are in the following.

- *HIVenvSweden*: HIV-1 sample of 136 patients from Sweden envelope glycoprotein (*env*) gene, V3 region. Thirteen HIV *env* genes used by Yang et al. [35] in developing models of variable selective pressures among sites (the NSsites models).

Table 2. Test Sets

	# of taxa	# of sites
HIVenvSweden	13	273
algae	8	900
hummt25	25	601
green	12	1314
rbcl55	55	1314
hasegawa5	14	232
mtprim9	9	888
exampleTipDate	17	1485
lysozymeSmall	7	390

- *algae*: 16s rDNA data.
- *hummt25*: Twenty five human D-loop sequences used in [34].
- *rbcl55*: Large subunit of RuBisCO gene from chloroplasts. Sequences of the chloroplast gene rbcL from a diversity of green plants, used in [17]. *green* extracted from rbcl55 consists of first 12 taxa of rbcl55.
- *hasegawa5*: Used by Hasegawa *et al.* [13].
- *mtprim9*: mtDNA primate dataset. A mitochondrial segment consisting 888 aligned sites from nine primate species [14], used by Yang [31] to test the discrete-gamma model and Yang [32] to test the auto-discrete-gamma models.
- *exampleTipDate*: Data set of 17 dengo viral strains sequenced at different dates from Andrew Rambaut's TipDate program. This was used for testing the TipDate models of [28].
- *lysozymeSmall*: Primate lysozyme genes of [22], used by Yang [33] in developing tests of positive selection along lineages. This is the "small data set" analyzed in that paper.

4.2 Performance

The main results are given in Table 3. The column "Basic order" shows the $\ln L$ scores by the usual random addition order, and the column "New order" shows the $\ln L$ scores by the addition order obtained by our GA. One can see that the results by "New order" significantly better than those of "Basic order."

Finally, we examine the effectiveness of the object function of Section 3.2. Since the fastDNAml itself requires rather high computational cost, it is impractical to use fastDNAml for fitness evaluation in GA. Although impractical, we replaced the object function by the $\ln L$ score of fastDNAml. This means that we run fastDNAml for evaluation whenever an offspring is created. Table 4 shows the $\ln L$ scores by the best and worst addition orders found by GA. Although some instances were independent of the addition orders (*mtprim9* and *lysozymeSmall*), the results overall shows that the order of addition greatly affects the qualities

Table 3. Comparison of Results

	Basic order	New order
HIVenvSweden	−1160.40239	**−1159.27680**
algae	−3159.07438	**−3158.20339**
hummt25	−1710.83504	**−1706.99035**
green	−8808.40925	**−8800.40369**
rbcl55	−28586.07304	**−28575.65960**
hasegawa5	−2682.91642	**−2682.67452**
mtprim9	−5243.41821	−5243.41821
exampleTipDate	−3869.32158	**−3869.25645**
lysozymeSmall	−924.97205	−924.97205

The figures in the table are the ln$L$ scores with HKY evolutionary model [13].

Table 4. Results of the Best and Worst Addition Orders

	Worst order	Best order
HIVenvSweden	−1164.16258	−1159.27680
algae	−3166.91330	−3158.20339
hummt25	−1754.03450	−1706.99035
green	−8840.43147	−8800.40369
rbcl55	−28662.63835	−28571.70704
hasegawa5	−2692.47175	−2682.67452
mtprim9	−5243.41821	−5243.41821
exampleTipDate	−3869.52773	−3869.25645
lysozymeSmall	−924.97205	−924.97205

of the resultant trees. It is surprising that except for one instance, *rbcl55*, the ln$L$ scores by the "Best order" are the same as those by the "New Order" in Table 3. This supports the effectiveness of the Min-relation order; we suggest to use it practically. The results of fastDNAml could be improved in this way by the proposed GA.

5 Conclusions

We proposed a hybrid genetic algorithm for optimizing the order of taxon addition in the fastDNAml. Since the performance of the fastDNAml is dependent on the order of taxon addition, we attempted to optimize the order using a genetic algorithm.

Although we improved the fastDNAml with attractive orders, there is still room for improvements. First of all, we need to study more about the relation between the distance among taxa and the taxon addition order.

It is also necessary to incorporate more problem-specific information into the local optimizations. Since a phylogenetic tree with high ln$L$ score often reveals new relationship between taxa, it is practically valuable to improve the algorithm. It is left for further study.

Acknowledgments. This work was partly supported by Optus Inc. and Brain Korea 21 Project. The RIACT at Seoul National University provided research facilities for this study.

References

1. J. H. Camin and R. R. Sokal. A method for deducing branching sequences in phylogeny. *Evolution*, 19:311–326, 1965.
2. D. Cavicchio. *Adaptive Search Using Simulated Evolution*. PhD thesis, University of Michigan, Ann Arbor, MI, 1970.
3. R. Chakraborty. Estimation of time of divergence from phylogenetic studies. *Canadian Journal of Genetics and Cytology*, 19:217–223, 1977.
4. D. H. Colless. The phylogenetic fallacy. *Systematic Zoology*, 16:289–295, 1967.
5. W. H. E. Day. Computational complexity of inferring phylogenies from dissimilarity matrices. *Bulletin of Mathematical Biology*, 49(4):461–467, 1987.
6. A. W. F. Edwards. The reconstruction of evolution. *Heredity*, 18:553, 1963.
7. A. W. F. Edwards and L. L. Cavalli-Sforza. Reconstruction of evolutionary trees. *Phenetic and Phylogenetic Classification*, pages 67–76, 1964.
8. J. Felsenstein. Evolutionary trees from DNA sequences: A maximum likelihood approach. *Journal of Molecular Evolution*, 17:368–376, 1981.
9. J. Felsenstein. PHYLIP - phylogeny inference package (version 3.2). *Cladstics*, 5:164–166, 1989.
10. D. Goldberg and R. Lingle. Alleles, loci and the traveling salesman problem. In *International Conference on Genetic Algorithms*, pages 154–159, 1985.
11. D. E. Goldberg. *Genetic algorithms in search, optimization and machine learning*. Addison-Wesley, Reading, MA, 1989.
12. R. L. Graham and L. R. Foulds. Unlikelihood that minimal phylogenetics for a realistic biological study can be constructed in reasonable computational time. *Mathematical Biosciences*, 60:133–142, 1982.
13. M. Hasegawa, H. Kishino, and T. Yano. Dating the human-ape splitting by a molecular clock of mitochondrial DNA. *Journal of Molecular Evolution*, 22:160–174, 1985.
14. K. Hayasaka, T. Gojobori, and S. Horai. Molecular phylogeny and evolution of primate mitochondrial DNA. *Molecular Biology and Evolution*, 5:626–644, 1988.
15. J. Holland. *Adaptation in natural and artificial systems*. University of Michigan Press, Ann Arbor, 1975.
16. K. Katoh, K. Kuma, and T. Miyata. Genetic algorithm-based maximum-likelihood analysis for molecular phylogeny. *Journal of Molecular Evolution*, 53:477–484, 2001.
17. P. O. Lewis. A genetic algorithm for maximum likelihood phylogeny inference using nucleotide sequence data. *Molecular Biology and Evolution*, 15(3):277–283, 1998.
18. W. H. Li. *Molecular Evolution*. Sinauer Associates, Sunderland MA, 1997.
19. F. G. Lobo and D. E. Goldberg. Decision making in a hybrid genetic algorithm. In *IEEE International Conference on Evolutionary Computation*, pages 121–125, 1997.

20. H. Matsuda. Protein phylogenetic inference using maximum likelihood with a genetic algorithm. In *Pacific Symposium on Biocomputing '96*, pages 512–523, 1996.
21. A. Meade, D. Corne, M. Pagel, and R. Sibly. Using evolutionary algorithms to estimate transition rates of discrete characteristics in phylogenetic trees. In *Congress on Evolutionary Computation*, pages 1170–1177, 2001.
22. W. Messier and C.-B. Stewart. Episodic adaptive evolution of primate lysozymes. *Nature*, 385:151–154, 1997.
23. M. Mitchell. *An introduction to genetic algorithms*. MIT Press, London, 1996.
24. M. Nei and S. Kumar. *Molecular Evolution and Phylogenetics*. Oxford University Press, New York, 2000.
25. J. Neyman. Molecular studies of evolution: a source of novel statistical problems. In *Statistical Decision Theory and Related Topics*, ed. S. S. Gupta and J. Yackel. New York: Academic Press, pages 1–27, 1971.
26. G. J. Olsen, H. Matsuda, R. Hagstrom, and R. Overbeek. fastDNAml: a tool for construction of phylogenetic trees of DNA sequences using maximum likelihood. *Computer Applications in the Biosciences*, 10(1):41–48, 1994.
27. R. D. M. Page and E. C. Holmes. *Molecular Evolution: A Phylogenetic Approach*. Blackwell Science, 1998.
28. A. Rambaut. Estimating the rate of molecular evolution: incorporating non-contemporaneous sequences into maximum likelihood phylogenies. *Bioinformatics*, 16(4):395–399, 2000.
29. J. M. Renders and H. Bersini. Hybridizing genetic algorithms with hill-climbing methods for global optimization: two possible ways. In *Proceedings of the First IEEE Conference on Evolutionary Computation*, pages 312–317, 1994.
30. D. Whitley, V. Gordon, and K. Mathias. Larmarckian evolution, the Baldwin effect and function optimization. In *International Conference on Evolutionary Computation*, Oct. 1994. *Lecture Notes in Computer Science*, 866:6-15, Springer-Verlag.
31. Z. Yang. Maximum likelihood phylogenetic estimation from DNA sequences with variable rates over sites: approximate methods. *Journal of Molecular Evolution*, 39:306–314, 1994.
32. Z. Yang. A space-time process model for the evolution of DNA sequences. *Genetics*, 139:993–1005, 1995.
33. Z. Yang. Likelihood ratio tests for detecting positive selection and application to primate lysozyme evolution. *Molecular Biology and Evolution*, 15:568–573, 1998.
34. Z. Yang and S. Kumar. New parsimony-based methods for estimating the pattern of nucleotide substitution and the variation of substitution rates among sites and comparison with likelihood methods. *Molecular Biology and Evolution*, 13:650–659, 1996.
35. Z. Yang, R. Nielsen, N. Goldman, and A.-M. K. Pedersen. Codon-substitution models for variable selection pressure at amino acid sites. *Genetics*, 155:431–449, 2000.

Multicriteria Network Design Using Evolutionary Algorithm

Rajeev Kumar and Nilanjan Banerjee

Department of Computer Science and Engineering
Indian Institute of Technology Kharagpur
Kharagpur, 721 302, India
rkumar@cse.iitkgp.ernet.in

Abstract. In this paper, we revisit a general class of multi-criteria multi-constrained network design problems and attempt to solve, in a novel way, with Evolutionary Algorithms (EAs). A major challenge to solving such problems is to capture possibly all the (representative) equivalent and diverse solutions. In this work, we formulate, without loss of generality, a bi-criteria bi- constrained communication network topological design problem. Two of the primary objectives to be optimized are network delay and cost subject to satisfaction of reliability and flow-constraints. This is a *NP-hard* problem so we use a hybrid approach (for initialization of the population) along with EA. Furthermore, the two-objective optimal solution front is not known *a priori*. Therefore, we use a multiobjective EA which produces diverse solution space and monitors convergence; the EA has been demonstrated to work effectively across complex problems of *unknown* nature. We tested this approach for designing networks of different sizes and found that the approach scales well with larger networks. Results thus obtained are compared with those obtained by two traditional approaches namely, the exhaustive search and branch exchange heuristics.

1 Introduction

Network design problems where even a single cost function or objective value (e.g., minimal spanning tree or shortest path problem) is optimized, are often NP-hard [1]. Many such uni-criterion network design problems are well studied and many heuristics/methods exist for obtaining exact/approximate solutions in polynomial-time [2]. But, in most real-life applications, network design problems generally require simultaneous optimization of multiple and often conflicting objectives, subject to satisfaction of some constraints. For example, topological design of communication networks, particularly mesh/wide area networks is a typical multiobjective problem involving simultaneous optimization of cost of the network and various performance criteria such as average delay of the network, throughput subject to some reliability measures and bandwidth/flow-constraints. The problem can be stated as: given a set of node locations and the traffic between the nodes, it is required to design the layout of links between the nodes while optimizing certain criteria e.g., overall cost, average per packet delay, reliability and provision for expansion. This requires optimization of conflicting factors, subject to various constraints. For example, reducing the packet delay could mean an increase in the link capacities, which will result in an increase in the network cost. Exploring the whole solution space for such a design problem is an NP hard problem [3]. Similar design problems exist for multicast routing of multimedia communication in constructing a minimal cost spanning/Steiner tree with given constraints on diameters [4].

Such multicriteria network design problems occur in many other engineering applications too. In VLSI design, the interconnect resistance increases significantly with deep micron technology. An increase in interconnect resistance increases interconnect delays thus making a dominant factor in timing analysis of VLSI circuits. The VLSI circuit design aims at finding minimum cost spanning/Steiner tree given delay bound constraints on source-sink connections [5]. Analogously, there exists the problem of degree/diameter- constrained minimum cost networks [6].

Many NP-hard bicriteria network design problems have been attempted and approximate solutions obtained using heuristics/methods, and verified in polynomial time, see - [6], [7] and [8]. For example, Ravi et al. [8] and Deo et al. [6] presented approximation algorithm by optimizing one criterion subject to a budget on the other. We argue that the use of heuristics may yield *single* optimized solutions in each objective-space, and may not yield many other equivalent solutions. Secondly, extending this approach to multi-criteria problems (involving more than two objectives/constraints) the techniques require improving upon more than one constraints. Thirdly and more importantly, such approaches may not yield all the representative optimal solutions. Most conventional approaches to solve network design problems start with a Minimum Spanning Tree (MST), and thus effectively minimizes the cost. With some variations induced by ϵ-constraint method, most other solutions obtained are located near the minimal-cost region of the Pareto-front, and thus do not form the complete Pareto-front.

In this work, we try to overcome the disadvantages of conventional techniques and single objective EAs. We use multiobjective EA to obtain a Pareto-front. For a wide-ranging review and a critical analysis of evolutionary approaches to multiobjective optimization - see [9] and [10]. There are many implementation of multiobjective EAs, for example, MOGA [11], NSGA [12], SPEA [13]) and PEAS [14]. These implementations achieve diverse and equivalent solutions by some diversity preserving mechanism, they do not talk about convergence. Any explicit diversity preserving method needs prior knowledge of many parameters and the efficacy of such a mechanism depends on successful fine-tuning of these parameters. In a recent study, Purshouse & Fleming [17] extensively studied the effect of sharing, along with elitism and ranking, and concluded that while sharing can be beneficial, it can also prove surprisingly *ineffective* if the parameters are not carefully tuned.

Some other recent studies have been done on combining convergence with diversity. Laumanns et al. [15] proposed an ϵ-dominance for getting an ϵ-approximate Pareto-front for problems whose optimal Pareto-set is *known*. Kumar & Rockett [16] proposed use of Rank-histograms for monitoring convergence of Pareto-front while maintaining diversity without any *explicit* diversity preserving operator. Their algorithm is demonstrated to work for problems of *unknown* nature. Secondly, assessing convergence does not need any *a priori* knowledge for monitoring movement of Pareto-front using rank-histograms.

In this work, we use their Pareto Converging Genetic Algorithm (PCGA) [16] which has been demonstrated to work effectively across complex problems and achieves diversity without needing *a priori* knowledge of the solution space. PCGA excludes any explicit mechanism to preserve diversity and allows a natural selection process to maintain diversity. Thus multiple, equally good solutions to the problem, are provided. PCGA assesses convergence to the Pareto-front which, by definition, is unknown in most real search problems, by use of rank-histograms.

We select topological design of communication network as a sample network problem domain. We present a novel approach to design a network with two minimization objectives of cost and delay subject to satisfaction of reliability and flow constraints. (In the past, EAs have been extensively used in *single* objective optimization for various communication network related design problems - we give a brief survey of such work in the next section.) The remainder of the paper is organized as follows. In section 2, we present the related work done for communication network design problem. We describe, in section 3, a suitable model for the representation of a communication network and its implementation. Then, we present results in section 4 along with a comparison with the conventional methods. Finally, we draw conclusions in section 5.

2 Related Work

Since Network Design Optimization is an NP-hard problem, heuristic techniques have been used widely for such design. Heuristic methods that have been used include techniques, such as branch exchange, cut saturation etc. For example Jan et al. developed a branch and bound based technique to optimize network cost subject to a reliability constraint [18]. Ersoy and Panwar developed a technique for the design of interconnected LAN and MAN networks to optimize average network delay [19]. Clarke and Anandalingam used a heuristic to design minimal cost and reliable network [20]. However, these being heuristics, they do not ensure that the solutions obtained are optimal. Some of these heuristics evaluate trees and thus a large number of possible solutions are left unexplored.

Linear and Integer Programming has been used to a limited extent for network optimization since the number of equations varies exponentially with the number of nodes [21]. Also, greedy randomized search procedures [22] and other meta heuristics have been used for combinatorial optimization.

EAs have been extensively used in single objective optimization for many communication network related optimization problems. For example, Baran and Laufer [23] presented an Asynchronous Team Algorithms (A-Team) implementation, in a parallel heterogeneous asynchronous environment, to optimize the design of reliable communication networks given the set of nodes and possible links. The proposed Team combines parallel GAs, with different reliability calculation approaches in a network of personal computers. Abuali et al. assigned terminal nodes to concentrator sites to minimize costs while considering maximum capacity [24]. Ko et al. used GA for design of mesh networks but the optimization was limited to optimizing the single objective of cost while keeping minimum network delay as a constraint [25]. Elbaum and Sidi used GA to design a LAN with the single objective of minimizing network delay [26]. Kumar et al. used GA for the expansion of computer networks while optimizing the single objective of reliability [27]. White et al. used GA to design Ring Networks optimizing the single objective of network cost [28]. Dengiz et. al [29] presented a EA with specialized encoding, initialization, and local search operators to optimize the design of communication network topologies.

Most approaches attempted to optimize just one objective. For some approaches, the problem is broken down into a number of subproblems, solved in sequence using some heuristics thereby possibly leading to locally optimal design. Ravi et al. [8] and Deo et al. [6] presented approximation algorithm by optimizing one criterion subject to a budget on the other. Since then, many polynomial-time algorithm have been developed for several NP-hard optimization problems arising in network design. Different connectivity requirements such as spanning trees, Steiner trees, generalized Steiner forests, and 2-connected networks have been considered.

However, a practical multiobjective optimization approach should *simultaneously* optimize multiple objectives subject to satisfiability of multiple constraints. In this work, we present a framework using EAs that simultaneously optimize multiple objectives and produces a set of non-dominated *equivalent* solutions that lie on (near-) optimal Pareto- front.

3 Design and Implementation

Problem Definition: Topological design of WANs involves determining the layout of links between nodes given the mean/peak inter node traffic such that certain parameters of the network are optimized. In the solution developed, the total network cost and average delay on links is minimized simultaneously to obtain a Pareto front of optimal non-dominated solutions.

Design Parameters: For design, we use the following network parameters: the total number of nodes in the network N, the distance matrix D_{ij} which gives the physical distance between nodes i and j in kms, the traffic matrix T_{ij} which gives the expected peak network traffic between nodes i and j in packets per second, the number of types of network equipment slabs available K, and the number of types of link slabs available M along with the link cost per unit distance and link capacity.

Objective Functions: We use two objective functions - cost and delay - each of which is approximated by the following formulation:

1. **Cost:**
   $$Cost = Costnodes + Costlinks + Costamps$$
   where,
   $Costnodes = \sum_i C_i;$ C_i = cost of the network equipment placed at node i
   $Costlinks = \sum_i \sum_j C_{ij};$ C_{ij} = cost of the link between node i and node j

   $Costamp = \dfrac{\sum_i \sum_j D_{ij} \times A}{L};$ L = maximum distance for which the signal is sustained without amplification, and A = cost of each amplifier unit.

2. **Average Delay:**
   $$AvgDelay = \frac{\sum_i \sum_j (Delay_{ij} \times LinkFlow_{ij})}{\sum_i \sum_j LinkFlow_{ij}}$$

   $LinkFlow_{ij} = \sum_k \sum_l Traffic_{kl}$ $\forall\, k, l$ nodes in the network such that the route from node k to node l includes the link (i, j). From queuing theory,

   $$Delay_{ij} = \frac{1}{Cap_{ij} - LinkFlow_{ij}}$$

   $Delay_{ij}$ is the link delay for packets flowing through link (i, j), and Cap_{ij} is the capacity of link (i, j). $LinkFlow_{ij}$ and $Delay_{ij}$ are 0 if there is no link between nodes i and j. $AvgDelay$ is ∞ if the network cannot handle the required traffic pattern with the existing capacities of the links and the routing policy adopted.

Constraints: Optimization of cost and delay functions are done subject to the following constraints:

1. **Flow Constraint:** Flow along a link (i, j) should not exceed the capacity of the link. Checking whether the total traffic along a link exceeds the capacity imposes this constraint. If it does, then the network is penalized.
2. **Reliability Constraint:** The network generated has to be reliable. The number of articulation points is a measure of the unreliability of the network. An articulation point of a graph is a vertex whose removal disconnects the graph. The number of articulation points is determined, and this constraint is imposed penalizing the network proportional to their number.

Routing Policy: To calculate the traffic through a particular link the routes between the nodes have to be known so that by superposition principle the total traffic on a link can be calculated. Routing is dynamic in real life and at any point the delays on the various links calculated from the traffic flowing through them gives the best route to be evaluated from the traffic matrix. For solving the design problem at least a rough static route has to be obtained. Dijsktra's shortest path algorithm is used for routing. The metric used for this purpose is the length of the link.

Encoding: In the encoding scheme chosen, every chromosome encodes a possible topology for interconnecting the given nodes; i.e., a chromosome represents a network, which is an individual in a set of potential solutions of the problem. This set of potential solutions constitutes a population. A constant length bit string representation was used to represent the chromosome. The chromosome consists of two portions; the first portion containing details of the network equipments at the nodes and the second portion consisting of details of the links. For instance, if there are T types of nodes, then $\lceil log_2 T \rceil$ bits are needed to encode a node. Thus the first portion of the chromosome consists of $\lceil log_2 T \rceil \times N$ bits. If a link is present between nodes 1 and 2 then the first bit position in the link portion is set to 1. Thus, the second portion of the chromosome consists of $\frac{N \times (N-1)}{2}$ bits. For example, we take 4 bits to encode up to 16 types of nodes. So, the first part of the chromosome contains $4 \times 4 = 16$ bits and the second part of the chromosome contains $\frac{4 \times (4-1)}{2} = 6$ bits.

The capacity of the link is then the first capacity value in the link slab that is greater than the minimum of the capacities of the NE s at the two node ends.

Initial Population: We use hybridization of EAs and conventional algorithm in generating the initial population. The following steps are used to generate the initial population. The network equipments (NE) at the nodes are randomly assigned and maintained in the chromosome. Assuming that the individual is fully connected, a minimal spanning tree is generated using Prim's algorithm. All co-tree links are then removed. A random number of links is then added from the co-tree set to the spanning tree. The number of links added is a random number in between one-third of the total number of links to half of the total number of links. This is done so that the initial population is not limited to spanning trees. This way we adopt a hybrid approach so that the time for exploitation and exploration of the search space is significantly reduced, and the number of lethals produced for large nets is minimized.

Fitness Evaluation: We use Pareto-rank based EA implementation. The Pareto rank [11] of each individual is equal to one more than the number of individuals dominating it in the multiobjective vector space. All the non-dominated individuals are assigned rank one. The values of the two objectives to be minimized (cost and average delay) are used to calculate the rank of the individual. Using the superposition principle the traffic on each individual node is calculated and hence the average delay for the network is calculated. Based on these two objectives the rank of the individual is calculated. In this work, we calculate fitness of an individual by $Fitness = \frac{1}{(Rank)^2}$.

Other Genetic Operators: We use Roulette wheel selection for selecting the parents. We divide chromosome in two parts for crossover. In the first part of the chromosome, initially the crossover point would lie at any position in the chromosome irrespective of the boundaries of the bits encoding. Node type values are not preserved to ensure maximum exploration. As the algorithm proceeds the probability of getting a crossover point within a node's NE boundary in the chromosome is constantly reduced so as to exploit the collected experience regarding optimal values of NE types so far. In this case only the existing NE types in the parents can be present in the children. In the link portion of the chromosome, since a single bit is used to code the presence or absence of the link, such considerations regarding tradeoff between exploration and exploitation do not arise. As a result, the crossover point is purely random. We use multi-point crossover; the number of crossover points depends on the problem-size. We use a simple bit-flipping mutation to further increase the exploration of the solution space.

Presence of Unconnected Components: As a result of the crossover and mutation operations, unconnected networks are generated as offspring. We do not completely eliminate the unconnected networks from further consideration. We maintain a pool of unconnected networks. This may give rise to fitter and connected components after further evolutions. This approach of maintaining unconnected, unfit individuals separately in the population is in accordance with the philosophy that unfit individuals can produce fit children.

Ensuring Convergence: For this we compute Intra-Island Rank-Histogram for each epoch of the genetic evolution and monitor the movement of Pareto-front. Since, this is a *hard* problem, it is likely that the problem may get trapped in local optima. To ensure a global (near-) optimal Pareto-front, we use a multi-island approach and monitors the Pareto-front using Inter-Island Rank histogram. The computation of Rank-histogram is analogous to that given in [16].

4 Results

We collected data of mass communication networks of different cities to carry out the simulation. We used the data which was used by the researchers in their previous work. We tested the algorithm for networks with up to 36 nodes, and convergence to an optimal Pareto front was observed. We conducted the experiments with many sets of random populations, and analyzed many sets of results. We also compared results with those obtained from other approaches namely exhaustive search and the Branch Exchange heuristic. In the following subsections, we include a few representative results.

Network of 10 Chinese Cities: The GA was run for the same problem as solved by Ko et al. [25]. In brief, the problem consisted of designing a packet switched mesh communication network among 10 major Chinese cities with realistic topology and traffic requirements. The design assumed a cost structure proportional to the distance among nodes and accounted for three different line rates: 6, 45 and 150 Mbps.

For a set of initial population of size 100, the solution space was found to improve very quickly up to the 40th epoch. Then the improvement was marginal. We carried the evolutions up to the 100th epoch. The rate of improvement was observed to be very slow; this was monitored by a rank-ratio histogram [16]. We include the initial population and the population at 60th epoch in Fig. 1.

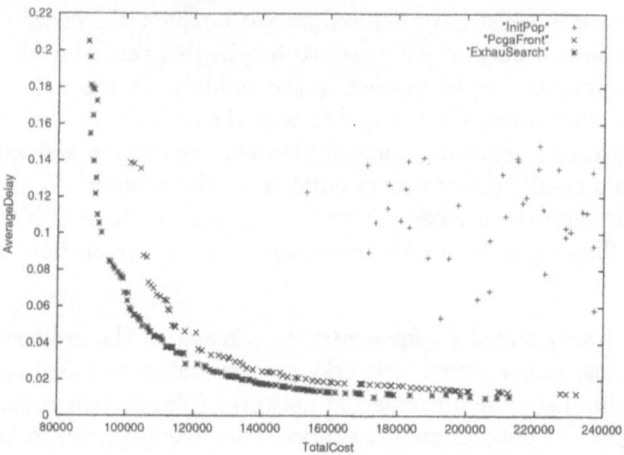

Fig. 1. 10 node network : (a) Initial population, (b) the converging Pareto-front obtained from EA and (c) the optimal Pareto-front obtained from exhaustive search.

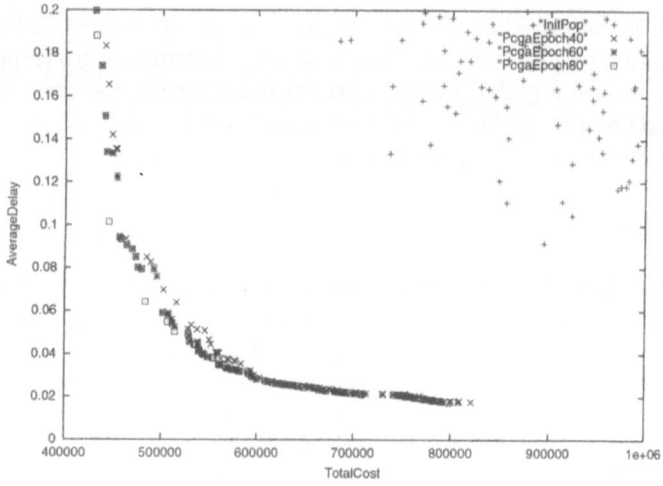

Fig. 2. A converging Pareto-front for a 21-node network. The population of size 100 is converging slowly to the Pareto-front during the later stages of evolution.

Network of 21 US Cities: Next, we tested our algorithm on a problem with larger number of nodes. This is a more complex than the earlier problem, so the improvement with epochs was slower.

Figure 2 shows the initial population and the non-dominated points obtained at epochs 40, 60 and 80. As seen from the plots, the movement of the Pareto Front is very-very marginal after the 40th epochs. However, a few new solutions were being added to the Pareto-front with evolutions in low-cost and high-delay region. We observed that finding unformly distributed diverse solution in this non-linear region was a difficult task. However, we obtained diversity in this region by running EA for longer epochs. Alternately, this could be done by adopting the multi-island approach and by assessing convergence using inter-island rank histogram [16].

Network of 36 European Cities: Finally, we ran EA for a problem with 36 nodes. This is much more complex than the previous two problems. It took much more computational resources; we started with a population of 250 size and we could get nearly converged solution space at some 60th epochs. This is shown in Fig. 3. The improvement was significant but with slower rate. The behavior of the population dynamics was quite analogous to the earlier results obtained with smaller and medium sized networks.

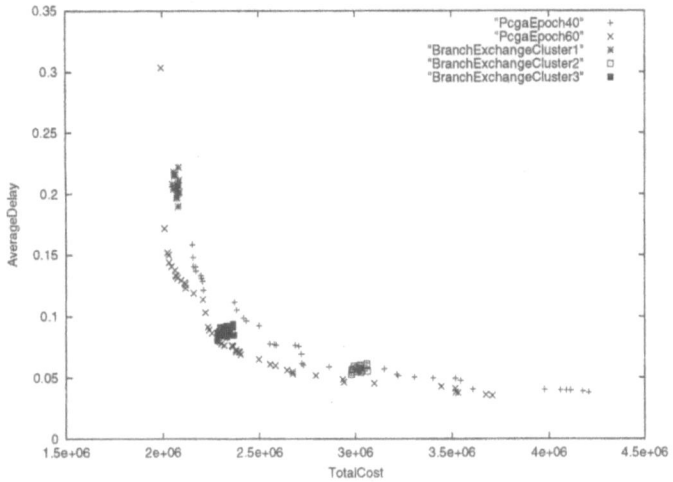

Fig. 3. A 36-node network: Comparison of solutions obtained by EA with the clusters obtained from Branch Exchange Heuristics.

Comparison: In order to show the relative merits of the EA approach we compared the results with those obtained by two of the commonly used conventional methods, namely, Exhaustive Search and Branch Exchange Heuristic.

An exhaustive search was done for all possible networks of size $N = 10$ nodes. All the possible networks were generated and evaluated, then all the non-dominated solutions in a given range were plotted against the results obtained with the Genetic Algorithm in Fig. 1. Since the problem is NP hard, the exhaustive search is of exponential complexity and is completely unfit for networks with more than 10 nodes. The complexity of the exhaustive search was found to be $O(2^{N^2})$. This is because there are $2^{{}^N C_2}$ graphs possible with N nodes. The deterministic solution is slightly superior to the results obtained by EA. This is expected because the deterministic Algorithm exhaustively searches all possible topologies. But it is also observed that the difference between the results obtained by Exhaustive Search and Genetic Algorithm is quite close. This gap is specific to a solution space which was obtained by a random sampling of the initial population; secondly this was not run to a *total* convergence. A multi-island approach as suggested in [16] is one of the possible solution to obtain a *superior* convergence. This is an area of further investigation.

The price paid for this marginal improvement obtained by Exhaustive Search over a single island Genetic Algorithm is the computation time involved. It was observed that for $N = 10$ node network the Exhaustive Search took more than 10 hours on a typical Intel Pentium P-IV, 1.7 MHz machine, whereas, the GA took a couple of minutes only. We could not compute the results for $N > 10$ nodes because of the exponential nature of the problem.

Another conventional method widely used for network optimization problems is Branch Exchange Heuristic. Many authors have used this heuristics to compare the results obtained by their algorithms. So we also compare the results obtained by EA with the clusters obtained by the Branch Exchange method. Here, we use an ϵ-constraint Branch Exchange to extend its use to multiobjective optimization. Different constraints have been put on any one objective function to obtain the solution in different regions of the Pareto-front. A few clusters are depicted in Fig. 3. As observed in Fig. 3 the results obtained by the Branch Exchange algorithm are comparable to a subset of the solutions obtained by EA but the diversity of the branch exchange is much less compared to that of EA. This is due to the fact that the branch exchange method considers only those network topologies that are spanning trees. However, such heuristics are unable to obtain most regions of the Pareto-front. This is a distinct advantage of EA in solving such *hard* problem.

5 Discussion and Conclusions

In this work, we demonstrated the solution of optimizing topologies of communication networks subject to their satisfying the twin objectives of minimum cost and delay along with two constraints. The solution to the network design problem is a set of optimal network topologies that are non-inferior with respect to each other. The multiple objectives to be optimized have not been combined into one and hence the general nature of the solution is maintained. These topologies are reliable in case of single link failures and it is guaranteed that the maximum packet load on any link will not exceed the link capacity. Thus the network is two edge connected and satisfies the constraints.

The algorithm has been test run on small as well as large networks. The initial population used in EA was taken from some hybridization of spanning tree and random topologies. The initial population and the final front were located far apart (Figs. 1 and 2). As a result much of the optimization was done by EA.

In most optimization problems like network design, it is crucial for the final solution-space to be diverse. As is observed from the results, EA achieves greater diversity in *polynomial time* as compared to other methods considered. A network designer having a range of network cost and packet delay in mind, can examine several optimal topologies simultaneously and choose one based on these requirements and other engineering considerations. The solutions obtained by traditional approaches do not show diversity. This is the primary advantage of using Pareto-rank based techniques to solve multiobjective optimization problems of such a hard nature.

Acknowledgements. This research is supported by Ministry of Human Resource Development(MHRD), Government of India project grant.

References

1. M. R. Garey and D. S. Johnson. *Computers and Interactability: A Guide to the Theory of NP-Completeness*, 1979. San Francisco, LA: Freeman.
2. D. Hochbaum (Ed.). *Approximation Algorithms for NP-Hard problems*, 1997. Boston, MA: PWS.
3. M. Gerla and L. Kleinrock. On the topological design of distributed computer networks. *IEEE Trans. Communications*, 25(1): 48–60, 1977.
4. V. P. Kompella, J. C. Pasquale, and G. C. Polyzos. Multicast routing for multimedia communication. *IEEE/ACM Trans. Networking*, 286–292, 1993.
5. M. Borah, R. M. Owens, and M. J. Irwin. An edge-based heuristic for Steiner routing. *IEEE Trans. Computer Aided Design of Integrated Circuits and Systems*, 13(12): 1563–1568, 1995.
6. N. Boldon, N. Deo, and N. Kumar. Minimum-weight degree-constrained spanning tree problem: Heuristics and implementation on an SIMD parallel machine. *Parallel Computing*, 22(3): 369–382, 1996.
7. M. V. Marathe, R. Ravi, R. Sundaram, S. S. Ravi, D. J. Rosenkrantz, and H. B. Hunt. Bicriteria network design problems. *J. Algorithms*, 28(1): 142–171, 1998.
8. R. Ravi, M. V. Marathe, S. S. Ravi, D. J. Rosenkrantz, and H. B. Hunt. Approximation algorithms for degree-constrained minimum-cost network design problems. *Algorithmica*, 31(1): 58–78, 2001.
9. C. A. C. Coello, D. A. Van Veldhuizen, and G. B. Lamont. *Evolutionary Algorithms for Solving Multi-Objective Problems*, 2002. Boston, MA: Kluwer.
10. K. Deb. *Multiobjective Optimization Using Evolutionary Algorithms*, 2001. Chichester, UK: Wiley.
11. C. M. Fonseca and P. J. Fleming. Multiobjective optimization and multiple constraint handling with evolutionary algorithms – Part I: a unified formulation. *IEEE Transactions on Systems, Man and Cybernetics-Part A: Systems and Humans*, 28(1): 26–37, 1998. 26–37.
12. K. Deb et al. A fast non-dominated sorting genetic algorithm for multiobjective optimization: NSGA-II. *Parallel Problem Solving from Nature*, PPSN-VI: 849–858, 2000.
13. E. Zitzler, M. Laumanns and and L. Thiele. SPEA2: Improving the strength Pareto evolutionary algorithm. *EUROGEN* 2001.
14. Knowles, J. D. and Corne, D. W. Approximating. *Evolutionary Computation*, 8(2): 149–172, 2000.

15. M. Laumanns, L. Thiele, K. Deo and E. Zitzler. Combining convergence and diversity in evolutionary multiobjective optimization. *Evolutionary Computation*, 10(3): 263–182, 2002.
16. R. Kumar and P. I. Rockett. Improved sampling of the Pareto-front in multiobjective genetic optimizations by steady-state evolution : a Pareto converging genetic algorithm. *Evolutionary Computation*, 10(3): 283–314, 2002.
17. R. C. Purshouse and P. J. Fleming. Elitism, sharing and ranking choices in evolutionary multi-criterion optimization. Research Report No. 815, Dept. Automatic Control & Systems Engineering, University of Sheffield, Jan. 2002.
18. R. H. Jan, F. J. Hwang, and S. T. Cheng. Topological optimization of a communication network subject to a reliability constraint. *IEEE Trans. Reliability*, 42(1): 63–69, 1993.
19. C. Ersoy and S. S. Panwar. Topological design of interconnected LAN/MAN Networks. *IEEE J. Select. Areas Communication*, 11(8): 1172–1182, 1993.
20. L. W. Clarke and G. Anandalingam. An integrated system for designing minimum cost survivable telecommunication networks. *IEEE. Trans. Systems, Man and Cybernetics- Part A*, 26(6): 856–862, 1996.
21. A. Atamturk and D. Rajan. Survivable network design: simultaneous routing of flows and slacks. Research Report, IEOR, University of California at Berkeley.
22. T. A. Feo and M. G. C. Resende. Greedy randomized adaptive search procedures. *Journal of Global Optimization*, 1995.
23. B. Baran and F. Laufer. Topological optimization of reliable networks using A-Teams. National Computer Center, National University of Asuncion, University Campus of San Lorenzo – Paraguay.
24. F. N. Abuali, D. A. Schnoenefeld, and R. L. Wainwright. Designing telecommunication networks using genetic algorithms and probabilistic minimum spanning Trees. In *Proc. 1994 ACM Symp. Applied Computing*, pp. 242–246, 1994.
25. K. T. Ko, K. S. Tang, C.Y. Chan, K. F. Man and S. Kwong. Using genetic algorithms to design mesh networks. *IEEE Computer*, 6–58, 1997.
26. R. Elbaum and M. Sidi. Topological design of local-area networks using genetic algorithms. *IEEE/ACM Trans. Networking*, 4(5): 766–777, 1996.
27. A. Kumar, R. M. Pathak, and Y.P. Gupta. Genetic-algorithm based reliability optimization for computer network expansion. *IEEE Trans. Reliability*, 44(1): 63–72, 1995.
28. A. R. P White, J. W. Mann, and G. D. Smith. Genetic algorithms and network ring design. *Annals of Operational Research*, 86: 347–371, 1999.
29. B. Dengiz, F. Altiparmak, and A. E. Smith. Local search genetic algorithm for optimal design of reliable networks. *IEEE Trans. Evolutionary Computation*, 1(3): 179–188, 1997.

Control of a Flexible Manipulator Using a Sliding Mode Controller with Genetic Algorithm Tuned Manipulator Dimension

N.M. Kwok[1] and S. Kwong[2]

[1] Faculty of Engineering, University of Technology, Sydney, NSW, Australia
ngai.kwok@eng.uts.edu.au
[2] Department of Computer Science, City University of Hong Kong
Hong Kong SAR, P.R. China
cssamk@cityu.edu.hk

Abstract. The tip position control of a single-link flexible manipulator is considered in this paper. The cross-sectional dimension of the manipulator is tuned by genetic algorithms such that the first vibration mode dominates the higher order vibrations. A sliding mode controller of reduced dimension is employed to control the manipulator using the rigid and vibration measurements as feedback. Control effort is shared dynamically between the rigid mode tracking and vibration suppression. Performance and effectiveness of the proposed reduced dimension sliding mode controller in vibration suppression and against payload variations are demonstrated with simulations.

1 Introduction

Flexible manipulators are robotic manipulators made of light-weight materials, e.g. aluminium. Because of the reduction in weight, lower power consumption and faster movement can be realised. Other advantages include the ease of setting up and transportation, as well as reduced impact destruction should the manipulator system went faulty. Light-weight manipulators can be applied in general industrial processes, e.g. pick and place, other applications could include the space shuttle on-board robotic arm. However, due to its light-weight, vibrations are inherent in the flexible manipulator that hinders its wide application.

Earlier research on the control of flexible manipulators could be found in [1] where a state-space model of the manipulator was proposed and the linear quadratic gaussian approach was used. Other research work included the use of an H_∞ controller in [2], the manipulator considered allow movements in both the horizontal and vertical planes. The use of deterministic control was found in [3], where a sliding mode controller was applied. The control effort was switched between the vibration mode errors, however, in an ad-hoc weighting basis. Recently, in [4], a sliding mode controller with reduced controller dimension, using only the rigid mode and first vibration mode as feedback, was demonstrated. The sliding surfaces were made adaptive to enhance the controller performance.

However, a relatively large number of switching in controller gains had to be incorporated in the controller. Apart from these advanced controller designs, intelligent controllers were also reported in [5], where a model-free fuzzy controller was used to control the flexible manipulator. In [6], a flexible manipulator was controlled by a neural network controller, trained with genetic learning. Genetic tuning of a Lyapunov based controller was also reported in [7], where stability was guaranteed by the Lyapunov design. Moreover, the controller performance was enhanced by genetic algorithm tuning. The work in [8] also showed that a sliding mode controller retaining only the rigid mode and first vibration mode as feedback was feasible. The approach adopted was that the control on the tip position and vibrations were dynamically weighted in a fuzzy-like manner. Diverging from the focus on controller designs, structural dimensions of the flexible manipulator was considered in [9], where a tapered beam resulted and the vibration frequency was increased for faster manoeuvre.

In this work, we consider the tip angular positional control of a single-link flexible manipulator. The manipulator is constructed in the form of a narrow beam of rectangular cross-section. At the hub is driven by a dc motor and the tip carries a payload. We use a sliding mode controller for its robustness against model uncertainties and parameter variations, e.g. payload variations. An inspection on the manipulator mathematical model will reflect that the manipulator is of infinite dimension, and the design and implementation of an infinite dimension controller is challenging. Here, we follow the work of [4] and [8], that only the rigid mode and first vibration mode are used as feedback and the controller dimension is therefore reduced. When using only the first vibration mode as one of the feedback, we need to ensure that it dominates all other higher order vibrations. Motivated by the work in [9], we propose to adjust the cross sectional dimension of the manipulator. In addition to the above requirement, we also aim to increase the vibration frequency for reduced vibration magnitude, while rejecting a large cross section area that increases the weight of the manipulator. We also see from the manipulator model that vibrations are proportional to the slope of the mode shapes at the hub. Therefore, in order to reduce vibration as a whole, we penalise large mode shape slopes. The above arguments lead to the formulation of a multi-objective optimisation problem. From [10], we propose to use the genetic algorithm to search for the optimal manipulator cross sectional dimension. The resultant manipulator characteristics will facilitate the design and implementation of a sliding mode controller of reduced dimension.

This paper is arranged as follows. The manipulator model will be developed in Section 2. Section 3 will present the sliding mode controller design. The genetic tuning of manipulator dimensions will be treated in Section 4. Simulation results will be presented in Section 5 and a conclusion will be drawn in Section 6.

2 System Modelling

The flexible manipulator considered here is a uniform aluminium beam with rectangular cross section and moves in the horizontal plane. One end of the

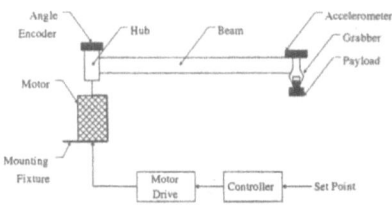
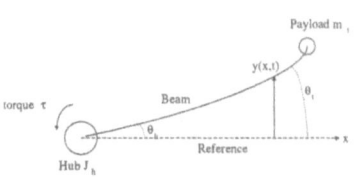

Fig. 1. System Configuration 　　　　**Fig. 2.** Parameter Definition

beam is fixed to the hub consisting of a dc motor and associated mounting fixtures. At the other end, the tip, is mounted a gribber for pick and place operation. The gribber and the load together form the payload. Angular sensors, e.g. accelerometers and angular encoders, are mounted at the tip and the motor shaft respectively. The manipulator set-up is shown in Fig. 1. The definitions of the manipulator parameters are shown in Fig. 2. Using the Eular-Bernoulli beam theory, we assume that shear and rotary inertia are negligible and vibrations being small as compared to the length of the manipulator, the equation of motion of the manipulator can be given as [2]

$$EI\frac{\partial^4 y}{\partial x^4} + \rho \frac{\partial^2 y}{\partial t^2} = 0 \qquad (1)$$

where E=Young's modulus, I=cross sectional moment of inertia, ρ=mass per unit length, y=displacement from reference
The boundary conditions are

$$y(0) = 0, \ EIy''(0) = \tau + J_h \ddot{\theta}_h, \ EIy'''(L) = m_t \ddot{y}(L), \ EIy''(L) = -J_t \ddot{y}'(L) \qquad (2)$$

where τ=motor torque, J_h=hub inertia, θ_h=hub angle, m_t=payload mass, J_t=payload inertia, L=manipulator length, *prime* stands for differentiation against x, *dot* stands for differentiation against time t
Using separation of variables, put

$$y(x,t) = \phi(x)\varphi(t) \qquad (3)$$

then we have

$$\ddot{\varphi} + \omega^2 \varphi = 0, \ \phi'''' - \beta^4 \phi = 0, \ \omega^2 = \beta^4 \frac{EI}{\rho} \qquad (4)$$

The general solution to Equ.4 is the mode shape given by

$$\phi(x) = A\sin\beta x + B\cos\beta x + C\sinh\beta x + D\cosh\beta x \qquad (5)$$

where β is the solution of the characteristic equation formed from the boundary conditions
Equ.5 can be expressed with coefficient A only, giving

$$\phi(x) = A(\sin\beta x + \gamma(\cos\beta x - \cosh\beta x) + \xi\sinh\beta x) \qquad (6)$$

where γ and ξ are functions of β and system parameters, and A is yet to be determined by normalisation

Using the assumed mode method, put

$$y(x,t) = \sum_{i=0}^{\infty} \phi_i(t)\varphi_i(t) = x\theta_h + \sum_{i=1}^{\infty} \phi_i(x)\varphi_i(t) \qquad (7)$$

The normalization coefficient A_i of the general solution is

$$A_i = \sqrt{\frac{J_{total}}{\int_0^L \rho\psi_i^2(x)dx + m_t\psi_i^2(L) + J_t\psi_i'^2(L) + J_h\psi_i'^2(0)}} \qquad (8)$$

where ψ_i is the expression inside the bracket in Equ.6

After normalisation using the orthogonal property and keeping consistence with the rigid mode, we have

$$\ddot{\varphi}_i + 2\zeta\omega_i\dot{\varphi}_i + \omega^2\varphi_i = \frac{\tau}{J_{total}}\phi_i'(0) \qquad (9)$$

where ζ is the material damping of small value, $\phi_i'(0)$ is the slope of the ith vibration mode shape at the hub, J_{total} is the total inertia making up of the hub, beam and the tip

With further manipulations, a state space system equation can now be written as

$$\begin{bmatrix} \dot{\theta}_h \\ \ddot{\theta}_h \\ \dot{\varphi}_1 \\ \ddot{\varphi}_1 \\ \vdots \\ \dot{\varphi}_n \\ \ddot{\varphi}_n \end{bmatrix} = \begin{bmatrix} 0 & 1 & . & . & . & . & . \\ 0 & 0 & . & . & . & . & . \\ . & . & 0 & 1 & . & . & . \\ . & . & -2\zeta\omega_1 & -\omega_1^2 & . & . & . \\ . & . & . & . & \ddots & . & . \\ . & . & . & . & . & 0 & 1 \\ . & . & . & . & . & -2\zeta\omega_n & -\omega_n^2 \end{bmatrix} \begin{bmatrix} \theta_h \\ \dot{\theta}_h \\ \varphi_1 \\ \dot{\varphi}_1 \\ \vdots \\ \varphi_n \\ \dot{\varphi}_n \end{bmatrix} + \begin{bmatrix} 0 \\ 1 \\ 0 \\ \phi_1'(0) \\ \vdots \\ 0 \\ \phi_n'(0) \end{bmatrix} \frac{\tau}{J_{total}} \qquad (10)$$

The motor dynamic equation is

$$J_h\ddot{\theta}_h = \tau = \frac{1}{R}k_t(V - k_b\dot{\theta}_h), \quad V = \frac{1}{k_t}R\tau + k_b\dot{\theta}_h \qquad (11)$$

where J_h is the hub inertia, τ is the torque generated by the motor, k_t is the torque constant, V is the voltage applied to the motor, k_b is the back-emf constant, R is the armature resistance

Let the controller generate the required driving torque τ, then we will use inverse dynamics to compute the motor voltage V. It is because the dc motor model has been well studied and its identification in real practice is not difficult. Finally, we will apply the derived control voltage to the motor to drive the flexible manipulator.

3 Controller Design

We will design the controller using only the rigid mode and first vibration mode as feedback signals. From the state space equation, Equ.10, we re-write the system differential equations

$$\ddot{\theta} = \frac{T_1}{J_{total}}, \quad \ddot{\varphi} = -2\zeta\omega\dot{\varphi} - \omega^2\varphi + \frac{T_2}{J_{total}}\phi' \qquad (12)$$

Note that we have dropped all subscripts for clarity. The control objective is to make

$$\theta \to \theta_d \quad \text{and} \quad \varphi \to 0 \quad \text{for} \quad t \to \infty$$

where θ_d is the desired hub angle
We then define the error variables as

$$\theta_e = \theta_d - \theta, \ \varphi_e = -\varphi \qquad (13)$$

The system differential equations become

$$\ddot{\theta}_e = \ddot{\theta}_d - \frac{T_1}{J_{total}}, \quad \ddot{\varphi}_e = -2\zeta\omega\dot{\varphi}_e - \omega^2\varphi_e - \frac{T_2}{J_{total}}\phi' \qquad (14)$$

Following the sliding mode design methods, we define the sliding surface as

$$s_1 = c_1\theta_e + \dot{\theta}_e \qquad (15)$$

where c_1 is the slope of the sliding surface
When sliding mode is attained, we have

$$\theta_e(t) = \theta_e(t_s)e^{-c_1 t} \quad for \quad t > t_s$$

where t_s is the time that sliding mode firstly occurred and $c_1 > 0$ then

$$\theta_e(t) \to 0 \quad as \quad t \to \infty$$

The sliding surface time derivative is

$$\dot{s}_1 = c_1\dot{\theta}_e + \ddot{\theta}_e = c_1\dot{\theta}_e + \ddot{\theta}_d - \frac{T_1}{J_{total}} \qquad (16)$$

Put

$$T_1 = (c_1\dot{\theta}_e + \ddot{\theta}_d + j_1 sign(s_1))J_{total} \qquad (17)$$

where $sign(x) = 1$ for $x > 0$ and $sign(x) = -1$ otherwise, and the sliding surface s_1 can be reached for $j_1 > 0$
Similarly, define the sliding surface s_2 as

$$s_2 = c_2\varphi_e + \dot{\varphi}_e \qquad (18)$$

The time derivative is

$$\dot{s}_2 = c_2\dot{\varphi}_e + \ddot{\varphi}_e = (c_2 - 2\zeta\omega)\dot{\varphi}_e - \omega^2\varphi_e - \frac{T_2}{J_{total}}\phi' \qquad (19)$$

Put

$$T_2 = ((c_2 - 2\zeta\omega)\dot{\varphi}_e - \omega^2\varphi_e + j_2 sign(s_2))\frac{J_{total}}{\phi'} \qquad (20)$$

where sliding surface s_2 can also be reached when $j_2 > 0$
Let the sliding mode controller output be

$$u_1 = \gamma_1 + k_1 sign(s_1), \quad u_2 = \gamma_2 + k_2 sign(s_2) \qquad (21)$$

where

$$\begin{aligned}\gamma_1 &= (c_1\dot{\theta}_e + \ddot{\theta}_d)J_{total}, \quad k_1 = j_1 J_{total} \\ \gamma_2 &= ((c_2 - 2\zeta\omega)\dot{\varphi}_e - \omega^2\varphi_e)\frac{J_{total}}{\phi'}, \quad k_2 = j_2\frac{J_{total}}{T_2}\end{aligned} \qquad (22)$$

Note that there is only one control torque from the motor but there are two control objectives; namely, $\theta \to \theta_d$ and $\varphi \to 0$. We have to assign weighting factors n_1 and n_2 applying to u_1 and u_2 respectively.

$$u = n_1 u_1 + n_2 u_2 = n_1\gamma_1 + n_1 k_1 sign(s_1) + n_2\gamma_2 + n_2 k_2 sign(s_2) \qquad (23)$$

In order to ensure that sliding surfaces are reached, that is, to make $s_1 \to 0$, $s_2 \to 0$; we also have to assign the values of k_1 and k_2. In so doing, we define a Lyapunov function

$$V = \frac{1}{2}(s_1^2 + s_2^2) > 0 \quad \text{for all} \quad s_1 \neq 0, s_2 \neq 0 \qquad (24)$$

The time derivative is required to be negative for reachability, that is

$$\dot{V} = s_1\dot{s}_1 + s_2\dot{s}_2 < 0 \qquad (25)$$

When we apply the weighted and aggregated control, Equ.23, to the plant, Equ.25 becomes

$$\begin{aligned}\dot{V} = &s_1(n_2(\gamma_1 - \gamma_2) - n_1 k_1 sign(s_1) - n_2 k_2 sign(s_2)) + \\ &s_2(n_1(\gamma_2 - \gamma_1) - n_1 k_1 sign(s_1) - n_2 k_2 sign(s_2))\end{aligned} \qquad (26)$$

From Equ.26, we see that there are two sliding surfaces s_1 and s_2 being interrelated to each other. Their combination will affect the Lyapunov function derivative. It was shown in [8] that the effect will be reduced if the values of k_1 and k_2 are given by
for $s_1 > 0$ and $s_2 > 0$

$$k_1 = max(\gamma_2 - \gamma_1, 0) + \epsilon_1, \quad k_2 = max(\gamma_1 - \gamma_2, 0) + \epsilon_2 \quad (27)$$

for $s_1 < 0$ and $s_2 < 0$

$$k_1 = max(\gamma_1 - \gamma_2, 0) + \epsilon_1, \quad k_2 = max(\gamma_2 - \gamma_1, 0) + \epsilon_2 \quad (28)$$

for $s_1 > 0$ and $s_2 < 0$

$$k_1 = max(\frac{n_2(\gamma_1 - \gamma_2)}{n_1}, 0) + \epsilon_1, \quad k_2 = 0 \quad (29)$$

for $s_1 < 0$ and $s_2 > 0$

$$k_1 = max(\frac{n_2(\gamma_2 - \gamma_1)}{n_1}, 0) + \epsilon_1, \quad k_2 = 0 \quad (30)$$

where ϵ_1 and ϵ_2 are small positive constants to compensate for model uncertainties and parameter variations
Regarding the sharing of control effort between regulation and vibration suppression, we state our control strategy in the following rule.

A larger control effect is to be applied for hub angle regulation when the hub angle error is large; when the hub angle is near the set point (small error), apply larger control effort to suppress the vibration mode.

We also observe from Equ.29 and Equ.30; that k_1 may go unbounded if $n_1 \to 0$. Therefore, we put

$$n_1 > n_2 \quad \text{with} \quad n_1 + n_2 = 1 \quad (31)$$

We also assign n_1 according to the control strategy in the rule stated above. Put

$$n_1 = \frac{1}{2}\left(1 + \frac{1}{1 + exp(-a(\left|\frac{\theta_e}{\theta_d}\right| - b))}\right), \quad n_2 = 1 - n_1 \quad (32)$$

where θ_e is the hub angle error, θ_d is the desired set point, a determines the slope of weight crossover, b determines the point where crossover in control weighting is to occur

4 Manipulator Dimension Tuning by Genetic Algorithm

The reduced dimension sliding mode controller developed in Section 3 depends critically on the fact that the first vibration mode magnitude dominates the higher order vibrations. When the tip displacement is given by Equ.7, what we can do to make y_1 dominates y_i, $i > 1$, is to make ϕ_1 and φ_1 to dominate the corresponding higher order variables. Referring to Equ.6, we see that the value of ϕ_i depends on the value of A_i and β_i. Also from Equ.10, the vibrations will be excited according to the gain $\phi'_i(0)$. We also see that in normalising the variable A_i in Equ.8, it is a function of the variable β_i. To sum up, the variable β_i plays a critical role in satisfying our requirements in making the first vibration mode dominate.

From a practical point of view, the hub parameters and the tip parameters are relatively fixed by the motor torque required and the task assigned. The length of the manipulator is also fixed by the workspace. A relatively free parameter, which bears an effect on β_i is the cross sectional area of the manipulator. Adjusting the height and width of the manipulator, we change the value of mass per unit length, cross sectional moment of inertia as well as the beam inertia. The effect of the manipulator dimension on β_i and in turn on ϕ_i is very non-linear. From the equation of motion in Equ.1 and boundary conditions in Equ.2, the determination of β_i in closed form analytical solution is very involved. Therefore, we turn to the use of genetic algorithms to search for the manipulator dimension that best satisfies our requirements.

Genetic algorithms are stochastic search methods based on the theory of evolution, the survival of the fittest, and the theory of schema [10]. When implementing the genetic algorithm, variables to be searched are coded in binary strings. For two such strings, parents, the binary bit patterns are exchanged in some bit position selected randomly. The resultant pair of strings, off-springs, become members of the population in the next generation. The exchange of binary bits, crossover, is conducted according to a crossover probability p_c. The parents with higher fitness have a higher chance for crossover. The above process exploits the search space but diversity should be explored. This is made possible by applying mutation to the off-springs. Mutation flips one of the bit in the string according to the probability p_m.

For every string in the population, fitness is evaluated and then the process repeats in the next generation. Termination criteria may be selected from tracking the improvement over the average fitness of a generation or according to the count of number of generations being processed.

Variations from the standard genetic algorithms are adopted here in tuning the manipulator dimensions. We will apply elitism, such that the string with the highest fitness value is stored irrespective of the processing of generations. It is because the last generation before termination of the algorithm may or may not contain the best string. We also adjust the crossover and mutation probability dynamically according to the improvement on the average fitness of a generation. If found improving, then p_c and p_m are reduced by a scale factor. While the average fitness is not improving, we increase p_c and p_m to gain more chances in finding a string of higher fitness. Our algorithm terminates when the average fitness is improving for several consecutive generations. The genetic algorithm becomes that described below.

```
initialise the population randomly
evaluate fitness and average
while not end of algorithm
   select parents by fitness
```

```
apply crossover and mutation
evaluate fitness and average
if improving then decrease pc and pm
else increase pc and pm
store the fittest string
end on fitness improved or generations count
```

The fitness function is defined as

$$fitness = f_r + f_1 + f_2 - f_3 - f_4 - f_5 + f_6 \qquad (33)$$

where f_r =reference datum keeping fitness> 0, $f_1 = 1$ when $\phi_i(L)$ are in descending order $f_1 = 0$ otherwise, $f_2 = \sum(\phi_i(L) - \phi_{i+1}(L))$ reward greater dominance of lower order vibrations, $f_3 = \sum |i \times \phi_i(L)|$ penalise weighted magnitude of vibrations at tip, $f_4 = \sum \phi_i'(0)$ penalise mode shape slopes at the hub, $f_5 = H_b W_b$ penalise for large cross-sectional beam area, $f_6 = \sum \omega_i$ reward higher vibration frequencies

5 Simulation

Simulations will be described in this section as well as the presentation of simulation results. Cases for comparisons include responses from open-loop and from the use of the proposed sliding mode controller. We will also present cases for initially un-tuned and the resulting tuned manipulator responses. The parameters used in the simulations are given below in Table 1.

Table 1. Simulation Parameters

beam initial height/width/length	0.04m/0.002m/0.8m
hub inertia	$1.06^{-4} kgm^2$
payload mass/inertia	$0.38kg/1.78^{-4} kgm^2$
no. of population/generation	10/20
probability p_c/p_m (initial)	0.9/0.02
scaling factor for p_c and p_m	1.1
termination on improvement	3 generations

5.1 Response with Un-tuned Dimensions

Using the initial manipulator parameters, an open-loop simulation was conducted for subsequent performance comparisons. The motor was fed with a voltage pulse, by some trial and error that brings the response to 0.2rad as the set-point for close-loop control. The result is shown in Fig. 3. The close-loop response using the sliding mode controller is shown in Fig. 4, where initial un-tuned dimensions were used. Comparing Figs. 3 and 4, we see that vibrations are effectively suppressed in the steady state with the close-loop control. However, vibration during the transient period is still observed.

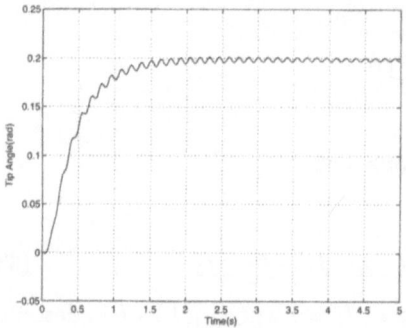
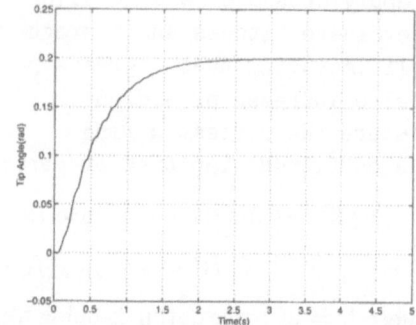

Fig. 3. Open-loop response, dimensions not tuned

Fig. 4. Close-loop response, dimensions not tuned

5.2 Genetic Algorithm Tuning

Tuning of the manipulator height and width was conducted according to the genetic algorithm described in Section 4. It is noted that the algorithm terminated before reaching the maximum number of generations with a relatively small number of populations. This observation shows the modification on the standard genetic algorithms with dynamic adjustment of the crossover and mutation probabilities were effective. Figures 5 and 6 shows the value of fitness function and the average fitness over the generations. It is also observed from the plots that the maximum fitness function though not increasing explicitly, the occurrences of low fitness were decreasing.

Results of mode shapes at tip, $\phi_i(L)$, mode shape slopes at the hub, ϕ'_i, and the vibration frequencies, ω_i, are tabulated in Table 2 for comparison between the initially un-tuned and the resulting tuned values. From the results, we see that the mode shape at the tip is arranged in descending order as required. The vibration frequency shows an increase from the tuned dimensions. An overall

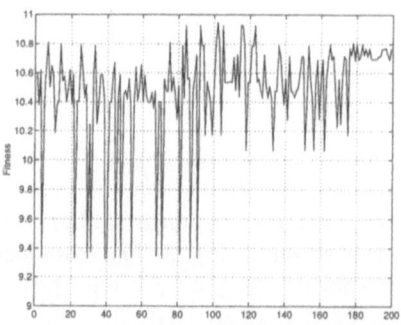
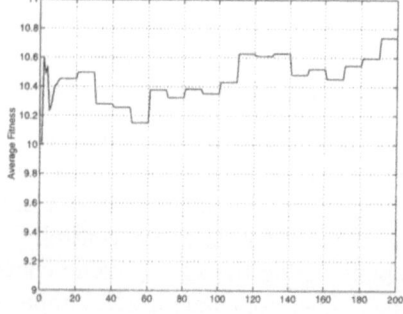

Fig. 5. Fitness Function

Fig. 6. Average fitness function

Table 2. Comparison of design variables resulted from initially un-tuned and tuned manipulator dimension

Results from un-tuned dimensions				Results from tuned dimensions			
mode	$.\phi_i(L)$	$.\phi'_i$	$.\omega_i$	mode	$.\phi_i(L)$	$.\phi'_i$	$.\omega_i$
1	-0.228	6.983	45.985	1	-0.211	7.421	68.366
2	0.118	17.242	164.929	2	0.102	18.074	243.026
3	0.012	23.389	320.977	3	0.000	24.191	465.226

optimal manipulator dimension results irrespective of the slight increase in mode shape slopes.

5.3 Responses with Tuned Dimensions

The tip responses with the use of tuned dimensions are shown in Figs. 7 and 8 below. The tuning of manipulator dimension using genetic algorithms is effective

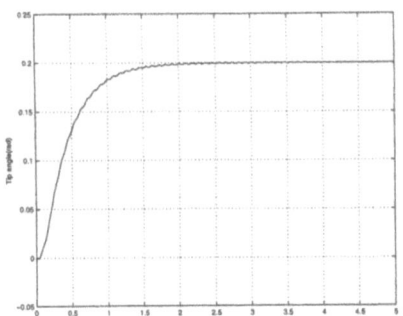

Fig. 7. Open-loop response, dimensions tuned

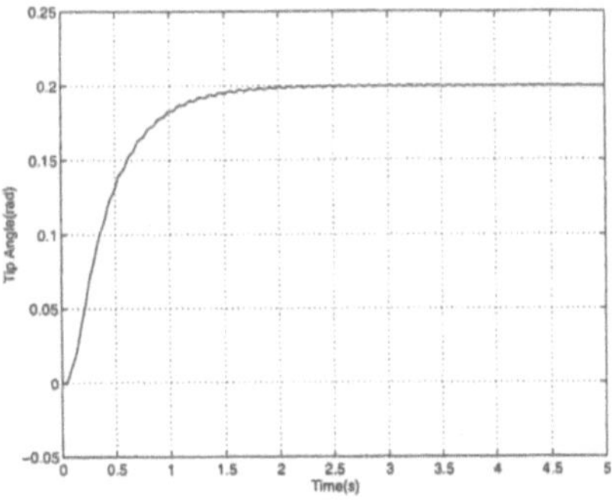

Fig. 8. Close-loop response, dimensions tuned

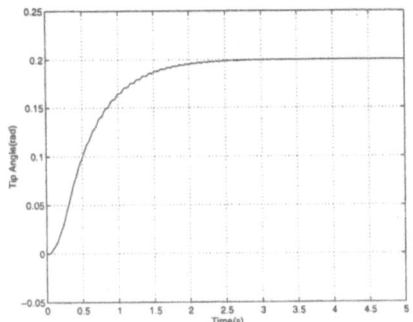
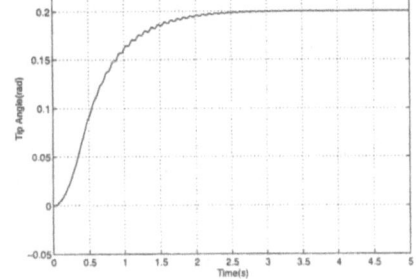

Fig. 9. Close-loop response, dimensions tuned, payload increased by 50%

Fig. 10. Close-loop response, dimensions tuned, payload increased by 100%

in reducing the vibration magnitudes even in the open-loop response. Figures 9 and 10 show the close-loop response when the payload is increased by 50% and 100% respectively. No significant degrade in response is observed.

6 Conclusion

A sliding mode controller was used to control the position of a single-link flexible manipulator. Only the rigid mode and first vibration mode were used as feedback signals that reduced the controller dimension. The manipulator dimensions were tuned using a genetic algorithm routine, with the application of elitism and adaptive crossover and mutation probabilities, had made the first vibration mode dominating the higher order vibrations and reduced the complexity in the implementation of the controller. Simulation results had demonstrated that the controller was effective in suppressing vibrations and achieving set-point regulation.

Acknowledgement. This work is supported by the City University Strategic Grant 7001416 in part.

References

1. R.H.Cannon, E.Schmitx: Initial Experiments on the End-point Control of a Flexible One-link Robot. Intl. J. of Robotics Research, Vol.3, No.3 (1984) 62–75
2. R.P.Sutton, G.D.Halikas, A.R.Plumer, D.A.Wilson: Modelling and H_∞ Control of a Single-link Flexible Manipulator. Proc. Instn. Mech. Engrs., Vol.213, Pt.1 (1999) 85–104
3. A.R.Ingole, B.Bandyopadhyay, R.Gorez: Variable Structure Control Application for Flexible Manipulators. Proc. 3rd IEEE Conf. Control Applications, Glasgow, UK (1994) 1311–1316
4. F.H.K.Fung, C.K.M.Lee: Variable Structure Tracking Control of a Single-link Flexible Arm Using Time-varying Sliding Surface. J. Robotic Sys., Vol.16, No.12 (1999) 715–726
5. J.X.Lee, G.Vukovich, J.Z.Sasiadek: Fuzzy Control of a Flexible Link Manipulator. Proc. American Control Conf., Baltimore, USA (1994) 568–574
6. S.Jain, P.Y.Peng, A.Tzes, F.Khorrami: Neural Network Designs With Genetic Learning for Control of a Single Link Flexible Manipulator. Proc. American Control Conf., Baltimore, USA (1994) 2570–2574
7. S.S.Ge, T.H.Lee, G.Zhu: Genetic Algorithm Tuning of Lyapunov-based Controllers: An Application to a Single-link Flexible Robot System. IEEE Trans. Ind. Elecn, Vol.43, No.5 (1996) 567–573
8. C.K.Lee, N.M.Kwok: Control of a Flexible Manipulator Using a Sliding Mode Controller With a Fuzzy-like Weighting Factor. Proc. 2001 IEEE Intl. Symposium on Ind. Elecn., Pusan, Korea (2001) 52–57
9. F.Y.Wang, J.L.Russell: Optimum Shape Construction of Flexible Manipulators With Total Weight Constraint. IEEE Trans. Sys. Man and Cybernetics, Vol.24, No.4 (1995) 605–614
10. K.F.Man, K.S.Tang, S.Kwong: Genetic Algorithms: Concepts and Applications. IEEE Trans. Ind. Elecn., Vol.43, No.5 (1996) 519–534

Daily Stock Prediction Using Neuro-genetic Hybrids

Yung-Keun Kwon and Byung-Ro Moon

School of Computer Science & Engineering, Seoul National University
Shilim-dong, Kwanak-gu, Seoul, 151-742 Korea
{kwon,moon}@soar.snu.ac.kr

Abstract. We propose a neuro-genetic daily stock prediction model. Traditional indicators of stock prediction are utilized to produce useful input features of neural networks. The genetic algorithm optimizes the neural networks under a 2D encoding and crossover. To reduce the time in processing mass data, a parallel genetic algorithm was used on a Linux cluster system. It showed notable improvement on the average over the buy-and-hold strategy. We also observed that some companies were more predictable than others.

1 Introduction

It is a topic of practical interest to predict the trends of financial objects such as stocks, currencies, and options. It is not an easy job even for financial experts because they are mostly nonlinear, uncertain, and nonstationary. There is no consensus among experts as to how well, if possible, financial time series are predictable and how to predict them.

Many approaches have been tried to predict a variety of financial time series such as portfolio optimization, bankruptcy prediction, financial forecasting, fraud detection, and scheduling. They include artificial neural networks [16] [19], decision trees [3], rule induction [5], Bayesian belief networks [20], evolutionary algorithms [9] [12], classifier systems [13], fuzzy sets [2] [18], and association rules [17]. Hybrid models combining a few approaches are also popular [15] [7].

Artificial neural networks (ANNs) have the ability to approximate well a large class of functions. ANNs have the potential to capture complex nonlinear relationships between a group of individual features and the variable being forecasted, which simple linear models are unable to capture. It is important in ANNs how to find the most valuable input features and how to use them. In [22], they used not only quantitative variables but also qualitative information to more accurately forecast stock prices. Textual information in articles published on the web was also used in [21]. In addition, the optimization of weights is another important issue in ANNs. The backpropagation algorithm is the most popular one for the supervised training of ANNs, but yet it is just a local gradient technique and there is room for further optimization.

In this paper, we try to predict the stock price using a hybrid genetic approach combined with a recurrent neural network. We describe a number of input variables that enable the network to forecast the next day price more accurately. The genetic algorithm operators provide diverse initial weights for the network.

The rest of this paper is organized as follows. In section 2, we explain the problem and present the objective. In section 3, we describe our hybrid genetic algorithm for predicting the stock price. In section 4, we provide our experimental results. Finally, conclusions are given in section 5.

2 Preliminaries

2.1 The Problem and Dataset

We attack the automatic stock trading problem. There are a number of versions such as intraday, daily, weekly, and monthly trading, depending on behavioral interval. In this work, we consider the daily trading.

In the problem, each record of dataset includes daily information which consists of the closing price, the highest price, the lowest price, and the trading volume. We name them at day t as $x(t)$, $x_h(t)$, $x_l(t)$, and $v(t)$, respectively. The trading strategy is based on the series of $x(t)$; if we expect $x(t+1)$ is higher than $x(t)$ we buy the stocks; if lower, we sell them; otherwise, we do not take any action. The problem is a kind of time-series data prediction that can be usually first tried with delay coordinates as follows:

$$x(t+1) = f(x(t), x(t-1), x(t-2), \ldots).$$

But, it is a simple and weak model. We transform the original time series to another that is more suitable for neural networks. First, $\frac{x(t+1)-x(t)}{x(t)}$ is used instead of $x(t+1)$ as the target variable. For the input variables, we use technical indicators or signals that were developed in deterministic trading techniques. To achieve it, we construct the model as follows:

$$\frac{x(t+1) - x(t)}{x(t)} = f(g_1, g_2, \ldots, g_m).$$

where g_k ($k = 1, \ldots, m$) is a technical indicator or signal.

2.2 The Objective

There can be a number of measures to evaluate the performance of the trading system. In our problem, we simulate the daily trading as close as possible to the actual situation and evaluate the profit. We have a cash balance C_t and stock S_t at day t ($t = 1, \ldots, N$). We start with C, i.e, $C_1 = C$ and $S_1 = 0$. Figure 1 shows the investing strategy and change of property at day $t+1$ according to the signal at day t of the trading system. In the strategy, the constant B is the upper bound of stock trade per day. We have the final property ratio P as follows:

$$P = \frac{C_N + S_N}{C_1 + S_1}.$$

```
if ( signal is SELL ) {
    C_{t+1} ← C_t + min(B, S_t)
    S_{t+1} ← S_t - min(B, S_t)
}
if ( signal is BUY ) {
    C_{t+1} ← C_t - min(B, C_t)
    S_{t+1} ← S_t + min(B, C_t)
}
S_{t+1} ← S_t × x_{t+1}/x_t
```

Fig. 1. Investing strategy and change of the property

3 The Suggested System

3.1 Processing Data

As mentioned, we have four daily data, x, x_h, x_l, and v, but we do not use them for the input variables as they are. We utilize a number of technical indicators being used by financial experts [10]. We describe some of them in the following:

- Moving average (MA)
  - The numerical average value of the stock prices over a period of time.
  - MA_S, MA_L : short-term and long-term moving average, respectively.
- Golden-cross and dead-cross
  - States that MA_S crosses MA_L upward and downward, respectively.
- Moving average convergence and divergence ($MACD$)
  - A momentum indicator that shows the relationship between MA_S and MA_L.
  - $MACD = MA_S - MA_L$
- Relative strength index (RSI)
  - An oscillator that indicates the internal strength of a single stock.
  - $RSI = 100 - \dfrac{100}{1 + U/D}$,

  U, D : An average of upward and downward price changes, respectively.
- Stochastics
  - An indicator that compares where a stock price closed relative to its price range over a given time period.
  - $\%K = \dfrac{x(t) - L}{H - L} \times 100$,

  H, L = The highest and the lowest price in a given time period.
  - $\%D$ = Moving average of $\%K$

We generate 64 input variables using the technical indicators. Figure 2 shows some representative variables. The others not shown in this figure include variables in the same forms as the above that use trading volumes in place of the prices. After generating the new variables, we normalize them by dividing by the maximum value of each variable. It helps the neural network to learn efficiently.

$X_1 = \dfrac{MA(t) - MA(t-1)}{MA(t-1)}$

$X_2 = \dfrac{MA_S(t) - MA_L(t)}{MA_L(t)}$

$X_3 = $ # of days since the last golden-cross

$X_4 = $ # of days since the last dead-cross

$X_5 = \dfrac{x(t) - x(t-1)}{x(t-1)}$

$X_6 = $ the profit while the stock has risen or fallen continuously

$X_7 = $ # of days for which the stock has risen or fallen continuously

$X_8 = MACD(t)$

$X_9 = \%K(t)$

$X_{10} = \%D(t)$

$X_{11} = \dfrac{x(t) - x_l(t)}{x_h(t) - x_l(t)}$

Fig. 2. Some examples of input variables

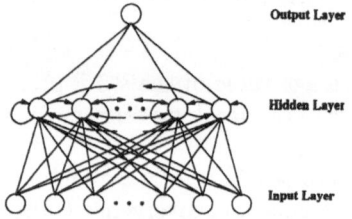

Fig. 3. The recurrent neural network architecture

3.2 Artificial Neural Networks

We use a recurrent neural network architecture which is a variant of Elman's network [4]. It consists of input, hidden, and output layers as shown in Fig. 3. Each hidden unit is connected to itself and also connected to all the other hidden units. The network is trained by a backpropagation-based algorithm.

It has 64 nodes in the input layer corresponding to the variables described in Section 3.1. Only one node exists in the output layer for $\frac{x(t+1)-x(t)}{x(t)}$.

3.3 Distribution of Loads by Parallelization

There have been many attempts to optimize the architectures or weights of ANNs. In this paper, we use a GA to optimize the weights. Especially, we parallelize the genetic algorithm since it takes much time to handle data over a long period of time. The structure of the parallel GA is shown in Fig. 4.

Parallel genetic algorithms can be largely categorized into three classes [1]: (i) global single-population master-slave GAs, (ii) single-population fine-grained,

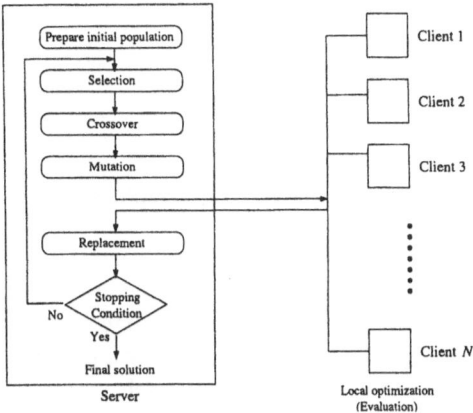

Fig. 4. The framework of the parallel genetic algorithm

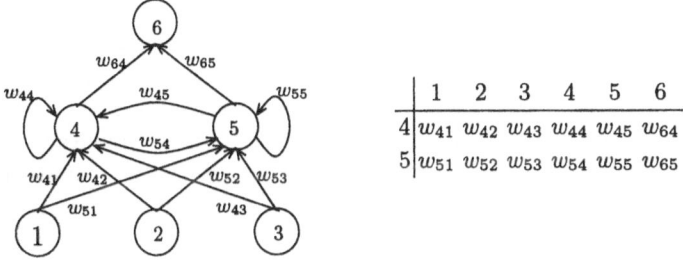

Fig. 5. Encoding in the GA

and (iii) multiple-population coarse-grained GAs. We take the first model for this work. In this neuro-genetic hybrid approach, the fitness evaluation is dominant in running time. To evaluate an offspring (a network) the backpropagation-based algorithm trains the network with training data.

In a measurement with *gprof*, the evaluation part took about 95% of the total running time. We distribute the load of evaluation to the clients (slaves) of a Linux cluster system. The main genetic parts locate in the server (master). When a new ANN is created by crossover and mutation, the GA passes it to one of the clients. When the evaluation is completed in the client, it sends the result back to the server. The server communicates with the clients in an asynchronous mode. This eliminates the need to synchronize every generation and it can maintain a high level of processor utilization, even if the slave processors operate at different speeds. This is possible because we use a steady-state GA which does not wait until a set of offspring is generated. All these are achieved with the help of MPI (Message Passing Interface), a popular interface specification for programming distributed memory systems. In this work, we used a Linux cluster system with 46 CPUs.

As shown in Fig. 4, the process in the server is a traditional steady-state GA. In the following, we describe each part of the GA.

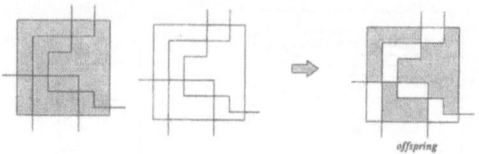

Fig. 6. An example of 2D geographical crossover

- *Representation:* Most GAs used linear encodings in optimizing ANN's weights [6] [14]. Recently, a two-dimensional encoding has proven to perform favorably [11]. We represent a chromosome by a two-dimensional weight matrix as shown in figure 5. In the matrix, each row corresponds to a hidden unit and each column corresponds to an input, hidden, or output unit. A chromosome is represented by $p \times (n+p+q)$ where n, p, and q are the numbers of input, hidden, output units, respectively. In this work, the matrix size is $20 \times (64 + 20 + 1)$.
- *Selection, crossover, and mutation:* Roulette-wheel selection is used for parent selection. The offspring is produced through geographic 2D crossover [8]. It is known to create diverse new schemata and reflect well the geographical relationships among genes. It chooses a number of lines, divides the chromosomal domain into two equivalent classes, and alternately copies the genes from the two parents as shown in figure 6. The mutation operator replaces each weight in the matrix with a low probability. All these three operators are performed in the server.
- *Local optimization:* After an offspring is modified by a mutation operator, it is locally optimized by backpropagation which helps the GA fine-tune around local optima. Local optimization is mingled with quality evaluation. As mentioned, it is performed in the client and the result is sent to the server.
- *Replacement and stopping criterion:* The offspring first attempts to replace the more similar parent to it. If it fails, it attempts to replace the other parent and the most inferior member of the population in order. Replacement is done only when the offspring is better than the replacee. The GA stops if it does not find an improved solution for a fixed number generations.

4 Experimental Results

We tested our approaches with 36 companies' stocks in NYSE and NASDAQ. We evaluated the performance for ten years from 1992 to 2001. We got the entire data from YAHOO (http://quote.yahoo.com). The GA was trained with two consecutive years of data and validated with the third year's. The solution was tested with the fourth year's data. This process was shifted year by year. Thus, totally 13 years of data were used for this work.

Table 1 shows the experimental results. The values mean the final property ratio P defined in section 2.2. In the table, the *Hold* strategy buys the stock at the first day and holds it all the year through. *RNN* and *GA* are the average results

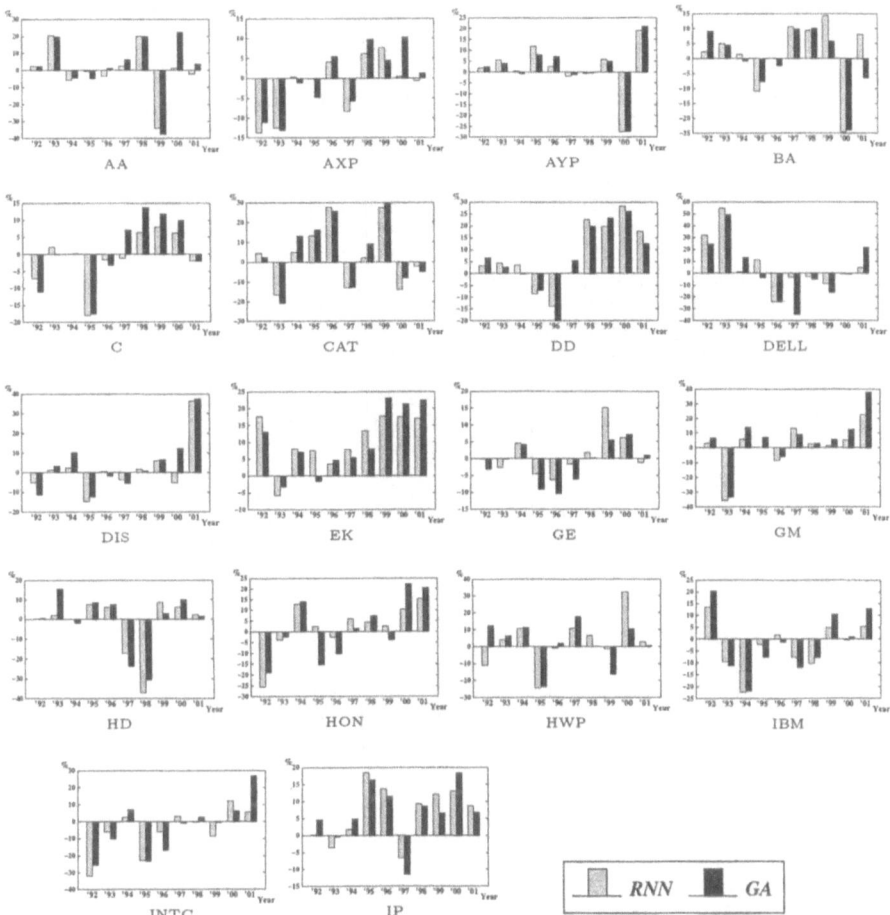

Fig. 7. Improvements of *RNN* and *GA* over *Hold* in the first half of the companies

by the recurrent neural network (20 trials) and the hybrid genetic algorithm (20 trials), respectively. For qualitative comparison, we summarized the relative performance in table 3. In the table, *Up*, *Down*, and *NC* represent the states of the stock market in each year. Since there are 36 companies tested for 10 years, we have 359 cases except one case with deficient data. The *Up* and *Down* mean that the closing price has risen or fallen over the year's starting price by 5% or more, respectively. *NC* means no notable difference. *Better*, *Worse*, and *Even* represent the relative performance of *RNN* and *GA* over the *Hold*. *Better* and *Worse* mean that P value of the learned strategy is at least 5% higher or lower than that of *Hold*, respectively. The GA performed better than the *Hold* in 153 cases, worse in 88 cases, and comparable in 118 cases. *RNN* and *GA* showed a significant performance improvement over *Hold* on the average. The *GA* considerably improved the performance of *RNN*.

Table 1. P values

Symbols	Strategies	1992	1993	1994	1995	1996	1997	1998	1999	2000	2001
AA	Hold	1.141	0.967	1.211	1.272	1.198	1.094	1.035	2.195	0.797	1.102
	RNN	1.168	1.166	1.140	1.262	1.154	1.120	1.243	1.446	0.806	1.078
	GA	1.167	1.159	1.157	1.211	1.212	1.161	1.243	1.373	0.975	1.143
AXP	Hold	1.205	1.185	1.116	1.419	1.316	1.624	1.144	1.549	0.992	0.686
	RNN	1.039	1.037	1.119	1.416	1.370	1.488	1.214	1.667	0.996	0.682
	GA	1.071	1.030	1.103	1.353	1.389	1.531	1.256	1.617	1.094	0.696
AYP	Hold	1.070	1.108	0.825	1.328	1.039	1.077	1.062	0.756	1.757	0.795
	RNN	1.089	1.168	0.829	1.484	1.065	1.057	1.055	0.800	1.274	0.946
	GA	1.096	1.152	0.818	1.433	1.113	1.065	1.058	0.795	1.276	0.960
BA	Hold	0.839	1.101	1.072	1.709	1.300	0.941	0.674	1.231	1.528	0.625
	RNN	0.858	1.156	1.087	1.523	1.303	1.040	0.737	1.408	1.152	0.676
	GA	0.915	1.151	1.063	1.579	1.270	1.034	0.742	1.304	1.163	0.586
C	Hold	1.238	1.656	0.817	1.878	1.441	1.803	0.940	1.576	1.273	1.000
	RNN	1.150	1.691	0.817	1.540	1.418	1.782	1.001	1.702	1.352	0.979
	GA	1.101	1.655	0.819	1.548	1.394	1.931	1.070	1.764	1.401	0.980
CAT	Hold	1.246	1.644	1.231	1.082	1.227	1.342	0.967	1.026	0.952	1.128
	RNN	1.303	1.374	1.291	1.225	1.567	1.169	0.986	1.308	0.819	1.103
	GA	1.275	1.304	1.390	1.256	1.542	1.173	1.054	1.331	0.875	1.071
DD	Hold	1.038	1.026	1.131	1.286	1.326	1.272	0.909	1.176	0.738	0.886
	RNN	1.071	1.071	1.172	1.178	1.142	1.273	1.116	1.408	0.948	1.043
	GA	1.105	1.052	1.127	1.196	1.063	1.342	1.089	1.452	0.931	0.997
DELL	Hold	2.514	0.530	1.661	1.787	2.877	3.346	3.461	1.372	0.344	1.553
	RNN	3.312	0.822	1.680	1.987	2.182	3.233	3.364	1.248	0.343	1.624
	GA	3.130	0.793	1.880	1.720	2.184	2.186	3.288	1.148	0.342	1.890
DIS	Hold	1.476	1.009	1.081	1.306	1.107	1.479	0.890	1.011	0.935	0.742
	RNN	1.399	1.020	1.106	1.115	1.112	1.424	0.906	1.072	0.887	1.012
	GA	1.308	1.043	1.191	1.148	1.087	1.397	0.897	1.077	1.050	1.019
EK	Hold	0.830	1.370	1.076	1.419	1.149	0.807	1.119	0.907	0.596	0.764
	RNN	0.976	1.290	1.163	1.526	1.190	0.869	1.267	1.067	0.701	0.894
	GA	0.938	1.326	1.153	1.396	1.202	0.850	1.206	1.117	0.724	0.936
GE	Hold	1.116	1.216	0.981	1.441	1.328	1.516	1.359	1.492	0.875	0.916
	RNN	1.114	1.184	1.024	1.375	1.244	1.492	1.382	1.718	0.928	0.905
	GA	1.082	1.215	1.021	1.309	1.192	1.423	1.361	1.575	0.937	0.924
GM	Hold	1.060	1.684	0.754	1.251	1.120	1.107	1.162	1.245	0.699	0.931
	RNN	1.091	1.086	0.797	1.251	1.024	1.252	1.192	1.260	0.737	1.143
	GA	1.131	1.124	0.858	1.340	1.052	1.206	1.196	1.314	0.786	1.284
HD	Hold	1.437	0.787	1.171	1.044	1.037	1.786	2.011	1.663	0.699	1.120
	RNN	1.440	0.802	1.171	1.123	1.098	1.478	1.266	1.805	0.742	1.145
	GA	1.442	0.909	1.148	1.132	1.114	1.363	1.401	1.711	0.768	1.135
HON	Hold	1.437	1.288	0.875	1.449	1.365	1.175	1.103	1.301	0.781	0.764
	RNN	1.071	1.239	0.986	1.483	1.330	1.244	1.151	1.333	0.860	0.881
	GA	1.167	1.257	0.996	1.227	1.228	1.192	1.182	1.251	0.956	0.921
HWP	Hold	1.219	1.141	1.265	1.698	1.207	1.282	1.068	1.704	0.641	0.679
	RNN	1.084	1.185	1.399	1.286	1.197	1.416	1.136	1.680	0.849	0.699
	GA	1.368	1.212	1.405	1.298	1.229	1.507	1.070	1.430	0.708	0.683
IBM	Hold	0.555	1.150	1.280	1.232	1.686	1.378	1.733	1.264	0.734	1.426
	RNN	0.630	1.040	0.996	1.204	1.715	1.272	1.555	1.324	0.732	1.502
	GA	0.669	1.022	1.003	1.139	1.668	1.216	1.599	1.397	0.741	1.611
INTC	Hold	1.721	1.416	1.041	1.839	2.224	1.114	1.664	1.440	0.714	1.012
	RNN	1.175	1.330	1.067	1.422	2.094	1.148	1.659	1.316	0.802	1.068
	GA	1.285	1.272	1.113	1.413	1.856	1.103	1.708	1.432	0.759	1.284
IP	Hold	0.946	1.026	1.105	1.042	1.067	1.082	0.932	1.318	0.711	1.023
	RNN	0.948	0.990	1.125	1.233	1.214	1.010	1.019	1.477	0.803	1.112
	GA	0.990	1.022	1.158	1.212	1.190	0.959	1.012	1.402	0.841	1.092

Table 1. Continued

Symbols	Strategies	1992	1993	1994	1995	1996	1997	1998	1999	2000	2001
JNJ	Hold	0.882	0.914	1.204	1.546	1.181	1.308	1.271	1.115	1.106	1.159
	RNN	0.881	0.921	1.211	1.554	1.184	1.313	1.411	1.150	1.314	1.162
	GA	0.913	0.912	1.185	1.490	1.074	1.434	1.426	1.275	1.376	1.179
JPM	Hold	0.959	1.065	0.814	1.436	1.194	1.156	0.957	1.133	1.363	
	RNN	1.003	1.051	0.822	1.369	1.091	1.234	0.961	1.223	1.404	N/A
	GA	0.976	1.032	0.830	1.381	1.242	1.234	0.960	1.101	1.345	
KO	Hold	1.047	1.060	1.163	1.449	1.383	1.290	1.004	0.839	1.079	0.775
	RNN	1.058	1.060	1.082	1.464	1.385	1.256	1.012	0.870	1.154	1.007
	GA	1.114	1.059	1.054	1.430	1.352	1.243	0.955	0.916	1.164	0.982
MCD	Hold	1.248	1.148	1.036	1.562	0.992	1.055	1.615	1.030	0.845	0.790
	RNN	1.268	1.256	1.197	1.579	1.087	1.050	1.483	1.114	0.845	0.934
	GA	1.254	1.245	1.227	1.584	1.092	1.044	1.358	1.179	0.855	0.966
MMM	Hold	1.056	1.062	1.011	1.255	1.321	0.968	0.894	1.265	1.263	0.992
	RNN	1.161	1.112	1.075	1.310	1.245	1.048	0.916	1.100	1.121	0.995
	GA	1.180	1.135	1.081	1.318	1.221	1.101	0.903	1.138	1.052	0.949
MO	Hold	0.959	0.755	1.000	1.594	1.214	1.219	1.160	0.446	1.971	0.993
	RNN	0.959	0.750	1.006	1.356	1.164	1.450	1.201	0.678	1.440	1.005
	GA	0.960	0.779	0.980	1.280	1.204	1.455	1.244	0.690	1.662	0.993
MRK	Hold	0.791	0.803	1.085	1.680	1.243	1.347	1.391	0.903	1.375	0.632
	RNN	0.790	0.803	1.239	1.465	1.310	1.263	1.430	0.981	1.360	0.975
	GA	0.817	0.868	1.196	1.260	1.258	1.270	1.433	0.988	1.377	0.950
MSFT	Hold	1.120	0.941	1.502	1.491	1.819	1.606	2.151	1.653	0.372	1.527
	RNN	1.153	0.941	1.343	1.418	1.694	1.603	1.894	1.705	0.382	1.564
	GA	1.270	1.031	1.343	1.293	1.898	1.438	1.920	1.793	0.407	1.435
NMSB	Hold	0.893	1.360	1.000	1.618	1.273	1.514	0.896	1.063	0.842	1.379
	RNN	7.049	4.855	4.603	4.896	3.012	3.280	2.174	1.530	2.485	1.571
	GA	6.773	5.081	4.843	4.777	3.021	3.437	2.126	1.623	2.440	1.524
ORCL	Hold	1.924	1.991	1.504	1.513	1.457	0.821	1.870	4.121	0.893	0.524
	RNN	1.445	1.840	1.647	1.517	1.667	1.024	1.416	1.576	0.916	0.528
	GA	1.377	1.558	1.613	1.491	1.554	0.974	1.540	1.633	0.977	0.735
PG	Hold	1.145	1.080	1.085	1.333	1.283	1.518	1.112	1.191	0.732	1.008
	RNN	1.136	1.069	1.115	1.399	1.224	1.455	1.111	1.398	1.115	1.151
	GA	1.137	1.058	1.091	1.354	1.237	1.417	1.113	1.395	1.010	1.188
RYFL	Hold	0.750	0.889	0.688	2.273	0.760	1.053	0.260	1.154	0.800	1.333
	RNN	8.658	12.406	16.073	7.267	7.670	5.034	1.770	3.627	3.983	5.753
	GA	9.037	12.718	15.765	6.346	8.001	5.191	1.827	2.950	3.967	5.655
SBC	Hold	1.121	1.137	0.979	1.437	0.901	1.424	1.404	0.897	1.068	0.778
	RNN	1.169	1.274	0.983	1.217	0.931	1.513	1.538	1.066	1.020	1.020
	GA	1.170	1.299	1.000	1.242	0.948	1.526	1.635	1.067	1.032	1.115
SUNW	Hold	1.165	0.851	1.245	2.512	1.196	1.551	2.170	3.398	0.665	0.484
	RNN	1.104	1.019	1.412	1.212	1.334	1.638	1.605	3.298	0.701	0.544
	GA	1.361	0.943	1.465	1.556	1.594	1.836	1.433	2.580	0.776	0.538
T	Hold	1.293	1.037	0.950	1.347	0.891	1.404	1.324	1.028	0.342	1.287
	RNN	1.031	0.987	0.950	1.188	1.102	1.336	1.317	1.207	0.395	1.655
	GA	1.095	0.981	0.950	1.169	1.102	1.219	1.231	1.123	0.444	1.417
UTX	Hold	0.920	1.228	1.048	1.491	1.400	1.114	1.477	1.157	1.204	0.859
	RNN	1.052	1.170	1.175	1.325	1.396	1.134	1.560	1.246	1.233	0.745
	GA	0.988	1.183	1.216	1.330	1.301	1.175	1.559	1.288	1.279	0.720
WMT	Hold	1.063	0.811	0.819	1.114	0.989	1.712	2.048	1.659	0.806	1.068
	RNN	1.060	0.813	0.771	1.035	1.038	1.262	1.166	1.700	0.943	1.076
	GA	1.107	0.857	0.791	1.158	1.097	1.265	1.217	1.771	0.990	1.111
XOM	Hold	1.023	1.039	0.951	1.330	1.220	1.258	1.174	1.077	1.140	0.882
	RNN	1.219	1.198	1.134	1.267	1.293	1.392	1.280	1.105	1.523	0.957
	GA	1.247	1.182	1.131	1.234	1.312	1.397	1.296	1.156	1.536	0.994

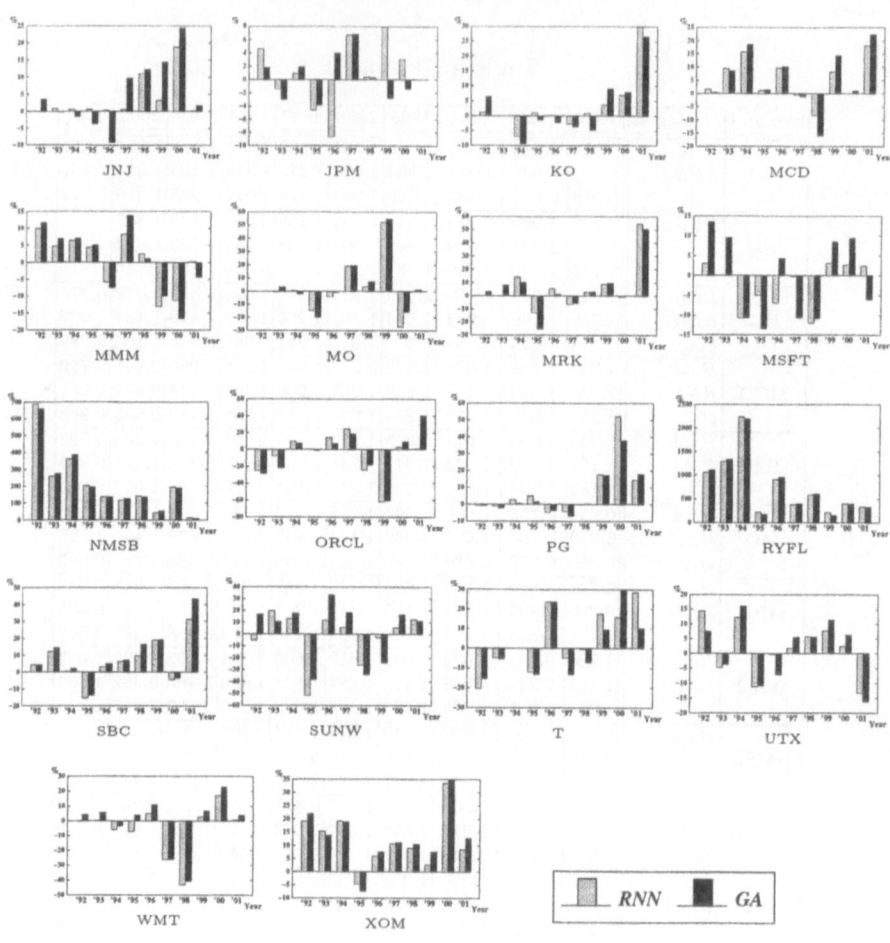

Fig. 7. Continued

Table 2. Relative performance of *RNN* and *GA* over *Hold*

(1) *RNN*				
	Better	Worse	Even	Total
Up	63	64	103	230
Down	55	3	28	86
NC	21	1	21	43
Total	139	68	152	359

(2) *GA*				
	Better	Worse	Even	Total
Up	79	76	75	230
Down	66	2	18	86
NC	28	2	13	43
Total	173	80	106	359

Figure 7 shows the relative performance graphically; the Y-axis represents the percentage improvement of *RNN* and *GA* over *Hold*. We can observe in the figure that some companies are more predictable than others. Practically, one can choose the stocks of companies that turned out to be more stably predictable by the GA.

5 Conclusion

In this paper, we proposed a genetic algorithm combined with a recurrent neural network for the daily stock trading. It showed significantly better performance than the "buy-and-hold" strategy with a variety of companies for recent ten years. For the input nodes of the neural network, we transform the stock prices and the volumes into a large number of technical indicators or signals. The genetic algorithm helps optimize the weights of neural network globally. This turned out to contribute to the performance improvement.

In the experiments, the proposed GA predicted better in some companies than in others. It implies that this work can be useful in portfolio optimization. Future study will include finding the stock trading strategy combined with portfolio. In addition, this approach is not just restricted to the stock market.

Acknowledgment. This work was partly supported by Optus Inc. and Brain Korea 21 Project. The RIACT at Seoul National University provided research facilities for this study.

References

1. E. Cantu-Paz. A survey of parallel genetic algorithms. *Calculateurs Parallels*, 10(2):141–171, 1998.
2. O. Castillo and P. Melin. Simulation and forecasting complex financial time series using neural networks and fuzzy logic. In *Proceedings of IEEE Conference on Systems, Man, and Cybernetics*, pages 2664–2669, 2001.
3. Y. M. Chae, S. H. Ho, K. W. Cho, D. H. Lee, and S. H. Ji. Data mining approach to policy analysis in a health insurance domain. *International Journal of Medical Informatics*, 62(2):103–111, 2001.
4. J. L. Elman. Finding structure in time. *Cognitive Science*, 14:179–211, 1990.
5. J. A. Gentry, M. J. Shaw, A. C. Tessmer, and D. T. Whitford. Using inductive learning to predict bankrupcy. *Journal of Organizational Computing and Electronic Commerce*, 12(1):39–57, 2002.
6. S. A. Harp, T. Samad, and A. Guha. Towards the genetic synthesis of neural networks. In *International Conference on Genetic Algorithm*, pages 360–369, 1989.
7. P. G. Harrald and M. Kamstra. Evolving artificial neural networks to combine financial forecasts. *IEEE Transactions on Evolutionary Computation*, 1(1):40–52, 1997.
8. A. B. Kahng and B. R. Moon. Toward more powerful recombinations. In *International Conference on Genetic Algorithms*, pages 96–103, 1995.
9. M. A. Kanoudan. Genetic programming prediction of stock prices. *Computational Economics*, 16:207–236, 2000.
10. P. J. Kaufman. *Trading Systems and Methods*. John Wiley & Sons, 1998.
11. J. H. Kim and B. R. Moon. Neuron reordering for better neuro-genetic hybrids. In *Proceedings of the Genetic and Evolutionary Computation Conference*, pages 407–414, 2002.
12. K. J. Kim. Genetic algorithms approach to feature discretization in artificial neural networks for the prediction of stock price index. *Expert Systems with Applications*, 19(2):125–132, 2000.

13. P. Y. Liao and J. S. Chen. Dynamic trading strategy learning model using learning classifier system. In *Proceedings of the Congresson Evolutionary Computation*, pages 783–789, 2001.
14. C. T. Lin and C. P. Jou. Controlling chaos by ga-based reinforcement learning neural network. *IEEE Transactions on Neural Networks*, 10(4):846–869, 1999.
15. K. N. Pantazopoulos, L. H. Tsoukalas, N. G. Bourbakis, Brün, and E. N. Houstis. Financial prediction and trading strategies using neurofuzzy approaches. *IEEE Transactions on Systems, Man, and Cybernetics-Part:B*, 28(4):520–531, 1998.
16. P. Tiño, C. Schittenkopf, and G. Dorffner. Financial volatility trading using recurrent neural networks. *IEEE Transactions on Neural Networks*, 12:865–874, 2001.
17. R. Veliev, A. Rubinov, and A. Stranieri. The use of an association rules matrix for economic modelling. In *International conference on neural information processing*, pages 836–841, 1999.
18. Y. F. Wang. Predicting stock price using fuzzy grey prediction system. *Expert Systems with Applications*, 22(1):33–38, 2002.
19. I. D. Wilson, S. D. Paris, J. A. Ware, and D. H. Jenkins. Residential property price time series forecasting with neural networks. *Knowledge-Based Systems*, 15(5):335–341, 2002.
20. R. K. Wolfe. Turning point identification and Bayesian forecasting of a volatile time series. *Computers and Industrial Engineering*, 15:378–386, 1988.
21. B Wuthrich, V. Cho, S. Leung, D. Permunetilleke, K. Sankaran, and J. Zhang. Daily stock market forecast from textual web data. In *IEEE International Conference on Systems, Man and Cybernetics*, pages 2720–2725, 1999.
22. Y. Yoon and G. Swales. Predicting stock price performance: a neural network approach. In *Proc. 24th Annual Hawaii International conference on System Sciences*, pages 156–162, 1991.

Finding the Optimal Gene Order in Displaying Microarray Data

Seung-Kyu Lee[1], Yong-Hyuk Kim[2], and Byung-Ro Moon[2]

[1] NHN Corp., 7th floor, Startower
737 Yoksam-dong, Kangnam-gu, Seoul, Korea
spin30@soar.snu.ac.kr
[2] School of Computer Science & Engineering, Seoul National University
Shilim-dong, Kwanak-gu, Seoul, 151-742 Korea
{yhdfly, moon}@soar.snu.ac.kr

Abstract. The rapid advances of genome-scale sequencing have brought out the necessity of developing new data processing techniques for enormous genomic data. Microarrays, for example, can generate such a large number of gene expression data that we usually analyze them with some clustering algorithms. However, the clustering algorithms have been ineffective for visualization in that they are not concerned about the order of genes in each cluster. In this paper, a hybrid genetic algorithm for finding the optimal order of microarray data, or gene expression profiles, is proposed. We formulate our problem as a new type of traveling salesman problem and apply a hybrid genetic algorithm to the problem. To use the 2D natural crossover, we apply the Sammon's mapping to the microarray data. Experimental results showed that our algorithm found improved gene orders for visualizing the gene expression profiles.

1 Introduction

The recent marvelous advances of genome-scale sequencing have provided us with a huge amount of genomic data. Microarrays [36,37], for example, have been used for revealing gene expression profiles for more than thousands of genes. In general, microarray data can be represented by a real-valued matrix; each row represents a gene and each column represents a condition, or experiment. If we let the matrix be X, the element X_{ij} represents the expression level of gene i for a given condition j. The microarray data are usually preprocessed with a clustering algorithm. The clustered microarray data, then, can be analyzed by biologists.

A number of algorithms for clustering gene expression profiles were proposed. Eisen et al. [10] applied hierarchical clustering [38] which has been a widely used tool [1,22,24,35]. It also has some variants [2,17]. Self-organizing maps (SOMs) [42,44] and k-means clustering [43] were also used for the same purpose. Ben-Dor et al. [3] developed an algorithm, cluster affinity search technique (CAST), which has a good theoretical basis. Merz and Zell [28] proposed a memetic algorithm for the problem formulated as finding the minimum sum-of-squares clustering

[48,9]. However, all the proposed algorithms were ineffective in visualizing the microarray data since they were not concerned about aligning genes within each cluster in a meaningful way. This raises the problem of finding the optimal order of genes for visualization.

Although there is no standard optimal criterion for evaluating which order is better than the others for visualization, placing genes with similar or the same expression profiles next to each other is considered to be natural and intuitive. Since finding the optimal order of microarray data is known to be NP-hard [5], evolutionary approaches such as genetic algorithms [18,13], memetic algorithms [29] are considered to be well suited for solving the problem.

Recently Tsai et al. [46] formulated the problem as the traveling salesman problem (TSP) and applied family competition genetic algorithm (FCGA). In the FCGA, the edge assembly crossover [30] was combined with the family competition concept [45] and neighbor-join mutation [47]. Using this consolidation, they showed that their formulation was effective in finding attractive gene orders for visualizing microarray data. However, they implicitly tried to minimize the distance between distant genes as well, which is less important for visualization.

In this paper, we propose a hybrid genetic algorithm for finding the optimal gene order of microarray data. We suggest a new variation of TSP formulation for this purpose. We use the 2D natural crossover [20,21], which is one of the state-of-the-art crossovers in the TSP literature. To use the 2D natural crossover, we need a 2D mapping of the data, which are virtually real-valued vectors, from a high-dimensional space into a two-dimensional Euclidean space. This mapping is necessary since the crossover exploits two-dimensional geographical information. We choose the Sammon's mapping [34] among several candidates.

Another important contribution of this paper is that we used a new formulation of TSP for the problem. In this model of TSP, relatively long edges in a tour are ignored for fitness evaluation. This is because reducing the length of long edges, which represents distant genes in relation, is not very meaningful for visualizing microarray data. We tested this idea on a spectrum of different rates of excluded edges.

The remainder of the paper is organized as follows. In Section 2, we summarize the traveling salesman problem and the Sammon's mapping. In Section 3, we describe our variation of TSP formulation for finding the optimal gene order in displaying microarray data. In Section 4, we explain our hybrid genetic algorithm in detail and present the experimental results in Section 5. Finally, we make our conclusions in Section 6.

2 Preliminaries

2.1 Traveling Salesman Problem

Given n cities and a distance matrix $D = [d_{ij}]$ where d_{ij} is the distance between city i and city j, the traveling salesman problem (TSP) is the problem of finding a permutation π that minimizes $\sum_{i=1}^{n-1} d_{\pi_i, \pi_{i+1}} + d_{\pi_n, \pi_1}$. In metric TSP the cities lie in a metric space (i.e., the distances satisfy the triangle inequality). In Euclidean

TSP, the cities lie in $\Re^d$ for some d; the most popular version is 2D Euclidean TSP where the cities lie in $\Re^2$. Euclidean TSP is a sub-case of metric TSP.

2.2 Sammon's Mapping

Sammon's mapping [34] is a mapping technique for transforming a dataset from a high-dimensional (say, m-dimensional) input space onto a low-dimensional (say, d-dimensional) output space (with $d < m$). The basic idea is to arrange all the data points on a d-dimensional output space in such a way that minimizes the distortion of the relationships among data points.

Sammon's mapping tries to preserve distances. This is achieved by minimizing an error criterion which penalizes the differences of distances between the points in the input space and the output space. Consider a dataset of n objects. If we denote the distance between two points x_i and x_j in the input space by δ_{ij} and the distances between x'_i and x'_j in the output space by δ'_{ij}, then Sammon's stress measure E is defined as follows:

$$E = \frac{1}{\sum_{i=1}^{n-1}\sum_{j=i+1}^{n}\delta_{ij}} \sum_{i=1}^{n-1}\sum_{j=i+1}^{n} \frac{(\delta_{ij} - \delta'_{ij})^2}{\delta_{ij}}.$$

The stress range is [0,1] with 0 indicating a lossless mapping. This stress measure can be minimized using any minimization technique. Sammon [34] proposed a technique called pseudo-Newton minimization, a steepest-descent method. The complexity of Sammon's mapping is $O(n^2 m)$. There were many studies about Sammon's mapping [8,33,31].

3 A New TSP Formulation

To visualize the microarray data, or gene expression profiles, in a meaningful way, it is natural and intuitive to align genes with similar expression profiles, or within the same group, close together. For genes with similar expression profiles to be aligned next to each other, it is useful to formulate the problem as the TSP.

For the TSP formulation, a distance measure is needed to quantify the similarity between gene expression profiles, which then defines the similarity between the genes themselves. Several distance measures were proposed to define the distance. They include Pearson correlation [1], absolute correlation [2], Spearman rank correlation [39], Kendall rank correlation [23], and Euclidean distance. In this paper, we choose the Pearson correlation as the distance measure.

Let $X = x_1, x_2, \ldots, x_k$ and $Y = y_1, y_2, \ldots, y_k$ be the expression levels of two genes X and Y, which were observed over a series of k conditions. The Pearson

---

[1] Karl Pearson (1857-1936) is considered to be the first to call the quantity a correlation efficient in 1896 [7]. It first appeared as a published form by Harris [16]. It is also referred to as Pearson product-moment correlation coefficient.
[2] absolute value of correlation

(a) with the naive TSP formulation   (b) with our new TSP formulation

Fig. 1. A comparison of good tours between the naive and our new TSP formulation.

correlation of the two genes X and Y is

$$s_{X,Y} = \frac{1}{k}\sum_{i=1}^{k}(\frac{x_i - \overline{X}}{\sigma_X})(\frac{y_i - \overline{Y}}{\sigma_Y})$$

where $\overline{X}$ and σ_X are the mean and the standard deviation of the expression levels, respectively. Then we define the distance between the genes X and Y by

$$D(X,Y) = 1 - s_{X,Y}$$

where $s_{X,Y}$ is the Pearson correlation.

Once the distance measure is defined, it is possible to formulate finding the optimal gene order for visualization with the TSP model. Each gene corresponds to a city in TSP, and the distance between two genes corresponds to the length of the edge between the two cities. Tsai et al. [46] reduced the problem of finding the optimal order of genes to the problem of finding the shortest tour of the corresponding TSP.

In the model, the fitness function is naturally defined to be

$$\sum_{i=1}^{n} D(g_{\pi_i}, g_{\pi_{i+1}})$$

where $g_{\pi_{n+1}} = g_{\pi_1}$, g_i denotes a gene, π denotes a gene order, n is the number of genes, and $D(g_i, g_j)$ is the distance between two genes g_i and g_j.

The above TSP formulation [46] aims at aligning genes with similar profiles close together. However, it also tries to minimize the distances between pairs of genes with not-very-similar profiles. When two genes are adjacent in a TSP tour, they are placed next to each other. If they have distant profiles, replacing a gene by a third one with a less distant profile has little meaning, as long as they are not considerably similar. Tsai et al.'s TSP formulation implicitly tries to reduce this type of edges as well.

To alleviate the problem, we propose a new fitness function. Our key idea is that we can improve the visualization results by excluding less meaningful edges in a tour from the fitness function. Figure 1 illustrates the motivation of our new TSP formulation. The length of the tour in Fig. 1(a) is shorter than that of the tour in Fig. 1(b), thus it is preferred in the naive TSP formulation. However, the tour in Fig. 1(b) can be a better tour in that it reflects the natural grouping, denoted by ellipses. Since an edge between two genes means that the two genes are placed next to each other in visualization, we can make the naturally-grouped genes placed next to each other if a meaningful tour like Fig. 1(b), possibly including long edges, is preferred. For such tours to be preferred, relatively long edges are excluded from the fitness function. The dotted edges in Fig. 1(b) represent the relatively long edges, thus they are not counted in our new TSP formulation. By excluding them, meaningful tours like Fig. 1(b) can be favored.

More formally, our variation of the TSP formulation defines the fitness function by

$$\sum_{i=1}^{n} D(g_{\pi_i}, g_{\pi_{i+1}}) \delta(g_{\pi_i}, g_{\pi_{i+1}})$$

where

$$\delta(i,j) = \begin{cases} 0 & \text{if } (i,j) \in L \\ 1 & \text{otherwise} \end{cases}$$

in which (i,j) is the edge connecting gene i and gene j, and L is the set of excluded edges.

We use the new fitness function only in selection and replacement. In other words, the other stages of GA except them still use the common TSP formulation which considers all the edges in a tour. This setting was settled down after some experiments.

4 A Hybrid Genetic Algorithm

A genetic algorithm hybridized with local optimizations is called a hybrid GA. A great many studies about hybridization of GAs were proposed [49,32].

- Sammon's Mapping
  Since the microarray data are virtually real-valued vectors in a high-dimensional space, we map them into the two-dimensional space in order to use the 2D natural crossover, which operates on chromosomes encoded by 2D *graphic images*. We chose the Sammon's mapping described in Section 2.2. Figure 2(a) shows a Sammon-mapped image from a small subset of real-field microarray data.
- Encoding
  Using the Sammon's mapping, we obtain a 2D Euclidean TSP instance using the distance information. We use the *graphic image itself* of a tour as a chromosome. This encoding was used in [20,21] and showed successful results on most TSP benchmarks.
- Initialization
  All the chromosomes are created at random. We set the population size to be 50 in our algorithm.
- Selection
  We use the tournament selection [14]. The tournament size is 2.

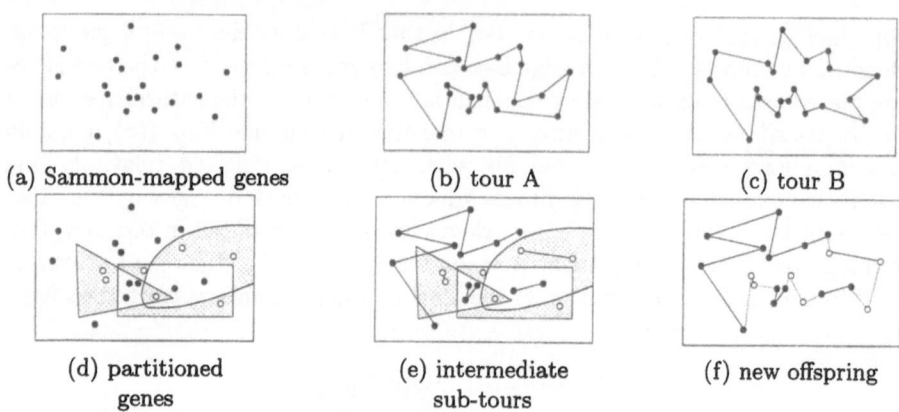

Fig. 2. A Sammon-mapped image and an example crossover on it

- Crossover
  We use the natural crossover [20,21]. The natural crossover draws *free curves* on the 2D space where genes are located. The curves divide the chromosomal positions into two disjoint partitions. Then we copy the genes in one partition from one parent to the offspring and those in the other partition from the other parent. Figures 2(b) through (f) show an example operation of the natural crossover on a Sammon-mapped chromosome.
- Mutation
  The double-bridge kick move, which is known to be effective from the literature [19,27], is used.
- Local Optimization
  We use the Lin-Kernighan (LK) algorithm [26], which is one of the most effective heuristics for TSP. The LK used here is an advanced version incorporating the techniques of *don't-look bit* [4] and *segment tree* [12] which cause dramatic speed-up.
- Replacement
  The replacement scheme proposed in [6] is used. The offspring tries to first replace the more similar parent, measured by Hamming distance [15], if it fails, then it tries to replace the other parent (replacement is done only when the offspring is better than one of the parents). If the offspring is worse than both parents, we replace the worst member of the population (GENITOR-style replacement [50]).
- Stopping Criterion
  The GA stops when one of the three conditions is satisfied: i) 80% of the population is occupied by solutions with the same quality, whose chromosomes are not necessarily the same, ii) the number of consecutive fails to replace the best solution reaches 200, or iii) the number of generations reaches 2000.

Table 1. Data Set

Data Set Name	Number of Genes	Number of Experiments
Cell cycle cdc15	782	24
Cell cycle	803	59
Yeast complexes	979	79

5 Experimental Results

5.1 Test Beds and Test Environment

We tested the proposed algorithm on three data sets, Cell Cycle cdc15, Cell Cycle, and Yeast Complexes. The first two data sets consist of about 800 genes each, which are cell cycle regulated in *saccharomyces cerevisia* with different numbers of experiments [40]. They are classified into five groups termed G1, S, S/G2, G2/M, and M/G1 by Spellman *et al.* [40]. Although it is controversial whether the group assignment does reflect the real grouping, it is known to be meaningful to some degree [46]. The final data set, Yeast Complexes, is from MIPS yeast complexes database [10]. All these three data sets can be found in [2] and downloaded at a web site.[3] Table 1 shows a brief description of each data set.

All programs were written in C++ language and run on Pentium III 866 MHz with Linux 2.2.14. They were compiled using GNU's *g++* compiler. We performed 100 runs for each experiment.

5.2 Performance

We denote by NNGA our proposed hybrid GA using the natural crossover and the new TSP formulation. The performances of the visualization results are evaluated by a score described in [46], which is defined by

$$Score = \sum_{i=1}^{n} G(g_{\pi_i}, g_{\pi_{i+1}})$$

where $g_{\pi_{n+1}} = g_{\pi_1}$, and

$$G(g_{\pi_i}, g_{\pi_j}) = \begin{cases} 1, \text{ if } g_{\pi_i} \text{ and } g_{\pi_j} \text{ are in the same group} \\ 0, \text{ if } g_{\pi_i} \text{ and } g_{\pi_j} \text{ are not in the same group} \end{cases}.$$

It is clear that a solution gets a higher score, under this scoring system, when more genes with the same group are aligned next to each other.

Figure 3 shows the scores found by NNGA when the percent of the excluded edges varies from 0% to 90% at intervals of 10%. The NNGA improved the

---
[3] http://www.psrg.lcs.mit.edu/clustering/ismb01/optimal.html

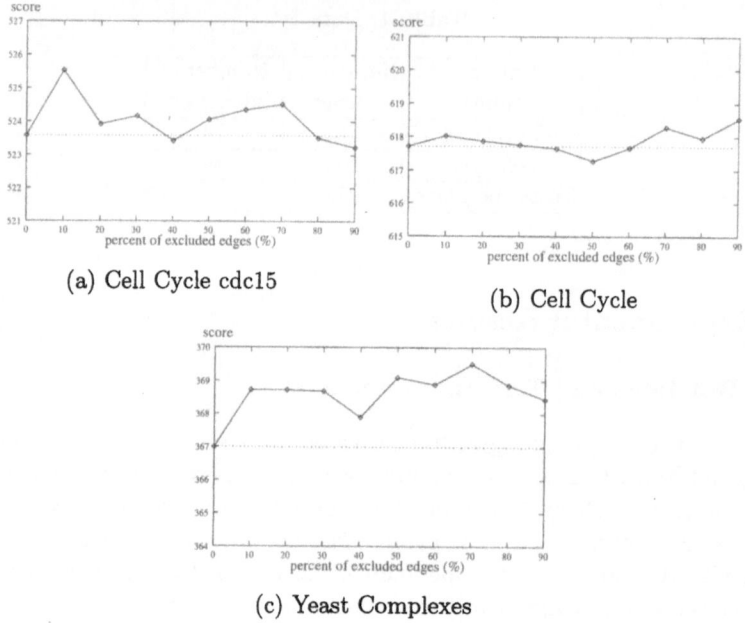

Fig. 3. Scores on a spectrum of different rates of excluded edges

results for most of the tested percents. In particular, for the Yeast Complexes data set, it showed improvement for all of the tested percents.

It is interesting that two peaks, not necessarily the highest ones, were observed at around 10% and 70% for all data sets. They imply that it is more favorable to exclude either a small or a large number of edges than do an intermediate number of edges in our experimental settings.

Table 2 compares the performance of our NNGA with state-of-the art algorithms for clustering gene expression profiles in terms of the score. The Single-, Complete-, and Average-linkage represent different versions of hierarchical clustering [10] and SOM [42] is a self-organizing map. We used the CLUSTER package [4] for the three versions of hierarchical clustering and the SOM. The NNGA with the new TSP formulation dominated the others.

It is more intuitive to inspect the visualized results than to just compare the scores between the algorithms for clustering gene expression profiles. To visualize them, we should assign a color to each expression level. We follow the typical red/green coloring scheme [41,10], while other schemes using different colors are available [41]. The red/green coloring scheme is as follows:

– Expression levels of zero are colored black, increasingly positive levels with reds of increasing intensity, and increasingly negative levels with greens of increasing intensity.
– Missing expression levels are usually colored gray.

---

[4] http://genome-www.stanford.edu/clustering

Table 2. Comparisons of NNGA with other algorithms in terms of best score

	Cell cycle cdc15	Cell cycle	Yeast Complexes
NNGA	**539**	**634**	**384**
FCGA	521	627	N.A.
Single-linkage	251	336	300
Complete-linkage	498	598	340
Average-linkage	500	581	331
SOM	461	578	306

N.A. : Not Available

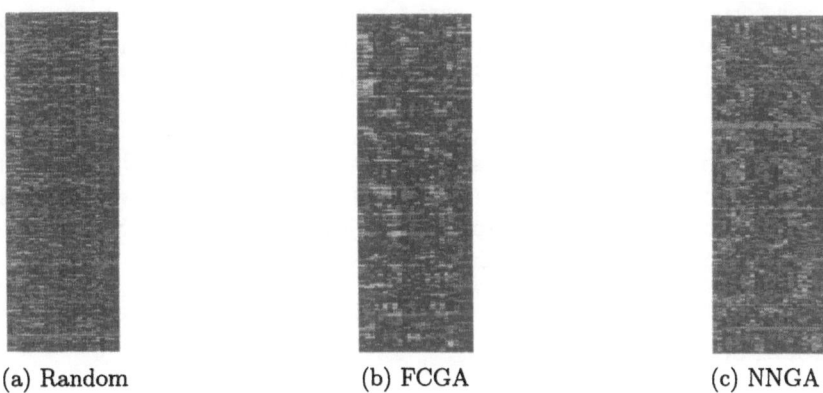

(a) Random   (b) FCGA   (c) NNGA

Fig. 4. Visualization results for Cell cycle cdc15

Figure 4 shows the visualization results for Cell cycle cdc15. In particular, Fig. 4(a) shows a random order, which is the original order and Fig. 4(b) and 4(c) show the best orders found by FCGA and NNGA, respectively. The NNGA shows a notable feature that clusters gene expression profiles with many missing data. One can find the clustered gray rows in Fig. 4(c).

6 Conclusions

We proposed a hybrid genetic algorithm for finding the optimal gene order in displaying the microarray data. To use the natural crossover, which exploits two-dimensional geographical information, we applied the Sammon's mapping to the data.

Furthermore, we improved the visualization results using our new TSP formulation. Our key idea is that reducing relatively long edges in a tour is less meaningful for visualization and it is thus advantageous to exclude the long edges from the fitness function. Experimental results showed that our idea improved the visualization results. Using the new fitness function, we could align more genes with the same group next to each other compared to state-of-the-art algorithms.

However, there is still a lot of work to give insight into the visualization of the microarray data. Since there have been no official criterion for evaluating visualization results, it is controversial to claim which one is better than the others in terms of a measure. We think that it is because most biologists analyze the results based on their limited visual intuition. We believe that examining what visualization is more meaningful to biologists is one of the most fundamental and demanding study.

The distance measure itself is also an interesting issue. While the Pearson correlation has been extensively used in the literature, it is not clear that the Pearson correlation is the best measure for defining the similarity between gene expression profiles. More elaborate distance measures such as an information-theoretic measure [11,25] are left for future studies.

Acknowledgments. The authors would like to thank Soonchul Jung and Huai-Kuang Tsai for invaluable discussions on this paper. This work was partly supported by Optus Inc. and Brain Korea 21 Project. The RIACT at Seoul National University provided research facilities for this study.

References

1. A. A. Alizadeh, M. B. Eisen, and et al. Distinct types of diffuse large b-cell lymphoma identified by gene expression profiling. *Nature*, 403(6769):503–511, 2000.
2. Z. Bar-Joseph, D. K. Gifford, and T. S. Jaakkola. Fast optimal leaf ordering for hierarchical clustering. *Bioinformatics*, 17:22–29, 2001.
3. A. Ben-Dor, R. Shamir, and Z. Yakhini. Clustering gene expresssion patterns. *Journal of Computational Biology*, 6:281–297, 1999.
4. J. L. Bentley. Experiments on traveling salesman problem. In *1st Annual ACM-SIAM Symposium on Discrete Algorithms (SODA '90)*, pages 129–133, 1990.
5. T. Biedl, B. Brejova, and et al. Optimal arrangement of leaves in the tree representing hierarchical clustering of gene expression data. Technical Report Technical Report CS-2001-14, Dept. of Computer Science, University of Waterloo, 2001.
6. T. N. Bui and B. R. Moon. Graph partitioning and genetic algorithms. *IEEE Transactions on Computers*, 45:841–855, 1996.
7. H. David. First (?) occurrence of common terms in mathematical statistics. *The American Statistician*, 49:121–133, 1995.
8. W. Dzwinel. How to make Sammon mapping useful for multidimensional data structures analysis. *Pattern Recognition*, 27(7):949–959, 1994.
9. A. Edwards and L. Cavalli-sforza. A method for cluster analysis. *Biometrics*, 21:362–375, 1965.
10. M. B. Eisen, P. T. Spellman, P. O. Brown, and D. Botstein. Cluster analysis and display of genome-wide expression patterns. In *Proceedings of the National Academy of Sciences*, pages 14863–14867, 1998.
11. A. M. Fraser. Reconstructing attractors from scalar time series: a comparison of singular system and redundancy criteria. *Physica D*, 34:391–404, 1989.

12. M. L. Fredman, D. S. Johnson, L. A. McGeoch, and G. Ostheimer. Data structures for traveling salesman. In *4th Annual ACM-SIAM Symposium on Discrete Algorithms (SODA '93)*, pages 145–154, 1993.
13. D. E. Goldberg. *Genetic algorithms in search, optimization and machine learning.* Addison-Wesley, Reading, MA, 1989.
14. D. E. Goldberg, K. Deb, and B. Korb. Do not worry, be messy. In *Proceedings of the Fourth International Conference on Genetic Algorithms*, pages 24–30, 1991.
15. R. Hamming. Error detecting and error correcting codes. *Bell systems Technical Journal*, 29(2):147–160, 1950.
16. J. Harris. The arithmetic of the product moment of calculating the coefficient of correlation. *American Nature*, 44:693–699, 1910.
17. J. Herrero, A. Valencia, and J. Dopazo. A hierarchical unsupervised growing neural network for clustering gene expression patterns. *Bioinformatics*, 17:126–136, 2001.
18. J. Holland. *Adaptation in natural and artificial systems.* University of Michigan Press, Ann Arbor, 1975.
19. D. S. Johnson. Local optimization and the traveling salesman problem. In *17th Colloquium on Automata, Languages, and Programming*, pages 446–461, 1990.
20. S. Jung and B. R. Moon. The natural crossover for the 2D Euclidean TSP. In *Genetic and Evolutionary Computation Conference*, pages 1003–1010, 2000.
21. S. Jung and B. R. Moon. Toward minimal restriction of genetic encoding and crossovers for the 2D Euclidean TSP. *IEEE Transactions on Evolutionary Computation*, 6(6):557–565, 2002.
22. S. Kawasaki, C. Borchert, and et al. Gene expression profiles during the initial phase of salt stress in rice. *Plant Cell*, 13(4):889–906, 2001.
23. M. Kendall. A new measure of rank correlation. *Biomerika*, 30:81–93, 1938.
24. A. B. Khodursky, B. J. Peter, and et al. DNA microarray analysis of gene expression in reponse to physiological and genetic changes that affect tryptophan metabolism in escherichia coli. In *Proceedings of the National Academy of Sciences*, pages 12170–12175, 2000.
25. W. Li. Mutual information functions versus correlation functions. *Journal of Statistical Physics*, 60:823–837, 1990.
26. S. Lin and B. Kernighan. An effective heuristic algorithm for the traveling salesman problem. *Operations Research*, 21(4598):498–516, 1973.
27. O. Martin, S. Otto, and E. Felten. Large-step Markov chains for the traveling salesman problem. *Complex Systems*, 5:299–236, 1991.
28. P. Merz and A. Zell. Clustering gene expression profiles with memetic algorithms. In *Proceedings of the 7th International Conference on Parallel Problem Solving from Nature*, pages 811–820, 2002.
29. P. Moscato. On evolution, search, optimization, genetic algorithms and martial arts: Towards memetic algorithms. Technical Report Technical Report C3P Report 826, Concurrent Computation Program, California Institute of Technology, 1989.
30. Y. Nagata and S. Kobayashi. Edge assembly crossover: A high-power genetic algorithm for the traveling saleman problem. In *7th International Conference on Genetic Algorithms*, pages 450–457, 1997.
31. E. Pekalska, D. De Ridder, R. P. W. Duin, and M. A. Kraaijveld. A new method of generalizing Sammon mapping with application to algorithm speed-up. In *Fifth Annual Conference of the Advanced School for Computing and Imaging*, pages 221–228, 1999.
32. J. M. Renders and H. Bersini. Hybridizing genetic algorithms with hill-climbing methods for global optimization: Two possible ways. In *Proceedings of the First IEEE Conference on Evolutionary Computation*, pages 312–317, 1994.

33. D. De Ridder and R. P. W. Duin. Sammon's mapping using neural networks: a comparision. *Pattern Recognition Letters*, 18(11–13):1307–1316, 1997.
34. J. W. Sammon, Jr. A non-linear mapping for data structure analysis. *IEEE Transactions on Computers*, 18:401–409, 1969.
35. R. Schaffer, J. Landgraf, and et al. Microarray analysis of diurnal and circadian-regulated genes in arabidopsis. *Plant Cell*, 13(1):113–123, 2001.
36. M. Schena, D. Shalon, R. W. Davis, and P. O. Brown. Quantitative monitoring of gene expresssion patterns with a complementary DNA microarray. *Science*, 270(5235):467–470, 1995.
37. D. Shalon, S. J. Smith, and P. O. Brown. A DNA microarray system for analyzing complex DNA samples using two-color fluorescent probe hybridization. *Genome Research*, 6(7):639–645, 1996.
38. R. R. Sokal and C. D. Michener. A statistical method for evaluating systematic relationships. *University of Kansas Science Bulletin*, 38:1409–1438, 1958.
39. C. Spearman. The proof and measurement of association between two things. *American Journal of Psychology*, 15:72–101, 1904.
40. T. S. Spellman, G. Sherlock, and et al. Comprehensive identification of cell cycle-regulated genes of the yeast *saccharomyces cerevisia* by microarray hybridization. *Molecular Biology of the Cell*, 9:3273–3297, 1998.
41. A. Sturn. Cluster analysis for large scale gene expression studies. Master's thesis, Graz University of Technology, Graz, Austria, 2001.
42. P. Tamayo, D. Slonim, and et al. Interpreting patterns of gene expresssion with self-organizing maps: Methods and application to hematopoietic differentiation. In *Proceedings of the National Academy of Sciences*, pages 2907–2912, 1999.
43. S. Tavazoie, J. D. Hughes, and et al. Systematic determination of genetic network architecture. *Nature Genetics*, 22:281–285, 1999.
44. P. Toronen, M. Kolehmainen, G. Wong, and E. Castren. Analysis of gene expression data using self-organizing maps. *FEBS Letters*, 451:142–146, 1999.
45. H. K. Tsai, J. M. Yang, and C. Y. Kao. A genetic algorithm for traveling salesman problems. In *Proceedings of the Genetic and Evolutionary Computation Conference (GECCO 2001)*, pages 687–693, 2001.
46. H. K. Tsai, J. M. Yang, and C. Y. Kao. Applying genetic algorithms to finding the optimal order in displaying the microarray data. In *Proceedings of the Genetic and Evolutionary Computation Conference (GECCO 2002)*, pages 610–617, 2002.
47. H. K. Tsai, J. M. Yang, and C. Y. Kao. Solving traveling salesman problems by combining global and local search mechanisms. In *Proceedings of the Congress on Evolutionary Computation (CEC 2002)*, pages 1290–1295, 2002.
48. J. Ward. Hierarchical grouping to optimize an objective function. *Journal of the American Statistical Association*, 58:236–244, 1963.
49. D. Whitley, V. Gordon, and K. Mathias. Larmarckian evolution, the baldwin effect and function optimization. In *International Conference on Evolutionary Computation*, Oct. 1994. *Lecture Notes in Computer Science*, 866:6–15, Springer-Verlag.
50. D. Whitley and J. Kauth. GENITOR: A different genetic algorithm. In *Proceedings of Rocky Mountain Conference on Artificial Intelligence*, pages 118–130, 1988.

Learning Features for Object Recognition

Yingqiang Lin and Bir Bhanu

Center for Research in Intelligent Systems
University of California, Riverside, CA, 92521, USA
{yqlin,bhanu}@vislab.ucr.edu

Abstract. Features represent the characteristics of objects and selecting or synthesizing effective composite features are the key factors to the performance of object recognition. In this paper, we propose a co-evolutionary genetic programming (CGP) approach to learn composite features for object recognition. The motivation for using CGP is to overcome the limitations of human experts who consider only a small number of conventional combinations of primitive features during synthesis. On the other hand, CGP can try a very large number of unconventional combinations and these unconventional combinations may yield exceptionally good results in some cases. Our experimental results with real synthetic aperture radar (SAR) images show that CGP can learn good composite features. We show results to distinguish objects from clutter and to distinguish objects that belong to several classes.

1 Introduction

In this paper, we apply genetic programming to synthesize composite features for object recognition. The basic task of object recognition is to identify the kinds of objects in an image, and sometimes the task may include to estimate the pose of the recognized objects. One of the key approaches to object recognition is based on features extracted from images. These features capture the characteristics of the object to be recognized and are fed into a classifier to perform recognition. The quality of object recognition is heavily dependent on the effectiveness of the features. However, it is difficult to extract good features from real images due to various factors, including noise. More importantly, there are many features that can be extracted. What are the appropriate features and how to synthesize composite features useful to the recognition from primitive features? The answers to these questions are largely dependent on the intuitive instinct, knowledge, experience and the bias of human experts.

In this paper, co-evolutionary genetic programming (CGP) is employed to generate a composite operator vector whose elements are synthesized composite operators for object recognition. A composite operator is represented by a binary tree whose internal nodes represent the pre-specified primitive operators and leaf nodes represent the primitive features, it is a way of combining primitive features. With each element evolved by a sub-population of CGP, a composite operator vector is cooperatively evolved by all the sub-populations. By applying composite operators, corresponding to each sub-population, to the primitive features extracted from images, we obtain composite feature vectors. These composite feature vectors are fed into a classifier for recognition. The primitive features are real numbers and designed by human experts based on the type of objects to be recognized. It is worth noting that the primitive operators and primitive features are decoupled from the CGP mechanism that generates composite operators. The users can tailor them to their own particular recognition task without affecting the other parts of the system. Thus, the method and the recognition system are flexible and can be applied to a wide variety of images.

2 Motivation and Related Research

- **Motivation:** The recognition accuracy of an automatic object recognition system is determined by the quality of the feature set. Usually, it is the human experts who design the features to be used in recognition. Handcrafting a set of features requires human ingenuity and insight into the characteristics of the objects to be recognized and in general, it is a very time consuming and expensive process due to the large number of features available and the correlations among them. Thus, automatic synthesis of composite features useful to the recognition from simple primitive features becomes extremely important. The process of synthesizing composite features can often be dissected into some primitive operations on primitive features. However, the ways of combining primitive features are almost infinite and human experts, relying on their knowledge, rich experience and limited by their speed and bias, can try only a small number of conventional combinations. Co-evolutionary genetic programming, on the other hand, may try many unconventional combinations and in some cases these unconventional combinations yield exceptionally good results. Also, the inherent parallelism of CGP and the speed of computers allow much more combinations to be considered by CGP compared to that by human experts and this greatly enhances the chances of finding good composite features.

- **Related Research:** Genetic programming (GP) has been used in image processing, object detection and recognition. Poli et al. [1] use GP to develop image filters to enhance and detect features of interest or to build pixel-classification-based segmentation algorithms. Bhanu and Lin [2] use GP for object detection and ROI extraction. Howard et al. [3] apply GP for automatic detection of ships in low resolution SAR imagery. Roberts and Howard [4] use GP to develop automatic object detectors in infrared images. Stanhope and Daida [5] use GP for the generation of rules for target/clutter classification and rules for the identification of objects. Unlike the work of Stanhope and Daida [5], the primitive operators in this paper are not logical operators, but operators that work on real numbers. They use GP to evolve logical expressions and the final outcome of the logical expressions determines the type of object under consideration (for example, 1 means target and 0 means clutter); we use CGP to evolve composite feature vectors for a Bayesian classifier and each sub-population is responsible for evolving a specific composite feature in the composite feature vector. The classifier evolved by GP in their system can be viewed as a linear classifier, but the classifier evolved by CGP here is a Bayesian classifier determined by the composite feature vectors learned from training images.

3 Technical Approach

In our CGP-based approach, individuals are composite operators and all possible composite operators form the huge search space, leading to the extreme difficulty in finding good composite operators unless one has a smart search strategy. The system we developed is divided into training and testing parts, which are shown in Fig. 1(a) and (b), respectively. During training, CGP runs on training images and evolves composite operators to obtain composite features. Since Bayesian classifier is completely determined by the composite feature vectors learned from training images, so both the composite features and the classifier are learned by CGP.

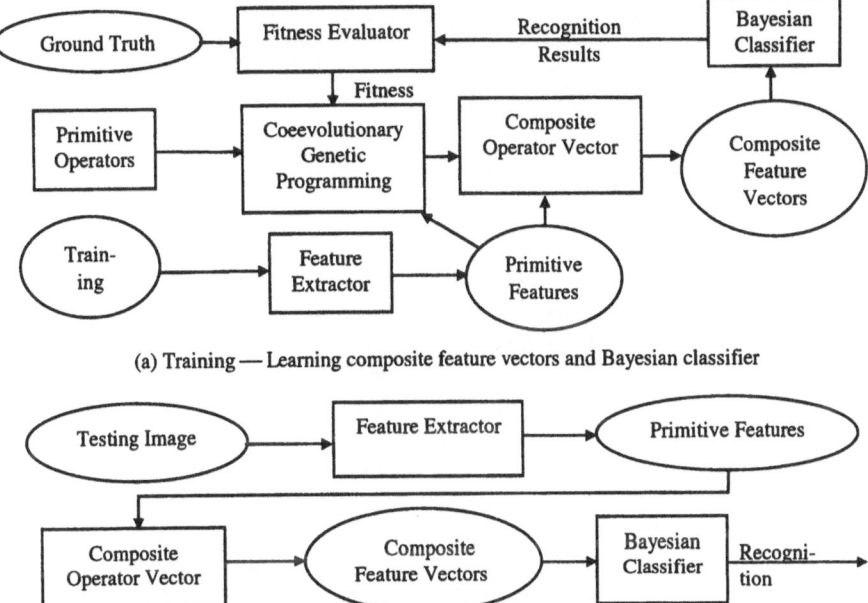

(a) Training — Learning composite feature vectors and Bayesian classifier

(b) Testing — Applying learned composite feature vectors and Bayesian classifier to a test image

Fig. 1. System diagram for object recognition using co-evolutionary genetic programming

3.1 Design Considerations

- **The Set of Terminals:** The set of terminals used in this paper are 20 primitive features used in [6]. The first 10 of them are designed by MIT Lincoln lab to capture the particular characteristics of synthetic aperture radar (SAR) imagery and are found useful for object detection. The other 10 features are common features used widely in image processing and computer vision. The 20 features are: (1) standard deviation of image; (2) fractal dimension and (3) weight rank fill ratio of brightest scatterers; (4) blob mass; (5) blob diameter; (6) blob inertia; (7) maximum and (8) mean values of pixels within blob; (9) contrast brightness of blob; (10) count; (11) horizontal, (12) vertical, (13) major diagonal and (14) minor diagonal projections of blob; (15) maximum, (16) minimum and (17) mean distances of scatterers from their centroid; (18) moment μ_{20}, (19) moment μ_{02} and (20) moment μ_{22} of scatters.

- **The Set of Primitive Operators:** A primitive operator takes one or two real numbers, performs a simple operation on them and outputs the result. Currently, 12 primitive operators shown in Table 1 are used, where a and b are real numbers and input to an operator and c is a constant real number stored in an operator.

Table 1. Twelve primitive operators

Primitive Operator	Description	Primitive Operator	Description
ADD (a, b)	Add a and b.	ADDC (a, c)	Add constant value c to a.
SUB (a, b)	Subtract b from a.	SUBC (a, c)	Subtract constant value c from a.
MUL (a, b)	Multiply a and b.	MUL (a, c)	Multiply a with constant value c.
DIV (a, b)	Divide a by b.	DIVC (a, c)	Divide a by constant value c.
MAX2 (a, b)	Get the larger of a and b.	MIN2 (a, b)	Get the smaller of a and b.
SQRT (a)	Return $\sqrt{a}$ if $a \geq 0$; otherwise, return $-\sqrt{-a}$.	LOG (a)	Return log(a) if $a \geq 0$; otherwise, return $-\log(-a)$.

- **The Fitness Measure:** the fitness of a composite operator vector is computed in the following way: apply each composite operator of the composite operator vector on the primitive features of training images to obtain composite feature vectors of training images and feed them to a Bayesian classifier. The recognition rate of the classifier is the fitness of the composite operator vector. To evaluate a composite operator evolved in a sub-population (see Fig. 2), the composite operator is combined with the current best composite operators in other sub-populations to form a complete composite operator vector where composite operator from the *ith* sub-population occupies the *ith* position in the vector and defines the fitness of the vector as the fitness of the composite operator under evaluation. The fitness values of other composite operators in the vector are not affected. When sub-populations are initially generated, the composite operators in each sub-population are evaluated separately without being combined with composite operators from other sub-populations. After each generation, the composite operators in the first sub-population are evaluated first, then the composite operators in the second sub-population and so on.

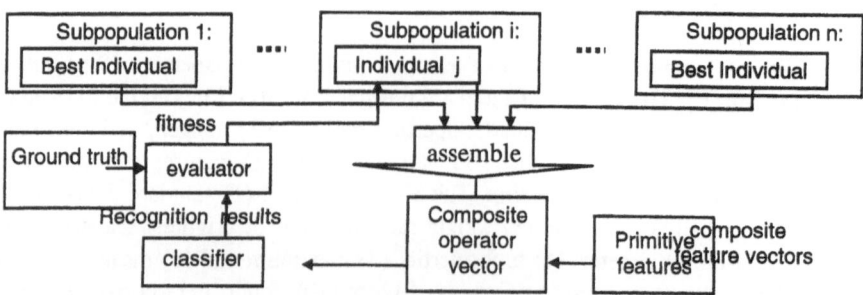

Fig. 2. Computation of fitness of *jth* composite operator of *ith* subpopulation

- **Parameters and Termination:** The key parameters are the number of sub-population N, the population size M, the number of generations G, the crossover and mutation rates, and the fitness threshold. GP stops whenever it finishes the specified number of generations or the performance of the Bayesian classifier is above the fitness threshold. After termination, CGP selects the best composite operator of each sub-population to form the learned composite operator vector to be used in testing.

3.2 Selection, Crossover, and Mutation

The CGP searches through the space of composite operator vectors to generate new composite operator vectors. The search is performed by selection, crossover and mutation operations. The initial sub-populations are randomly generated. Although sub-populations are cooperatively evolved (the fitness of a composite operator in a sub-population is not solely determined by itself, but affected by the composite operators from other sub-populations), selection is performed only on composite operators within a sub-population and the crossover is not allowed between two composite operators from different sub-populations.

- **Selection:** The selection operation involves selecting composite operators from the current sub-population. In this paper, we use tournament selection. The higher the fitness value, the more likely the composite operator is selected to survive.

- **Crossover:** Two composite operators, called parents, are selected on the basis of their fitness values. The higher the fitness value, the more likely the composite operator is selected for crossover. One internal node in each of these two parents is randomly selected, and the two subtrees rooted at these two nodes are exchanged between the parents to generate two new composite operators, called offspring. It is easy to see that the size of one offspring (i.e., the number of nodes in the binary tree representing the offspring) may be greater than both parents if crossover is implemented in such a simple way. To prevent code bloat, we specify a maximum size of a composite operator. If the size of one offspring exceeds the maximum size, the crossover is performed again until the sizes of both offspring are within the limit.

- **Mutation:** To avoid premature convergence, mutation is introduced to randomly change the structure of some composite operators to maintain the diversity of sub-populations. Candidates for mutation are randomly selected and the mutated composite operators replace the old ones in the sub-population. There are three mutations invoked with equal probability:
  1. Randomly select a node of the composite operator and replace the subtree rooted at this node by another randomly generated binary tree
  2. Randomly select a node of the composite operator and replace the primitive operator stored in the node with another primitive operator randomly selected from the primitive operators of the same number of input as the replaced one.
  3. Randomly selected two subtrees of the composite operator and swap them. Of course, neither of the two sub-trees can be the sub-tree of the other.

3.3 Generational Co-evolutionary Genetic Programming

Generational co-evolutionary genetic programming is used to evolve composite operators. The GP operations are applied in the order of crossover, mutation and selection. Firstly, two composite operators are selected on the basis of their fitness values for crossover. The two offspring from crossover are kept aside and won't participate in the following crossover operations on the current sub-population. The above process is repeated until the crossover rate is met. Then, mutation is applied to the composite operators in the current sub-population and the offspring from crossover. Finally, selection is applied to select some composite operators from the current sub-population and combine them with the offspring from crossover to get a new sub-population of the same size as the old one. In addition, we adopt an elitism replacement method that keeps the best composite operator from generation to generation.

- **Generational Co-evolutionary Genetic Programming:**

0. *randomly generate N sub-populations of size M and evaluate each composite operator in each sub-population individually.*
1. *for gen = 1 to generation_num do*
2.   *for i = 1 to N do*
3.   *keep the best composite operator in sub-population P_i.*
4.   *perform crossover on the composite operators in P_i until the crossover rate is satisfied and keep all the offspring from crossover.*
5.   *perform mutation on the composite operators in P_i and the offspring from crossover with the probability of mutation rate.*
6.   *perform selection on P_i to select some composite operators and combine them with the composite operators from crossover to get a new sub-population P_i' of the same size as P_i.*
7.   *evaluate each composite operator C_j in P_i'. To evaluate C_j, select the current best composite operator in each of the other sub-population, combine C_j with those N-1 best composite operators to form a composite operator vector where composite operator from kth sub-population occupy the kth position in the vector (k=1, ..., N). Run the composite operator vector on the primitive features of the training images to get composite feature vectors and use them to build a Bayesian classifier. Feed the composite feature vectors into the Bayesian classifier and let the recognition rate be the fitness of the composite operator vector and the fitness of C_j.*
8.   *let the best composite operator from P_i replace the worst composite operator in P_i' and let $P_i = P_i'$*
9.   *Form the composite operator vector consisting of the best composite operators from corresponding sub-populations and evaluate it. If its fitness is above the fitness threshold, goto 10.*
       endfor // loop 2
     endfor // loop 1
10. *select the best composite operator from each sub-population to form the learned composite operator vector and output it.*

4 Experiments

Various experiments are performed to test the efficacy of CGP in generating composite features for object recognition. In this paper, we show some selected examples. All the images used in experiments are real synthetic aperture radar (SAR) images and they are divided into training and testing images. 20 primitive features are extracted from each SAR image. CGP runs on primitive features from training images to generate a composite operator vector and a Bayesian classifier. The composite operator vector and the Bayesian classifier are tested against the testing images. It is to be noted that the ground truth is used only during training. The parameters of CGP used throughout the experiments are sub-population size (50), number of generations (50), fitness threshold (1.0), crossover rate (0.6) and mutation rate (0.05). The maximum size of composite operators is 10 in experiment 1 and 20 in experiment 2. The constant real number c stored in some primitive operators is from −20 to 20. For the purpose of objective comparison, CGP is invoked ten times for each experiment with the same set of parameters and the same set of training images. Only the average performances are used for comparison

- **Experiment 1 – Distinguish Object from Clutter:** From MSTAR public real SAR images, we generate 1048 SAR images containing objects and 1048 SAR images containing natural clutter. These images have size 120×120 and are called object images and clutter images, respectively. An example object image and clutter image are shown in Fig. 3, where white spots indicate scatterers with high magnitude. 300 object images and 300 clutter images are randomly selected as training images and the rest are used in testing.

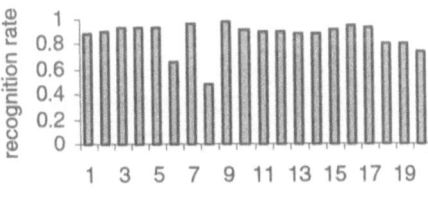

(a) A typical object image   (b) A typical natural clutter image

Fig. 3. Example target and clutter SAR images   **Fig. 4.** Recognition rates of 20 primitive features

First, the efficacy of each primitive feature in discriminating the object from clutter is examined. Each primitive feature from training images is used to train a Bayesian classifier and the classifier is tested against the same kind of primitive features from the testing images. The results are shown in Fig. 4. Feature contrast brightness of blob (9) is the best one with recognition rate 0.98. To show the efficacy of CGP in synthesizing effective composite features, we consider three cases: only the worst two primitive features (blob inertia (6) and mean values of pixels within blob (8)) are used by CGP; five bad primitive features (blob inertia (6), mean values of pixels within blob (8), moments μ_{20} (18), μ_{02} (19) and μ_{22} (20) of scatters) are used by CGP; 10 common features (primitive features 11 to 20) not specifically designed to deal with SAR images are used by CGP during synthesis. The number of sub-populations is 3, which means the dimension of the composite feature vector is 3. The results are shown in Fig. 5, where the horizontal coordinates are the number of primitive features used in synthesis and the vertical coordinates are the recognition rate. The bins on the left show the training results and those on the right show the testing results. The numbers above the bins are the average recognition rates over all ten runs. Then the number of sub-population is increased from 3 to 5. The same 2, 5 and 10 primitive features are used as building blocks by CGP to evolve composite features. The experimental results are shown in Fig. 6. Table 2 shows the maximum and minimum recognition rates in these experiments, where tr and te mean training and testing and max and min stand for the maximum and minimum recognition rates.

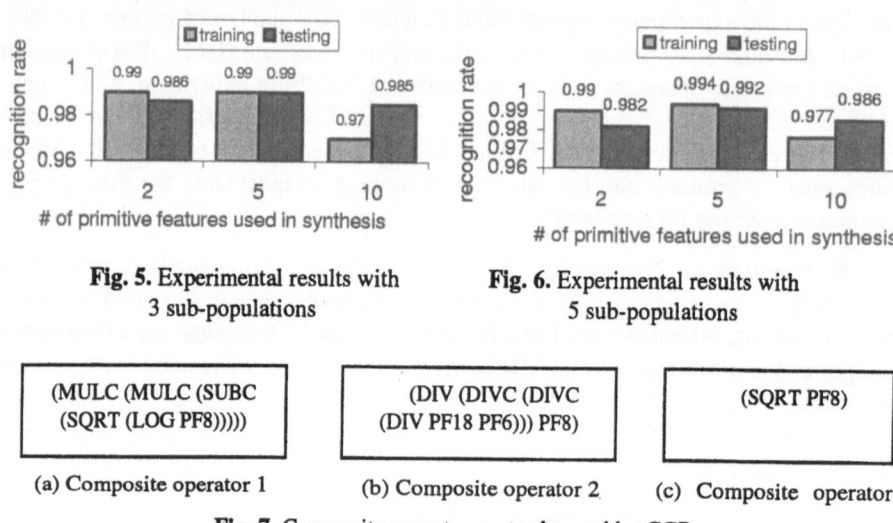

Fig. 5. Experimental results with 3 sub-populations

Fig. 6. Experimental results with 5 sub-populations

(a) Composite operator 1: (MULC (MULC (SUBC (SQRT (LOG PF8)))))

(b) Composite operator 2: (DIV (DIVC (DIVC (DIV PF18 PF6))) PF8)

(c) Composite operator: (SQRT PF8)

Fig. 7. Composite operator vector learned by CGP

Table 2. The maximum and minimum recognition rate

	3 sub-populations						5 sub-populations					
	2		5		10		2		5		10	
	Tr	Te	Tr	Te	Tr	Te	Tr	Te	Tr	Te	Tr	Te
Max	0.992	0.991	0.995	0.995	0.978	0.989	0.992	0.984	0.995	0.995	0.983	0.992
Min	0.988	0.979	0.978	0.974	0.965	0.979	0.988	0.98	0.993	0.987	0.972	0.979

From Figs. 5 and 6, it is obvious that composite feature vectors synthesized by CGP are very effective in distinguishing object from clutter. They are much better than the primitive features from which they are built. Actually, if both features 6 and 8 jointly form 2-dimensional primitive feature vectors for recognition, the recognition rate is 0.668; if features 6, 8, 18, 19, and 20 jointly form 5-dimensional primitive feature vectors, the recognition rate is 0.947; if all the last 10 primitive features are used, the recognition rate is 0.978. The average recognition rates of composite feature vectors are better than all the above results. Figure 7 shows the composite operator vector evolved by CGP maintaining 3 sub-populations in the 6th run when 5 primitive features are used, where PFi means the primitive feature i and so on.

- **Experiment 2 – Recognize Objects:** Five objects (BRDM2 truck, D7 bulldozer, T62 tank, ZIL131 truck and ZSU anti-aircraft gun) are used in the experiments. For each object, we collect 210 real SAR images under 15°-depression angle and various azimuth angles between 0° and 359° from MSTAR public data. Fig. 8 shows one optical and one SAR image of each object. From Fig. 8, we can see that it is not easy to distinguish SAR images of different objects. Since SAR images are very sensitive to azimuth angles and training images should represent the characteristics of the object under various azimuth angles, 210 SAR images of each object are sorted in the ascending order of their corresponding azimuth angles and the first, fourth, seventh, tenth SAR images and so on are selected for training. Thus, for each object, 70 SAR images are used in training and the rest are used in testing.

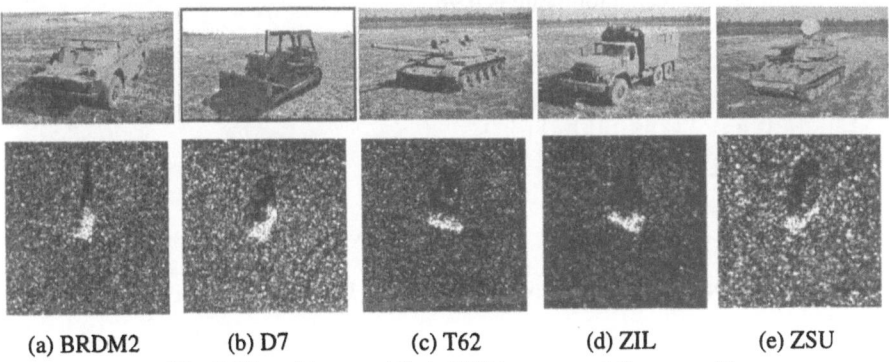

(a) BRDM2   (b) D7   (c) T62   (d) ZIL   (e) ZSU

Fig. 8. Five objects and their SAR images used for recognition

1) **Discriminate three objects:** CGP synthesizes composite features to recognize three objects: BRDM2, D7 and T62. First, the efficacy of each primitive feature in discriminating these three objects is examined. The results are shown in Fig. 9. Feature mean values of pixels within blob (8) is the best primitive feature with recognition 0.73. Three series of experiments are performed in which CGP maintains 3, 5 and 8 sub-populations to evolve 3, 5 and 8-dimensional composite features, respectively. The primitive features used in the experiments are all the 20 primitive features and the last 10 primitive features. The maximum size of the composite operators is 20. The experimental results are shown in Figs. 10, 11 and 12, where 10f and 20f mean primitive features 11 to 20 and all the 20 primitive features, respectively. The bins on the left show the training results and those on the right show the testing results. The numbers above the bins are the average recognition rates over all ten runs. Table 3 shows the maximum and minimum recognition rates in these experiments.

From Figs. 10, 11 and 12, it is clear that composite feature vectors synthesized by CGP are effective in recognizing objects. They are much better than primitive features used by CGP to synthesize composite features. Actually, if all 20 primitive features are used jointly to form 20-dimensional primitive feature vectors for recognition, the recognition rate is 0.96. This result is a little bit better than the average performance shown in Fig 10 (0.94), but the dimension of the feature vector is 20. However, the dimension of composite feature vectors in Figs 10 and 11 are just 3 and 5 respectively. If we increase the dimension of composite feature vector from 3 to 5 and 8, CGP results are better. If the last 10 primitive features are used, the recognition rate is 0.81. From these results, we can see that the effectiveness of the primitive features has an important impact on the effectiveness of the composite features synthesized by CGP. With effective primitive features in hand, CGP will synthesize better composite features. Fig. 13 shows the composite operator vector evolved by CGP maintaining 5 sub-populations in the 10th run when all 20 primitive features are used. The size of the first and second composite operators is 20. The size of the third composite operator is 9 and the size of the last one is 15. The fourth composite operator is very simple, just selects the primitive feature 11. The primitive features used by the synthesized composite operator vector are primitive features 2, 3, 4, 5, 6, 7, 8, 11, 12, 14, 18, 19, 20. If all these 13 primitive features directly form a 13-dimensional primitive feature vector for recognition, the recognition rate is 0.96.

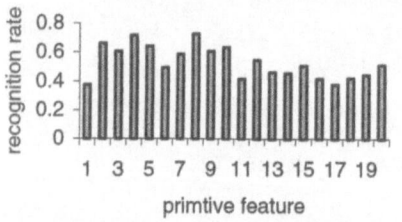

Fig. 9. Recognition rates of 20 primitive features

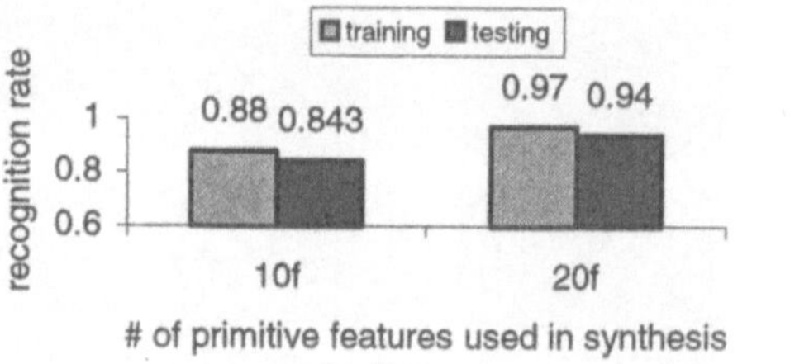

Fig. 10. Recognition rate with 3 sub-populations

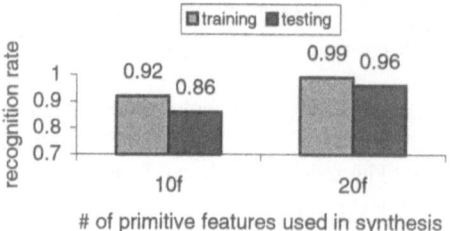

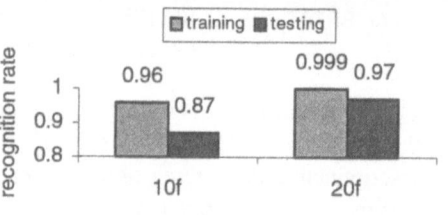

Fig. 11. Recognition rate with 5 sub-populations

Fig. 12. Recognition rate with 8 sub-populations

(DIV (MULC (SUB (SUB (DIVC (SQRT PF6)) (MULC (SUB PF18 (MULC (SUB PF18 (SQRT PF4)))))) (SQRT PF6))) (MIN2 PF12 PF19))

(a) Composite operator 1

(DIV (MULC (ADD (ADDC (MULC (MUL (MIN2 (ADDC (DIV PF20 PF4)) PF14) PF3))) (LOG (ADDC (DIV PF20 PF4))))) (DIVC PF4))

(b) Composite operator 2

(DIV (MIN2 (SUBC (SUBC PF11)) (MAX2 PF7 PF8)) PF8)

(c) Composite operator 3

(PF11)

(d) Composite operator 4

(LOG (ADDC (LOG (DIV (SUBC (LOG (DIV (SUBC (LOG PF5)) (SUBC PF5)))) (MUL PF2 PF5)))))

(e) Composite operator 5

Fig. 13. Composite operator vector learned by CGP for 5 sub-populations

Table 3. The maximum and minimum recognition rate

	3 sub-populations				5 sub-populations				8 sub-populations			
	10f		20f		10f		20f		10f		20f	
	Tr	Te	Tr	Te	Tr	Te	Tr	Te	Tr	Te	Tr	Te
Max	0.91	0.86	0.98	0.97	0.94	0.88	0.995	0.97	0.98	0.9	1.0	0.98
Min	0.85	0.83	0.95	0.92	0.91	0.84	0.98	0.94	0.95	0.85	0.995	0.95

2) Discriminate Five Objects: With more objects added, the recognition becomes more difficult. This can be seen from Fig. 14, which shows the efficacy of each primitive feature in discriminating these five objects. Feature blob mass (4) is the best primitive feature with recognition 0.49. If all 20 primitive features are used jointly to form 20-dimensional primitive feature vectors for recognition, the recognition rate is 0.81; if only the last 10 primitive features are used, the recognition rate is 0.62. This number is much lower, since the last 10 features are common features and are not designed with the characteristics of SAR images taken into consideration.

Two series of experiments are performed in which CGP maintains 5 and 8 sub-populations to evolve 5 and 8-dimensional composite features for recognition. The primitive features used in the experiments are all the 20 primitive features and the last 10 primitive features. The maximum size of composite operators is 20. The experimental results are shown in Fig. 15. The left two bins in columns 10f and 20f correspond to 5 sub-populations and the right two bins correspond to 8 sub-populations. The bins showing the training results are to the left of those showing testing results. The numbers above the bins are the average recognition rates over all ten runs. Table 4 shows the maximum and minimum recognition rates in these experiments.

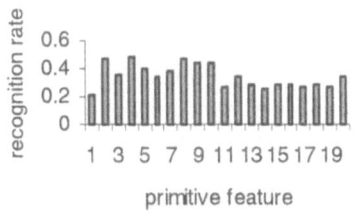

Fig. 14. Recognition rates of 20 primitive features

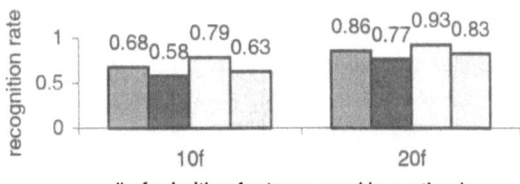

Fig. 15. Recognition rate with 5 (left two bins) and 8 (right two bins) sub-populations

Table 4. The maximum and minimum recognition rate

	5 sub-population				8 sub-population			
	10f		20f		10f		20f	
	Tr	Te	Tr	Te	Tr	Te	Tr	Te
Max	0.71	0.63	0.88	0.8	0.80	0.65	0.94	0.85
Min	0.65	0.55	0.83	0.73	0.75	0.62	0.91	0.80

From Fig.15, we can see that when the dimension of the composite feature vector is 8, the performance of the composite features is good and it is better than using all 20 (0.81) or 10 (0.62) primitive features from which the composite features are built. When the dimension of the composite feature vector is 5, the recognition is not satisfactory when using just 10 common features as building blocks. Also, when the dimension is 5, the average performance is a little bit worse than using all 20 or 10 primitive features, but the dimension of composite feature vector is just one-fourth or half of the number of primitive features, saving a lot of computational burden in recognition. When all 20 primitive features are used and CGP has 8 sub-populations, the composite operators in the best composite operator vector evolved have sizes 19, 1, 16, 19, 15, 7, 16 and 6, respectively. The primitive features used by the synthesized composite operator vector are primitive features 2, 3, 4, 5, 8, 9, 10, 11, 12, 13, 14, 15, 16, 18, 19 and 20. If all these 16 primitive features directly form a 16-dimensional primitive feature vector for recognition, the recognition rate is 0.80, which is lower than the average performance of the composite feature vector shown in Fig. 15.

- **Discussion:** The above experiments show that CGP is a viable tool to synthesize effective composite features from primitive features for object recognition and the learned composite features outperform the primitive features or any combination of primitive features from which they are evolved. The effectiveness of composite features learned by CGP is dependent on the effectiveness of primitive features. The usefulness of CGP is that it can evolve composite features that are more effective than the primitive ones upon which they are evolved. To achieve the same recognition rate, the number of composite features needed is smaller than the number of primitive features needed (one-fourth or half), thus, reducing the computational expenses during run-time recognition.

5 Conclusions

In this paper, CGP is used to synthesize composite features for object recognition. Our experimental results using real SAR images show that CGP can evolve composite features that are more effective than the primitive features upon which they are built. To achieve the same recognition performance of primitive features, fewer composite features are needed and this reduces the computational burden during recognition. However, primitive features still have a significant impact on the effectiveness of the evolved composite features. How to let CGP evolve effective composite features using general primitive features is the focus of our future research.

Acknowledgment. This research is supported by the grant F33615-99-C-1440. The contents of the information do not necessarily reflect the position or policy of the U. S. government.

References

1. R. Poli, "Genetic programming for feature detection and image segmentation," in *Evolutionary Computation*, T.C. Forgarty (Ed.), pp. 110–125, 1996.
2. B. Bhanu and Y. Lin, "Learning composite operators for object detection," *Proc. Genetic and Evolutionary Computation Conference*, pp. 1003–1010, July, 2002.
3. D. Howard, S.C. Roberts, and R. Brankin, "Target detection in SAR imagery by genetic programming," *Advances in Engg. Software*, vol. 30, no. 5, pp. 303–311, May 1999.

4. S.C. Roberts and D. Howard, "Evolution of vehicle detectors for infrared line scan imagery," *Proc. Evolutionary Image Analysis, Signal Processing and Telecommunications, First European Workshops*, pp. 110–125, Springer-Verlag, 1999.
5. S.A. Stanhope and J. M. Daida, "Genetic programming for automatic target classification and recognition in synthetic aperture radar imagery," *Proc. Conference. Evolutionary Programming VII*, pp. 735–744, 1998.
6. B. Bhanu and Y. Lin, "Genetic algorithm based feature selection for target detection in SAR images," *Image and Vision Computing*, 2003.

An Efficient Hybrid Genetic Algorithm for a Fixed Channel Assignment Problem with Limited Bandwidth

Shouichi Matsui, Isamu Watanabe, and Ken-ichi Tokoro

Communication & Information Research Laboratory (CIRL)
Central Research Institute of Electric Power Industry (CRIEPI)
2-11-1 Iwado-kita, Komae-shi, Tokyo 201-8511, Japan
{matsui,isamu,tokoro}@criepi.denken.or.jp

Abstract. We need an efficient channel assignment algorithm for increasing channel re-usability, reducing call-blocking rate and reducing interference in any cellular systems with limited bandwidth and a large number of subscribers. We propose an efficient hybrid genetic algorithm for a fixed channel assignment problem with limited bandwidth constraint. The proposed GA finds a good sequence of codes for a virtual machine that produces channel assignment. Results are given which show that our GA produces far better solutions to several practical problems than existing GAs.

1 Introduction

The channel assignment problem (CAP), or frequency assignment problem (FAP) is a very important problem today, but is a difficult, NP-hard problem. The radio spectrum is a limited natural resource used in a variety of private and public services, the most well known example can be found in cellular mobile phone systems, or personal communication services (PCS). To facilitate this expansion the radio spectrum allocated to a particular service provider needs to be assigned as efficiently and effectively as possible.

Because the CAP is a very important problem in the real world and an NP-hard problem, a number of heuristic algorithms have been proposed (e.g., [6]). To achieve the optimal solution of fixed channel assignment problems, most proposed algorithms try to minimize the amount of necessary channels while satisfying a set of given constraints (e.g., [2,3,5,8,9,10,11,19]).

However, the total number of available channels, or bandwidth of frequencies, are given and fixed in many situations. Minimizing the bandwidth becomes meaningless for such applications. To address the problem, Jin et al. proposed a new cost model [7] in which the available number of channels are given and fixed and the electro-magnetic compatibility constraints, and demand constraint are relaxed. They also proposed genetic algorithms to solve the problems [4,7]. But, their algorithms use naïve representation, and the search space is too large in scale, therefore the results obtained using these algorithms are not good enough.

We propose a new hybrid genetic algorithm for the problem. Because the proposed algorithm combines GA with a local search, it can be thought of as a memetic algorithm [13]. The proposed GA is tested using a set of standard benchmark problems, and the performance is far better than the previously proposed GAs [4,15] for the problem [4]. The proposed GA can obtain far better solutions than the previously proposed GAs.

2 Fixed Channel Assignment with Limited Bandwidth

2.1 Notation

Let us consider a cellular system that consists of N cells, and each cell is numbered from 1 to N. A compatibility matrix is a symmetric $N \times N$ matrix $C = (c_{ij})$ with nonnegative integer elements. The value c_{ij} prescribes the minimum frequency separation required between frequencies assigned to cell i and cell j, i.e., if f_i^k and f_j^l are the frequencies assigned to cell i and j respectively, then the following condition $|f_i^k - f_j^l| \geq c_{ij}$ should be satisfied for all i and j.

Radio frequencies are assumed to be evenly spaced, therefore they can be identified with the positive integers. Let $X_{i,j}$ be the variable that takes 1 when the j-th mobile that stays in cell i wishes to own a frequency, and n_i be the number of mobiles that stay in cell i, and let T be the number of channels that each frequency provides under TDMA (Time Division Multiple Access). And let $X_i = \sum_{j=1}^{n_i} X_{i,j}$ be the random variable of required channels in cell i, and let μ_i and σ_i be the expected number and the standard deviation of X_i (no matter the request is failed or success). In general, the load of each cell is maintained by the system, therefore μ_i and σ_i can be estimated from the long-term history of base stations' load data.

2.2 Damage of Blocked Calls

Call attempts may fail at a busy station because there are no available channels, and failed calls are called *blocked calls*. The more blocked calls, the more damage caused to the system. Thus, an ideal channel assignment should guarantee that the total amount of blocked calls in all cells be as low as possible.

Let H_i be the number of assigned channels to cell i, then the expected number of blocked calls in cell i is $\sum_{j=H_i+1}^{n_i} P(X_i = j)(j - H_i)$. The objective here is to minimize the cost $\sum_{i=1}^{N} \sum_{j=H_i+1}^{n_i} P(X_i = j)(j - H_i)$. As Horng et al. showed, we can assume the random variable X_i ($1 \leq i \leq N$) is close to the normal random variable with parameters μ_i and σ_i [4]. Therefore, as an approximation of $\sum_{j=x+1}^{n_i} P(X_i = j)(j - x)$, we can use

$$I_E(x) = \frac{1}{\sqrt{2\pi}\sigma_i} \int_x^\infty (y-x) \exp\left\{-\frac{1}{2}\left(\frac{y-\mu_i}{\sigma_i}\right)^2\right\} dy$$

$$= \frac{1}{\sqrt{2\pi}}\sigma_i \exp\left\{-\frac{1}{2}\left(\frac{x-\mu_i}{\sigma_i}\right)^2\right\} + \frac{1}{2}(\mu_i - x)\mathrm{erfc}\left(\frac{x-\mu_i}{\sqrt{2}\sigma_i}\right),$$

where $\mathrm{erfc}(x)$ is the complementary error function defined as

$$\text{erfc}(x) = 1 - \text{erf}(x) = \frac{2}{\sqrt{\pi}} \int_x^\infty \exp(-t^2) dt.$$

2.3 Damage from Interference

Interference can occur between a pair of transmitters if the interfering signal is sufficiently strong. Whether a transmitter pair has the potential to interfere depends on many factors, e.g., distance, terrain, power, or antenna design. The higher the potential for interference between a transmitter pair is, the larger the frequency separation required. For example, if two transmitters are sufficiently geographically separated then the frequency can be re-used, i.e., the same frequency can be assigned. At the other extreme if two transmitters are located at the same site then they may require, say, a five-frequency separation.

Violating frequency separation constraint, or EMC (Electro-Magnetic Compatibility) constraint, would bring some degree of disadvantage to the mobiles that experience interference. And the disadvantage would be in proportion to the degree of interference that depends on the frequency distance (i.e., how many Hz between them) and the power it suffered. The degree of damage is defined as follows. Let p be the assigned frequency to cell i, and q be the one to cell j, then the damage caused by interference from this assignment $f(i,j,p,q)$ is defined as follows.

$$f(i,j,p,q) = \begin{cases} 0 & \text{if } |p-q| \geq c_{ij}, \\ f_{i,p} f_{j,q} I_C(c_{ij} - |p-q|) & \text{if } |p-q| < c_{ij} \text{ and } i = j, \\ f_{i,p} f_{j,q} I_A(c_{ij} - |p-q|) & \text{otherwise.} \end{cases}$$

where $f_{i,p} = 1$ if frequency p is assigned to cell i, otherwise $f_{i,p} = 0$, and I_C and I_A are two strictly increasing functions.

2.4 Objective Function

The objective of the problem is to minimize the total damage, the sum of the cost of blocked calls and the cost of interference, therefore the problem is defined as follows.

$$\text{Min } O = \sum_{i=1}^{N} \sum_{j=1}^{N} \sum_{p=1}^{Z} \sum_{q=1}^{Z} f(i,j,p,q) + \alpha \sum_{i=1}^{N} I_E(TF_i),$$

subject to $f_{i,p} = 0$ or 1 for $1 \leq i \leq N$ and $1 \leq p \leq Z$ where $F_i = \sum_{p=1}^{Z} f_{i,p}$ for $1 \leq i \leq N$, Z is the allowable number of frequencies, and α is the relative weight of the damage of blocked calls to the damage from interference.

2.5 Related Works

Because CAP (FAP) is an important and very difficult problem to solve exactly, GA based algorithms for the minimum span frequency assignment problem (MS-FAP) have been proposed (e.g., [2,3,5,8,9,10,11,14,17,19]).

The performance of the GAs for MSFAP that represent possible assignment directly as a bit-string or a sequence of integers is not good enough, and the permutation based GAs are reported to show good performance [10,11,19]. In these GAs, an assignment order of transmitters is represented by a permutation and an assignment is carried out using a sequential algorithm. The scheme has overcome the weakness of the previous two schemes and the performance has improved [10,11,19], and a GA with an adaptive mutation rate and a new initialization method was developed and showed very good performance [11]. The performance of the permutation based GAs are high for MSFAP, but they are designed to find an assignment without violating compatibility constraints. Therefore, they cannot be used for the problem shown in this section.

The formulation shown above, which is quite different from MSFAP, is first proposed and a GA for solving the problem was developed by Jin et al. [7], and an improved version was proposed by Horng et al. [4] and Park et al. [15]. However they use naïve representation of $N \times Z$ matrix that is bad coding and they use simple GA. Rothlauf et al. [16] showed that we should use well-designed GA in the case of bad codings, therefore the performance of previously proposed GAs is not good enough.

3 The Proposed Algorithm

We propose a new algorithm for the FAP formulated in the previous section. The main idea of the proposed GA is that sequences of codes of a virtual machine that performs assignment are encoded as chromosomes, and the GA searches for a good code sequence that minimizes the total damage cost.

The genetic algorithm (GA) used here is an example of 'steady state', overlapping populations GA [12]. The proposed GA is outlined in Figure 1.

3.1 How Many Frequencies Are Necessary for a Cell?

The second term of the objective function decreases as we increase the number of assigned frequencies (F_i), but the first term, interference cost, would increase. Let us consider a cost function

$$D_i(F_i) = \sum_{p=1}^{Z} \sum_{q=1}^{Z} f(i,i,p,q) + \alpha D_B(TF_i),$$

which consists of the blocking cost and interference cost within a cell i only considering the co-site channel separation constraint.

We can find a good candidate of F_i using $D_i(F_i)$. The frequency separation in a cell decreases as we increase F_i, and the interference cost increases, whereas $D_B(TF_i)$ decreases. Therefore we can find F_i that minimize $D_i(F_i)$ by evaluating $D_i(F_i)$ for $F_i = 1, 2, \cdots, \bar{F}_i$, i.e., the optimal F_i^* can be defined as

$$F_i^* = \operatorname*{argmin}_{F_i \in \{1,2,\cdots,\bar{F}_i\}} D_i(F_i).$$

```
Procedure GA
BEGIN
Initialize:
  (1) Calculate F_i^* (i = 1, 2, ···, N).
  (2) Generate N individuals (N being the population size), and randomly generate
      mutation rate N_m^i.
  (3) Produce N assignments and store each assignment.
  (4) Store best-so-far.
LOOP
  (1) Generate offspring: Generate N_n offspring.
      (1-1) Select two parents by roulette wheel rule;
      (1-2) Apply crossover with the probability of P_c, and generate two offspring;
            when no crossover is applied, then two parents will be the offspring;
      (1-3) Apply adaptive mutation to offspring and generate one offspring by
            Adaptive-Mutation;
      (1-4) If the offspring is better than best-so-far then replace best-so-far.
  (2) Selection: Select best N individuals from the pool of old N and new N_n
      individuals.
UNTIL stopping condition satisfied.
Print best-so-far.
END
```

Fig. 1. Outline of the proposed GA.

Because $D_B(x)$ rapidly decreases and becomes close to zero when $x \geq \mu_i + 5\sigma$, we can set $\bar{F}_i = \lceil \mu_i + 5\sigma \rceil$. The minimal frequency separation S_i^m of each cell i is calculated by

$$S_i^m = \min\{\lfloor (Z-1)/(F_i^* - 1) \rfloor, c_{ii}\}.$$

We can use $TGT = \sum_{i=1}^{N} D_i(F_i^*)$ as a target of total cost.

3.2 Virtual Machine

Let us consider a virtual machine that assigns the frequency one by one to cells according to a given code. The cost function $D_i(F_i^*)$ is minimized by assigning all frequencies with the separation of S_i^m when $(Z-1)\mathbf{mod}(F_i^* - 1) = 0$. When $(Z-1)\mathbf{mod}(F_i^* - 1) \neq 0$, some frequencies must be assigned with the separation $S_i^m + 1$. Because $D_i(F_i)$ does not consider the compatibility constraint between cells i and j, the above assignment does not minimize the total cost, therefore some assignments should be with the separation that does not violate the inter-cell constraints.

With these observations, the virtual machine must have at least 3 codes or operations that are shown in Table 1. The assignment order to cells is important, so an instruction of the virtual machine is defined as a pair of integers, namely (cell_number, action).

The virtual machine assigns a frequency f that is determined by the Table 1 to cell cell_number when $1 \leq f \leq Z$, and do nothing in other cases. When

Table 1. Action specification.

Action	Assigning frequency
0	frequency with separation of S_i^m
1	frequency with separation of $S_i^m + 1$
2	minimum usable frequency

a frequency f is assigned to cell i, the set of usable frequencies of all cells are updated according to the compatibility matrix C. In this step, we use S_i^m instead of c_{ii}.

3.3 Chromosome

A chromosome of the GA is a sequence of the instructions. A single instruction is a pair of integers, namely (p, a) where $1 \leq p \leq N$, and $a \in \{0, 1, 2\}$.

An instruction (p, a) assigns a frequency to cell p, therefore a valid chromosome must have F_i^* times of instructions whose cell number is i for each cell i, and the length of the sequence becomes $L = \sum_i^n F_i^*$. Thus a chromosome is expressed as a sequence $\{(p_1, a_1), (p_2, a_2), \cdots, (p_L, a_L)\}$.

3.4 Local Search

After the assignment by the virtual machine we could improve the assignment by a local search. If a cell has a frequency f_p that violates the interference constraints, and there is a usable frequency f_q, then we can replace f_p by f_q, and the replacement always reduces the total cost. After this local search, the chromosome is modified to reflect the modification.

The modification algorithm changes the order of instructions in the chromosome and the action of the genes according to the result of the local search. The basic ideas are as follows;

- If the frequency is generated by the local search, the corresponding instruction is moved towards the tail.
- The action that assigns the minimum usable frequency should be kept unchanged.

The modification algorithm shown in Figure 2 is applied to each cell in decreasing order of the interference damage of each cell.

3.5 Crossover

We use a crossover operator that does not generate an invalid chromosome. If we ignore the action part of the genes, a chromosome is a permutation of cell numbers. Because cells have multiple channel demands, a permutation contains

```
Procedure Modification
BEGIN
Initialize:
  (1) Let two sequences of S₁ and S₂ be empty.
  (2) Sort assigned frequencies in ascending order for each cell, and result be
      (f_{i,1}, f_{i,2}, ···, f_{i,F_i}). And also let f_{i,0} ← 1 − S_i^m.
Scan genes from head to tail:
LOOP
  (1) Let i be the cell number, and a be the action of current gene that corresponds
      to the k-th assignment to cell i.
  (2) Calculate channel separations s ← f_{i,k} − f_{i,(k−1)}.
  (3) If (s > S_i^m + 1) then t ← 2 else t ← s − S_i^m.
  (4) If t = 2 then
        If f_{i,k} is generated by the local search
        then append instruction (i, t) to S₂,
        else append instruction (i, t) to S₁.
      else
        If a = 2 then append instruction (i, a) to S₁,
        else append instruction (i, t) to S₁.
UNTIL all genes are scanned.
Return the concatenation of S₁ and S₂.
END
```

Fig. 2. Modification algorithm.

multiple occurrence of the same cell numbers. Therefore, the candidates are Generalized Order Crossover (GOX) [1], Generalized Partially Mapped Crossover (GPMX) [1], Precedence Preservation Crossover (PPX) [1], and modified PMX (mPMX) [11].

By comparing the performance of the four operators, we decided to use GPMX as the crossover operator. Because GPMX attained the best average, and the difference between the best solution attained by GPMX and the other operators is small, we decided to use GPMX. The comparisons is shown in 'Experiments and Results' section.

3.6 Mutation

The mutation is done in 2 steps. In the first step, instructions are mutated by swap mutation, i.e., randomly chosen instruction at position p and q are swapped, i.e., (n_p, a_p) and (n_q, a_q) is swapped.

In the second step, the action part of randomly chosen genes are mutated by flip mutation, i.e., the value of the action part of a randomly chosen gene g_i is replaced by a randomly chosen integer in the range $[0, 2]$.

The mutation operator does not generate any invalid chromosomes, i.e., offsprings are always valid sequence of code for the virtual machine.

Self-Adaptive Mutation Rate. The efficiency of GA depends on two factors, namely the maintenance of suitable working memory, and quality of the

Procedure Adaptive-Mutation
BEGIN

Initialization: Let L be the length of chromosome, o_1, o_2 be the offspring, and N_m^1, N_m^2 be the product of L and the mutation rate of o_1, o_2 respectively. Let $U()$ be a function that returns random variable uniformly distributed in the range of $[0, 1)$.

Update Mutation Rate:
  if $U() < (N_m^1/L)$ then
    if $U() < 1/2$ then $N_m^{1\prime} \leftarrow N_m^1 + 1$ else $N_m^{1\prime} \leftarrow N_m^1 - 1$
  else $N_m^{1\prime} \leftarrow N_m^1$
  if $U() < (N_m^2/L)$ then
    if $U() < 1/2$ then $N_m^{2\prime} \leftarrow N_m^2 + 1$ else $N_m^{2\prime} \leftarrow N_m^2 - 1$
  else $N_m^{2\prime} \leftarrow N_m^2$

Apply Mutation: Apply mutation to offspring o_1 with the rate of $N_m^{1\prime}/L, N_m^{2\prime}/L$ and the results be o_1^1, o_1^2. And apply mutation to offspring o_2 with the rate of $N_m^{1\prime}/L, N_m^{2\prime}/L$ and the results be o_2^1, o_2^2. Individuals inherit the mutation rate that generated them.

Selection: Select the best individual from $o_1^1, o_1^2, o_2^1, o_2^2$.
END

Fig. 3. Adaptive mutation and reproduction method.

match between the probability density function generated and the landscape being searched. The first of these factors will depend on the choice of population size and selection algorithm. The second will depend on the action of the reproductive operators, i.e., crossover and mutation operators, and the set of associative parameters on the current population [18].

We reported that the performance of their permutation based GA for MSFAP heavily depends on the mutation rate [10], and we also showed that an adaptive mutation rate mechanism improved the performance greatly [11]. Therefore, we have built an individual level self-adaptive mutation rate mechanism into our algorithm.

Proposed Mutation Scheme. The mutation rate r_m^i for each individual i is coded as an integer N_m^i in the range of $[N_m^L, N_m^U]$, and the mutation rate r_m^i is defined as $r_m^i = N_m^i/L$. The values of N_m^L, N_m^U used in the numerical experiments will be discussed later.

In the mutation step, the mutation rate N_m^i is first modified according to the value of itself. The N_m^i is incremented or decremented with the mutation rate of r_m^i, and the result $r_m^{i\prime}$ is used as the mutation rate for the individual i. The mutation and the resulting offsprings are created by the algorithm shown in Figure 3.

3.7 Selection

The selection of the GA is roulette wheel selection. The selection probability p_1 for the first parent is calculated for each genotype based on its rank. The selection probability p_2 for the second parent is calculated based on its fitness[1].

3.8 Other Genetic Operators

We use sigma truncation for the scaling function. The fitness function F' is defined as

$$F' = F - (\mu_F - 2 \times \sigma_F),$$

where $F = 1/O$, μ_F is the average, and σ_F is the standard deviation of F over the population.

The GA evolves population to maximize the F'.

4 Experiments and Results

4.1 Benchmark Problems

We tested the algorithm using the problem proposed by Horng [4]. The eight benchmark problems were examined in our experiments. Three compatibility matrices C_3, C_4, C_5 and three communication load tables D_1, D_2, D_3, are combined to make eight problems [4].

The interference cost function $I_C(x) = 5^{x-1}$ and $I_A(x) = 5^{2x-1}$, the weight of blocking cost $\alpha = 1000$, and the value of T (number of channels that each frequency provides under TDMA) was set to 8^2.

4.2 Parameters

The population size for each problem was 100, the maximal number of generation was 1000, The GA terminates if the variance of fitness of the population becomes 0, and does not change during contiguous 200 generations. The crossover ratio (P_c) is 0.8, and the number of new offsprings (N_n) is equal to the population size. The lower bound and the upper bound of the mutation rate is set $N_m^L = 3$ and $N_m^U = \lfloor 0.1 \times L \rfloor$.

---

[1] Valenzuela et al. used similar asymmetric selection probability in their GA for MSFAP [19]. In their GA, the first parent is selected deterministically in sequence, but the second parent is selected in a roulette wheel fashion with the probability based on its rank.

[2] According to private communications with an author of the paper, there are typos in the paper [4]. The typos are the communication load table of Problem 1, and the weight α. And there is no description of the value of T. We used the correct data provided by the author. The results by Park et al.[15] used the same corrected data.

Table 2. Comparison of simulation results

Prob	TGT	Horng et al.[4]		Park et al.[15]			Our GA		
		Best	Ave	Best	Ave	CPU[†]	Best	Ave	CPU[‡]
P1	3.7e-4	203.4	302.6	0.4	0.5	65504	3.7e-4	3.7e-4	75.3
P2	4.1	271.4	342.5	27.9	30.9	88692	4.1	4.1	68.4
P3	4.1	1957.4	2864.1	63.1	79.3	89918	5.8	7.1	196.0
P4	231	906.3	1002.4	675.8	684.1	95585	243.8	248.3	69.4
P5	231	4302.3	4585.4	1064.1	1092.5	87905	535.8	659.8	255.4
P6	190	4835.4	5076.2	1149.8	1227.3	35790	552.1	692.2	249.7
P7	2232	20854.3	21968.4	5636.7	5831.8	37323	3243,5	3537.9	249.4
P8	22518	53151.7	60715.4	41883.0	41967.5	135224	27714.3	29845.4	813.8

†: CPU seconds on Pentium III 750MHz, ‡: CPU seconds on Pentium III 933MHz.

4.3 Results

We tested the performance of the GA by running 100 times for each problem. The results are shown in Table 2. Comparison with the results by Horng et al.[4] and by Park et al.[15] and the target of total cost defined in the subsection 3.1 are also given. The column P denotes the problem name, and TGT denotes the target of total cost. The column 'Best' shows the minimum cost found in the 100 runs, the column 'Ave' shows the average cost over the 100 runs, and the column 'CPU' shows the computational time for a single run.

Table 2 shows that our GA performs very well, and outperforms the previous GAs [4,15] for all cases. The cost obtained by our GA is very small compared to the others. Our GA can find a very good assignment with the same cost of the target for P1 and P2, and the cost is very close to the target for P3 and P4.

The GA can obtain assignments without any violation of the EMC constraints for the problems P1, P2, and P3. The performance improvement is significant for all problems.

The CPU time is very short compared with the GA by Park et al., it is almost 1/100 for all problems. Our GA not only runs faster, but also the solution is better than the GA by Park et al.

Comparison of Crossover Operators. We ran the proposed GA with the same parameters shown above, only changing the crossover operators. Table 3 shows the comparison of crossover operators.

Table 3 shows the following.

- GPMX attained the best average solution for all eight problems.
- The best solution of P1, P2, P3, and P4 is the same among the four operators. GPMX attained the best solution for P8, PPX attained the best solution for P6 and P7, mPMX attained the best solution for P5.
- All four operators attained better results than that of Horng et al. [4] or Park et al. [15].

Table 3. Comparison of crossover operators

Prob	GOX Best	GOX Ave	GPMX Best	GPMX Ave	PPX Best	PPX Ave	mPMX Best	mPMX Ave
P1	3.7e-4	3.7e-4	3.7e-4	3.7e-4	3.7e-4	3.7e-4	3.7e-4	3.7e-4
P2	4.1	4.1	4.1	4.1	4.1	4.1	4.1	4.1
P3	5.8	7.1	5.8	7.1	5.8	8.0	5.8	7.7
P4	243.8	248.8	243.8	248.3	243.8	249.7	243.8	248.6
P5	548.5	713.7	535.8	659.8	558.0	690.0	526.7	691.4
P6	608.0	766.1	552.1	692.2	539.2	723.6	574.4	766.6
P7	3309.7	3693.4	3243.5	3537.9	3233.3	3667.8	3273.3	3651.2
P8	30308.6	31655.8	27714.3	29845.4	28842.5	30392.1	29040.2	30637.1

5 Conclusions

We have presented here a new efficient hybrid genetic algorithm for fixed channel assignment problems with limited bandwidth constraint. The algorithm uses the GA to find a good sequence of codes for a virtual machine that executes the assignment task, and to improve the performance, an adaptive mutation rate mechanism is developed.

The proposed GA is tested using a set of benchmark problems, and the performance is superior to existing GAs. The proposed GA can obtain very good solutions that were unable to be found using the previously proposed GAs for the problem.

We believe that our approach, finding good code sequence for a virtual machine by GA, with the adaptive mutation rate mechanism can be applied to other real world problems.

Acknowledgments. The authors are grateful to Mr. Ming-Hui Jin of National Central University, Chungli, Taiwan for providing us data for the experiment.

References

1. Bierwirth, C., Mattfeld, D.C., and Kopfer, H.: On permutation representations for scheduling problems, *Proc. 4th International Conference on Parallel Problem Solving from Nature—PPSN IV*, pp.310–318, 1996.
2. Crompton, W., Hurley, S., and Stephens, N.M.: Applying genetic algorithms to frequency assignment problems, *Proc. SPIE Conf. Neural and Stochastic Methods in Image and Signal Processing*, vol.2304, pp.76–84, 1994.
3. Cuppini, M.: A genetic algorithm for channel assignment problems, *Eur. Trans. Telecommun.*, vol.5, no.2, pp.285–294, 1994.
4. Horng, J.T., Jin, M.H., and Kao, C.Y.: Solving fixed channel assignment problems by an evolutionary approach, *Proc. of Genetic and Evolutionary Computation Conference 2001 (GECCO-2001)*, pp.351–358, 2001.

5. Hurley, S. and Smith, D.H.: Fixed spectrum frequency assignment using natural algorithms, *Proc. of Genetic Algorithms in Engineering Systems: Innovations and Applications*, pp.373–378, 1995.
6. Hurley, S., Smith,D.H., and Thiel, S.U.: FASoft: a system for discrete channel frequency assignment, *Radio Science*, vol.32, no.5, pp.1921–1939, 1997.
7. Jin, M.H., Wu, H.K., Horng,J.Z., and Tsai, C.H.: An evolutionary approach to fixed channel assignment problems with limited bandwidth constraint, *Proc. IEEE Int. Conf. Commun. 2001*, vol.7, pp.2100–2104, 2001.
8. Kim, J.-S., Park,S.H., Dowd, P.W., and Nasrabadi, N.M.: Comparison of two optimization techniques for channel assignment in cellular radio network, *Proc. of IEEE Int. Conf. Commun.*, vol.3, pp.850–1854, 1995.
9. Lai, K.W. and Coghill,G.G.: Channel assignment through evolutionary optimization, *IEEE Trans. Veh. Technol.*, vol.45, no.1, pp.91–96, 1996.
10. Matsui, S. and Tokoro, K.: A new genetic algorithm for minimum span frequency assignment using permutation and clique, *Proc. of Genetic and Evolutionary Computation Conference 2000 (GECCO-2000)*, pp.682–689, 2000.
11. Matsui, S. and Tokoro, K.: Improving the performance of a genetic algorithm for minimum span frequency assignment problem with an adaptive mutation rate and a new initialization method, *Proc. of Genetic and Evolutionary Computation Conference 2001 (GECCO-2001)*, pp.1359–1366, 2001.
12. Mitchell, M.:*An Introduction to Genetic Algorithms*, MIT Press, 1996.
13. Moscate, P: On evolution, search, optimization, genetic algorithms and martial arts: towards memetic algorithms, Caltech Concurrent Computation Program, C3P Report 826, 1989.
14. Ngo, C.Y. and Li, V.O.K.: Fixed channel assignment in cellular radio networks using a modified genetic algorithm, *IEEE Trans. Veh. Technol.*, vol.47, no.1, pp.163–172, 1998.
15. Park, E. J., Kim, Y. H., and Moon, B. R., Genetic search for fixed channel assignment problem with limited bandwidth, *Proc. of Genetic and Evolutionary Computation Conference 2002 (GECCO-2002)*, pp.1172–1179, 2002.
16. Rothlauf, F., Goldberg, D.E., and Heinzl, A.: Bad coding and the utility of well-designed genetic algorithms, *Proc. of Genetic and Evolutionary Computation Conference 2000 (GECCO-2000)*, pp. 355–362, 2000.
17. Smith, D.H., Hurley, S., and Thiel, S.U.: Improving heuristics for the frequency assignment problem, *Eur. J. Oper. Res.*, vol.107, no.1, pp.76–86, 1998.
18. Smith, J.E.: *Self Adaptation in Evolutionary Algorithms*, Ph.D thesis, Univ. of the West England, Bristol, 1998.
19. Valenzuela, C., Hurley, S., and Smith, D.: A permutation based algorithm for minimum span frequency assignment, *Proc. 5th International Conference on Parallel Problem Solving from Nature—PPSN V*, Amsterdam, pp. 907–916, 1998.

Using Genetic Algorithms for Data Mining Optimization in an Educational Web-Based System

Behrouz Minaei-Bidgoli and William F. Punch

Genetic Algorithms Research and Applications Group (GARAGe)
Department of Computer Science & Engineering
Michigan State University
2340 Engineering Building
East Lansing, MI 48824
{minaeibi,punch}@cse.msu.edu
http://garage.cse.msu.edu

Abstract. This paper presents an approach for classifying students in order to predict their final grade based on features extracted from logged data in an education web-based system. A combination of multiple classifiers leads to a significant improvement in classification performance. Through weighting the feature vectors using a Genetic Algorithm we can optimize the prediction accuracy and get a marked improvement over raw classification. It further shows that when the number of features is few; feature weighting is works better than just feature selection.

1 Statement of Problem

Many leading educational institutions are working to establish an online teaching and learning presence. Several systems with different capabilities and approaches have been developed to deliver online education in an academic setting. In particular, Michigan State University (MSU) has pioneered some of these systems to provide an infrastructure for online instruction. The research presented here was performed on a part of the latest online educational system developed at MSU, the *Learning Online Network with Computer-Assisted Personalized Approach* (*LON-CAPA*).

In LON-CAPA[1], we are involved with two kinds of large data sets: 1) educational resources such as web pages, demonstrations, simulations, and individualized problems designed for use on homework assignments, quizzes, and examinations; and 2) information about users who create, modify, assess, or use these resources. In other words, we have two ever-growing pools of data.

We have been studying data mining methods for extracting useful knowledge from these large databases of students using online educational resources and their recorded paths through the web of educational resources. In this study, we aim to an-

---

[1] See http://www.lon-capa.org

E. Cantú-Paz et al. (Eds.): GECCO 2003, LNCS 2724, pp. 2252–2263, 2003.
© Springer-Verlag Berlin Heidelberg 2003

swer the following research questions: Can we find *classes* of students? In other words, do there exist groups of students who use these online resources in a *similar* way? If so, can we identify that class for any individual student? With this information, can we *help* a student use the resources better, based on the usage of the resource by other students in their groups?

We hope to find similar patterns of use in the data gathered from LON-CAPA, and eventually be able to make predictions as to the most-beneficial course of studies for each learner based on their present usage. The system could then make suggestions to the learner as to how to best proceed.

2 Map the Problem to Genetic Algorithm

Genetic Algorithms have been shown to be an effective tool to use in data mining and pattern recognition. [7], [10], [6], [16], [15], [13], [4]. An important aspect of GAs in a learning context is their use in pattern recognition. There are two different approaches to applying GA in pattern recognition:

1. Apply a GA directly as a classifier. Bandyopadhyay and Murthy in [3] applied GA to find the decision boundary in N dimensional feature space.

2. Use a GA as an optimization tool for resetting the parameters in other classifiers. Most applications of GAs in pattern recognition optimize some parameters in the classification process. Many researchers have used GAs in feature selection [2], [9], [12], [18]. GAs has been applied to find an optimal set of feature weights that improve classification accuracy. First, a traditional feature extraction method such as Principal Component Analysis (PCA) is applied, and then a classifier such as k-NN is used to calculate the fitness function for GA [17], [19]. Combination of classifiers is another area that GAs have been used to optimize. Kuncheva and Jain in [11] used a GA for selecting the features as well as selecting the types of individual classifiers in their design of a Classifier Fusion System. GA is also used in selecting the prototypes in the case-based classification [20].

In this paper we will focus on the second approach and use a GA to optimize a combination of classifiers. Our objective is to *predict* the students' final grades based on their web-use features, which are extracted from the homework data. We design, implement, and evaluate a series of pattern classifiers with various parameters in order to compare their performance on a dataset from LON-CAPA. Error rates for the individual classifiers, their combination and the GA optimized combination are presented.

2.1 Dataset and Class Labels

As test data we selected the student and course data of a LON-CAPA course, PHY183 (Physics for Scientists and Engineers I), which was held at MSU in spring semester 2002. This course integrated 12 homework sets including 184 problems, all of which are online. About 261 students used LON-CAPA for this course. Some of students dropped the course after doing a couple of homework sets, so they do not have any final grades. After removing those students, there remained 227 valid samples. The grade distribution of the students is shown in Fig 1.

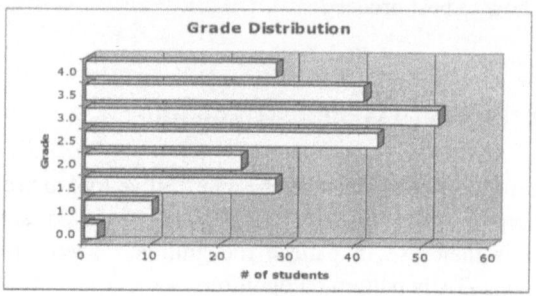

Fig. 1. Graph of distribution of grades in course PHY183 SS02

We can group the students regarding their final grades in several ways, 3 of which are:

1. Let the 9 possible class labels be the same as students' grades, as shown in table 1
2. We can label the students in relation to their grades and group them into three classes, *"high"* representing grades from 3.5 to 4.0, *"middle"* representing grades from 2.5 to 3, and *"low"* representing grades less than 2.5.
3. We can also categorize the students with one of two class labels: *"Passed"* for grades higher than 2.0, and *"Failed"* for grades less than or equal to 2.0, as shown in table 3.

Table 1. Selecting 9 class labels regarding to students' grades in course PHY183 SS02

Class	Grade	Student #	Percentage
1	0.0	2	0.9%
2	0.5	0	0.0%
3	1.0	10	4.4%
4	1.5	28	12.4%
5	2.0	23	10.1%
6	2.5	43	18.9%
7	3.0	52	22.9%
8	3.5	41	18.0%
9	4.0	28	12.4%

Table 2. Selecting 3 class labels regarding to students' grades in course PHY183 SS02

Class	Grade	Student #	Percentage
High	Grade >= 3.5	69	30.40%
Middle	2.0 < Grade < 3.5	95	41.80%
Low	Grade <= 2.0	63	27.80%

Table 3. Selecting 2 class labels regarding to students' grades in course PHY183 SS02

Class	Grade	Student #	Percentage
Passed	Grade > 2.0	164	72.2%
Failed	Grade <= 2.0	63	27.80%

We can predict that the error rate in the first class grouping should be higher than the others, because the sample size of the grades over 9 classes differs considerably. It is clear that we have less data for the first three classes in the training phase, and so the error rate would likely be higher in the evaluation phase.

2.2 Extractable Features

An essential step in doing classification is selecting the features used for classification. Below we discuss the features from LON-CAPA that were used, how they can be visualized (to help in selection) and why we normalize the data before classification.

The following features are stored by the LON-CAPA system:

1. Total number of correct answers. (Success rate)
2. Getting the problem right on the first try, vs. those with high number of tries. (Success at the first try)
3. Total number of tries for doing homework. (Number of attempts before correct answer is derived)
4. Time spent on the problem until solved (more specifically, the number of hours until correct. The difference between time of the last successful submission and the first time the problem was examined). Also, the time at which the student got the problem correct relative to the due date. Usually better students get the homework completed earlier.
5. Total time spent on the problem regardless of whether they got the correct answer or not. (Difference between time of the last submission and the first time the problem was examined).
6. Participating in the communication mechanisms, vs. those working alone. LON-CAPA provides online interaction both with other students and with the instructor. Where these used?
7. Reading the supporting material before attempting homework vs. attempting the homework first and then reading up on it.

8. Submitting a lot of attempts in a short amount of time without looking up material in between, versus those giving it one try, reading up, submitting another one, and so forth.
9. Giving up on a problem versus students who continued trying up to the deadline.
10. Time of the first log on (beginning of assignment, middle of the week, last minute) correlated with the number of tries or number of solved problems. A student who gets all correct answers will not necessarily be in the successful group if they took an average of 5 tries per problem, but it should be verified from this research.

In this paper we focused on the first six features in the PHY183 SS02 dataset that we have chosen for the classification experiment.

2.3 Classifiers

Pattern recognition has a wide variety of applications in many different fields, such that it is not possible to come up with a single classifier that can give good results in all the cases. The optimal classifier in every case is highly dependent on the problem domain. In practice, one might come across a case where no single classifier can classify with an acceptable level of accuracy. In such cases it would be better to pool the results of different classifiers to achieve the optimal accuracy. Every classifier operates well on different aspects of the training or test feature vector. As a result, assuming appropriate conditions, combining multiple classifiers may improve classification performance when compared with any single classifier [4].

The scope of this survey is restricted to comparing some popular non-parametric pattern classifiers and a single parametric pattern classifier according to the error estimate. Six different classifiers using the LON-CAPA datasets are compared in this study. The classifiers used in this study include *Quadratic Bayesian classifier, 1-nearest neighbor (1-NN), k-nearest neighbor (k-NN), Parzen-window, multi-layer perceptron (MLP),* and *Decision Tree.*[2] These classifiers are some of the common classifiers used in most practical classification problems. After some preprocessing operations were made on the dataset, the error rate of each classifier is reported. Finally, to improve performance, a combination of classifiers is presented.

2.4 Normalization

Having assumed in Bayesian and Parzen-window classifiers that the features are normally distributed, it is necessary that the data for each feature be normalized. This ensures that each feature has the same weight in the decision process. Assuming that the given data is Gaussian distributed, this normalization is performed using the mean and standard deviation of the training data. In order to normalize the training data, it is necessary first to calculate the sample mean μ, and the standard deviation σ of

---

[2] The first five classifiers are coded in MATLAB™ 6.0, and for the decision tree classifiers we have used some available software packages such as C5.0, CART, QUEST, and CRUISE.

each feature, or column, in this dataset, and then normalize the data using the equation (1)

$$x_i = \frac{x_i - \mu}{\sigma} \qquad (1)$$

This ensures that each feature of the training dataset has a normal distribution with a mean of zero and a standard deviation of one. In addition, the kNN method requires normalization of all features into the same range. However, we should be cautious in using the normalization before considering its effect on classifiers' performances.

2.5 Combination of Multiple Classifiers (CMC)

In combining multiple classifiers we want to improve classifier performance. There are different ways one can think of combining classifiers:
- The simplest way is to find the overall error rate of the classifiers and choose the one which has the least error rate on the given dataset. This is called an *offline CMC*. This may not really seem to be a CMC; however, in general, it has a better performance than individual classifiers.
- The second method, which is called *online CMC*, uses all the classifiers followed by a vote. The class getting the *maximum votes* from the individual classifiers will be assigned to the test sample. This method intuitively seems to be better than the previous one. However, when tried on some cases of our dataset, the results were not better than the best result in previous method. So, we changed the rule of majority vote from *"getting more than 50% votes"* to *"getting more than 75% votes"*. This resulted in a significant improvement over offline CMC.

Using the second method, we show in table 4 that CMC can achieve a significant accuracy improvement in all three cases of 2, 3, and 9-classes. Now we are going to use GA to find out that whether we can maximize the CMC performance.

3 Optimizing the CMC Using a GA

We used GAToolBox[3] for MATLAB to implement a GA to optimize classification performance. Our goal is to find a population of best weights for every feature vector, which minimize the classification error rate.

The feature vector for our predictors are the set of six variables for every student: Success rate, Success at the first try, Number of attempts before correct answer is derived, the time at which the student got the problem correct relative to the due date, total time spent on the problem, and the number of online interactions of the student both with other students and with the instructor.

---

[3] Downloaded from http://www.shef.ac.uk/~gaipp/ga-toolbox/

We randomly initialized a population of six dimensional weight vectors with values between 0 and 1, corresponding to the feature vector and experimented with different number of population sizes. We found good results using a population with 200 individuals. The GA Toolbox supports binary, integer, real-valued and floating-point chromosome representations. Real-valued populations may be initialized using the Toolbox function *crtrp*. For example, to create a random population of 6 individuals with 200 variables each: we define boundaries on the variables in *FieldD* which is a matrix containing the boundaries of each variable of an individual.

```
FieldD = [ 0 0 0 0 0 0;  % lower bound
           1 1 1 1 1 1]; % upper bound
```

We create an initial population with Chrom = crtrp(200, FieldD), So we have for example:
```
Chrom = 0.23 0.17 0.95 0.38 0.06 0.26
        0.35 0.09 0.43 0.64 0.20 0.54
        0.50 0.10 0.09 0.65 0.68 0.46
        0.21 0.29 0.89 0.48 0.63 0.89
        ..................
```

We used the simple genetic algorithm (SGA), which is described by Goldberg in [9]. The SGA uses common GA operators to find a population of solutions which optimize the fitness values.

3.1 Recombination

We used "*Stochastic Universal Sampling*" [1] as our selection method. A form of stochastic universal sampling is implemented by obtaining a cumulative sum of the fitness vector, *FitnV*, and generating N equally spaced numbers between 0 and sum(FitnV). Thus, only one random number is generated, all the others used being equally spaced from that point. The index of the individuals selected is determined by comparing the generated numbers with the cumulative sum vector. The probability of an individual being selected is then given by

$$F(x_i) = \frac{f(x_i)}{\sum_{i=1}^{N_{ind}} f(x_i)} \quad (2)$$

where $f(x_i)$ is the fitness of individual x_i and $F(x_i)$ is the probability of that individual being selected.

3.2 Crossover

The crossover operation is not necessarily performed on all strings in the population. Instead, it is applied with a probability Px when the pairs are chosen for breeding. We selected $Px = 0.7$. There are several functions to make crossover on real-valued matrices. One of them is *recint*, which performs intermediate recombination between pairs of individuals in the current population, OldChrom, and returns a new population after mating, NewChrom. Each row of OldChrom corresponds to one individual. *recint* is a function only applicable to populations of real-value variables. Intermediate recombination combines parent values using the following formula [14]:

$$Offspring = parent1 + Alpha \times (parent2 - parent1) \qquad (3)$$

Alpha is a Scaling factor chosen uniformly in the interval [-0.25, 1.25]

3.3 Mutation

A further genetic operator, mutation is applied to the new chromosomes, with a set probability *Pm*. Mutation causes the individual genetic representation to be changed according to some probabilistic rule. Mutation is generally considered to be a background operator that ensures that the probability of searching a particular subspace of the problem space is never zero. This has the effect of tending to inhibit the possibility of converging to a local optimum, rather than the global optimum.

There are several functions to make mutation on real-valued population. We used *mutbga*, which takes the real-valued population, OldChrom, mutates each variable with given probability and returns the population after mutation, *NewChrom = mutbga(OldChrom, FieldD, MutOpt)* takes the current population, stored in the matrix OldChrom and mutates each variable with probability by addition of small random values (size of the mutation step). We considered 1/600 as our mutation rate. The mutation of each variable is calculated as follows:

$$Mutated\ Var = Var + MutMx \times range \times MutOpt(2) \times delta \qquad (4)$$

where delta is an internal matrix which specifies the normalized mutation step size; MutMx is an internal mask table; and MutOpt specifies the mutation rate and its shrinkage during the run. The mutation operator *mutbga* is able to generate most points in the hypercube defined by the variables of the individual and the range of the mutation. However, it tests more often near the variable, that is, the probability of small step sizes is greater than that of larger step sizes.

3.4 Fitness Function

During the reproduction phase, each individual is assigned a fitness value derived from its raw performance measure given by the objective function. This value is used in the selection to bias towards more fit individuals. Highly fit individuals, relative to the whole population, have a high probability of being selected for mating whereas less fit individuals have a correspondingly low probability of being selected. The error rate is measured in each round of cross validation by dividing "the total number of misclassified examples" into "total number of test examples". Therefore, our *fitness function* measures the error rate achieved by CMC and our objective would be to maximize this performance (minimize the error rate).

4 Experiment Results

Without using GA, the overall results of classifiers' performance on our dataset, regarding the four tree-classifiers, five non-tree classifiers and CMC are shown in the Table 4. Regarding individual classifiers, for the case of 2-classes, kNN has the best performance with 82.3% accuracy. In the case of 3-classes and 9-classes, CART has the best accuracy of about 60% in 3-classes and 43% in 9-Classes. However, considering the combination of non-tree-based classifiers, the CMC has the best performance in all three cases. That is, it achieved 86.8% accuracy in the case of 2-Classes, 71% in the case of 3-Classes, and 51% in the case of 9-Classes.

Table 4. Comparing the Error Rate of all classifiers on PHY183 dataset in the cases of 2-Classes, 3-Classess, and 9-Classes, using 10-fold cross validation, **without GA**

Classifier		Performance %		
		2-Classes	3-Classes	9-Classes
Tree Classifier	C5.0	80.3	56.8	25.6
	CART	81.5	**59.9**	**33.1**
	QUEST	80.5	57.1	20.0
	CRUISE	81.0	54.9	22.9
Non-tree Classifier	Bayes	76.4	48.6	23.0
	1NN	76.8	50.5	29.0
	KNN	**82.3**	50.4	28.5
	Parzen	75.0	48.1	21.5
	MLP	79.5	50.9	-
	CMC	**86.8**	**70.9**	**51.0**

For GA optimization, we used 200 individuals in our population, running the GA over 500 generations. We ran the program 10 times and got the averages, which are shown, in table 5. In every run 500×200 times the fitness function is called in which we used 10-fold cross validation to measure the average performance of CMC. So every classifier is called 3×10^6 times for the case of 2-classes, 3-classes and 9-classes. Thus, the time overhead for fitness evaluation is critical. Since using the MLP in this process took about 2 minutes and all other four non-tree classifiers (Bayes, 1NN, 3NN, and Parzen window) took only 3 seconds, we omitted the MLP from our classifiers group so we could obtain the results in a reasonable time.

Table 5. Comparing the CMC Performance on PHY183 dataset Using GA and without GA in the cases of 2-Classes, 3-Classess, and 9-Classes, 95% confidence interval.

	Performance %		
Classifier	2-Classes	3-Classes	9-Classes
CMC of 4 Classifiers without GA	83.87 ± 1.73	61.86 ± 2.16	49.74 ± 1.86
GA Optimized CMC, Mean individual	94.09 ± 2.84	72.13 ± 0.39	62.25 ± 0.63
Improvement	10.22 ± 1.92	10.26 ± 1.84	12.51 ± 1.75

The results in Table 5 represent the mean performance with a two-tailed t-test with a 95% confidence interval. For the improvement of GA over non-GA result, a P-value indicating the probability of the Null-Hypothesis (There is no improvement) is also given, showing the significance of the GA optimization. All have p<0.000, indicating significant improvement. Therefore, using GA, in all the cases, we got more than a 10% mean individual performance improvement and about 12 to 15% mean individual performance improvement. Fig. 2 shows the best result of the ten runs over our dataset. These charts represent the population mean, the best individual at each generation and the best value yielded by the run.

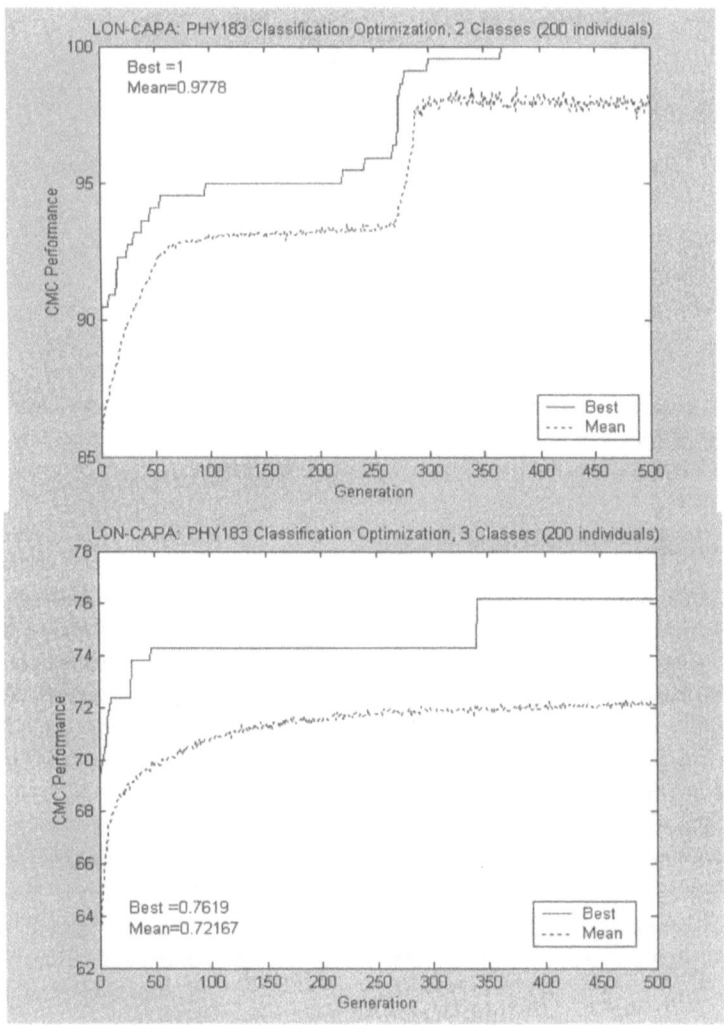

Fig. 2. Graph of GA Optimized CMC performance in the case of 2, and 3-Classes

Finally, we can examine the individuals (weights) for features by which we obtained the improved results. This feature weighting indicates the importance of each feature for making the required classification. In most cases the results are similar to Multiple Linear Regressions or tree-based software that use statistical methods to measure feature importance. Table 6 shows the importance of the six features in the 3-classes case using the Entropy splitting criterion. Based on entropy, a statistical property called *information gain* measures how well a given feature separates the training examples in relation to their target classes. Entropy characterizes *impurity* of an arbitrary collection of examples S at a specific node N. In [5] the impurity of a node N is denoted by $i(N)$ such that:

$$Entropy(S) = i(N) = -\sum_{j} P(\omega_j) \log_2 P(\omega_j) \tag{5}$$

where $P(\omega_j)$ is the fraction of examples at node N that go to category ω_j.

Table 6. Feature Importance in 3-Classes Using Entropy Criterion

Feature	Importance %
Total_Correct_Answers	100.00
Total_Number_of_Tries	58.61
First_Got_Correct	27.70
Time_Spent_to_Solve	24.60
Total_Time_Spent	24.47
Communication	9.21

The GA results also show that the "Total number of correct answers" and the "Total number of tries" are the most important features for the classification. The second column in table 6 shows the percentage of feature importance.

5 Conclusions and Future Work

Four classifiers were used to segregate the students. A combination of multiple classifiers leads to a significant accuracy improvement in all 3 cases. Weighing the features and using a genetic algorithm to minimize the error rate improves the prediction accuracy at least 10% in the all cases of 2, 3 and 9-Classes. In cases where the number of features is low, the feature weighting worked much better than feature selection. The successful optimization of student classification in all three cases demonstrates the merits of using the LON-CAPA data to *predict* the students' final grades based on their features, which are extracted from the homework data.

We are going to apply Genetic Programming to produce many different combinations of features, to extract new features and improve prediction accuracy. We plan to use Evolutionary Algorithms to classify the students and problems directly as well. We also want to apply Evolutionary Algorithms to find Association Rules and Dependency among the groups of problems *(Mathematical, Optional Response, Numerical, Java Applet, and so forth)* of LON-CAPA homework data sets.

Acknowledgements. This work was partially supported by the National Science Foundation under ITR 0085921.

References

1. Baker, J. E.: Reducing bias and inefficiency in the selection algorithm, Proceeding ICGA 2, Lawrence Erlbuam Associates, Publishers, (1987) 14–21
2. Bala J., De Jong K., Huang J., Vafaie H.: and Wechsler H. Using learning to facilitate the evolution of features for recognizing visual concepts. Evolutionary Computation 4(3) - Special Issue on Evolution, Learning, and Instinct: 100 years of the Baldwin Effect. (1997)
3. Bandyopadhyay, S., and Muthy, C.A.: Pattern Classification Using Genetic Algorithms, Pattern Recognition Letters, Vol. 16, (1995) 801–808
4. De Jong K.A., Spears W.M. and Gordon D.F.: Using genetic algorithms for concept learning. Machine Learning 13, (1993) 161–188
5. Duda, R.O., Hart, P.E., and Stork, D.G.: Pattern Classification. 2nd Edition, John Wiley & Sons, Inc., New York NY. (2001)
6. Falkenauer E.: Genetic Algorithms and Grouping Problems. John Wiley & Sons, (1998)
7. Freitas, A.A.: A survey of Evolutionary Algorithms for Data Mining and Knowledge Discovery,See: www.pgia.pucpr.br/~alex/papers. A chapter of: A. Ghosh and S. Tsutsui. (Eds.) Advances in Evolutionary Computation. Springer-Verlag, (2002)
8. Goldberg, D.E.: Genetic Algorithms in Search, Optimization, and Machine Learning, MA, Addison-Wesley, (1989)
9. Guerra-Salcedo C. and Whitley D.: Feature Selection mechanisms for ensemble creation: a genetic search perspective. In: Freitas AA (Ed.) Data Mining with Evolutionary Algorithms: Research Directions, Technical Report WS-99-06. AAAI Press, (1999)
10. Jain, A. K.; Zongker, D.: Feature Selection: Evaluation, Application, and Small Sample Performance, IEEE Transaction on Pattern Analysis and Machine Intelligence, Vol. 19, No. 2, February (1997)
11. Kuncheva, L.I., and Jain, L.C.: Designing Classifier Fusion Systems by Genetic Algorithms, IEEE Transaction on Evolutionary Computation, Vol. 33 (2000) 351–373
12. Martin-Bautista MJ and Vila MA. A survey of genetic feature selection in mining issues. Proceeding Congress on Evolutionary Computation (CEC-99), Washington D.C., July (1999) 1314–1321
13. Michalewicz Z.: Genetic Algorithms + Data Structures = Evolution Programs. 3rd Ed. Springer-Verlag, (1996)
14. Muhlenbein and Schlierkamp-Voosen D.: Predictive Models for the Breeder Genetic Algorithm: I. Continuous Parameter Optimization, Evolutionary Computation, Vol. 1, No. 1, (1993) 25–49
15. Park Y and Song M.: A genetic algorithm for clustering problems. Genetic Programming 1998: Proceeding of 3rd Annual Conference, Morgan Kaufmann, (1998), 568–575.
16. Pei, M., Goodman, E.D. and Punch, W.F.: Pattern Discovery from Data Using Genetic Algorithms, Proceeding of 1st Pacific-Asia Conference Knowledge Discovery & Data Mining (PAKDD-97) Feb. (1997)
17. Pei, M., Punch, W.F., and Goodman, E.D.: Feature Extraction Using Genetic Algorithms, Proceeding of International Symposium on Intelligent Data Engineering and Learning'98 (IDEAL'98), Hong Kong, Oct. (1998)
18. Punch, W.F., Pei, M., Chia-Shun, L., Goodman, E.D.: Hovland, P., and Enbody R. Further research on Feature Selection and Classification Using Genetic Algorithms, In 5th International Conference on Genetic Algorithm, Champaign IL, (1993) 557–564
19. Siedlecki, W., Sklansky J., A note on genetic algorithms for large-scale feature selection, Pattern Recognition Letters, Vol. 10, (1989) 335–347
20. Skalak D. B.: Using a Genetic Algorithm to Learn Prototypes for Case Retrieval an Classification. Proceeding of the AAAI-93 Case-Based Reasoning Workshop, Washigton, D.C., American Association for Artificial Intelligence, Menlo Park, CA, (1994) 64–69

Improved Image Halftoning Technique Using GAs with Concurrent Inter-block Evaluation

Emi Myodo, Hernán Aguirre, and Kiyoshi Tanaka

Dept. of Electrical and Electronic Engineering, Faculty of Engineering
Shinshu University, 4-17-1 Wakasato, Nagano 380-8553, Japan
{myodo,ahernan,ktanaka}@iplab.shinshu-u.ac.jp

Abstract. In this paper we propose a modified evaluation method to improve the performance of an image halftoning technique using GAs. We design the algorithm to avoid noise in the fitness function by evolving all image blocks concurrently, exploiting the inter-block correlation, and sharing information between neighbor image blocks. The effectiveness of the method when the population and image block size are reduced, and the configuration of selection and genetic operators are investigated in detail. Simulation results show that the proposed method can remarkably reduce the entire processing time to generate high quality bi-level halftone images.

1 Introduction

Recently the application of evolutionary algorithms (EAs) to real-world problems has been rapidly increasing. Signal processing is one of the areas in which methods using EAs are steadily being developed [1]. In this work, we focus on the image halftoning problem using genetic algorithms (GAs). In the halftoning problem a N-gray tone image must be portrayed as a n-gray tone image, where $n < N$. GAs have been used for halftoning in two ways. One approach seeks to evolve (or co-evolve) filters, which are applied to the input N-gray tone image to generate a halftone image [2,3,4]. The second approach searches directly for the optimum halftoning image having as a reference the input N-gray tone image. Our work fit in the latter approach.

Kobayashi et al. [5] first proposed a direct search GA based halftoning technique to generate bi-level halftone images. This technique divides the input images into non-overlapping blocks and uses a simple GA with a specialized two dimensional crossover to search the corresponding optimum binary patterns (high gray level precision and high spatial resolution). The method's major advantages are that (i) it can generate images with a specific desired combination of gray level precision and spatial resolution, and (ii) it generates bi-level halftone images with quality higher than traditional schemes such as ordered dithering, error diffusion, and so on [6]. However, the method uses a substantial amount of processing time and computer memory. Recently, Aguirre et al. [7] have proposed an improved GA (GA-SRM) to overcome the drawbacks of [5]. GA-SRM applies varying mutation parallel to crossover and background mutation, putting the operators in a cooperative-competitive stand with each other by subjecting their offspring to extinctive selection [8,9]. The halftoning technique using GA-SRM [7], compared to the conventional GA based technique [5], can achieve a 98% reduction in population size and a 70%-85% reduction in processing time while generating high quality bi-level halftone images. Additional related work can be found in [10], where the halftoning problem is treated as a multiobjective optimization problem to generate simultaneously halftone images with various combinations of gray level precision and special resolution. Also, in [11] the bi-level halftoning technique is extended to a multi-level halftoning technique using GAs. The gains in performance achieved by GA-SRM alone are substantial, yet the direct GA-SRM based halftoning method still requires much more processing time than traditional methods, such ordered dithering and error diffusion, and enhancing efficiency is required.

The GA based halftoning methods mentioned above evolve all image blocks independently from each other. A side effect of this is that the evaluation function becomes approximate for the pixels close to the boundaries between image blocks, which introduces false optima and delays the search. In this paper we improve further GA based halftoning with a modified evaluation method that contemplates inter-block correlation between neighbor blocks evolving all image blocks concurrently. The effectiveness of the method when the population and image block size are reduced, and the configuration of selection and genetic operators are investigated in detail. Simulation results verify that the modified scheme further accelerates the search speed to generate high quality halftone images, reducing processing time to less than $\frac{1}{100}$ of the time required by the conventional GA based halftoning method proposed in [5].

2 Image Halftoning Scheme Using GA

2.1 Individual Representation

An input image is first divided into non-overlapping blocks D consisting of $r \times r$ pixels to reduce the search space of solutions [5,7]. The GA uses an individual x with a $r \times r$ two dimensional representation for the chromosome. In the case of bi-level halftoning each element of the chromosome $x(i,j) \in \{0,1\}$. Figure 1 illustrates the division of the image in blocks and an example of individual x corresponding to a current block D.

2.2 Evaluation

We evaluate chromosomes with two kinds of evaluation criteria. (i) One is high gray level precision (local mean gray levels close to the original image), and (ii) the other is high spatial resolution (appropriate contrast near edges) [5]. The bi-level image halftoning technique calculates a gray level precision error by

$$E_m = \sum_{(i,j) \in D} \frac{1}{r^2} |g(i,j) - \hat{g}(i,j)| \tag{1}$$

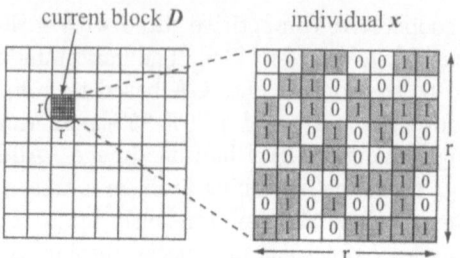

Fig. 1. Image division and individual representation($r = 8$)

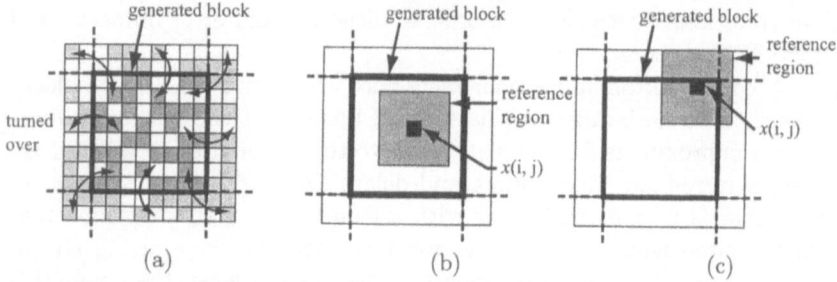

Fig. 2. (a) shows a current generated block x with its binary pattern copied around block boundaries for gray level estimation. (b) and (c) show examples of generated pixels and their reference region to calculate gray level estimation.

where $g(i,j)$ $(i,j=0,1,\cdots,r\text{-}1)$ is the gray level of the (i,j)-th pixel in the input image block, and $\hat{g}(i,j)$ is the estimated gray level associated to the (i,j)-th pixel of the generated halftone block $(x(i,j))$. To obtain $\hat{g}(i,j)$ a reference region around $x(i,j)$ is convoluted by a gaussian filter that models the correlation among pixels. In order to reduce discontinuity around block boundaries, the pixel pattern of x is copied around the boundary regions, as shown in Fig. 2a, and used to calculate the gray level estimation $\hat{g}(i,j)$. Figure 2b and Fig. 2c illustrate two examples of $x(i,j)$ and its reference region to calculate $\hat{g}(i,j)$. In Fig. 2b the reference region lies within the generated block while in Fig. 2c it exceeds the block boundaries and includes also the copied pixels.

In order to preserve the edge information of the input image well, the spatial resolution error in the bi-level image halftoning technique is calculated by

$$E_c = \sum_{(i,j) \in D} \frac{1}{r^2} |G(i,j) - B(i,j)| \qquad (2)$$

$$G(i,j) = g(i,j) - \bar{g}(i,j)$$

$$B(i,j) = (x(i,j) - \frac{1}{2})N$$

where $G(i,j)$ is the difference between the gray level $g(i,j)$ of the (i,j)-th pixel in the input image block and its neighboring local mean value $\bar{g}(i,j)$.

The two errors E_m and E_c are combined into one single objective function as

$$E = \alpha_m E_m + \alpha_c E_c \quad (3)$$

where α_m and α_c are weighting parameters of E_m and E_c, respectively. The chromosome's fitness is assigned by

$$F = E_{max} - E \quad (4)$$

where E_{max} is the error associated with the worst chromosome in a population. The GA is used to search for optimum compromise between (i) and (ii) with the above fitness function.

2.3 Genetic Operators and Selection

The improved GA-SRM for the halftoning problem [7] is based on a model of GA that puts genetic operators with complementary roles in a cooperative-competitive stand with each other [8,9]. The main features of the model are (i) two genetic operators to create offspring: Self-Reproduction with Mutation (SRM) that puts emphasis on mutation, and Crossover and Mutation (CM) that puts emphasis on recombination, (ii) an extinctive selection mechanism, and (iii) an adaptive mutation schedule that varies SRM's mutation rates from high to low values based on SRM's own contribution to the population. In [7], CM uses the same two dimensional crossover as [5] and SRM is provided with an Adaptive Dynamic Block (ADB) mutation schedule, in which the size of mutation block is dynamically adjusted. Extinctive selection is implemented with (μ, λ) proportional selection.

3 Proposed Method

3.1 Problems

As indicated in 2.2, the GA based halftoning schemes in [5] and [7] evolve the image blocks independently from each other, copy the binary pattern of the current block around the boundary regions, and use that information to evaluate the quality of the generated block. Due to the expected high correlation between neighboring pixels in an image, the pixels copied around the boundaries of the generated block aim to reduce discontinuities between blocks. However, these pixels are not the true information of the generated neighbor blocks. Although mathematically the same fitness function is used for every pixel, from an information standpoint the conventional GA based halftoning method induce a kind of approximate fitness function [12] for the pixels close to the boundary regions, which introduces false optima. This can mislead the algorithm and greatly affect its search speed. Note that if the area of image block is reduced the number of pixels evaluated with the approximate function (e.g. Fig. 2c) will increase while the number of pixels evaluated with the true fitness function (e.g. Fig. 2b) will decrease, negatively affecting the quality of the generated image and delaying processing time. The noise introduced by the approximated function is a real obstacle for further reduction of processing time.

3.2 Modified Evaluation Method

To have a fitness function that models the halftoning problem with higher fidelity, we make use of the inter-block correlation between neighbor blocks in the evaluation, linking each block with its neighbor blocks and sharing some genetic information between them. A GA is allocated to each block and each GA evolves its population of possible solutions concurrently. In this process the best individuals $x_{u,v}^{*(t-1)}$ in the neighbor populations are generationally referred and used to calculate the fitness values for individuals $x_{k,l}^{(t)}$ in the current population, as shown in Fig. 3. With this procedure of information sharing between populations we can supplement incomplete information in the evaluation process of [5,7] expecting that it would contribute to reduce processing time, improve the image quality around block boundaries, and allow further reductions of block size. Parallel implementations can be realized with the required number of processing units, linking at most 8 neighbor units. In this work the parallel GA is simulated as concurrent processes in a serial machine.

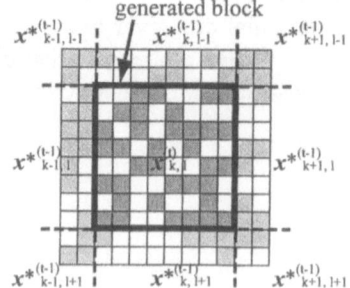

Fig. 3. A current block and connected neighbor blocks for gray level estimation

4 Results and Discussion

4.1 Experimental Setup

In this paper, we apply the proposed method to a canonical GA (cGA) [5] and GA-SRM [7]. To test the alogorithms we use SIDBA's benchmark images in our simulation. Unless stated otherwise, results presented here are for image "Lenna". The size of the original image is 256×256 pixels with $N = 256$ gray levels and the generated images are bi-level halftone images ($n = 2$). The image block size is $r \times r = 16 \times 16$ and the population size is 200. The weighting

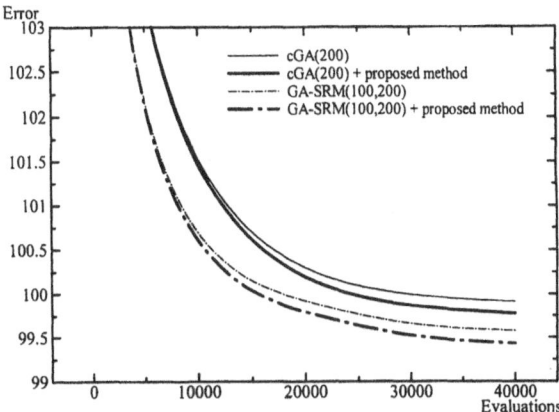

Fig. 4. Performance by cGA(200) and GA-SRM(100,200) with proposed and conventional method

Table 1. Comparison of number of evaluations

method	cGA(200)	GA-SRM(100,200)
conventional	1.000T	0.510T
proposed	0.695T	0.430T

parameters in Eq. (3) are set to $\alpha_m = 0.2$ and to $\alpha_c = 0.8$. In the case of the canonical GA the crossover probability is set to $p_c = 1.0$ and mutation probability to $p_m = 0.001$, similar to [5]. In our simulation we consider the result (error value of Eq. (3)) achieved by cGA with this settings after $T = 40,000$ evaluations as a reference value for image quality [5]. In the case of GA-SRM, for CM the crossover and mutation probabilities are set to $p_c = 1.0$ and $p_m^{(CM)} = 0.001$, respectively. For SRM, ADB mutation schedule with bit swapping mutation is used [7]. The balance for offspring creation is $\lambda_{CM} : \lambda_{SRM} = 1 : 1$, and the ratio between the number of offspring and the number of parents is $\mu : \lambda = 1 : 2$.

4.2 Effects in Conventional Schemes

First, we show the effect of the proposed method in cGA and GA-SRM with large populations. Figure 4 plots the error reduction over the generations by cGA(200), which denotes a canonical GA with a population size of 200 individuals, and by GA-SRM(100,200), which denotes GA-SRM with a parent population of $\mu = 100$ and an offspring populaion of $\lambda = 200$. From Fig. 4 we can see that both cGA and GA-SRM can achieve higher image quality with the proposed method than with the conventional evaluation method. Table 1 shows the number of evaluations needed to reach the reference value for image quality. We can reduce the number of evaluations about 31% in cGA and 16% in GA-SRM. Note that GA-SRM even with the conventional evaluation method is faster than cGA with the proposed evaluation method.

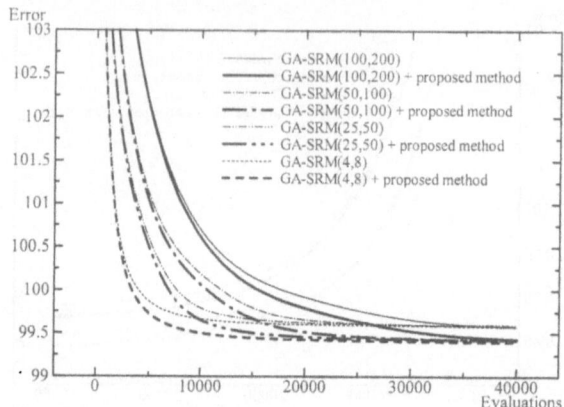

Fig. 5. Performance by GA-SRM with proposed and conventional method in different population sizes

Table 2. Effect of population size reduction

method	GA-SRM			
	(100, 200)	(50, 100)	(25, 50)	(4, 8)
conventional	0.510T	0.330T	0.211T	0.115T
proposed	0.430T	0.290T	0.185T	0.094T

4.3 Effect in Population Size Reduction

Second, since memory is an important issue in this application we study the effect on performance of reducing the population size. Figure 5 plots the error transition over the evaluations by GA-SRM with the conventional and proposed method. From Fig. 5 we can see that GA-SRM with the proposed method using smaller population sizes accelerates the search speed without deteriorating the final image quality. Table 2 shows the number of evaluations needed to reach the reference value for image quality.

4.4 Effects in Block Size Reduction

Next, we study the effect of reducing the size of the image block fixing the population size to $(\mu, \lambda) = (4, 8)$ in GA-SRM. Here, the mutation probability for CM is set to $p_m^{(CM)} = \frac{1}{r \times r}$ [13], because this value for mutation rate causes better performance in combination with extinctive selection. Figure 6 plots the error reduction over the evaluations for "Lenna" and Table 3 shows the number of evaluations needed to reach the image quality reference value for "Lenna" and other benchmark images. Note that with the proposed method we can further accelerate the search by reducing the block size to be evolved and still keep high image quality. For example, in case of Lenna and $r \times r = 4 \times 4$ the proposed method needs only 240 evaluations to achieve the image quality reference value

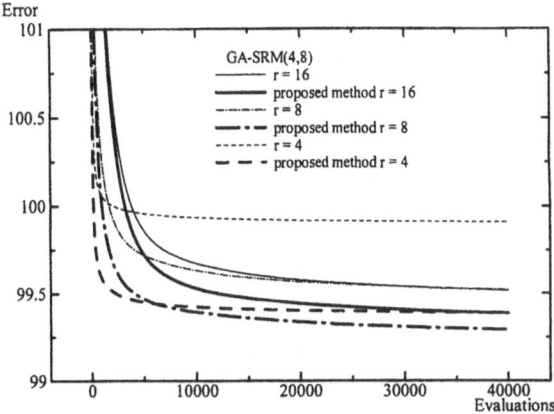

Fig. 6. Performance by GA-SRM(4,8) with proposed and conventional method using different block sizes

(the same image quality obtained by cGA after 40,000 evaluations)[†], which means less than $\frac{1}{100}$ of the processing time compared with the original scheme [5]. Running software implementations of the algorithms in a Pentium IV processor (2GHz), it takes about 7 seconds to generate one image of 256 × 256 pixels that reaches the reference value for image quality. Processing time increases in proportion to image pixels. From Table 3 note also that we could consistently observe similar behavior for other benchmark images. On the other hand, the conventional method considerably delays the search and cannot achieve visually satisfactory images for various test images. As mentioned in 3.2, in this paper the parallel implementation of the GA is simulated as concurrent processes in a serial machine. A true parallel implementation will benefit from a further reduction in processing time due to the distribution of work to several processors.

Figure 7 shows the original image "Lenna" and several generated images by traditional methods, such ordered dithering and error diffusion, and GA based halftoning methods with the conventional and proposed evaluation method.

4.5 Inter-block Information Sharing Gap

The effect on image quality and processing time of using a generational gap to perform the information sharing was also examined [14]. From our experiments we found that a *Gap* of 10 generations reduces substantially the frequency needed to share the inter-block information and still achieves a very significant reduction in processing time (only 1% of the processing time is needed compared to the original cGA method [5]). For values of *Gap* greater than 10 the search is delayed and no significant gain is observed reducing the number of times the information is updated.

[†] Here we compare average number of evaluation per pixel between different schemes, which is a good coincident with the actual processing time.

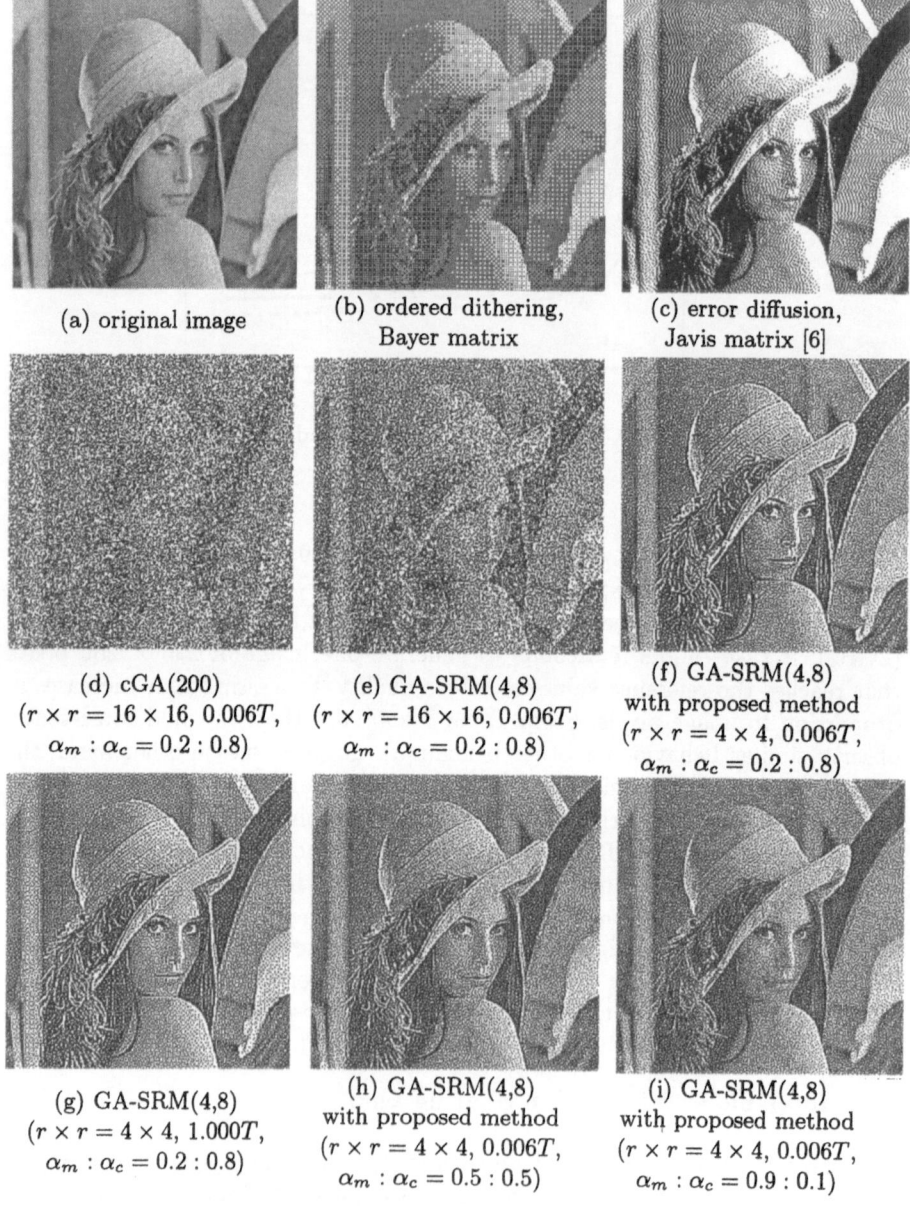

Fig. 7. Original "Lenna" and output images

Table 3. Effect of block size reduction for "Lenna" and other benchmark images

image	method	GA-SRM(4,8)		
		16 × 16	8 × 8	4 × 4
Lenna	conventional	0.112T	0.054T	0.84T
(256 × 256)	proposed	0.090T	0.029T	0.006T
Aerial	conventional	0.129T	0.062T	—
(256 × 256)	proposed	0.095T	0.032T	0.006T
Airplane	conventional	0.105T	0.056T	—
(256 × 256)	proposed	0.088T	0.030T	0.005T
Girl	conventional	0.108T	0.053T	—
(256 × 256)	proposed	0.093T	0.027T	0.005T
Moon	conventional	0.133T	0.059T	0.073T
(256 × 256)	proposed	0.100T	0.031T	0.005T
Title	conventional	0.127T	0.050T	0.018T
(256 × 256)	proposed	0.117T	0.043T	0.010T
Woman	conventional	0.120T	0.058T	0.130T
(256 × 256)	proposed	0.092T	0.031T	0.006T

— : never achieved the reference fitness value by cGA(200) with T=40,000 evaluations

5 GA Configuration in Different Block Sizes

In 4.4 we showed that by using the proposed method the block size can be reduced from $r \times r = 16 \times 16$ to $r \times r = 4 \times 4$ to generate high quality bi-level halftone images. The block size determines the size of search space. Smaller blocks imply an easier optimization problem and simpler optimization methods could be used. However, smaller blocks would also require more hardware resources, especially if we want to have a true parallel implementation of the algorithm. On the other hand, bigger blocks would increase the difficulty of finding the optimum and enhanced optimization methods would be required. This trade-off is an important issue that should be considered during the implementation of this application.

In this section we study the configuration of selection and genetic operators in GA for different block sizes. Figure 8 shows results by several EAs set with an offspring population of 8. M indicates only mutation with $p_m = \frac{1}{r \times r}$, CM is crossover ($p_c = 1.0$) followed by background mutation ($p_m = \frac{1}{r \times r}$), and SRM is the adaptive varying mutation operator. If the values of (μ, λ) are indicated, e.g. (4,8), the algorithm uses (μ, λ) proportional selection. Otherwise, the algorithm uses proportional selection.

From these figures, we can see that GA-SRM (CM-SRM(4,8)) is the most reliable method in all block sizes to generate a high quality halftone image rapidly. But to make the implementation of this problem easier in certain block size, other methods can be used. For example, when halftone images are generated in block size $r \times r = 4 \times 4$, we can use M(4,8) method without disadvantages because the performance of M(4,8) is as good as CM-SRM(4,8) as shown in Fig. 8a.

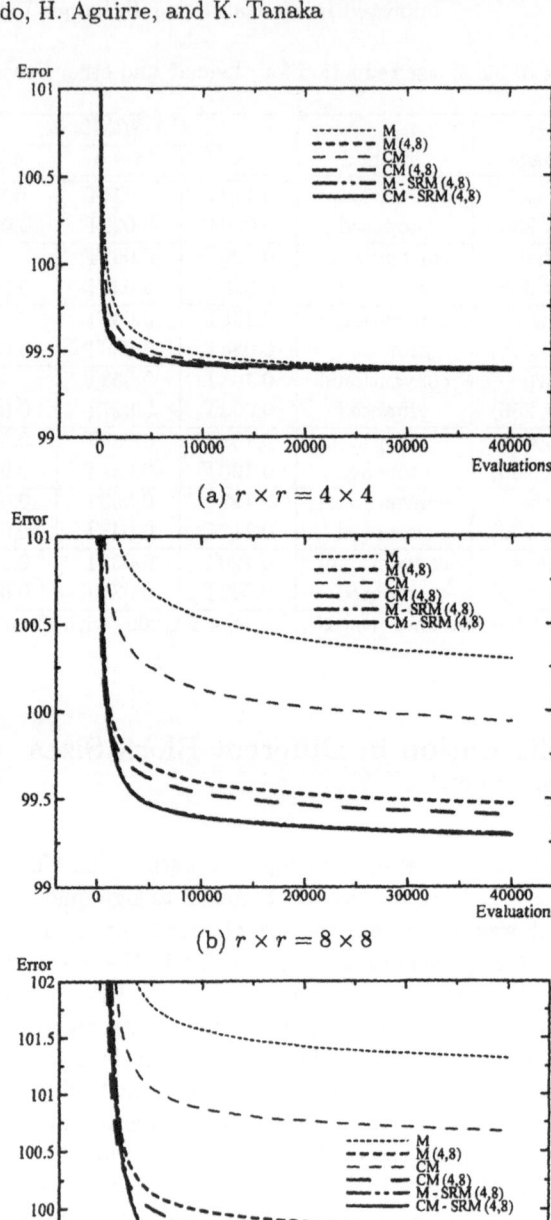

Fig. 8. Configuration of selection and genetic operators in GA for different block sizes

6 Conclusions

In this paper we have presented a modified evaluation method to improve the performance of an image halftoning technique using GAs. The proposed algorithm avoid noise in the fitness function by evolving all image blocks concurrently, exploiting the inter-block correlation, and sharing information between neighbor image blocks. The effectiveness of the method when the population and image block size are reduced, and the configuration of selection and genetic operators were investigated in detail. We could verify that the proposed method can remarkably reduce the processing time to less than $\frac{1}{100}$ of the time required by the conventional GA based halftoning method without deteriorating high image quality. The proposed method also generates images with less noise around block boundaries.

As future work, we are planning to extend this method to multi-level and color halftoning using multiobjective optimization techniques [10].

References

1. K. F. Man, K. S. Tang, S. Kwong and W. A. Halong, *Genetic Algorithms for Control and Signal Processing*, Springer-Verlag, 1997.
2. J. Newbern and V. M. Bove, Jr., "Generation of Blue Noise Arrays by Genetic Algorithm", *Proc. SPIE Human Vision and Electronic Imaging II*, vol.3016, pp.441–450, 1997.
3. J. T. Alander, T. Mantere, and T. Pyylampi, "Threshold Matrix Generation for Digital Halftoning by Genetic Algorithm Optimization", *Proc. SPIE Intelligent Systems and Advanced Manufacturing: Intelligent Robots and Computer Vision XVII: Algorithms, Techniques, and Active Vision*, vol.3522, pp.204–212, 1998.
4. J. T. Alander, T. Mantere, and T. Pyylampi, Digital Halftoning Optimization via Genetic Algorithms for Ink Jet Machine, *Developments in Computational Mechanics with High Performance Computing*, CIVIL-COMP Press, Edinburgh, UK, pp.211–216, 1999.
5. N. Kobayashi and H. Saito, "Halftoning Technique Using Genetic Algorithm", *Proc. IEEE ICASSP'94*, Vol.5, pp.105–108, 1994.
6. R. Ulichney, *Digital Halftoning*, MIT Press, Cambridge, 1987.
7. H. Aguirre, K. Tanaka and T. Sugimura, "Accelerated Halftoning Technique Using Improved Genetic Algorithm with Tiny Populations", *Proc. IEEE Intl. Conf. on Systems, Man, and Cybernetics*, pp.905–910, 1999.
8. H. Aguirre, K. Tanaka and T. Sugimura, "Cooperative Model for Genetic Operators to Improve GAs", *Proc. IEEE Intl. Conf. on Information, Intelligence and Systems*, pp.98–106, 1999.
9. H. Aguirre, K. Tanaka, T. Sugimura and S. Oshita, "Cooperative-Competitive Model for Genetic Operators: Contributions of Extinctive Selection and Parallel Genetic Operators", *Proc. Late Breaking Papers at Genetic and Evolutionary Computation Conference 2000*, pp.6–14, 2000.
10. H. Aguirre, K. Tanaka and T. Sugimura, "Halftone Image Generation with Improved Multiobjective Genetic Algorithm", *Proc. First Intl. Conf. on Evolutionary Multi-Criterion Optimization*, Lecture Notes in Computer Science, Springer, vol.1993, pp.501–515, 2001.

11. T. Umemura, H. Aguirre, K. Tanaka, "Multi-level Halftone Image Generation with Genetic Algorithms," *Proc. WSEAS Multiconference on Applied and Theoretical Mathematics*, pp.202–207, 2001.
12. Y. Jin, M. Olhofer, and B. Sendhoff, "A Framework for Evolutionary Optimization With Approximate Fitness Functions", *IEEE Trans. Evolutionary Computation*, vol.6, no.5, pp.481–494, 2002.
13. T. Bäck, "Optimal mutation rates in genetic search", *Proc. 5th Int'l Conf. on Genetic Altorithms*, Morgan Kaufmann, pp.2–8, 1993.
14. E. Myodo, H. Aguirre, K. Tanaka, "Improved Image Halftoning Scheme Using GAs with a Modified Evaluation Method " *IEEE-EURASIP Int'l Workshop on Signal and Image Processing*, to appear, 2003.

Complex Function Sets Improve Symbolic Discriminant Analysis of Microarray Data

David M. Reif[1], Bill C. White[1], Nancy Olsen[2], Thomas Aune[2], and Jason H. Moore[1]

[1]Program in Human Genetics, Department of Molecular Physiology and Biophysics
Vanderbilt University, Nashville, TN, USA
{Reif,BWhite,Moore}@phg.mc.Vanderbilt.edu
[2]Program in Human Genetics, Department of Medicine
Vanderbilt University, Nashville, TN, USA 37232-0700
{Nancy.Olsen,Thomas.Aune}@Vanderbilt.edu

Abstract. Our ability to simultaneously measure the expression levels of thousands of genes in biological samples is providing important new opportunities for improving the diagnosis, prevention, and treatment of common diseases. However, new technologies such as DNA microarrays are generating new challenges for variable selection and statistical modeling. In response to these challenges, a genetic programming-based strategy called symbolic discriminant analysis (SDA) for the automatic selection of gene expression variables and mathematical functions for statistical modeling of clinical endpoints has been developed. The initial development and evaluation of SDA has focused on a function set consisting of only the four basic arithmetic operators. The goal of the present study is to evaluate whether adding more complex operators such as square root to the function set improves SDA modeling of microarray data. The results presented in this paper demonstrate that adding complex functions to the terminal set significantly improves SDA modeling by reducing model size and, in some cases, reducing classification error and runtime. We anticipate SDA will be an important new evolutionary computation tool to be added to the repertoire of methods for the analysis of microarray data.

1 Introduction

Biomedicine is at a critical inflection point in the relationship between the amount of data it is possible to collect and our understanding of that data. For the first time in history, we are undergoing an information explosion and an understanding implosion. That is, new technologies are making it possible to collect data at a much faster rate than we can understand it. For example, DNA microarray technology facilitates the simultaneous measurement of the expression levels of tens of thousands of genes in biological samples [1]. As a result of this explosion of genetic information, bioinformatics and computational biology are faced with two important challenges. First, what are the most appropriate statistical and computational modeling approaches for relating gene expression data with clinical endpoints? Second, what are the most appropriate computational search strategies for identifying combinations of gene expression variables from an effectively infinite search space?

In response to these challenges, Moore et al. [2,3] have developed a machine learning strategy called symbolic discriminant analysis (SDA) for the automatic selection of gene expression variables and mathematical functions for statistical modeling of clinical endpoints. One advantage of this approach is that it uses the parallel search features of genetic programming [4], or GP, that are desirable for microarray data analysis [5]. Another important advantage is that no *a priori* assumptions are made about the functional form of the statistical model. This is important if the relationships between gene expression variables and clinical endpoints are complex and nonlinear, with interactions playing a more important role than the independent main effects of each gene. It is anticipated that such complex interactions among genes are common and play an important role in susceptibility to multifactorial diseases [6,7]. Indeed, applications of SDA to identifying patterns of gene expression that are associated with human leukemia and autoimmune diseases has revealed nonadditive relationships between gene expression variables that would not have been identified using a parametric statistical approach such as linear discriminant analysis [2,3,8].

The initial development of SDA has focused on a function set consisting of only four arithmetic functions. The goal of the present study is to evaluate whether adding complex functions such as square root to the terminal set improves SDA modeling. Adding complex functions will be considered an improvement if 1) they significantly decrease classification error of SDA models, 2) they significantly reduce the size of SDA models in terms of the overall number of nodes and the node depth, and/or 3) they significantly reduce the computational time required to complete a certain number of GP generations. For this study, we evaluate the addition of complex functions using real microarray data from human autoimmune disease that has been previously analyzed using SDA with just arithmetic functions [3]. We begin in Section 2 with an overview of the SDA approach. In Section 3, we provide an overview of the microarray data. In Section 4, we describe the details of the GP. In Section 5, we describe the experimental design and statistical analysis. A summary and discussion of the results are presented in Sections 6 and 7 respectively. The results presented in this paper demonstrate that adding complex functions to the terminal set significantly reduces the size of SDA models and, in some cases, reduces the GP runtime and classification error.

2 An Overview of Symbolic Discriminant Analysis

2.1 Introduction to Symbolic Discriminant Analysis

An important limitation of parametric statistical approaches such as linear discriminant analysis and logistic regression is the need to pre-specify the functional form of the model. To address this limitation, Moore et al. [2,3] developed SDA for automatically identifying the optimal functional form and coefficients of discriminant functions that may be linear or nonlinear. This is accomplished by providing a list of mathematical functions and a list of explanatory variables that can be used to build discriminant scores. Similar to symbolic regression [4], GP is utilized to perform a parallel search for a combination of functions and variables that optimally discriminate between two endpoint groups. GP permits the automatic discovery of symbolic discriminant functions that can take any form defined by the mathematical operators provided. GP builds symbolic discriminant functions using expression trees. Each expression tree has a mathematical function at the root node and all other nodes. Terminals in the expression tree are comprised of gene expression variables and constants. The primary advantage of this approach is that the functional form of the statistical model does not need to be pre-specified. This is important for the identification of combinations of expressed genes whose relationship with the endpoint of interest may be non-additive or nonlinear [6,7].

In its first implementation, SDA used leave one out cross-validation (LOOCV) to estimate the classification and prediction error of SDA models [2]. With LOOCV, each subject is systematically left out of the SDA analysis as an independent data point (i.e. the testing set) used to assess the predictive accuracy of the SDA model. Thus, SDA is run on a subset of the data (i.e. the training set) comprised of n-1 subjects. The model that classifies subjects in the training set with minimum error is selected and then used to predict the group membership of the single independent testing subject. This is repeated for each of the possible training sets yielding n SDA models. Moore et al. [2] selected LOOCV because it is an unbiased estimator of model error [9]. However, it should be noted that LOOCV may have a large variance due to similarity of the training datasets [9,10] and the relatively small sample sizes that are common with microarray experiments. It is possible to reduce the variance using perhaps 5-fold or 10-fold cross-validation. However, these procedures may lead to biased estimates and may not be practical when the sample size is small [9].

In this first implementation of SDA, models were selected that had low classification and prediction errors as evaluated using LOOCV. The end result of this initial implementation of SDA is an ensemble of models from separate cross validation divisions of the data. In fact, over k runs of n-fold cross validation, there are a maximum of kn possible models generated assuming a 'best' model for each interval is identifiable. This is a common result when GP is used with cross-validation methods because of its stochastic elements. In response to this, Moore et al. [3] and Moore [8] developed a strategy for SDA modeling that involves evaluating the consistency with which each gene was identified across each of the LOOCV trials. This new statistic is referred to as cross validation consistency (CVC) and is similar to the approach taken by Ritchie et al. [11] for the identificationi of gene-gene interactions in epidemiological study designs. The idea here is that genes that are important for discriminating between biological or clinical endpoint groups should consistently be identified regardless of the LOOCV dataset. The number of times each gene is identified is counted and this value compared to the value expected 5% of the time by chance were the null hypothesis of no association true. This empirical decision rule is established by permuting the data 1,000 or more times and repeating SDA analysis on each permuted dataset as described above. In this manner, a list of statistically significant genes derived from SDA can be compiled.

Once a list of statistically significant genes or variables is compiled, a symbolic discriminant function that maximally describes the entire dataset can then be derived.

This is accomplished by rerunning SDA multiple times on the entire dataset using only the list of statistically significant candidate genes identified in the CVC analysis. A symbolic discriminant function that maximizes the distance between distributions of symbolic discriminant scores between the two endpoint groups is selected. In this manner, a single 'best' symbolic discriminant function can be identified and used for prediction in independent datasets. This modeling process is designed to limit false-positive results and provide an objective way of dealing with different GP results in different cross validation divisions of the data. Moore [8] has suggested this approach may be useful for other GP-based modeling strategies.

2.2 Advantages of Symbolic Discriminant Analysis

As discussed by Moore et al. [3], there are two important advantages of SDA over traditional multivariate methods such as linear discriminant analysis [12-14]. First, SDA does not pre-specify the functional form of the model. For example, with linear discriminant analysis, the discriminant function must take the form of an additive linear equation. This limits the models to linear additive functions of the explanatory variables. With SDA, the basic mathematical building blocks are defined and then flexibly combined with explanatory variables to derive the best discriminant function.

The second advantage of SDA is the automatic selection of variables from a list of thousands. Traditional model fitting involves stepwise procedures that enter a variable into the model and then keep it in the model if it has statistically significant marginal or independent main effect [15]. Interaction terms are only evaluated for those variables that are already in the model. This deals with the combinatorial problem of selecting variables, however, variables whose effects are primarily through interactions with other variables will be missed. This may be an unreasonable assumption for most complex biological systems. The SDA approach employs a parallel machine learning approach to selecting variables that permits interactions to be modeled in the absence of marginal effects. For example, both of the SDA models of human autoimmune disease identified by Moore et al. [3] involve multiplicative relationships among the gene expression variables (see Figure 1 below). No one variable contributes to the discriminant function independently of the others. For comparison, Moore et al. [3] analyzed the data using stepwise LDA [14]. Only one of the genes was identified by both SDA and LDA, and this was one of the genes that had a statistically significant main effect in a univariate analysis. Thus, SDA identified a set of genes that was not identified by LDA.

2.3 Disadvantages of Symbolic Discriminant Analysis

Although SDA has several important advantages over traditional multivariate statistical methods, there are several disadvantages [3]. First, there is no guarantee that GP will find the optimal solution. Heuristic searches tend to sacrifice finding an optimal solution in favor of tractability [16]. Implementing an evolutionary algorithm in parallel certainly improves the chances of finding an optimal solution [17], but it is not a certainty. This is due to the stochastic nature of GP. The initial populations of solutions are randomly generated and the recombination and mutation occurs at ran-

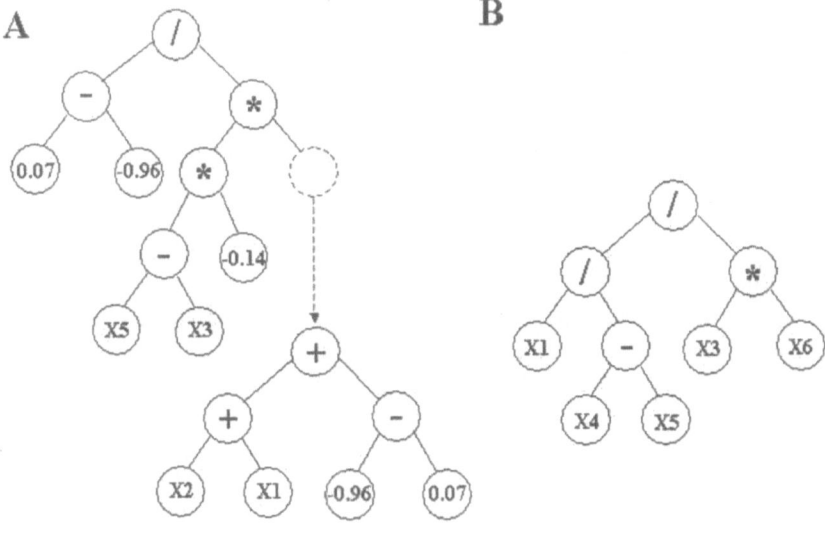

Fig. 1. Symbolic discriminant functions of gene expression variables in the form of expression trees and mathematical equations for discriminating rheumatoid arthritis from normal (A) and systemic lupus erythematosus from normal (B) [3].

dom positions in the binary expression trees. Further, there may be a stochastic component to how the highest fit individuals are selected. For these reasons, evolutionary algorithms should be run multiple times with multiple parallel populations.

A second disadvantage of this approach is the computational requirement. LDA can be performed on a standard desktop workstation while SDA requires a parallel computer cluster for optimal performance. When the number of genes to be evaluated is large, the power of a parallel computer is required. Although such systems are fairly inexpensive to build, the time investment to establish and manage a parallel computing farm may be prohibitive to some.

A third disadvantage is the complexity of the symbolic discriminant functions obtained. An attractive feature of LDA is the simplicity of the models, which facilitates interpretation. SDA has the potential to generate rather large models.

The goal of the present study is to evaluate whether adding more complex functions to the terminal set overcomes some of the disadvantages of SDA described by Moore et al. [3]. Specifically, does SDA with complex functions find better models faster? Also, are the SDA models identified smaller?

3 An Overview of the Microarray Data Used

A detailed description of the data is provided by Maas et al. [20]. Briefly, gene expression was measured in peripheral blood mononuclear cells (PBMC) from normal individuals (n=12) and patients with rheumatoid arthritis (RA) (n=7) or systemic lupus erythematosus (SLE) (n=9) using cDNA microarrays. Normal individuals (i.e. controls) were examined before and after routine immunization with flu vaccine to allow comparison of normal immune response genes to those that are differentially regulated in autoimmune disease. Reproducibility of the experimental method was first established by performing four hybridizations to separate microarrays using the same RNA sample. Data were normalized against a common control and linear regression analysis was utilized to estimate reproducibility. All hybridizations were highly reproducible with R^2 values ranging from 0.87 to 0.99. In the present study, we used SDA to identify symbolic discriminant functions that differentiate RA from normal and SLE from normal.

4 Details of the Genetic Programming (GP) Strategy

We used genetic programming or GP [4] in the present study to optimize the selection of gene expression variables and mathematical operators for building symbolic discriminant functions. We implemented the GP optimization of SDA using the lil-gp software package [18] modified to operate in parallel using the parallel virtual machine (PVM) message-passing library. The parallel GP was run on two processors of a 110-processor Beowulf-style parallel computer system running the Linux operating system. Each node has two Pentium III 600Mhz processors, 256 Mb RAM, a network card, and a 10 Gb hard drive. A total of two demes were used each consisting of 100 individuals for total population size of 200. We allowed the GP to run a total of 100 iterations with migration between each population every 25 iterations. A recombination frequency of 0.9 was used along with a reproduction rate of 0.1. Table 1 summarizes the GP parameters.

Table 1. The GP parameter settings for the SDA analyses.

Objective	Identify optimal SDA models
Fitness function	Classification error
Number of runs	100 per dataset
Stopping criteria	Classification error = 0
Population size	200
Number of demes	2
Generations	100
Selection	Fitness proportionate
Crossover probability	0.9
Reproduction probability	0.1

5 Experimental Design and Data Analysis

SDA was run with and without complex functions in two study designs. In the first study design, we compared normal samples and rheumatoid arthritis (RA) samples. In the second study design, we compared normal samples and systemic lupus erythematosus (SLE) samples. For each of the two study designs, we ran SDA a total of 100 times with just the basic arithmetic operators in the terminal set (+, -, *, /). Next, we ran SDA a total of 100 times for each comparison with a set of complex functions in addition to the basic arithmetic operators in the terminal set. The complex function set included square, square root, logarithm, exponential, absolute value, sine, and cosine. The goal of the statistical analysis was to determine whether the addition of more complex functions to the terminal set 1) significantly decreases the classification error of SDA models as assessed by leave one out cross validation (LOOCV), 2) significantly reduces the size of SDA models in terms of the overall number of nodes and the node depth, and/or 3) significantly reduces the computational time required to complete a certain number of GP generations. Across the 100 runs in each study design, we specifically compared the average classification error as assessed by LOOCV, the average runtime of the GP, the average node count, and the average node depth. In addition, we compared the average number of gene expression variables, constants, and mathematical functions in the best SDA models. A one-tailed Student's t-test was used to test the null hypothesis that complex functions do not decrease classification error, reduce SDA model size, or reduce computation time. When the assumptions of normality or equal variance were violated, a nonparametric Wilcoxon rank-sum test was carried out. All results were considered statistically significant at the 0.05 level.

6 Results

Table 2 summarizes the average classification error, GP runtime, node count, node depth, number of gene expression variables, number of constants, and number of mathematical functions in SDA models with and without complex functions for the RA vs. normal comparison. Table 3 summarizes the same information for the SLE vs. normal comparison. For both comparisons, the majority of measures were significantly improved with the addition of complex functions to the terminal set. For the RA vs. normal comparison, only classification error was not improved. For the SLE vs. normal comparison, only GP runtime was not improved. The most striking difference is that the SDA models are significantly smaller when complex functions are used.

Table 2. Comparison of mean SDA performance measures for modeling with and without complex functions in the terminal set (RA vs. normal).

Performance Measure	Arithmetic Functions		Complex Functions		p-value
	n	mean	n	mean	
Classification error	100	<.01	100	<.01	.5656
GP runtime (sec.)	100	67.04	100	66.80	<.0001
Node count	100	37.79	100	17.31	<.0001
Node depth	100	6.71	100	7.64	<.0001
Number of variables	100	4.83	100	2.37	<.0001
Number of constants	100	17.21	100	4.65	<.0001
Number of functions	100	16.21	100	10.17	<.0001

Table 3. Comparison of mean SDA performance measures for modeling with and without complex functions in the terminal set (SLE vs. normal).

Performance Measure	Arithmetic Functions		Complex Functions		p-value
	n	mean	n	mean	
Classification error	100	.06	100	.03	<.0001
GP runtime (sec.)	100	73.05	100	73.09	.5372
Node count	100	40.46	100	24.51	<.0001
Node depth	100	6.95	100	9.72	<.0001
Number of variables	100	5.51	100	4.10	<.0001
Number of constants	100	18.21	100	6.28	<.0001
Number of functions	100	17.21	100	14.07	<.0001

7 Discussion

Symbolic discriminant analysis (SDA) was developed as an attempt to deal with the challenges of selecting subsets of gene expression variables and features that facilitate the classification and prediction of biological and clinical endpoints [2,3,8]. Motivation for the development of SDA came from the limitations of traditional parametric approaches such as linear discriminant analysis and logistic regression. Application of SDA to high-dimensional microarray data from several human diseases has demonstrated that this approach is capable of identifying biologically relevant genes from among thousands of candidates. Further, because SDA makes no assumptions about the functional form of the model, it is capable of modeling complex nonlinear relationships between gene expression variables and clinical endpoints [2,3,8]. The initial studies using SDA used only the four basic arithmetic functions in the terminal set. The goal of the present study is to determine whether adding more complex functions such as logarithm and square root improve SDA modeling of microarray data. Adding complex function was considered an improvement if 1) they significantly decrease the classification error of SDA models as assessed by leave one out cross validation, 2) they significantly reduce the size of SDA models in terms of the overall number of nodes and the node depth, and/or 3) they significantly reduce the computational time required to complete a certain number of GP generations. The results presented in this paper demonstrate that adding complex functions to the terminal set significantly improves SDA modeling by reducing model size and, in some cases, reducing classification error and runtime. Thus, the primary conclusion of this study is that a richer set of functions should be included in the terminal set.

The results of this study address several of the primary disadvantages of the SDA approach outlined in Section 2.3 and by Moore et al. [3]. First, the finding that adding complex functions reduces model size and runtime is important since a primary concern is that the search space is rugged and effectively infinite [5]. Further, there is no guarantee that GP will identify an optimal solution. Thus, anything that can be done to reduce the size of the search space without compromising the flexibility and power of the modeling approach is desirable. The availability of complex functions in the terminal set reduces the search space that would have been necessary to construct the same complex functions from the basic arithmetic operators. In both comparisons, the average number of nodes in the expression trees of the best SDA models was reduced by approximately 50%.

The results of this study also address the disadvantage of increased computational time associated with SDA modeling. It is certainly true that SDA is more computationally intensive than methods such as LDA. In the RA versus normal comparison, the GP runtime was significantly less when complex functions were used. A time saving of even seconds for a single run can be very important when computational methods such as permutation testing are used. Indeed, Moore et al. [3] propose permutation testing as a strategy for carrying our formal hypothesis testing with the SDA approach. This entails running SDA 1000 or more times on randomized datasets to create an empirical distribution of the test statistic being used under the null hypothesis of no association. Thus, a time savings of even one second per run would save 1000 seconds or approximately 16.7 minutes. A time savings of one minute would save 16.7 hours. These time savings could be very important, especially if computational resources are at a minimum or if many such runs need to performed on many different datasets.

The final disadvantage addressed by this study is model complexity. As with any GP-based modeling procedure, SDA models can be quite large and complicated. We demonstrated that adding complex functions significantly reduced model size in both comparisons. Smaller models, even with more complex functions, are easier to interpret because there are fewer variable interrelationships that need to be interpreted. Indeed, in both comparisons, the average number of variables included in the optimal SDA models was reduced by approximately 25-50% with the addition of complex functions. Further, the average number of constants and functions was also significantly reduced.

In the present study, we only carried out the very first few steps in the SDA modeling process outlined in Section 2 and by Moore et al. [3] in order to compare the features of the models during cross validation. In future studies, we will carry out the full SDA analysis of this dataset and others using the more complex functions. This will allow us to compare the final model obtained with those found previously (see

Figure 1 for example). Additionally, it will be very important to assess the prediction error of models with complex functions. This will be possible when multiple, independently collected datasets are available from the same study. We anticipate SDA models with complex functions will have a lower prediction error because the overall models are likely to be smaller and simpler. However, this is a working hypothesis that still needs to be tested.

As suggested by Moore and Parker [5], GP-based modeling strategies are expected to have a large impact on the statistical and computational analysis of high-dimensional datasets derived from high-throughput technologies such as DNA microarrays. The inherent parallel or beam search strategy used by GP may be necessary for traversing effectively infinite rugged search spaces. Indeed, GP is finding its way into a number of bioinformatics and computational biology studies [19]. This study significantly improves SDA modeling through the addition of complex functions to the terminal set. We anticipate SDA will continue to be an important methodology for the analysis of microarray data.

Acknowledgements. This work was supported by generous funds from the Robert J. Kleberg, Jr. and Helen C. Kleberg Foundation. This work was also supported by the Arthritis Foundation and National Institutes of Health grants AR02027, AR41943, DK58749, HL68744, CA90949, CA95103, and CA98131.

References

1. Schena, M., Shalon, D., Davis, R.W., Brown, P.O.: Quantitative monitoring of gene expression patterns with a complementary DNA microarray. Science 270 (1995) 467–470
2. Moore, J.H., Parker, J.S., Hahn, L.W.: Symbolic discriminant analysis for mining gene expression patterns. In: De Raedt, L., Flach, P. (eds) Lecture Notes in Artificial Intelligence 2167, pp 372–81, Springer-Verlag, Berlin (2001)
3. Moore, J.H., Parker, J.S., Olsen, N., Aune, T. Symbolic discriminant analysis of microarray data in autoimmune disease. Genetic Epidemiology 23 (2002) 57–69
4. Koza, J.R.: Genetic Programming: On the Programming of Computers by Means of Natural Selection. The MIT Press, Cambridge London (1992)
5. Moore, J.H., Parker, J.S.: Evolutionary computation in microarray data analysis. In: Lin, S. and Johnson, K. (eds): Methods of Microarray Data Analysis. Kluwer Academic Publishers, Boston (2001)
6. Templeton, A.R.: Epistasis and complex traits. In: Wade, M., Brodie III, B., Wolf, J. (eds.): Epistasis and Evolutionary Process. Oxford University Press, New York (2000)
7. Moore, J.H., Williams, S.M.: New strategies for identifying gene-gene interactions in hypertension. Annals of Medicine 34 (2002) 88–95
8. Moore, J.H.: Cross validation consistency for the assessment of genetic programming results in microarray studies. In: Raidl, G. et al. (eds) Lecture Notes in Computer Science 2611, in press, Springer-Verlag, Berlin (2003).
9. Hastie, T., Tibshirani, R., Friedman, J.: The Elements of Statistical Learning: Data Mining, Inference, and Prediction. Springer, New York (2001)

10. Devroye, L., Gyorfi, L., Lugosi, G.: A Probabilistic Theory of Pattern Recognition. Springer-Verlag, New York (1996)
11. Ritchie, M.D., Hahn, L.W., Roodi, N., Bailey, L.R., Dupont, W.D., Plummer, W.D., Parl, F.F. and Moore, J.H.: Multifactor dimensionality reduction reveals high-order interactions among estrogen metabolism genes in sporadic breast cancer. American Journal of Human Genetics 69 (2001) 138–147
12. Fisher, R.A.: The Use of Multiple Measurements in Taxonomic Problems. Ann. Eugen. 7 (1936) 179–188
13. Johnson, R.A., Wichern, D.W.: Applied Multivariate Statistical Analysis. Prentice Hall, Upper Saddle River (1998)
14. Huberty, C.J.: Applied Discriminant Analysis. John Wiley & Sons, Inc., New York (1994)
15. Neter, J., Wasserman, W., Kutner, M.H.: Applied Linear Statistical Models, Regression, Analysis of Variance, and Experimental Designs. 3rd edn. Irwin, Homewood (1990)
16. Langley, P.: Elements of Machine Learning. Morgan Kaufmann Publishers, Inc., San Francisco (1996)
17. Cantu-Paz, E.: Efficient and Accurate Parallel Genetic Algorithms. Kluwer Academic Publishers, Boston (2000)
18. http://garage.cps.msu.edu/software/software-index.html
19. Fogel, G.B., Corne, D.W.: Evolutionary Computation in Bioinformatics. Morgan Kaufmann Publishers, Inc., San Francisco (2003)
20. Maas, K., Chan, S., Parker, J., Slater, A., Moore, J.H., Olsen, N., and Aune, T.M.: Cutting edge: molecular portrait of human autoimmunity. Journal of Immunology 169 (2002) 5–9

GA-Based Inference of Euler Angles for Single Particle Analysis

Shusuke Saeki[1], Kiyoshi Asai[2], Katsutoshi Takahashi[2], Yutaka Ueno[2], Katsunori Isono[3], and Hitoshi Iba[1]

[1] Graduate School of Frontier Science, The University of Tokyo, Hongo 7-3-1 Bunkyo-ku, Tokyo, 113-8656, Japan
{saeki,iba}@miv.t.u-tokyo.ac.jp

[2] Computational Biology Research Center, National Institute of Advanced Industrial Science and Technology, Aomi 2-41-6, Koutou-ku, Tokyo 135-0064, Japan
{asai-cbrc,takahashi-k,yutaka.ueno}@aist.go.jp

[3] INTEC Web and Genome Informatics Corporation, 1-3-3 Shinsuna, Koto-ku, Tokyo 136-0075, Japan
isono@isl.intec.co.jp

Abstract. Single particle analysis is one of the methods for structural studies of protein and macromolecules developed in image analysis on electron microscopy. Reconstructing 3D structure from microscope images is not an easy analysis because of the low resolution of images and lack of the directional information of images in 3D structure. To improve the resolution, different projections are aligned, classified and averaged. Inferring the orientations of these images is so difficult that the task of reconstructing 3D structures depends upon the experience of researchers. But recently, a method to reconstruct 3D structures is automatically devised [6]. In this paper, we propose a new method for determining Euler angles of projections by applying Genetic Algorithms (i.e., GAs). We empirically show that the proposed approach has improved the previous one in terms of computational time and acquired precision.

1 Introduction

Structural analysis of proteins is currently conducted using primarily NMR and X-ray methods. However, each of these has limitations: NMR is applicable only to relatively small proteins, and X-ray analysis is constrained by the difficulty and limitations of crystallization.

Recently, a technique called single-particle analysis has been recognized as a viable method for analyzing three-dimensional protein structures that are difficult to analyze through other methods [2].Using this technique, a protein is frozen for observation by electron microscopy, and the three-dimensional structure is then determined from images of the protein in various orientations. This technique does not require crystallization of the protein and therefore very large proteins can be analyzed. However, the presence of noise in the raw micrographs causes the protein images to be barely recognizable, making resolution of the three-dimensional structure problematic due to the extremely small size of the proteins and their fragility under the electron beam.

In order to successfully obtain a protein image from noisy photomicrography, resolution must be enhanced by classifying multiple two-dimensional images according to orientation and then overlaying the images to determine the unknown three-dimensional structure [4]. This requires very complex image analysis techniques. Typically, a few tens of averaged images are obtained from 10,000 to 100,000 raw projection images, and from these, the three-dimensional structure is determined.

This method has finally become practical due to improvements in micrographic techniques and the recent rapid progress in computational capabilities [9,8]. In this paper, we propose a new algorithm based on a GA to address an especially difficult image analysis problem involved in the single-particle technique, i.e. obtaining a clear three-dimensional structural projection from images of proteins at various orientations.

2 Euler Angle Determination Problem in Single-Particle Analysis

2.1 Common-Line Method

Reconstructing the three-dimensional structures requires determination of the projection angle of each image. By obtaining three-dimensional positional relationships between images, practical techniques used for CT scan analysis can be applied to obtain a density distribution of the three-dimensional structure, which then allows the density distribution to be displayed by assigning a certain threshold value. In the three-dimensional density function, the particle orientation with respect to the image plane is represented by the Euler angle. As shown in Fig. 1, the Euler angle is composed of three fundamental angles:¿,Å and Á.

The basis of the Euler angle representation of the projection angle of each image is the common-line method [11,3], which holds well because micrographic images are transparent projections.

> In the three-dimensional density function, the profile that is projected onto a straight line at a certain angle is common to two projection images, and can be obtained by projecting either projection image onto a straight line at a certain angle.

> When there are three projection images, the projection angle of the three projection images is determined by the profile (common line) that matches the given angle.

Even when the common line for two projected images is determined, the three-dimensional projection direction cannot be expressed by a single set of parameters, as one rotational degree of freedom remains (Fig. 2a). At least three projection images are required in order to express the three-dimensional projection direction using a single set of parameters (Fig. 2b) [5].

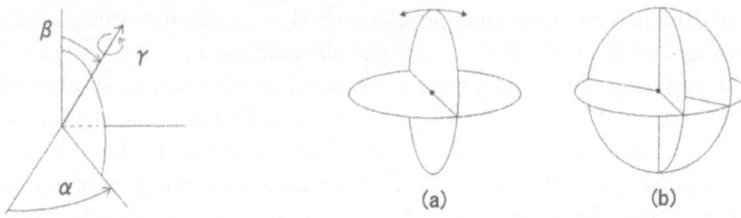

Fig. 1. Euler Angles ¿,À,Á

Fig. 2. Projections of a 3D Object.

2.2 Actual Common-Line Determination

Figure 3 illustrates the actual procedure used to determine the common line from sample projection images A and B. First, the profile, which is a three-dimensional density function called a sinogram, is obtained and then projected on a straight line at a certain angle (Fig. 3, left). The profile can be calculated to obtain the projection on the straight line while varying the angle of each of the projection images.

Next the sinogram of projection image A (angle θ_a) is compared to the sinogram of projection image B (angle θ_b). The index of similarity is expressed as the root mean square deviation (RMSD). By representing the sinograms of projections A and B as $snA(\theta_a, x)$, $snB(\theta_b, x)$, respectively, the RMSD value is expressed generally as follows:

$$RMSD(\theta_a, \theta_b) = \sqrt{\int_x \{snA(\theta_a, x) - snB(\theta_b, x)\}^2 \, dx}$$

The profile obtained by acquiring RMSD values by varying θ_a and θ_b, respectively, from 0 to 360 is called the cross sinogram, as shown on the right in Fig. 3.

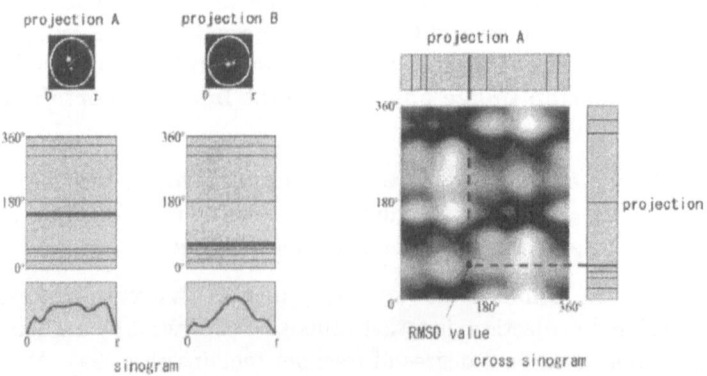

Fig. 3. Sinogram and Cross Sinogram.

2.3 Inference of the Euler Angle

If the Euler angles are given, RMSD values can be calculated. Therefore, the inference of the Euler angles based on common-line method boils down to the following optimization problem:

> One of the projection images is regarded as a reference image and its Euler angle is fixed as (0, 0, 0). Determine the Euler angles of the rest of the projection images to minimize the sum of RMSD values of all projection image pairs.

Therefore, if N is the number of projection images, this optimization problem is a function of $3(N-1)$ parameters. Judging from the cross sinogram of Fig. 3, the sum total of functions of RMSD values has a fairly complicated shape, which we believe would result in a huge number of local solutions, implying the need for a highly efficient search method.

3 Search Method Using GA

3.1 Real-Coded GA

GA is a population-based search method by means of simulated evolution in a computer. The parameters of the optimization problem (the optimization problem function) are coded as genes, and the solution is searched by repeating the operations of selection, crossover and mutation.

Binary or gray representation is generally used for GA coding; however, for real-coded handling, a problem exists in that the phase structures of the genotype and the phenotype are too dissimilar. Thus, in recent years, the real-coded GA, which uses directly the actual number of vectors for coding, has been proposed. One example of the successful practical application of real-coded GA is the lens design system [7]. As the crossover method, Unimodal Normal Distribution Crossover (UNDX) is used in the current experiment. The UNDX generates offsprings around the line segment connecting two parents. It was shown that the UNDX can efficiently optimize some benchmark functions with strong epistasis among parameters compared to other crossover methods such as BLX-¿. Euler angles which we are going to optimize is considered to be the parameters with such a feature.

3.2 Sequential Additive Search Method

When applying the real-coded GA to optimize all Euler angles simultaneously, the precise three-dimensional reconstruction was not successful. Because the search space was simply too large, our investigations were steered toward one of the innumerable existing local solutions.

Therefore, consideration was given to introducing heuristics that would satisfy the problem. If a set of at least three images is provided, the optimized value

1. Three images are optimized by real-coded GA. The Euler angle of the initial population is taken at random.
2. One image to be added is selected from among the remaining images.
3. The initial Euler angle of the original set of images is treated as the best solution of the previous optimization, and the Euler angle of the additional image generates a population as a random value.
4. Optimization is performed by real-coded GA.
5. Further optimization is performed by the steepest-descent method.
6. If an image to be added remains, return to step 2. If no images remains, finish.

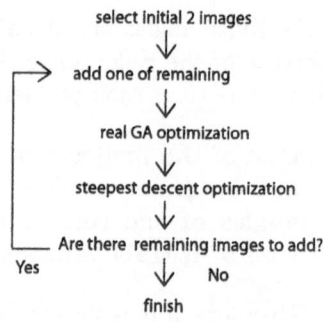

Fig. 4. Flow of Optimization

among these images can be obtained. Thus, the following method of sequential additive optimization was considered.

Figure 4 illustrates the flow of sequential additive search by GA. The best solution acquired by GA is further optimized by the steepest-descent method, because even though the precision remains unchanged, the required computation time is reduced to approximately one-tenth by halting the search at the proper generation. The best local solution is obtained by force from the steepest-descent method rather than searching through lengthy generations using GA.

Figure 5 shows the initialization details for this method. By including the best solution of the previous optimization as an initial individual, a more optimal initial population can be generated than for the case in which all parameters are initially random. Since the crossover method is UNDX [7], the best solution portion of the previous optimization of an individual can be changed, except by mutation. By this method, we believe that an incredibly large search space can be searched efficiently.

A problem associated with this method is that the quality of the optimization solution depends on the order of the additional images. The order of image addition can produce approximately a 10^n variation in RMSD value. Trying all possible orders is not realistic because the number of cases increases exponentially with the number of images. For the heuristic, a method that tries all possibilities of addition at each step while adding images one-by-one and then selects having the lowest RMSD value, is considered. However, with this additional method, because a GA is used for optimization, some unevenness remains in the quality of the solution at certain levels for each trial. To stabilize the quality of the solution, a number of solutions are retained as "candidates" for the order of image addition during the search (Fig. 6).

1. Randomly select two sets of images as the number of candidates.
2. Add one image from those remaining in each set, and perform optimization and verification. Check all additional possibilities.
3. From the new image set obtained by the above addition, select the number of candidates in sequence toward higher evaluation, and designate this as the next image set.
4. If an image to be added remains, return to step 2. If no images remain, finish.

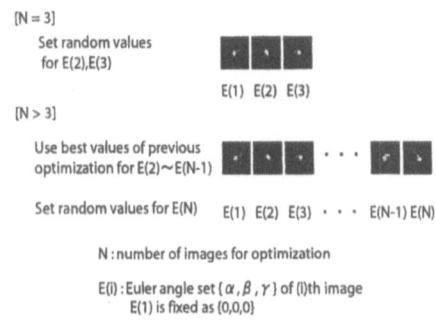

Fig. 5. Initialization of GA Optimization.

1. Randomly select two sets of images as the number of candidates.
2. Add one of the remaining images to each set, and perform optimization and verification. Check all additional possibilities.
3. From the new image set obtained by the addition, select the number of candidates in sequence toward higher evaluation, and designate this as the next image set.
4. If an image to be added remains, return to step 2. If no image remain, finish.

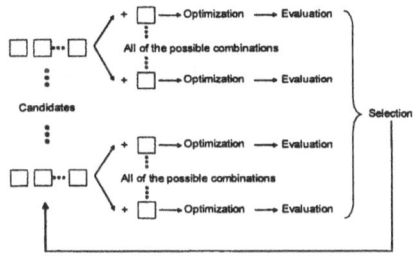

Fig. 6. Flow of Image Addition.

4 Experimental Results with the Three-Dimensional Structures Reconstructed from Projection Images

4.1 Experimental Projection Images

In order to evaluate the method by GA, an experiment was performed using a protein having a known three-dimensional structure. The protein used for this experiment was myosin, which generates power in muscle tissue. Figure 7 shows the three-dimensional structure of myosin and twenty images acquired by projection from different directions. Each pixel of the image takes a real value from 0.0 to 256.0. The objective was to determine the Euler angles of these 20 projection images.

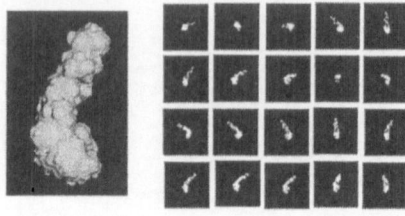

Fig. 7. Projection of Myosin.

4.2 Validity of Sequential Additive Search

Reconstruction of the three-dimensional structure was performed using the GA-based method described earlier. The image size is 65 x 65 pixels. In order to investigate the validity of sequential additive search by GA, comparison of a result with global search by GA(i.e. real-coded GA holding a vector of 57 variables from the start) was performed. The parameters for global search by GA are shown listed in Table 1. The relationship between the number of generations and the RMSD values in global search by GA is shown in Fig. 8. The best search, the worst search, and the average in ten trial are displayed. The last result was further optimized by the steepest descent method. RMSD values were saturated around 1.0 in a certain amount of generation.

Table 1. The Parameters for Global Search.

Population	100
Crossover	0.8
Mutation	0.00175
Generation	15000

Table 2. The Parameters for Sequential Additive Search(N : number of images).

Population	50
Crossover	0.8
Mutation	$0.3/\{3(N-1)\}$
Generation	$25N$
Repeat	5

The parameters for sequential additive search by GA are listed in Table 2. The number of mutations was selected to be proportional to the gene length (i.e., the number of parameters). In addition, the number of generations was selected so as to be proportional to the gene length. The search process was repeated five times, and the best solution among the obtained solutions was selected, thus scattering of the solution was minimized. We found that the solution quality tended to be better by performing five trials (resulting in five times the population) than by making the number of individuals in the population five times larger, although the computational cost was approximately the same. Because the population approaches uniformity with the evolution of individuals, we believe that various evolved individuals can be ultimately obtained if the population number is larger.

A result of seqential additive search is shown in Figs. 9, 10, and 11. The search by GA was tried 10 times to each number of candidates, in order to investigate the stability of the solution. Figure 9 plots the best value, the average value and the worst value to the number of candidates. We found that average value of RMSD values decreases when the number of candidates increased. Moreover, it turned out that the best value of RMSD values also decreases slightly to the increase in the number of candidates. Variance of the RMSD values to the number of candidates is shown in Fig. 11. The tendency for the variance of the solution to decrease exponentially to the number of candidates was observed.

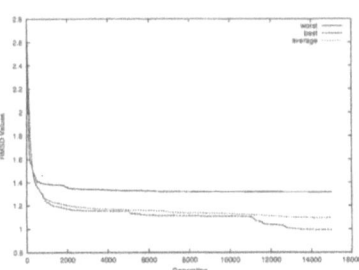

Fig. 8. RMSD Values v.s. Generations.

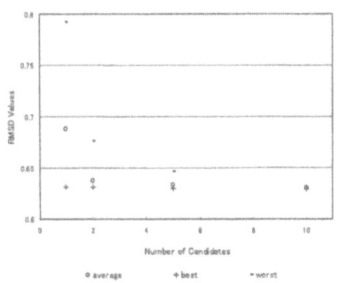

Fig. 9. RMSD Values v.s. Number of Candidates.

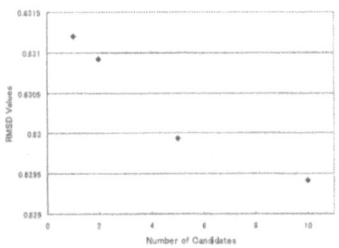

Fig. 10. Best of RMSD Values

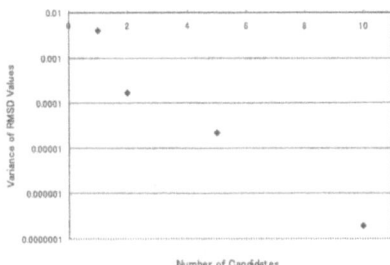

Fig. 11. Variance v.s. Number of Candidates.

Figure 12 shows the reconstructed three-dimensional structure obtained using the derived Euler angles. Figure 12(a) is a solid reconstruction based on the projection direction when the projection image was generated, so the purpose of this experiment was to reconstruct this structure. Figure 12(b) is a result of sequential additive search by GA with ten candidates of the solutions. Figure 12c is a result of global search by GA. Figure 12b is obviously closer to the target structure Fig. 12a as compared with Fig. 12c. When the result of global search is 0.92 in a RMSD value, the result of sequential additive search is 0.63 in a RMSD value. It can be said that sequential additive search is superior to global search in this problem.

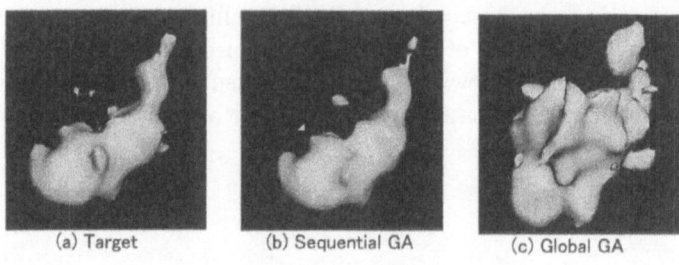

Fig. 12. Reconstructed 3D Structures.

4.3 Effect of Noise and Blur

In order to investigate the applicability to real micrographic images, some noise and blur were added and their effects were examined. Gaussian noise was applied, and blur was added using a Gaussian filter. Gaussian noise is a model of real noise in raw micrographic images, whereas Gaussian filter can simulate blur phenomena caused by the clustering operations. Figure 13 shows the images with noise and blur. The level in Fig. 13 is the Gaussian noise average level parameter, and half of a standard deviation was added. The noise factor and intensity follow a Gaussian distribution. SN in Fig. 13 is the parameter of the Gaussian filter representing the standard deviation of the angle at which pixels are blurred on a circle centered on the image. When SN is zero, no filtering is applied. The Gaussian noise was added first, followed by the Gaussian filter.

Reconstruction was performed by sequential additive search by GA with 50 candidates. The parameters for GA are the same as Table 2. The image size used this experiment is 96 x 96 pixels in order to show that our proposed method is applicable also to the image of other resolution. Figure 15 shows the target the three-dimensional structure reconstructed using the correct Euler angles. The arrangement of the three-dimensional structure corresponds to that of the projection image in Fig. 13. Figure 14 is a solid reconstruction based on the projection direction when the projection image was generated, so the purpose of this experiment was to reconstruct this structure. When no noise was added, the reconstruction was almost perfect. We observed a tendency for the reconstructed three-dimensional structure to be smoothed when the Gaussian filter was applied. However, when the Gaussian noise level was set to ten, we were still able to clearly discern the tail structure of myosin when using the Gaussian filter. We believe that the influence of Gaussian noise was reduced

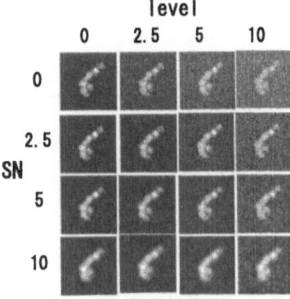

Fig. 13. Projections with Gaussian Noise and Gaussian Filter

by the blurring. It could be claimed that a certain degree of degradation of the images is a valid sacrifice for this noise reduction when the global three-dimensional structure is of interest.

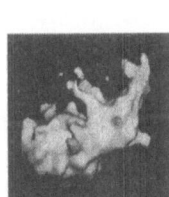

Fig. 14. Target 3D Structure.

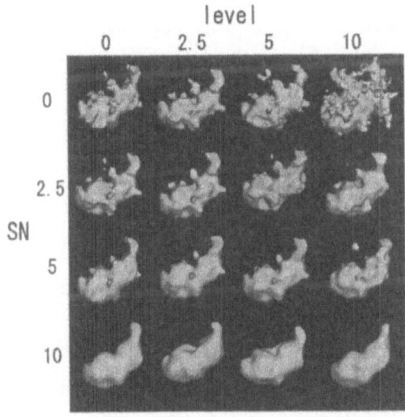

Fig. 15. Reconstructed 3D Structures.

4.4 Comparison with the Previously Proposed Method

There was previously proposed method called spot detection method. This method is based on a sequential additive search method identical to the present method, except that the solution candidate is searched by detecting a comparatively large area (spot) of mutual correlation of cross sinograms [12,10]. For added images, the ¿ and Á components of the Euler angle are obtained from *the spot information of the cross sinogram* between the standard image and the added image, and À is estimated from the overall relationship.

Table 3. Comparison with Other Methods.

	RMSD for inference	RMSD with target 3D structure	Time(hour)
(a) sequential GA	0.03136	5.11	30
(b) spot detection	0.03160	5.12	110
(c) global GA	0.04358	11.35	30

Table 3 lists the results for the experimental RMSD values and the required computation times for this method along with those of various other methods. The microprocessor used for the present experiments was an Athlon XP1800+. Table 3(a) shows the case using fifty candidates in the present method. Table 3(b) is, as far as we know, the only method proposed in the past to solve this problem. Table 3(c) shows the results of searching a wide space via GA. Projection images with noise (level 10, sn 10) are used because acutual microscopic images contain a great deal of noise. RMSD values of 3D structures between the target and the acquired ones by these methods are compared to investigate the solution precision. The proposed method(a) is superior with respect to both computation time and solution precision, as compared to the spot searching method(b). Compared to these two methods, the wide-space search via GA(c) showed greatly reduced precision.

5 Discussion

Only the spot-detection method has been known as the single particle analysis so far, for the sake of inferring the Euler angles solely from the projection images, i.e., not by using the symmetrical information of 3D structures. The proposed GA-based method gave better performance to the noisy projection images, in terms of the computational time and the acquired precision. One reason seems to be that our approach is relatively independent from the reference images so that it can explore the search space more widely. On the other hand, in case of noise-free projection images, the spot detection method may be superior, especially when we handle the higher resolution. This is because the spot information can provide a more correct common line with the higher resolution. When applying to the real data, we inevitably have to cope with the noisy projection, which will make our method more practical and effective.

Our GA-based method is a stochastic search so that different solutions are acquired from one run to another. However, the RMSD values by our method were much smaller, i.e., better, for all the runs. In general, many parameters (e.g., see Table 2) have to be effectively tuned for the sake of efficient GA-base search. We have tried two types of resolutions, i.e., 65x65 pixels and 96x96 pixels, for the projected images. As a result of several comparative experiments, we have confirmed that only the number of candidates should be changed to give the satisfactory performance to these cases. In other words, most of the other GA parameters need not be considered for the further tuning to cope with different experimental set-ups.

Our future reseach concerns are on reconstructing three-dimensional structures from projection images acquired by clustering operations.Clustering and averaging algorithms for the robust image-processing have been worked on [1]. We are currently working on the application to clustered and averaged projection images of a simulation and we plan to use actual micrographic images.

6 Conclusion

We found that three-dimensional structures could be restored from projection images by the method proposed in the present paper. In addition, noise was added to the projection images, and its effect on the reconstructed solid body was investigated. In processing, projection images that are judged to be oriented in the same direction are clustered and then overlaid to reduce noise by sacrificing image clarity through fading to a certain level. The experimental results suggest that the proposed type of processing is viable for discerning large target structures.

The present method was found to be superior compared to previous methods when noise is included in the images. Because actual micrographic images contain a great deal of noise, the present method is considered to be more practicable than previous methods. Introducing heuristics that correspond to the characteristics of the problems involved in searching via GA narrows the search space. Conversely however, for difficult problems, the present method can still search the space effectively, and thus is quite useful.

In the above described experiments, we restored a three-dimensional structure directly, based on projection images of a simulation having added noise. However, we must first investigate reconstruction of three-dimensional structures based on projection images acquired by clustering operations. Ultimately, we intend to apply the present method to actual micrographic images.

References

1. K.Asai, Y.Ueno, C.Sato K.Takahashi   "Clustering and Averaging of Images in Single-Particle Analysis"   Genome Informatics 11, 151–160 2000.
2. J.Frank "Three-dimentional Electron Microscopy of Macromolecular Assemblies" Academic Press, London 1996.
3. G. Harauz   "Exact filters for general geometry three dimensional reconstruction" Optik 73, 146–156 1986.
4. van Heel, et al   "Multivariate statistical classification of noisy images"   Ultramicroscopy 13, 165–183 1984.
5. van Heel, et al   "Angular reconstruction : a posteriori assignment of projection directions for 3D reconstruction"   Ultramicroscopy 38, 241–251 1987.
6. van Heel, et al.   "A new generation of the IMAGIC image processing system" J.Struct. Biol. 116, 17–24 1996.

7. I.Ono, S.Kobayashi "A Real-coded Genetic Algorithm for Function Optimization Using Unimodal Normal Distribution Crossover, Proc. of 7th Int. Conf. on Genetic Algorithms" Proc. of 7th Int. Conf. on Genetic Algorithms 246–253 1997.
8. S.Saeki, K.Asai, K.Takahashi, Y.Ueno, K.Isono, H.Iba "Inference of Euler Angles for Single Particle Analysis by Using Genetic Algorithms" Genome Informatics 12, 151–160 2002.
9. C.Sato, Y.Ueno, K.Asai, K.Takahashi, M.Sato, A.Engel, Y.Fujiyoshi "The voltage-sensitive sodium channel is a bell-shaped molecule with several cavities" *Nature*, Vol.409, No.6823, pp. 1047–1051, 2001.
10. K.Takahashi, Y.Ueno, K.Asai "Euler Angle Decision Method between Transfer Images of a Protein Molecule for Single Particle Analysis" The Biophysical Society of Japan, 39th Annual Meeting, 1P041, 2001.(in Japanese)
11. Penczek, et al. "Three-dimentional reconstruction of single particles embedded in ice" Ultramicroscopy 40, 33–53 1992.
12. Y.Ueno, K.Takahashi, K.Asai, C.Sato "BESPA : software tool for Three-Dimensional Structure Reconstruction from Single Particle Images of Proteins" Genome Informatics 10, 241–242 1999.

Mining Comprehensible Clustering Rules with an Evolutionary Algorithm

Ioannis Sarafis[1], Phil Trinder[1], and Ali Zalzala[2,3]

[1] School of Mathematical and Computer Sciences, Heriot-Watt University
Riccarton Campus, Edinburgh, EH14 4AS Scotland, United Kingdom
{I.Sarafis,P.W.Trinder}@hw.ac.uk
[2] School of Engineering & Physical Sciences, Heriot-Watt University
Riccarton Campus, Edinburgh, EH14 4AS Scotland, United Kingdom
[3] School of Engineering, American University of Sharjah, P.O. 26666, Sharjah, UAE
A.Zalzala@hw.ac.uk

Abstract. In this paper, we present a novel evolutionary algorithm, called NOCEA, which is suitable for Data Mining (DM) clustering applications. NOCEA evolves individuals that consist of a variable number of non-overlapping clustering rules, where each rule includes d intervals, one for each feature. The encoding scheme is non-binary as the values for the boundaries of the intervals are drawn from discrete domains, which reflect the automatic quantization of the feature space. NOCEA uses a simple fitness function, which is radically different from any distance-based criterion function suggested so far. A density-based merging operator combines adjacent rules forming the genuine clusters in data. NOCEA has been evaluated on challenging datasets and we present results showing that it meets many of the requirements for DM clustering, such as ability to discover clusters of different shapes, sizes, and densities. Moreover, NOCEA is independent of the order of input data and insensitive to the presence of outliers, and to initialization phase. Finally, the discovered knowledge is presented as a set of non-overlapping clustering rules, contributing to the interpretability of the results.

1 Introduction

Clustering is a common data analysis task that aims to partition a collection of objects into homogeneous groups, called clusters [9]. Objects assigned to the same cluster exhibit high similarity among themselves and are dissimilar to objects belonging to other clusters. The challenging field of DM clustering led to the emergence of various clustering algorithms [7]. Evolutionary Algorithms (EAs) are optimization techniques that have been inspired from the biological evolution of species [4], [11]. NOCEA (*Non-Overlapping Clustering with an Evolutionary Algorithm*) employs the powerful search mechanism of EAs to meet some of the requirements for DM clustering such as discovery of clusters with different shapes, sizes, and densities, independency of the order of input data, insensitivity to the presence of outliers and to initialization phase and interpretability of the results [7].

2 Related Work

There are four basic types of clustering algorithms: *partitioning algorithms, hierarchical algorithms, density-based algorithms* and *grid-based algorithms*. Partitioning algorithms construct a partition of N objects into a set of k clusters [9]. Hierarchical algorithms create a hierarchical decomposition of the database that can be presented as dendrogram [13]. Density-based algorithms search for regions in the data space that are denser than a threshold and form clusters from these dense regions [8]. Grid-based algorithms quantize the search space into a finite number of cells and then operate on the quantized space [1]. EAs have been proposed for clustering, because they avoid local optima and are insensitive to the initialization [3], [6], [12].

3 NOCEA Clustering Algorithm

3.1 Bin Quantization

Let $A = \{A_1, ..., A_d\}$ be a set of bounded domains and $S = A_1 \times ... \times A_d$ is a d-dimensional numerical space. The input consists of a set of d-dimensional *patterns* $P = \{p_1, ..., p_k\}$, where each p_i is a vector containing d numerical values, $p_i = [\alpha_1, ..., \alpha_d]$. The jth component of vector p_i is drawn from domain A_j. NOCEA's clustering mechanism based on a statistical decomposition of the feature space into a multi-dimensional grid. Each dimension is divided into a finite number of intervals, called *bins*. The number of bins m_j and the bin width h_j for jth dimension are dynamically computed by projecting the patterns in this dimension and then applying the statistical analysis described below. Initially, each dimension is divided into four segments, namely A, B, C and D. Segments A, B, C and D represent the intervals [a, Q1), [Q1, median), [median, Q3) and [Q3, b], respectively. Note that, a, b are the left and right bounds of A_j, while *median, Q1* and *Q3* are the median, first and third quartiles of patterns in jth dimension, respectively. The bin width h for each segment is then computed using the following formula [2]:

$$h = 3.729 * \sigma * n^{-\frac{1}{3}} \qquad (1)$$

where, n is the number of patterns inside this segment, while σ denotes the standard deviation of patterns lying in this segment. The number of bins for each segment is derived by dividing the length of the segment by the corresponding bin width h. The total number of bins m_j for the jth dimension is computed by forming the sum of bins from the four segments. Finally, the bin width h_j for the entire jth dimension is obtained by dividing the length of A_j domain by m_j. Of course it would be more realistic if each segment keeps the corresponding bin width leading to non-uniform grids. Despite the robustness of non-uniform grids we adopt the uniform approach because of the extra complexity introduced by the first and the way that infeasible solutions are repaired (see section 3.6).

3.2 Individual Encoding

The proposed encoding scheme is a non-binary, rule-based representation, containing a variable number of non overlapping rules. Each rule comprises d genes, where each gene corresponds to an interval involving one feature. Each ith gene, where i=1,..,d of a rule, is subdivided into two fields: *left boundary* (lb_i) and *right boundary* (rb_i), where lb_i and rb_i denotes the lower and upper value of the ith feature in this rule. The boundaries are drawn from discrete domains, reflecting the grid-based decomposition of the feature space. Note that two rules are non-overlapping if there exists at least one dimension where there is no intersection between the corresponding genes.

3.3 Fitness Function

In our clustering context, the fitness function is greedy with respect to the number of patterns that are covered by the rules of individuals. In particular, NOCEA aims to maximize the *coverage C*, which is defined to be the fraction of *total patterns* N_{total} that are covered by the rules of the individuals:

$$C = \max\left(\frac{\sum_{i=1}^{k} N_i}{N_{total}}\right) \quad (2)$$

where, k denotes the number of rules and N_i the number of patterns in ith rule. The above fitness function is suitable for comparing individuals that have different number of rules. Two individuals can have exactly the same performance with radically different genetic material (e.g. number, shape and size of rules). Thus, the fitness landscape may have multiple optima and as long as there is no any kind of bias towards a certain type of solution(s) (e.g. solution(s) with a particular number of rules, size or shape) equation 2 contributes to the diversity among the population members. Another important characteristic of the above fitness function is the fact that its theoretical lowest and highest values are always known, that is, the fitness of an individual can be between zero and one. Theoretically, as the level of noise increases the distance between the performance of the best individual(s) and one increases.

3.4 Recombination Operator

In a typical EA the recombination operator is used to exploit known solutions by exchanging genetic material between good individuals. Taking into consideration the requirement for evolving individuals without overlapping rules elaborate one and two point crossover operators have been developed. The crossover operator is applied to two parents generating two offsprings, from which only one can survive. Firstly, a number j is randomly drawn from the domain [1..d]. Recall, that d denotes the dimensionality of the feature space. If m_j is the number of bins corresponding to jth dimension, the crossover points *cp1* and *cp2* are determined by randomly picking one bin from the interval $[0, (m_j - 1)]$. Two offsprings are generated by exchanging the rules lying in the region from cp1 to cp2, between

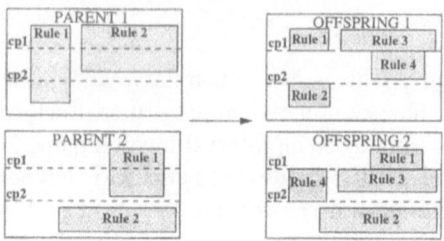

Fig. 1. An example of two-point crossover operation

the parents. An example of two-point crossover is depicted in Fig. 1. Note that if the number of crossover points is one then the above mentioned crossover operation reduces to one-point recombination. The proposed crossover operator can lead to the generation of offsprings that are not of the same length as their parents. This is because the offsprings must contain no overlapping rules, which in turn requires the splitting of any rule that intersects with the d-dimensional region between *cp1* and *cp2*. Clearly, for very low dimensional datasets and high recombination rates, two or even one point crossover can be disruptive in terms of breaking a relatively large rule into a number of smaller rules. Obviously, the greater the number of rules the greater the computational complexity as far as the application of genetic operators is concerned. Moreover, for the gain of simplicity in describing the clusters, fewer rules are always preferable. On the other hand, the length-changing characteristic of the crossover operator can contribute to increased diversity among the population members and this can be useful especially in cases with arbitrary-shaped clusters. To merely cope with the disruptive effect of crossover operator, we apply the following heuristic in the reported experiments. Instead of arbitrarily selecting the crossover points in *j*th dimension from the interval $[0, (m_j - 1)]$, these points are now fixed and they correspond to the bins containing Q1, Q3 or median in *j*th dimension. Theoretically, the disruptive effect of crossover is reduced as the dimensionality of the dataset increases, because the probability of intersection between the crossover points and rules decreases. As a consequence, for very low dimensional datasets (2D or 3D) it is reasonable to set the ratio of crossover in relatively small values or even to deactivate the recombination operator.

3.5 Mutation Operator

The evolutionary search in our system is mainly based on an elaborate mutation operator, which although alters the genetic material randomly, the generated individuals do not contain overlapping rules. The mutation operator has two functionalities: a) to grow and b) to shrink an existing rule.

Growing-Mutation. The growing-mutation aims to increase the size of existing rules in an attempt to cover as many patterns as possible but with a minimum number of rules. It is particularly useful in cases where there are large and convex clusters that can be easily captured by a relatively small number of large rules.

These large rules can be generated starting from smaller ones and by applying the growing-mutation operator over the generations. It is reasonable to focus on the discovery of as few and as large rules as possible, because these kind of rules contribute to the high interpretability of the clustering output. Additionally, the combined functionalities of the growing-mutation and the repair operator (see section 3.6) can lead to the discovery of new interesting regions (isolated clusters) within the d-dimensional feature space. Lets assume that the jth dimension of a d-dimensional rule R undergoes growing-mutation. We can compute the maximum value $rb_{j(max)}$ that rb_j can be extended to the right without causing overlapping as follows:

1. Sort in ascending order all the rules that their left bound is greater than the right bound of rule R in jth dimension.
2. If the sorted list is empty, set $rb_{j(max)}$ to $(m_j - 1)$ and exit. Otherwise, proceed to step 3.
3. Pick the next element RN of the list. If there is intersection in at least one dimension (excluding the jth) between the rules R and RN, set $rb_{j(max)}$ equal to the left bound of RN in jth dimension minus one and terminate the loop. If there is no intersection, proceed to the next element.

The derivation of the minimum value $lb_{j(min)}$ that lb_j can be extended to the left without causing overlapping is the dual of computing $rb_{j(max)}$. Obviously, the allowable range of modification in the boundaries of rules due to growing mutation is solely determined by the relative position of rules within the d-dimensional feature space. The left and right bounds of each rule in jth dimension are randomly mutated within the intervals $[lb_{j(min)}, lb_j)$ and $(rb_j, rb_{j(max)}]$, respectively.

Shrinking-Mutation. This type of mutation as its name implies shrinks an existing rule. Each time, the bound of a rule undergoes shrinking-mutation, the corresponding bound shrinks by one bin. The intuition behind this small modification is that for high dimensional datasets where the existence of isolated rules is increased, an arbitrary shrinking-mutation operation can cause the elimination of these rules and as a consequence additional generations may be required to re-discover these promising regions of the feature space. The shrinking-mutation operator is particularly useful to perform local fine-tuning and to facilitate the precise capturing of non-convex clusters. This is done by allowing adjacent rules to be re-arranged within the d-dimensional grid in order to cover as many patterns as possible.

Balancing Shrinking and Growing. The ratio of shrinking to growing operations r_{sg} should be set to relatively small values (e.g. 0.05). This is mainly done to bias the evolutionary search to discover new interesting regions in the large d-dimensional feature space, rather than trying to perform fine local tunning. Furthermore, due to the replacement strategy (see section 3.8) that allows the surviving of individuals for only a certain number of generations, high values for r_{sg} may be an obstacle for NOCEA to converge in an optimal solution.

3.6 Repair Operator

Often the search space S consists of two disjoint subsets of feasible and infeasible solutions, F and U, respectively [11]. Infeasible solutions are those that violate at least one constraint of the problem and we have to deal with them very carefully, because their presence in the population influence other parts of the evolutionary search [11]. In the following we introduce some basic definitions in order to formalize the notion of feasibility in our clustering context. The *selectivity* s_{ij} of ith bin in jth dimension is defined to be the number of patterns lying inside this bin over the total number of patterns covered by this rule. We define as *Selectivity Control Chart (SCC)* in jth dimension a diagram that has a centerline CL, and an upper (UCL) and a lower (LCL) control limits that are symmetric about the centerline. Additionally, measurements corresponding to selectivity of bins in jth dimension are plotted on the chart. CL, UCL and LCL are given by equation 3.

$$\begin{aligned} \text{UCL} &= CL * (1+t) \\ \text{CL} &= \tfrac{1}{b_j} \\ \text{LCL} &= CL * (1-t) \end{aligned} \quad (3)$$

where, b_j denotes the number of bins in jth dimension covered by the rule and t is a tolerance threshold that controls the sensitivity of SCC chart in detecting shifts in the selectivity of bins. Consider the two dimensional feature space shown in Fig. 2a, where each dimension has been partitioned into a number of bins. Figures 2c and 2d illustrate the SCC charts for dimensions x and y of a rule R1 shown in Fig. 2a, respectively. If the patterns covered by R1 are projected to x-axis, there are two distinct and well separated regions. These regions can be easily detected by examining the corresponding SCC chart shown in Fig. 2c, where the selectivity of the third bin is below LCL. In contrast, the selectivity for each bin in the y-axis is within the control limits and as a consequence NOCEA detects a single region in y-axis. In our approach, for each dimension of a rule R we construct a SCC diagram. If there exist points below LCL, then the corresponding bins are relatively sparse with respect to the rest and can be discarded. On the other hand, if there are no points below LCL but there exists at least one bin with selectivity greater than UCL, then NOCEA reduces t gradually by subtracting a small constant number, until at least one point falls below LCL. If neither of the above conditions are true, the distribution of patterns across this dimension can be considered as uniform and as a consequence all bins are kept. The removal of sparse bins described above creates a set of intervals for each dimension. The original rule is replaced by a set of new rules, which are formed by combining one interval from each dimension. A rule R is said to be *solid*, if the selectivity of all bins for each dimension is between the corresponding LCL and UCL. An individual that contains only solid rules is a feasible solution in our clustering context. Finally, if the selectivity of a rule, that is the fraction of total patterns covered by this rule, is below a user-defined threshold t_{sparse}, then this rule is said to be *sparse* and is eliminated. Theoretically, for relatively large values of t, regardless the underlying distribution of patterns, each rule will be solid. In such a case, NOCEA can not produce homogeneous rules because

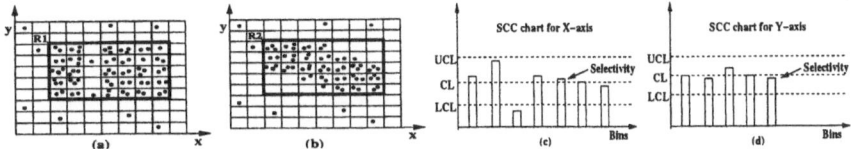

Fig. 2. Visualization of SCC charts

in essence there is no mechanism to stop rules from growing arbitrarily. On the other hand, relatively small values for t can cause over-triggering of the repair operator resulting in the generation of many small rules, which in turn add extra complexity to the application of the genetic operators. To merely cope with the problem of fine tunning the parameter t, we suggest a step-size adaptation mechanism that increases t over the generations according to:

$$t_i = t_{min} + (t_{max} - t_{min}) * \left(1 - \left(\frac{i}{T}\right)^b\right) \qquad (4)$$

where, b is a user-defined parameter, t_i denotes the value of t in ith generation, T is the total number of generations, while t_{min} and t_{max} are input parameters determining the minimum and maximum values of t, respectively.

A major drawback of SCC charts is the fact that consider each dimension independently. Thus, a SCC chart can only detect discontinuities occurring in a particular dimension as long as there is at least one bin in this dimension that is entirely very sparse. For instance, consider the rule R2 that is shown in Fig. 2b. Assuming relatively large values for t the corresponding SCC charts, which are not shown here, can not detect the very sparse regions located in the bottom-left and top-right of rule R2. To merely cope with partially sparse bins we apply the following heuristic to each rule, periodically (e.g. every 40 generations). Initially, a dimension that contains at least six bins is randomly selected. Then the parent rule R is split in that dimension in the middle and forms two touching rules R' and R''. Both these rules undergo the repairing procedure described above. If these rules remain intact as far as their size is concerned, another dimension which has not previously been examined and which contains at least six bins is randomly chosen. However, if the repair operator modifies at least one of these rules, the parent rule R is replaced with the set of rules that has been produced after repairing both rules, R' and R''. Note that we only consider dimensions containing at least six bins to avoid local abnormalities in the derived rules R' and R''. For the same reason the repair operator is invoked using t_{max} as the value for the tolerance threshold t.

3.7 Merging Rules

As soon as NOCEA converges, a merging procedure that combines adjacent rules in order to form the genuine clusters in data, is activated. The merging procedure is based on the concept of density and it is different from other distance-based

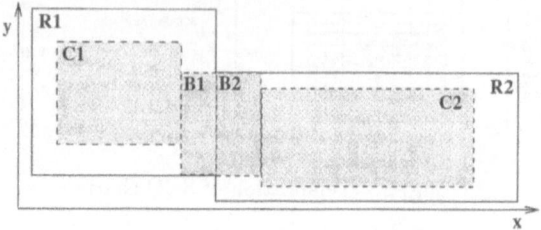

Fig. 3. An example of adjacent rules (R1 and R2).

merging approaches [5], [10]. Two rules are adjacent if they have a common face, i.e there are d-1 dimensions where there is an intersection in at least one bin between the corresponding genes and there is one dimension where the left bound of one rule is adjacent to the right bound of the other one. For instance, rules R1 and R2 shown in Fig. 3 are adjacent because there is an intersection between their corresponding genes in y-axis while in x-axis the right and left bound of rules R1 and R2, respectively, are touching. For each pair of adjacent rules R1 and R2 we compute three density metrics, namely, d_{C1}, d_{C2}, d_{B1} and d_{B2}. These density metrics correspond to the density of patterns within the d-dimensional regions $C1$, $C2$, $B1$ and $B2$, respectively, which are shown in Fig. 3. $C1$ and $C2$ represents the central regions while $B1$ and $B2$ the border regions within the rules R1 and R2, respectively. NOCEA assigns a pair of adjacent rules R1 and R2 to the same cluster as long as all the following conditions are satisfied:

$$\left(\frac{\min(d_{C1},d_{B1})}{\max(d_{C1},d_{B1})} \geq t_d\right)(1), \quad \left(\frac{\min(d_{B1},d_{B2})}{\max(d_{B1},d_{B2})} \geq t_d\right)(2), \quad \left(\frac{\min(d_{C2},d_{B2})}{\max(d_{C2},d_{B2})} \geq t_d\right)(3) \quad (5)$$

where t_d is a user-defined parameter. In essence, conditions (1), (2) and (3) ensure that the variation in the density across the path (shadowed region in Fig. 3) which connect the central regions of the two rules is not very high and as a consequence these two rules belong to the same cluster.

3.8 Setting Parameters

The mutation ratio is set to 1.0, while the probability of mutating the boundaries of rules is 0.05. The shrinking to growing ratio r_{sg} is set to 0.05. We used one-point crossover with ratio 0.1. The selection strategy that is used in our experiments is a k-fold tournament selection with tournament size k=4. The replacement strategy implements best of all scheme by merging the current and offspring populations and selecting the best individuals. However, to avoid premature convergence in infeasible optima, each individual is allowed to survive only for a certain number of generations (e.g. 5). Moreover, for every 40 generations all parents are replaced by the offsprings in order to eliminate all individuals containing partially sparse bins. Usually, the repairing of such rules cause a small reduction in the coverage of the entire individual, which can be easily observed from the fitness diagrams shown in Fig. 5b, 5d, and 5f. The only stopping criterion used in the reported experiments is the maximum number of generations 200, while the population size is 50. Each population member is randomly initialized with a single d-dimensional rule. The parameters t_{min} and t_{max} and b, which control the adaptation of the tolerance threshold t are set to 0.4, 0.65 and 1.5, respectively. The threshold t_{sparse} for eliminating sparse rules is 0.05.

4 Experimental Results

In this section we report the experimental results derived by running NOCEA against three synthetically generated datasets that contain patterns in 2-dimensional feature space as depicted in Fig. 4. These datasets are particularly challenging because they contain clusters of different sizes, shapes, densities and orientations. Furthermore, there are special artifacts such as streaks running across clusters and outliers that are randomly scattered in the features space. The first (DS1) and second (DS2) datasets consist of 8000 and 10000 patterns respectively and were used to evaluate the performance of a well-known clustering algorithm, called CHAMELEON [10]. The third dataset, DS3 has 100000 2-dimensional patterns and was used by another popular clustering algorithm called CURE [5]. These datasets are public available under the URL: http://www.macs.hw.ac.uk/~ceeis/gecco03/ds.htm. Figure 5 illustrates the rules and clusters that NOCEA found for the three datasets. Rules belonging to the same cluster are assigned the same number. Furthermore, for each dataset there is a fitness diagram where the coverage of the best and worst individuals as well as the mean coverage of the entire population in each generation are plotted. The experiments were conducted in a workstation running Windows 2000 with a 600MHz Intel Pentium III processor, 256MB of DRAM and 17GB of IDE disk. The evaluation time per generation for DS1, DS2 and DS3 is 1.5, 1.8 and 16 seconds, respectively. It is not guaranteed that NOCEA converges to the same set of rules using the same seeds. However, the union (or merging) of the derived rules always produces the same set of clusters, regardless the seeds used to initialize the individuals. It can be observed from Fig. 5 that NOCEA has the following desirable properties:

a) Discovery of Non-convex Clusters. NOCEA has the ability to discover arbitrary-shaped clusters, with different sizes and densities. The density-based merging procedure combines correctly adjacent rules forming the genuine clusters in data.

b) No a priori Provision of the Number of Clusters. Unlike other clustering algorithms such as k-means [9], NOCEA does not require the provision of the number of clusters a-priory, because the merging procedure discover the correct number of clusters on the fly.

c) Simplicity of Fitness Function. NOCEA has a simple fitness function which differs radically from distance-based criterion functions used in partitioning and hierarchical clustering algorithms [9]. Unlike other hierarchical and par-

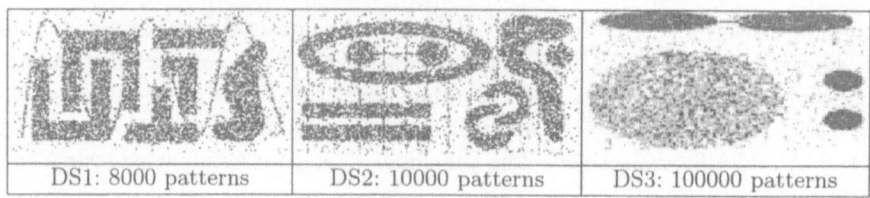

Fig. 4. The three datasets used in our experiments

titioning algorithms, NOCEA is not biased on splitting large clusters into smaller ones in order to minimize some distance criterion function [5].

d) Handling Outliers. Although NOCEA is greedy with respect to the number of patterns that are covered by the rules of the individuals, the fitness diagrams shown in Fig. 5 indicate that the maximum theoretical value of coverage 1.0, was never reached (assuming an amount of noise). This is due to the combined functionalities of the repair operator which do not allow a sparse region to be a member of a rule and to the procedure that eliminates very sparse rules. Thus, NOCEA is relatively insensitive to the presence of noise.

e) Data Order and Initialization Independency. The form of fitness function which is not distance-based and the randomized search of NOCEA ensures independency to the order of input data and to the initialization phase.

f) Interpretability of the Results. Whereas other clustering techniques describe the derived partitions by labeling the patterns with an identifier corresponding to the cluster that they have been assigned [7], NOCEA presents the discovered knowledge in the form of non-overlapping IF-THEN clustering rules, which have the advantage of being high-level, symbolic knowledge representation.

5 Conclusions and Future Work

In this paper, we have presented a novel evolutionary algorithm called NOCEA, which is suitable for DM clustering applications. NOCEA evolves individuals that consist of a variable number of non-overlapping clustering rules, where each rule includes d intervals, one for each feature. The encoding scheme is non-binary as the values for the boundaries of the intervals are drawn from discrete domains. These discrete domains reflect the dynamic quantization of the feature space, which is based on information derived by analyzing the distribution of patterns in each dimension. We use a simple fitness function, which is radically different from any distance-based criterion function suggested so far. A density-based merging procedure combines adjacent rules forming the correct number of clusters on the fly. Experimental results reported in section 4 indicate that the specific fitness function, together with, the elaborate genetic operators allow NOCEA to meet some of the requirements for DM clustering. NOCEA is an evolutionary algorithm and as a consequence does not easily scale up comparing to other hill-climbing clustering techniques [7]. However, EA are highly parallel

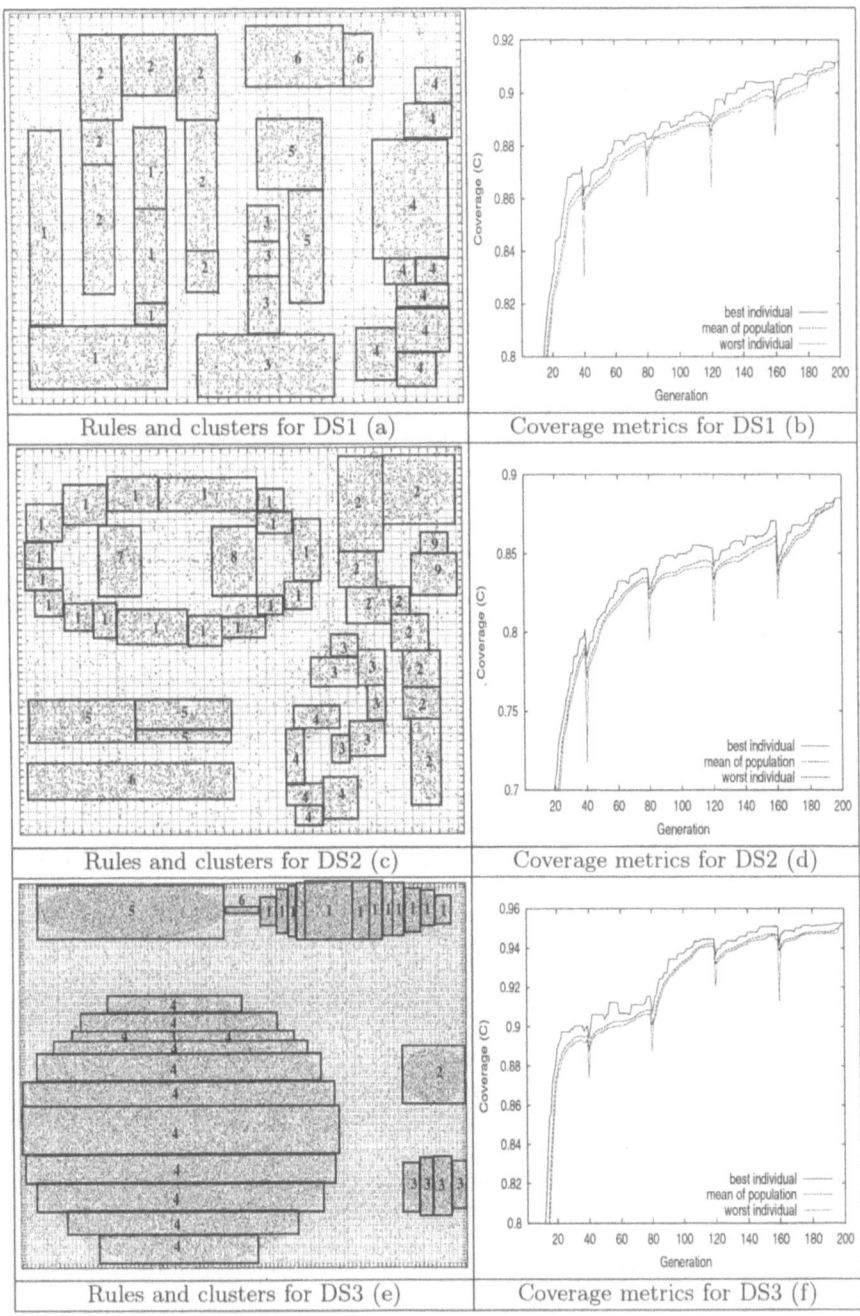

Fig. 5. NOCEA's experimental results

procedures and ongoing work attempts to investigate the improvements in the efficiency of NOCEA using various schemes of parallelism. Additionally, ongoing work attempts to overcome the problem of splitting large rules into smaller ones when the crossover points intersect with rules, by extending the functionality of the recombination operator. Another possible extension can be the development of a generalization operator that can be used to reduce the complexity of the cluster descriptors. In particular this procedure will take as argument a set of clustering rules corresponding to a particular cluster and will produce a more general descriptor for it.

References

1. R. Agrawal, J. Gehrke, D. Gunopulos, and P. Raghavan. Automatic subspace clustering of high dimensional data for data mining applications. In *Proc. 1998 ACM-SIGMOD Int. Conf. Management of Data*, pages 94–105, 1998.
2. Scott D. W. *Multivariate Density Estimation*. Wiley, New York, 1992.
3. A.A. Freitas. *Data Mining and Knowledge Discovery with Evolutionary Algorightms*. Springer-Verlag, August 2002.
4. D. E. Goldberg. *Genetic Algorithms in Search, Optimization, and Machine Learning*. Addison-Wesley, 1989.
5. S. Guha, R. Rastogi, and K. Shim. CURE: An efficient clustering algorithm for large databases. In *Proceedings of the ACM SIGMOD International Conference on Management of Data (SIGMOD-98)*, volume 27,2 of *ACM SIGMOD Record*, pages 73–84. ACM Press, June 1–4 1998.
6. L. O. Hall, I. B. Özyurt, and J. C. Bezdek. Clustering with a genetically optimized approach. *IEEE Trans. on Evolutionary Computation*, 3(2):103–112, 1999.
7. M. Han, J.and Kamber. *Data Mining: Concepts and Techniques*. Morgan Kaufmann Publishers, August 2000.
8. A. Hinneburg and D. A. Keim. An efficient approach to clustering in large multimedia databases with noise. In *Proc. 1998 International Conference on Knowledge Discovery and Data Mining (KDD-98)*, pages 58–65. AAAI Press, 1998.
9. Jain, A.K., Dubes, R.C: . *Algorithms for Clustering Data*. Prentice-Hall, Englewood Cliffs, NJ, 1988.
10. G. Karypis, E-H. Han, and V. Kumar. Chameleon: Hierarchical clustering using dynamic modeling. *IEEE Computer*, 32(8):68–75, 1999.
11. Z. Michalewicz. *Genetic Algorithms + Data Structures = Evolution Programs*. Springer-Verlag, 1996. Third Edition.
12. I. Sarafis, A. Zalzala, and P. Trinder. A genetic rule-based data clustering toolkit. In *Proc. Congress on Evolutionary Computation (CEC), Honolulu, USA*, 2002.
13. T. Zhang, R. Ramakrishnan, and M. Livny. BIRCH: an efficient data clustering method for very large databases. In *Proceedings of the ACM SIGMOD International Conference on Management of Data*, volume 25, 2, pages 103–114. ACM Press, 1996.

Evolving Consensus Sequence for Multiple Sequence Alignment with a Genetic Algorithm

Conrad Shyu and James A. Foster

Initiatives for Bioinformatics and Evolutionary Studies (IBEST)
Department of Computer Science
University of Idaho, Moscow, Idaho 83843, USA
{tsemings,foster}@cs.uidaho.edu

Abstract. In this paper we present an approach that evolves the consensus sequence [25] for multiple sequence alignment (MSA) with genetic algorithm (GA). We have developed an encoding scheme such that the number of generations needed to find the optimal solution is approximately the same regardless the number of sequences. Instead it only depends on the length of the template and similarity between sequences. The objective function gives a sum-of-pairs (SP) score as the fitness values. We conducted some preliminary studies and compared our approach with the commonly used heuristic alignment program Clustal W. Results have shown that the GA can indeed scale and perform well.

1 Introduction

Living things diverge from common ancestors through changes in deoxyribonucleic acid (DNA) and millions of years of evolution [6]. DNA indeed plays a fundamental role in the processes of life in various aspects. It contains the template for the synthesis of proteins, which are crucial molecules for life. Moreover, DNA is essential to life because it functions as a medium to transmit information from one generation to another [10]. Evidently the most important regions in DNA are generally conserved to ensure survival. Sequence alignment is commonly used to detect and quantify similarities in DNA or protein sequences. Alignments of biological sequences generated by computational algorithms are routinely used as a basis for inference about sequences whose structures or functions are not well known [7]. The most common approach is to find the best-scoring algorithm between a pair of sequences where the score records aligning similar residues and penalizes substitutions and gaps. The best-scoring alignment is commonly found by the dynamic programming (DP) algorithms, such as Smith-Waterman and Needleman-Wunsch algorithms [14, 23]. DP algorithms guarantee a mathematically optimal alignment for the given evolutionary model; however, the complexity of DP algorithms grows exponentially as the length and number of sequences increase. Specifically multiple sequence alignments (MSA) with DP have been shown to be NP-hard [19]. Several heuristic approaches, such as Clustal W [20], are frequently used to approximate the optimal alignments. In this paper, we present an approach that utilizes the guided search in GA to evolve the most probable *consensus sequence* [25] for MSA.

The design of GA is derived from the most commonly used DP algorithms for sequence alignments. In addition, we have developed an encoding scheme such that the search complexity does not depend on the number of sequences. The search complexity instead depends on the length of the consensus sequence and similarity between sequences. The scheme encodes each possible matching nucleotide at given column with binary masks. This compact representation greatly reduces the space requirement as well as the search complexity. The GA constructs the final alignment through the backtracking process, which is identical to the one found in DP algorithms [14]. The objective or evaluation function gives the sum-of-pairs (SP) scores to determine the fitness of each chromosome in the population. SP score has been widely used to detect and quantify similarities between sequences; however it does not provide any probabilistic or biological justifications [7]. To further improve the performance of GA, we have devised a sequence profiling formulation that reduces the complexity for calculating the SP scores. We have compared our approach to the most commonly used heuristic alignment program Clustal W and demonstrated that GA can indeed perform and scale well. In most cases, the GA outperformed Clustal W and produced better alignments.

2 Sequence Alignment

There are diverse motivations behind the alignment of biological sequences. Genetic sequences are inherited from common ancestors through millions of years of evolution. Therefore, it is of interest to trace evolutionary history of mutation and other evolutionary changes through sequencing [2, 6]. Alignment of biological sequences in this context is generally understood as comparisons based on the criteria of evolution. For example, the number of mutations, insertions, and deletions of residues necessary to transform one DNA sequence into another is a measure of phylogeny or evolutionary relatedness [3, 13]. On the other hand, a comparison may pinpoint regions of common origin, which may in turn coincide with regions of similar structure or function [10]. A pairwise sequence alignment is a technique of arranging two sequences, so that the residues in certain positions are deemed to have a common evolutionary origin. In other words, if the same residue occurs in both sequences at the same position then it may have been conserved during the course of evolution. On the other hand, if two resides differ then it is generally assumed that they may have derived from a common ancestor. Homologous sequences, those related by common descent, may have different lengths, which is generally explained through insertions or deletions [6, 10].

DP has been commonly used to align two sequences because it guarantees a mathematically optimal alignment. MSA, on the other hand, is simply an extension of pairwise sequence alignment. MSA is the process of aligning three or more sequences simultaneously to bring as many similar residues into register as possible. The resulting alignments are commonly interpreted into two contexts; (a) to find regions that define a conserved pattern or domain; and (b) to derive the possible phylogeny or evolutionary relationships among the sequences [13]. The presence of similar domains in several similar sequences implies a similar biochemical function or structural fold that may be used as the basis for further experimental investigation. Simultaneous alignment of three or more sequences with DP, however, poses a difficult algorithmic challenge.

2.1 Dynamic Programming (DP)

Dynamic programming (DP) is a commonly used method for solving sequential or multi-stage decision problems and is recursive in nature. The essence of DP is the principle of optimality [15, 17]. DP has long been used to solve varieties of discrete optimization problems such as scheduling, string-editing, packaging, and inventory management [12]. It views a problem as a set of interdependent sub-problems. DP solves sub-problems and uses the results to solve larger sub-problems. The solution to a sub-problem is expressed as a function of solutions to one or more sub-problems at the preceding levels [7]. In other words, DP expresses the problem in a recurrent formulation. To make optimal decisions for the next and all future states, DP only needs to know the current decision. This is also known as the *Markovian property*. For a process to be Markovian the future must depend only on the present state, and past should not have any effect on the future [7, 12]. The term *programming* in the name actually refers to the mathematical rules that can be easily followed to solve a problem; it has nothing to do with writing a computer program. DP is known to be an efficient programming technique for solving certain combinatorial problems. It is particular important in bioinformatics [17], as it is the basis of sequence alignments for comparing DNA and protein sequences. The following figure shows the recurrent formulation of DP for sequence alignment.

$$F(i, j) = \max \begin{cases} F(i-1, j-1) + s(x_i, y_j), \\ F(i-1, j) - d, \\ F(i, j-1) - d. \end{cases}$$

Fig. 1. This recurrent equation is applied repeatedly to fill the matrix of $F(i, j)$ values. This particular formulation gives the global alignment of two sequences. $F(i, j)$ is the maximum of three previous values, namely $F(i-1, j-1)$, $F(i-1, j)$, and $F(i, j-1)$. The value $s(x_i, y_j)$ is the score for aligning the characters x_i and y_j and d is the penalty for inserting a gap

For pairwise sequence alignments, for example, DP first begins with the construction of an alignment matrix $F(i, j)$ with the indexes (i, j) for the two sequences S_x and S_y. The matrix is initialized with $F(0, 0)=0$. The value of $F(i, j)$ is the score of the best alignment from the first character x_1 to the character x_i of sequence S_x and the first character y_1 to the character y_j of S_y. There are three possible ways that x_i and y_j can be aligned; (a) x_i can align with y_j, which gives a match or mismatch; (b) x_i is aligned with a gap; or (c) y_j is aligned to a gap. Since the matrix is built recursively, in order to calculate $F(i, j)$, the previous states $F(i-1, j-1)$, $F(i-1, j)$, and $F(i, j-1)$ must be known beforehand [7].

2.2 Sum-of-Pairs (SP) Score

Carrillo and Lipman [5] first introduced the sum-of-pairs (SP) score function, which defines the scores of a multiple alignment of N sequences as the sum of the scores of the $N(N-1)/2$ pairwise alignments [5, 7]. Although SP score function has been widely used to evaluate MSA, it doesn't really provide any biological or probabilistic justification. Each sequence is scored as if it is descended from the $N-1$ other sequences instead of a single ancestor. As a result, evolutionary events are often overestimated. The problem worsens as the number of sequences increases [7]. A weighted SP score function has been proposed to partially compensate the problem [1, 8]. Moreover, despite the simplicity of the SP score function, its sheer running time and space consumption makes it impractical even for modestly sized sets of short sequences. Specifically it has been shown that the problem of computing MSA with optimal SP score is NP-hard [22]. Several fast approximations and divide-and-conquer approaches [18] have been proposed to overcome the computational complexity. The following figure shows the mathematical formulation of the weighted SP score function.

$$w(M) = \sum_{1 \leq p < q \leq k} \left(a_{p,q} \times \sum_{j=1}^{N} s(m_{pj}, m_{qj}) \right)$$

Fig. 2. The SP function, $w(M)$, sums all the pairwise substitution scores in the columns for the sequence pairs p and q. Each column is evaluated with a scoring matrix. The substitution scoring function, $s(m_p, m_q)$, defines all possible alignments for nucleotides p_j and q_j. The function $s(m_p, m_q)$ gives the score of the alignment at column j for sequence p and q. The weight, $a_{p,q}$, is intended to balance the overestimation problem in the SP score function [1, 7]

2.3 Clustal W

Clustal W is a commonly used progressive alignment program for biological sequences. It is based on a heuristic algorithm and therefore cannot always find the optimal alignments. Clustal W exploits the fact that homologous sequences are evolutionarily related. It builds up multiple alignments progressively with a series of pairwise alignments moving from the leaves upward in a guide tree that estimates the phylogeny of the sequences [9]. Clustal W first aligns regions of identical or highly conserved residues and gradually adds in more distance ones [21]. This approach is sufficiently fast and allows Clustal W to alignment virtually any number of sequences. Although Clustal W doesn't always find the optimal alignments; however, in most cases those alignments at least give a good starting points for further automatic or manual refinement. This type of alignment is generally useful for the study of identifying regions that are highly conserved. The alignments can be further improved through sequence weighting, positions-specific gap penalties and choice of weight matrix [20]. Clustal W nonetheless suffers two major problems, the local maxima and the choice of alignment parameters.

The local maxima problem stems from the nature of the progressive alignment strategy. As the algorithm follows the guide tree and merges sequences together, the solution is never guaranteed to be globally optimal, as defined by some overall measure of alignment quality [9, 16, 20]. Any misaligned regions made early in the alignment process cannot be corrected later as new information from other sequences is introduced. This problem is frequently a result of an incorrect branching order in the guide tree. The only way to correct this is to use an iterative or stochastic sampling procedure such as bootstrapping [20]. The choice of alignment parameters is another problem in Clustal W. If parameters are not chosen appropriately, alignments will not converge to a globally optimal solution [7, 20]. For closely related sequences, any reasonable scoring matrices should work fine because matches usually receive the most weights [11]. Therefore, when matches dominate an alignment, almost any weight matrices will find a good solution. However, when aligning more divergent sequences, scores for gaps and mismatches become narrow and critical because they occur more frequently. Moreover for highly conserved sequences, the range of gap penalties that will find the correct or best possible solution can be very broad. As more and more divergent sequences are added, however, the exact values for gap penalties become critical for success [20]. Our observations have confirmed that this is actually a common problem in most MSA algorithms. Statistically as the number of sequences increases, the expected number of matches in each column also increases. For example, the probability of finding a matching nucleotide in the column of ten sequences is much higher than that of three sequences. If the gap penalty is too low, alignments will generally contain excessive amounts of gaps. It is in general difficult to justify why one scoring matrix is better than the others [7].

3 Design of GA

3.1 Consensus Sequence

The *consensus sequence* [25] is a unique as well as the most interesting and important feature of our GA approach. It is essentially a compact formulation to represent all possible alignments for virtually any given numbers of sequences [4]. The consensus sequence borrows the idea from biology that sometimes it is necessary for certain positions in a sequence to be made ambiguous when some residues simply cannot be resolved during laboratory experiments. A sequence with ambiguity codes is actually a mix of sequences, each having one of the nucleotides defined by the ambiguity at that position. For example, if an R is encountered in the sequence, then the sequences in the assortment will have either an adenine or a guanine at that position. The ambiguity enables conserved sequences to be condensed into one single representation. The following figure lists the most commonly used ambiguity codes defined by the Nomenclature Committee of the International Union of Biochemistry (IUB) and the next figure shows a hypothetical alignment with five sequences and illustrates how the ambiguity codes are used to derive the consensus sequence. It is assumed that the optimal alignment is already known for a given evolutionary model. The consensus sequence in essence is a condensed sequence with ambiguity codes that shows what nucleotides are allowed in each column.

Symbol	Description	Symbol	Description
A	Adenine	R	Purines (A, G)
C	Cytosine	Y	Pyrimidines (C, T)
T	Thymine	K	Keto (T, G)
G	Guanine	M	Amino (A, C)
W	A or T	B	C, G, or T
S	C or G	D	A, G, or T
H	A, C, or T	V	A, C, or G
N, X	A, G, C, or T	-	Gap symbol

Fig. 3. The most commonly used DNA ambiguity codes are defined by the International Union of Biochemistry (IUB). The presence of ambiguity generally indicates that some residues can not be resolved during the laboratory experiments. Ambiguity codes also enable sequences to be represented in a more condensed form

3.2 Design of Encoding Scheme

To further enhance the design, our chromosomes are broken into four pieces according to the nucleotide they represent. In other words, our GA uses four parallel chromosomes to represent four different nucleotides. Each chromosome encodes the relative occurrences and locations of a nucleotide and is only evolved with the chromosome that encodes the same nucleotide. The fitness of the entire chromosomes is determined by how well they fit together to derive the final alignment. The geometry of the parallel chromosomes is very similar to the four dimensional hyper-plane. Each dimension is evolved and optimized separately and independently.

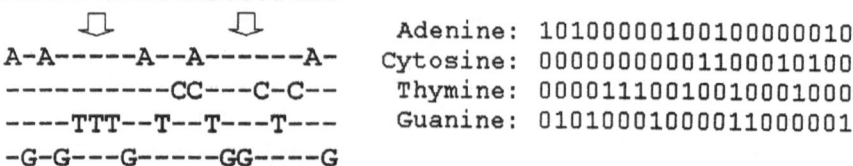

Fig. 4. Sequences are split up into four subsequences according to the nucleotides. Each subsequence only encodes the information where such a nucleotide can be found. Since each nucleotide is individually encoded, any possible ambiguities can be fully represented. Further, chromosomes are represented as binary strings, where 1 signals the existence of such a nucleotide at the given location while 0 signals the absence

The length of chromosomes is difficult to determine precisely. It depends on the evolutionary model as well as the similarity of the sequences. If sequences are highly conserved, chromosomes can be relatively short. Intuitively any two randomly generated sequences will have at least 25% similarity. For this study, we assumed that sequences have at least 50% similarity to be biologically significant. Therefore, we arbitrarily defined the length of the chromosome to be 1.5 times longer than the longest sequences. For most of our studies, this assumption worked fine. Furthermore, for implementation convenience, each chromosome is further divided into smaller blocks called loci. Biologically, a locus is a block of alleles where genes can be found. The length of a locus is determined by the length of an integer on the hardware platform. The number of loci depends on the length of the chromosome.

3.3 Crossover and Mutation Operations

For this research, we implemented a simple one-point crossover. A point is randomly selected in each locus for each chromosome and alleles are exchanged between two parent chromosomes to form an offspring. An offspring is produced at each generation and then competes with the population. Since each chromosome has separate string for the four nucleotides, the crossover points are chosen separately. Furthermore, we have implemented a bias function for selecting the parent chromosomes from the entire population. The bias function is like an "unfair" randomly number generator. It is essentially a quadratic equation that randomly generates a series of numbers with bias toward the lower indexes. Since the initial population has been sorted in the descending order according to the fitness values, consequently the individuals with higher fitness values are more likely to be selected. Mutation is an important operator that prevents the population from stagnating at local optima [24]. In our implementation, the mutation is only applied to the newly created offspring chromosomes. The GA first calculates the expected number of mutations for each locus in the chromosomes with a random factor. It then iteratively picks random locations on each locus for each of the four parallel chromosomes and changes the alleles. The mutation operator randomly flips the alleles independently on each locus with the binary XOR operator. It inverts the alleles from 0 to 1 or 1 to 0.

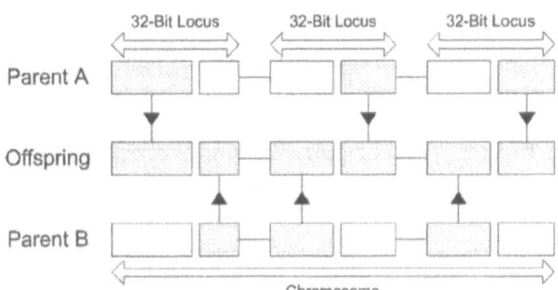

Fig. 5. The crossover operator retrieves the alleles from two parent chromosomes to create an offspring. A point is randomly selected on each locus for each chromosome. The offspring receives approximately a half of alleles from each parent. The length of a locus is 32-bit, which corresponds to the length of an integer on our machines

3.4 Objective Function

The objective function measures the quality of MSA. Therefore, ideally the better the score the more biologically relevant the multiple alignments are. The substitution costs are evaluated using a predefined substitution matrix. The matrix assigns every possible substitution or conservation according to its biological likeliness. We used the nucleic scoring matrix defined by IUB that each match receives 10 points and mismatch 0. The gap penalty is 10.0 for opening and 0.2 extending a gap. The alignment with the highest score is considered a potentially optimal solution. In addition, the objective function subtracts the fitness value with both the mismatch and gap scores multiplied by the numbers of nucleotides that are missing in the alignment for the chromosomes that do not include all nucleotides. Our experiments have confirmed that this strategy worked quite well. The calculation of SP scores for N sequences takes $O(M \times N^2)$ time [4, 26] where M is the average length of the sequences. To further improve the GA performance, we have devised a sequence profiling technique that simplifies the calculations of SP scores.

```
AGAGTTTGATC  A  TGGCTCAG
A-AGTTT--TC  C  TGGCTCAG
AGAG--TGATC  A  TGGCTCAG
AGAG--TGATC  A  TGGC-CAG
AG-GTTTGATC  C  TGG--CAG
```

$$W_{match}(M) = \sum_{i=\{A,C,T,G\}} \binom{i}{2} \times S(i,i)$$

$$W_{mismatch}(M) = \sum_{i=\{A,C,T,G\}} \sum_{j} i \times j \times S(i,j)$$

A	C	T	G	-
3	2	0	0	0

$$W_{gap}(M) = \sum_{i=\{A,C,T,G\}} i \times i_{gap} \times S(i,-)$$

Fig. 6. The figure shows the sequence profile for the column in the red color. The profiling process simply accumulates the frequencies or occurrences of each nucleotide on a given column. The process simplifies the calculations of the SP scores into three smaller tasks and reduces the complexity. Matches are only possible when two identical nucleotides are aligned together. Therefore, the score for matches is the sum of all possible combinations of identical nucleotides multiplied by the matching scores $S(i, i)$ from the substitution matrix. The number of mismatches is the sum of all combinations between two different nucleotides. There are only six such combinations. The gap penalties are the sum of all arrangements between each nucleotide and gaps

The objective function first computes the *profile* of the sequences for each column. A profile is simply the occurrences or frequencies of each nucleotide. The profiling process accumulates the occurrences of each nucleotide and reduces the calculations of SP scores into three smaller tasks. Matches are only possible when two identical nucleotides are aligned together. Therefore, the matching score is simply the sum of all possible combinations of the same nucleotides multiplied by the match score in the substitution matrix. The number of mismatches, on the other hand, is the sum of all the combinations of two different nucleotides. There are only six such combinations. The number of gap alignments is derived from the sum of the frequencies of each nucleotide multiplied by the number of gaps. The result is then multiplied by the gap penalty to obtain the overall gap score.

3.5 Alignment Construction

The construction of the final alignment is very similar to that of the DP algorithm [14, 23]. The GA derives the alignment from the last nucleotide to the first and the chromosomes are accordingly decoded backward. If a nucleotide is permitted at a given column, then it is consumed and added in the final alignment. The process moves on to the preceding ones. Otherwise a gap is inserted into the alignment. If no nucleotides are ever used in the column, then the allele is skipped. Alleles that are not used to derive the final alignment are considered the non-coding regions. One of the very interesting and important features of the scheme is that the alleles that are used to derive the alignment do not have to be consecutive. In addition, two different chromosomes can potentially give the same alignments. In other words, the alignment construction process picks the "appropriate" alleles as it moves along. The chromosomes do not have to encode exact bit patterns for the alignments. This makes every allele in the chromosome a potential solution for the alignment. Experiments have confirmed that the GA discovered the optimal alignment, as defined by the substitution matrix, quickly and effectively. The alignments produced by the GA are at least as good as the ones obtained from Clustal W. If the GA is allowed to continue evolving, better alignments are very likely to be found. As the number of sequences increases, the effectiveness of the GA begins to surface. Our experiments have shown that regardless the number of sequences being aligned; the GA performed extremely well and produced alignments with competitive scores.

```
Adenine:  1010100011110100110110000101101 0
Cytosine: 0101011110100110101100111010000 1
Thymine:  1111111101000011001011101110000
Guanine:  0101001000111011011010100000011 0

                AGA GTT T GA TCA-TG G CTC A G
                A-A GTT T -- TC-CTG G CTC A G
                AGA G-- T GA TCA-TG G CTC A G
                AGA G-- T GA TCA-TG G C-C A G
                AG- GTT T GA TC-CTG G --C A G
```

Fig. 7. The figure shows how the final alignment is constructed from the chromosomes. The alignment is derived in the reverse order. The spaces in between are the alleles that did not match up any of the nucleotides in the sequences. If a particular nucleotide does not present in the sequence, a gap is inserted. Chromosomes do not have to encode the exact bit patterns for the alignments. The alignment process simply picks the "appropriate" alleles

4 Experiment Settings

The objective of our experiments was to demonstrate that the GA could scale better as well as produce competitive alignments. For the purpose of this study, we assumed that Clustal W always gave the optimal scores. Sequences were first aligned with Clustal W and the scores were used as the stopping condition for the GA. We applied the standard IUB nucleic scoring matrix and used the gap penalties identical to that of Clustal W. For this study, we have gathered 20 short random DNA sequences of an average length of 60 base pairs. Sequences were manually verified to have at least 50% similarity. Each chromosome was about 96 bits long and had three loci. The mutation rate was 0.0625 and the average expected number of mutations on each locus was about one. The GA began with a randomly generated population of 64 individuals. The population was first evaluated and sorted in the descending order according to the fitness values. At each generation, the bias function randomly picked two individuals from the population that served as the parent chromosomes. The crossover operator exchanged alleles from two parent chromosomes and created an offspring. Mutation was applied to the offspring repeatedly until the fitness is higher than both parent chromosomes. The offspring then competed with the entire population and removed the individual with the lowest fitness. We gradually increased the number of sequences in each trial. Due to the stochastic nature of GA, all trials were performed at least three times in order to obtain more reliable results.

5 Experiment Results and Discussions

Experiments show promising results for our GA approach. In most cases, the GA outperformed Clustal W and produced better alignments. The number of generations needed to find the optimal solutions remained approximately the same even though the quantity of sequences increased. This is tribute to the fact that the GA was able to utilize the guided search effectively and found the optimal alignments.

During the course of experiments, we have tried various chromosome lengths in order to understand how they affect the performance of the GA. Notably the performance dropped dramatically when the length of sequences reached to 300 base pairs. Further investigation is still needed in order to gain a better understanding. The exact length of the consensus sequence is difficult to determine because it largely depends on the similarity of sequences and the evolutionary model. Our observations revealed that if the consensus sequence is too short, GA frequently failed to converge. On the other hand, if it is too long, the progress becomes extremely slow. In addition, we have confirmed that the SP scoring function was never a good measurement for MSA. If the gap is not heavily penalized, the same score can be easily achieved with more matches but excessive amounts of gaps. The relative difference in score between the correct and incorrect alignments decreases as the number of sequences increases. Clearly this is very counterintuitive and not realistic. The relative difference should increase when more sequences are introduced into the alignment.

Table 1. This table summarizes the numbers of generations needed to find the optimal alignments, at least as good as Clustal W, with various amounts of sequences

Trial	Number of Sequences / Generations					
	10	12	14	16	18	20
1	29,044	36,286	32,225	25,304	35,893	42,805
2	20,835	31,012	22,244	35,447	27,701	46,888
3	26,080	39,906	26,141	32,720	43,989	44,452

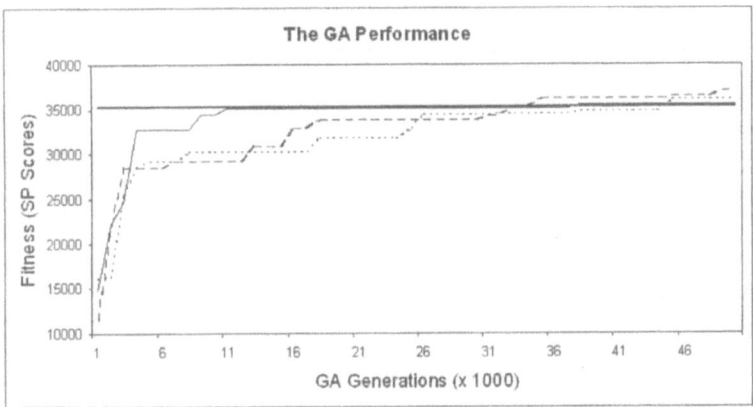

Fig. 8. The figure shows the GA performance on a set of 18 sequences. The GA typically found a good alignment within 50,000 generations. If the GA is allowed to continue evolving, alignments with even higher scores are very likely to be found. The horizontal line indicates the score found by Clustal W, which is about 35,294

6 Future Works

The consensus sequence with GA showed very promising performance and results. For future work, we would like to extend this approach to align protein sequences and implement statistical scoring techniques. In addition, we plan to incorporate the weighting scheme and analyze the impact on the GA performance. We are currently investigating an approach that incorporates several statistical and simulation techniques to try to quantify the significance of alignment scores.

Acknowledgements. Shyu was partially funded by a grant from Proctor and Gamble and Foster was part-ially funded for this research by NIH NCRR 1P20 RR16448.

References

1. Altschul, S.F., Carroll, R.J., and Lipman, D. Weights for data related by a tree. Journal of Molecular Biology, **207**: 647–653 (1989).
2. Altschul, S.F. and Lipman, D. Trees, stars, and multiple sequence alignment. SIAM Journal of Applied Mathematics, **49**: 197–209 (1989).
3. Altschul, S.F., Gish, W., Miller, W., Myers, E. W., and Lipman, D. Basic local alignment search tool. Journal of Molecular Biology, 215 (3):403–410 (1990).
4. Day, W.H. and McMorris, F.R. The computation of consensus patterns in DNA sequence. Mathematical and Computational Model. **17**, 49–52 (1993).
5. Carrillo, H. and Lipman, D. The multiple sequence alignment problem in biology. SIAM Journal of Applied Mathematics, **48**: 1073–1082 (1988).
6. Carroll, S.B., Grenier, J.K., and Weatherbee, S.D. From DNA to diversity: molecular genetics and the evolutionary of animal designs. Malden, MA: Blackwell Science (2001).

7. Durbin, R., Eddy, S., Krogh, A., and Mitchison, G. Biological sequence analysis: probabilistic models of proteins and nucleic acids. Cambridge, UK: Cambridge University (1998).
8. Fogel, D.B. and Corne, D.W. (ed.). Evolutionary Computation in Bioinformatics. San Fran-cisco, CA: Morgan Kaufmann Publishers (2003).
9. Feng, D. and Doolittle, R.F. Progressive sequence alignment as a prerequisite to correct phylogenetic trees. Journal of Molecular Evolutionary, **25**: 351–360 (1987).
10. Graur, D. and Li, W.H. Fundamental of Molecular Evolution, 2nd ed. Sunderland, MA: Sinauer Associates (2000).
11. Wang, L., and Gusfield, D. Improved approximation algorithms for tree alignment. Journal of Algorithms, **25**: 255–273 (1998)
12. Gusfield, D. Algorithms on strings, trees and sequences: computer science and computational biology. New York, NY: Cambridge University Press (1997).
13. Hall, B.G. Phylogenetics trees made easy: a how-to manual for molecular biologists. Sunderland, MA: Sinauer Associates (1997).
14. Needleman, S.B. and Wunsch, C.D. A general method applicable to the search for similarities in the amino acid sequence of two proteins. J. of Mol. Biol. **48**: 443–453 (1970).
15. Sean, R.E. A memory-efficient dynamic programming algorithm for optimal alignment of sequence to an RNA secondary structure. BMC Bioinformatics, **3**: 13 (2002).
16. Sauder, J., Arther, J., and Dunbrack, R. Large-scale of comparison of protein sequence alignment algorithms with structure alignments. Proteins: structures, function, and genetics, **40**: 6–32 (2000).
17. Setubal, J. and Meidanis, J. Introduction to computational molecular biology. Boston, MA: PWS Publishing (1997).
18. Stoye, J., Perry, S.W., and Dress, A.W.M. Improving the divide-and-conquer approach to sum-of-pairs multiple sequence alignment. Applied Mathematical Literature, vol. 10, no. 2, pp. 67–73 (1997).
19. Thomsen, R., Fogel, G.B., and Kirnk, T. A Clustal alignment improver using evolutionary algorithms. Congress on Evolutionary Computation, vol. 1; p. 121–126 (2002).
20. Thompson, J.D., Higgins, D.G., and Gibson, T.J. Clustal W: improving the sensitivity of progressive multiple sequence alignment through sequence weighting, position specific gap penalties and weight matrix choice. Nucleic Acids Research, **22**: 4673–4680 (1994).
21. Thompson, J. D., Gibson, T.J., Plewniak, F., Jeanmougin, F., and Higgins, D.G. The Clustal X windows interface: flexible strategies for multiple sequence alignment aided by quality analysis tools. Nucleic Acids Research, **24**: 4876–4882 (1997).
22. Wang, L. and Jiang, T. On the complexity of multiple sequence alignment. Journal of Computational Biology, **1**: 337–348 (1994).
23. Waterman, S.M. and Eggert, M. A new algorithm for best subsequence alignments with application to tRNA-rRNA comparisons. J. of Molecular Biology, **197**: 723–725 (1987).
24. Whitley, D. A genetic algorithm tutorial. Statistics and Computing, vol. **4**: 65–85 (1994).
25. Keith, J.M., Adams, P., Bryant, D. Kroese, D.P., Mitchelson, K.R., Cochran, D.A.E., and Lala, G.H. A simulated annealing algorithm for finding consensus sequences. Bioinformaics, vol. 18, no. 11, p. 1494–1499 (2002).
26. Wang, L., Jiang, T. and Gusfield, D. A more efficient approximation scheme for tree align-ment. SIAM Journal of Computational Biology. **30**: 283–299 (2000).

A Linear Genetic Programming Approach to Intrusion Detection

Dong Song, Malcolm I. Heywood, and A. Nur Zincir-Heywood

Dalhousie University, Faculty of Computer Science
6040 University Avenue, Halifax, NS, B3H 1W5, Canada
{dsong,mheywood,zincir}@cs.dal.ca

Abstract. Page-based Linear Genetic Programming (GP) is proposed and implemented with two-layer Subset Selection to address a two-class intrusion detection classification problem as defined by the KDD-99 benchmark dataset. By careful adjustment of the relationship between subset layers, over fitting by individuals to specific subsets is avoided. Moreover, efficient training on a dataset of 500,000 patterns is demonstrated. Unlike the current approaches to this benchmark, the learning algorithm is also responsible for deriving useful temporal features. Following evolution, decoding of a GP individual demonstrates that the solution is unique and comparative to hand coded solutions found by experts.

1 Introduction

The Internet, as well as representing a revolution in the ability to exchange and communicate information, has also provided greater opportunity for disruption and sabotage of data previously considered secure. The study of intrusion detection systems (IDS) provides many challenges. In particular the environment is forever changing, both with respect to what constitutes normal behavior and abnormal behavior. Moreover, given the utilization levels of networked computing systems, it is also necessary for such systems to work with a very low false alarm rate [1]. In order to promote the comparison of advanced research in this area, the Lincoln Laboratory at MIT, under DARPA sponsorship, conducted the 1998 and 1999 evaluation of intrusion detection [1]. As such, it provides a basis for making comparisons of existing systems under a common set of circumstances and assumptions [2]. Based on binary TCP dump data provided by DARPA evaluation, millions of connection statistics are collected and generated to form the training and test data in the Classifier Learning Contest organized in conjunction with the 5th ACM SIGKDD International Conference on Knowledge Discovery & Data Mining 1999 (KDD-99). The learning task is to build a detector (i.e. a classifier) capable of distinguishing between "bad" connections, called intrusions or attacks, and "good" or normal connections. There were a total of 24 entries submitted for the contest [3,4]. The top three winning solutions are all variants of decision trees. The winning entry is composed from 50×10 C5 decision trees fused by cost-sensitive bagged boosting [5]. The second placed entry consisted of a decision forest containing 755 trees [6]. The third placed entry consisted of two layers of voting decision trees augmented with human security expertise [7].

In this work, the first interest is to explore a Genetic Programming approach to produce computer programs with a much less complex structure, compared with the above data mining approaches, yet yielding satisfactory performance on the KDD-99 test set. The motivation being to provide transparent solutions that execute in real time with modest computational resources. Second, the training system must scale well on a comparatively large training data set (the 10% KDD-99 training set consists of approximately half a million patterns) whilst providing a small enough computational footprint to complete training in a matter of hours on a personal computer. Third, genetic programs are translated into human readable format and are compared with human reasoning. This will facilitate a better understanding of what constitutes a good rule as well as gaining the confidence of the more technically motivated users. Finally, only the most basic feature set is to be utilized, as opposed to the 41 connection features used in the decision tree solutions. To this end, a Page-based Linear Genetic Program with Dynamic Subset Selection and Random Subset Selection is proposed. Forty trials are conducted and results are compared with KDD-99 winning entries. One individual was simplified by removing structural introns and analyzed. The solution was found to be comparative to the type of rules extracted by domain experts on the same data set.

In the following text, Section 2 summarizes the properties associated with the KDD-99 IDS data set. Section 3 details the Genetic Programming (GP) approach taken, with a particular emphasis on the methodology used to address the size of the data set. Section 4 describes parameter settings and evaluates experiment results. Finally, conclusions and future directions discussed in Section 5.

2 Intrusion Detection Problem

From the perspective of the Genetic Programming (GP) paradigm KDD-99 posted several challenges [1, 2]. The amount of data is much larger than normally the case in GP applications. The entire training dataset consists of about 5,000,000 connection records. However, KDD-99 provided a concise training dataset – which is used in this work – and appears to be utilized in the case of the entries to the data-mining competition [3-7]. Known as "10% training" this contains 494,021 records among which there are 97,278 normal connection records (i.e. 19.69 %). This will be addressed by using a hierarchical competition between patterns, based on the Dynamic Subset Selection technique [8].

Each connection record is described in terms of 41 features and a label declaring the connection as either normal, or as a specific attack type. Of the 41 features, only the first eight (of nine) "Basic features of an Individual TCP connection"; hereafter referred to as 'basic features' are employed by this work. The additional 32 *derived* features, fall into three categories,

Content Features: Domain knowledge is used to assess the payload of the original TCP packets. This includes features such as the number of failed login attempts;

Time-Based Traffic Features: These features are designed to capture properties that mature over a 2 second temporal window. One example of such a feature would be the number of connections to the same host over the 2 second interval;

Host-Based Traffic Features: Utilize a historical window estimated over the number of connections – in this case 100 – instead of time. Host based features are therefore designed to assess attacks, which span intervals longer than 2 seconds.

In this work, none of these additional features are employed, as they appear to almost act as flags for specific attack behaviors. Our interest is on assessing how far the GP paradigm would go on 'basic features' alone.

Thirdly, the training data encompasses 24 different attack types, grouped into one of four categories: User to Root; Remote to Local; Denial of Service; and Probe. Naturally, the distribution of these attacks varies significantly, in line with their function – 'Denial of Service,' for example, results in many more connections than 'Probe'. Table 1 summarizes the distribution of attack types across the training data. Test data, on the other hand, follows a different distribution than in the training data, where this has previously been shown to be a significant factor in assessing generalization [3]. Finally, the test data added an additional 14 attack types not included in the training data.

Table 1. Distribution of Attacks

Data Type	Training	Test
Normal	19.69%	19.48%
Probe	0.83%	1.34%
DOS	79.24%	73.90%
U2R	0.01%	0.07%
R2L	0.23%	5.2%

Given that this work does not make use of the additional 32 derived features, it is necessary for the detector to derive any temporal properties associated with the current pattern, $x(t)$. To this end, as well as providing the detector with the labeled feature vector for the current pattern, $[x(t), d(t)]$, the detector is also allowed to address the previous 'n' features at some sampling interval (modulo(n)).

3 Methodology

In the case of this work a form of Linearly-structured GP (L-GP) is employed [9-12]. That is to say, rather than expressing individuals using the tree like structure popularized by the work of Koza [13], individuals are expressed as a linear list of instructions [9]. Execution of an individual therefore mimics the process of program execution normally associated with a simple register machine. That is, instructions are defined in terms of an opcode and operand (synonymous with function and terminal sets respectively) that modify the contents of internal registers {R[0],...,R[k]}, memory and program counter [9]. Output of the program is taken from register R[0] on completion of program execution (or some appropriate halting criterion [11]). Moreover, in an attempt to make the action of the crossover operator less destructive, the Page-based formulation of L-GP is employed [12]. In this case, an individual is described in terms of a number of *pages*, where each page has the *same* number of *instructions*. Crossover is limited to the exchange of *single* pages between two parents, and appears to result in concise solutions across a range of benchmark regression and classification problems. Moreover, a mechanism for dynamically changing page size was introduced, thus avoiding problems associated with the *a priori* selection of a specific number of instructions per page at initialization. Mutation operators take two forms. In the first case the 'mutation' operator selects an instruction for modification with uniform probability and performs an Ex-OR with a second instruction, also created with uniform probability. If the ensuing instruction represents a legal instruction the new instruction is accepted, otherwise the process is repeated. The second mutation operator 'swap' is designed to provide sequence modification. To do so, two instructions are selected within the same individual with uniform probability and their positions exchanged.

As indicated in the introduction, the specific interest of this work lies in identifying a solution to the problem of efficiently training with a large dataset (around half a million patterns). To this end, we revisit the method Dynamic Subset Selection [8] and extend it to the case of a hierarchy of subset selections. There are at least two aspects to this problem: the cost of fitness evaluation – the inner loop, which dominates the computational overheads associated with applying GP in practice; and the overhead associated with datasets that do not fit within RAM alone. In this work, a hierarchy is employed in which the data set is first partitioned into blocks small enough for retention in RAM, whilst a competition is initiated between training patterns within a selected block. Such a scheme also naturally matches the design methodology for computer memory hierarchies [14]. The selection of blocks is performed using Random Subset Selection (RSS) – layer 1. Dynamic Subset Selection (DSS) enforces a competition between different patterns – layer 2.

3.1 Subset Selection

First layer. The KDD-99 10% training data set was divided into 100 blocks with 5,000 connection records per block. The size of such blocks is defined to ensure that, when selected, they fit within the available RAM. Blocks are randomly selected with uniform probability. Once selected, a history of training pressure on a block is used to set up the number of iterations performed at the next layer in DSS. This is performed in proportion to the performance of the best-case individual. Thus, iterations of DSS, I, in block, b, at the current instance, i, is

$$I_b(i) = I_{(max)} \times E_b(i-1) \qquad (5)$$

where $I_{(max)}$ is the maximum number of subsets selected on a block; and $E_b(i-1)$ is the number of misclassifications of the best individual on the previous instance, i, of block, b. Hence, $E_b(i) = 1 - [\text{hits}_b(i) / \#\text{connections}(b)]$, where $\text{hits}_b(i)$ is the hit count over block 'b' for the best case individual identified over the last DSS tournament at iteration 'i' of block 'b'; and #connections(b) is the total number of connections in block 'b'.

Second Layer. A simplified DSS is deployed in this layer. That is, fixed probabilities are used to control the weighting of selection between age and difficulty. For each record in the DSS subset, there is a 30% (70%) probability of selecting on the basis of age (difficulty). Thus, a greater emphasis is always given to examples that resist classification. DSS utilizes a subset size of 50, with the objective of reducing the computational complexity associated with a particular fitness evaluation. Moreover, in order to further reduce computation, the performance of parent individuals on a specific subset is retained. After 6 tournaments the DSS subset will be reselected.

DSS Selection [8]. In the RSS block, every pattern is associated with an age value, which is the number of DSS selections since last selection, and a difficulty value. The difficulty value is the number of individuals that were unable to recognize a connection correctly the last time that the connection appeared in the DSS subset. Connections appear in a specific DSS stochastically, where there is 30% (70%) probability to select by age (difficulty). Roulette wheel selection is then conducted on the whole RSS block, with respect to age (difficulty). After the DSS subset is filled, age and difficulty of selected connections are reset. For the rest, age is increased by 1 and difficulty remains unchanged.

3.2 Parameterization of the Subsets

The low number of patterns actually seen by a GP individual during fitness evaluation, relative to the number of patterns in the training data set, may naturally lead to 'over fitting' on specific subsets. Our general objective was therefore to ensure that the performance across subsets evolved as uniformly as possible. The principle interest is therefore to identify the stop criterion necessary to avoid individuals that are sensitive to the composition of a specific Second Level subset.

To this end, a single experiment is conducted in which 2,000 block selections are made with uniform probability. In the case of each block selection, there are 400 DSS selections. Before selection of the next block takes place, the best performing individual (with respect to sub-set classification error) is evaluated over *all* patterns within the block, let this be the block error at selection i, or $E_b(i)$. A sliding window is then constructed consisting of 100 block selection errors, and a linear least-squares regression performed. The gradient of each linear regression is then plotted, Figure 1 (1900 points). A negative trend implies that the block errors are decreasing whereas a positive trend implies that the block errors are increasing (the continuous line indicates the trend). It is now apparent, that after the first 750 block selections, the trend in block error has stopped decreasing. After 750 selections, oscillation in the block gradients appears, where this becomes very erratic in the last 500 block selections.

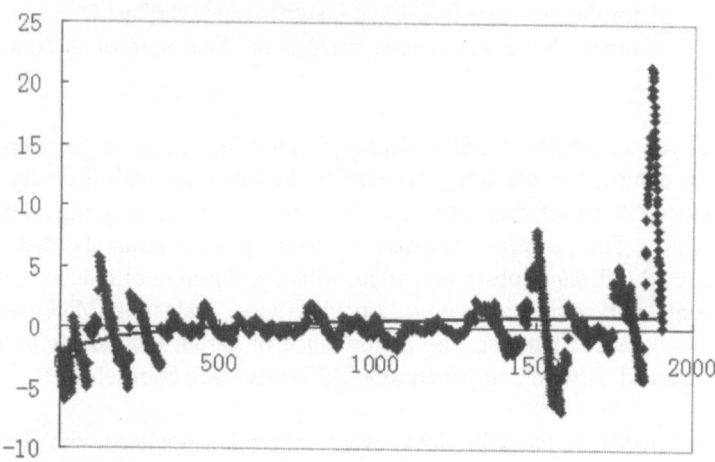

Fig. 1. Gradient of block error using best case DSS individual. X axis represents tournament and Y represents slope of best fitting line on 100 point window

On the basis of these observations, the number of DSS selections per block is limited to 100 (from 400) – with the objective of further reducing any tendency to prematurely specialize – where the principle cost is in a higher number of block selections, 1000 in this case.

3.3 Structural Removal of Introns

Introns are program pieces, which have no influence to the output, but appear to be a factor in the evolution of solutions. Moreover, two forms of introns are often distinguished: structural introns and semantic introns. Structural introns manipulate variables that are not used for the calculation of the outputs at that program position. Whereas, semantic introns manipulate variables on which the state of the program is invariant [15]. In this work, structural introns are detected using following pseudo code, initiated once evolution is complete, with the last reference to R[0] (the register *a priori* defined as the output) as the input argument.

```
markExon(reg, i)

{..for (destination in the i^th instruction != reg; i--)
    if (i = 0) exit;
    mark i^th instruction as exon
    markExon(oprand1, i-1)
    markExon(oprand2, i-1)    .... }
```

4 Experiment

The following experiments are based on 40 runs using Dynamic Page-based L-GP. Runs differ only in their choice of a random seed initializing the population. Table 2 lists the common parameter settings for all runs. The total number of records in training and test set is listed in Table 3. The method used for encouraging the identification of temporal relationships and composing the instruction set is defined as follows.

Sequencing Information. As indicated above, only the 8 basic features of each connection are used, corresponding to: Duration; Protocol; Service; normal or error status of the connection (Flag); number of data bytes from source to destination (DST); number of data bytes from destination to source (SRC); LAND (1 if connection is from/to the same host/port, 0 otherwise); and number of "wrong" fragments (WRONG). This implies that GP is required determine the temporal features of interest itself. To do so, for each 'current' connection record, $x(t)$, GP is permitted to index the previous 32 connection records relative to the current sample, modulo 4. Thus, for each of the eight basic TCP/IP features available in the KDD-99 dataset, GP may index the 8 connection records $[(t), (t-4), \ldots (t-32)]$, where the objective is to provide the label associated with sample 't'.

Table 2. Parameter Settings for Dynamic Page based Linear GP

Parameter	Setting
Population Size	125
Maximum number of pages	32 pages
Page size	8 instructions
Maximum working page size	8 instructions
Crossover probability	0.9
Mutation probability	0.5
Swap probability	0.9
Tournament size	4
Number of registers	8
Instruction type 1 probability	0.5
Instruction type 2 probability	4
Instruction type 3 probability	1
Function set	$\{+, -, *, /\}$
Terminal set	$\{0, \ldots, 255\} \cup \{i_0, \ldots, i_{63}\}$
RSS subset size	5000
DSS subset size	50
RSS iteration	1000
DSS iteration (6 tournaments/ iteration)	100
Wrapper function	0 if output <=0, otherwise 1
Cost function	Increment by 1 for each misclassification

Table 3. Distribution of Normal and Attacks

Connection	Training	Test
Normal	97249	60577
Attacks	396744	250424

Instruction Set. A 2-address format is employed in which provision is made for: up to 16 internal registers, up to 64 inputs (Terminal Set), 5 opcodes (Functional Set) – the fifth is retained for a reserved word denoting end of program – and an 8-bit integer field representing constants (0-255) [12]. Two mode bits toggle between one of three instruction types: opcode with internal register reference; opcode with reference to input; target register with integer constant. Extension to include further inputs or internal registers merely increases the size of the associated instruction field. The output is taken from the first internal register.

Training was performed on a Pentium III 1GHz platform with a 256M byte RAM under Windows 2000. The 40 best individuals within the last tournament are recorded and translated simplified as per Section 3.3. Note that 'best' is defined with respect to the cost function used during training, Table 2. Performance of these cases is then expressed in terms of false positive (FP) and detection rates, estimated as follows,

$$Detection\,Rate = 1 - \frac{\#False\,Negatives}{Total\,Number\,of\,Attacks} \quad (6)$$

$$False\,Positive\,Rate = \frac{\#False\,Positives}{Total\,Number\,of\,Normal\,Connections} \quad (7)$$

Figure 3 summarizes the performance of all 40 runs in terms of FP and Detection rate on both training and test data. Of the forty cases, three represented degenerate solutions (not plotted). That is to say, they basically classified everything as normal (i.e. only 20% of the training classifications would be correct) or attack (roughly 80% of the training connections would be correct). Outside of the case of the three degenerate cases, it is apparent that solution classification accuracy is consistently achieved. Table 4 makes a direct comparison between KDD-99 competition winners, verses the corresponding GP cases.

Structural removal of introns, Section 3.3, resulted in a 5:1 decrease in the average number of instructions per individual (87 to 17 instructions). With the objective of identifying what type of rules were learnt, the GP individual with best Detection Rate from Table 4 was selected for analysis. Table 5 lists the individual following removal of the structural introns.

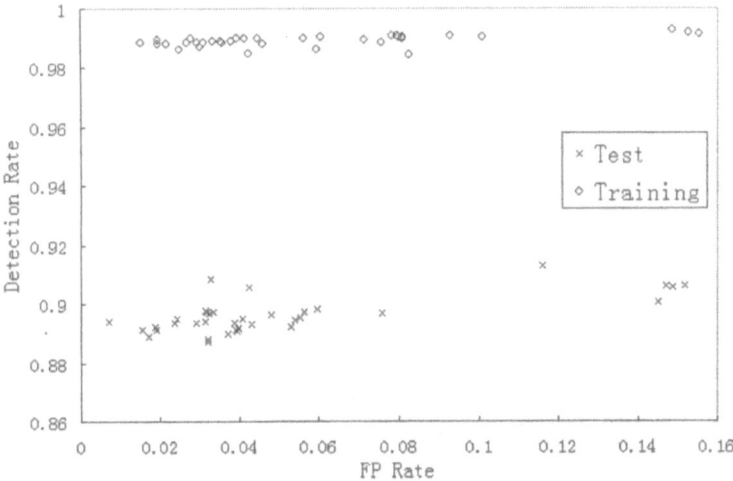

Fig. 3. FP and detection rate of 40 runs on KDD-99 test data set

Table 4. Comparison with KDD-99 winning entries

Parameter	Detection Rate	FP rate
Winning entry	0.908819	0.004472
Second place	0.915252	0.00576
Best GP – FP rate	0.894096	0.006818
Best GP – Detection rate	0.908252	0.032669

Table 5. Anatomy of Best Individual

Opcode	Destination	Source
LOD	R[0]	20
SUB	R[0]	Input[2][5]
MUL	R[0]	Input[0][1]
DIV	R[0]	Input[0][4]
SUB	R[0]	Input[2][5]
SUB	R[0]	Input[6][5]
DIV	R[0]	Input[0][4]

Table 6 summarizes performance of the individual over a sample set of the connection types in terms of connections types seen during training (24 different types) and connections types only during test (14 different types). Of particular interest here is that high classification accuracy is returned for connection types, which are both frequent and rare, where it might be assumed that only the connections with many examples might be learnt.

Table 6. Error rates on test data for top 16 attacks by individual with Best Detection Rate

Seen connection type	% Misclassified	Total Examples	Unseen connection type	% Misclassified	Total Examples
Neptune	0	58,001	Udpstorm	0	2
Portsweep	0	354	Prosstable	3.03	759
Land	0	9	Saint	5.978	736
Nmap	0	84	Mscan	8.452	1,053
Smurf	0.077	164,091	Httptunnel	15.823	158
Satan	3.552	1,633	Phf	50	2
Normal	3.267	60,577	Apache2	65.491	794

Re-expressing this individual analytically, below, indicates that the statistics of the number of bytes from the responder and the byte ratio responder-originator are utilized. This enables the individual to identify that the attacking telnet connections in the DARPA dataset are statistically different from the normal telnet connections. Moreover, not only telnet connections can be classified by this way. Such a rule never misses an attack of "Neptune", "portsweep", "land", "nmap", "udpstorm". It also provided 'good' performance on "smurf", "processtable", "normal", "satan", "saint", "mscan" and "httptunnel". For "Neptune," there are many half open tcp connections, without any data transfer. In "smurf," there are many echo replies to victim, but no echo requests from victim. In "http tunnel," the attacker defines attacks on the http protocol, which is normal, but the actual data exchange ratio makes it different from normal traffic. Currently, only [16] argued that telnet connection can be differentiated by a rule of the form discovered here. It has been suggested that attacks be formulated with such a rule in mind, [17], but without explicitly proposing using this statistic. Thus GP in this case has actually provided a unique generic rule for the detection of multiple attack types.

$$Output = \frac{\frac{(20 - Input[2][5]) \times Input[0][1]}{Input[0][4]} - Input[2][5] - Input[6][5]}{Input[0][4]}$$

where $Input[j][i]$ indexes the ith input feature at temporal location $t - 4 \times j$; and 't' ($= 0$) is the current connection.

5 Conclusion

A Page-based Linear Genetic Programming system with DSS and RSS was implemented and tested on the KDD'99 benchmark dataset, a problem involving a training dataset of half a million patterns. To do so, a *hierarchy* of data subset selections is introduced such that GP only perceives 50 of the total training set patterns at any one time. Moreover, such a hierarchy is designed to utilize the memory hierarchy commonly employed in computer architectures. As such the ensuing system completes each trial in approximately 15 minutes on a modest laptop-computing platform (1Ghz Pentium III, 256 Mbyte RAM) or 10 hours for 40 trials.

In addition, only the 'basic' *connection* features are employed, with GP deriving the necessary temporal features itself. Performance approaches that of data-mining solutions based on all 41 features, whilst solution transparency is also supported and verified, enabling the user to learn from the solutions provided. Note however, that the principle design interest of this work was to demonstrate that GP could be applied to data-driven learning problems on large datasets. The resulting GP classifier represents an anomaly detector, providing a binary decision boundary: normal or attack. Extensions to include the classification of different attack types would involve training additional detectors on the subset of patterns labeled as attack. The ensuing hierarchy of detectors would provide an additional (attack) class label at each level.

Future work is expected to include a dynamic cost function; with the objective of adjusting at run time the relative weighting associated with different attack types. Moreover, the function set at present is purely analytical. Of interest would be the significance of conditional statements or modular code within this problem context.

Acknowledgements. This research was partially supported by NSERC Discovery Grants of Drs. Heywood and Zincir-Heywood.

References

1. Lippmann R.P., Fried D.J., Graf I., Haines J.W., Kendall K.R., McClung D., Weber D., Webster S.E., Wyschogrod D., Cunningham R.K., Zissman M.A.: Evaluating Intrusion Detection Systems: the 1998 DARPA Off-Line Intrusion Detection Evaluation. Proceedings of the 2000 DARPA Information Survivability Conference and Exposition, 2 (2000)
2. McHugh J.: Testing Intrusion Detection Systems: A Critique of the 1998 and 1999 DARPA Intrusion Detection System Evaluations as Performed by Lincoln Laboratory. ACM Transactions on Information and System Security. 3(4), (2000) 262–294
3. Elkan C.: Results of the KDD'99 Classifier Learning Contest. SIGKDD Explorations. ACM SIGKDD. 1(2), (2000) 63–64
4. Wenke L., Stolfo S.J., Mok K.W.: A data mining framework for building intrusion detection models. Proceedings of the 1999 IEEE Symposium on Security and Privacy (1999) 120–132
5. Pfahringer B.: Winning the KDD99 Classification Cup: Bagged Boosting. SIGKDD Explorations. ACM SIGKDD. 1(2) (2000) 65–66
6. Levin I.: KDD-99 Classifier Learning Contest LLSoft's Results Overview. SIGKDD Explorations. ACM SIGKDD. 1(2) (2000) 67–75
7. Vladimir M., Alexei V., Ivan S.: The MP13 Approach to the KDD'99 Classifier Learning Contest. SIGKDD Explorations. ACM SIGKDD. 1(2) (2000) 76–77
8. Gathercole C., Ross P.: Dynamic Training Subset Selection for Supervised Learning in Genetic Programming. Parallel Problem Solving from Nature III. Lecture Notes in Computer Science, Vol. 866. Springer-Verlag, Berlin (1994) 312–321

9. Cramer N.L.: A Representation for the Adaptive Generation of Simple Sequential Programs. Proceedings of the International Conference on Genetic Algorithms and Their Application (1985) 183–187
10. Nordin P.: A Compiling Genetic Programming System that Directly Manipulates the Machine Code. In: Kinnear K.E. (ed.): Advances in Genetic Programming, Chapter 14. MIT Press, Cambridge, MA (1994) 311–334
11. Huelsbergen L.: Finding General Solutions to the Parity Problem by Evolving Machine-Language Representations. Proceedings of the 3rd Conference on Genetic Programming. Morgan Kaufmann, San Francisco, CA (1998) 158–166
12. Heywood M.I., Zincir-Heywood A.N.: Dynamic Page-Based Linear Genetic Programming. IEEE Transactions on Systems, Man and Cybernetics – PartB: Cybernetics. 32(3) (2002), 380–388
13. Koza J.R.: Genetic Programming: On the Programming of Computers by Means of Natural Selection. MIT Press, Cambridge, MA (1992)
14. Hennessy J.L., Patterson D.A.: Computer Architecture: A Quantitative Approach. 3rd Edition. Morgan Kaufmann, San Francisco, CA (2002)
15. Brameier M., Banzhaf W.: A Comparison of Linear Genetic Programming and Neural Networks in Medical Data Mining. IEEE Transactions on Evolutionary Computation, 5(1) (2001) 17–26
16. Caberera J.B.D., Ravichandran B., Mehra R.K.: Statistical traffic modeling for network intrusion detection. Proceedings of the 8th International Symposium on Modeling, Analysis and Simulation of Computer and Telecommunication Systems (2000) 466–473
17. Kendall K.: A Database of Computer Attacks for the Evaluation of Intrusion Detection Systems. Master Thesis. Massachusetts Institute of Technology (1998)

Genetic Algorithm for Supply Planning Optimization under Uncertain Demand

Tezuka Masaru and Hiji Masahiro

Hitachi Tohoku Software, Ltd.
2-16-10, Honcho, Aoba ward, Sendai City, 980-0014, Japan
{tezuka,hiji}@hitachi-to.co.jp

Abstract. Supply planning optimization is one of the most important issues for manufacturers and distributors. Supply is planned to meet the future demand. Under the uncertainty involved in demand forecasting, profit is maximized and risk is minimized. In order to simulate the uncertainty and evaluate the profit and risk, we introduced Monte Carlo simulation. The fitness function of GA used the statistics of the simulation. The supply planning problems are multi-objective, thus there are several Pareto optimal solutions from high-risk and high-profit to low-risk and low-profit. Those solutions are very helpful as alternatives for decision-makers. For the purpose of providing such alternatives, a multi-objective genetic algorithm was employed. In practice, it is important to obtain good enough solutions in an acceptable time. So as to search the solutions in a short time, we propose Boundary Initialization which initializes population on the boundary of constrained space. The initialization makes the search efficient. The approach was tested on the supply planning data of an electric appliances manufacturer, and has achieved a remarkable result.

1 Introduction

Manufacturers and distributors deal with a number of products. Supply planning problems are to decide the quantity, the type, and the due time of each product to supply as a replenishment. The supply plan is decided depending on the demand forecast of the products, and the forecasted demand involves uncertainty. If the demand exceeds the supply, opportunity losses occur while excess supply increases inventory level, and may result in dead stocks. In order to supply products, several resources are consumed to produce, and deliver the products. Materials, production machines, and transportation are the resources. The availability of the resources is limited, thus the supply quantity of the products also has a limit. Under the resource constraints and the uncertainty of demands, the supply plan is made to maximize the profit.

Traditionally, inventory management approach has been used to create supply plans[1]. This approach decides the supply plan to minimize the stockout rate, or opportunity loss rate. However, in practice, this approach has a problem. This approach does not take account of the relationship between profit and risk. For example, under the resource constraints, we can expect more profit by reducing the opportunity loss of the product with high gross margin than by doing so with low gross margin. However, the demand of the high gross margin product is often uncertain, while the demand of regular products with low gross margin is relatively steady and the risk is low. Thus the best mixture of high risk and low risk products is important to achieve a certain amount of profit with minimizing risk. In other words, portfolio management needs to be introduced in supply planning problems. It is desired to optimize profit and risk of supply planning problems under uncertain demand.

There is also an operational problem. Supply planning problems are multi-objective. Though the maximization of profit and the minimization of risk are required simultaneously, those objectives are in trade-off relationship. Therefore, the problem has some Pareto optimal solutions from high-risk and high-profit to low-risk and low-profit. For decision-makers, such alternative solutions are very helpful because what they should do is to simply select one solution, which fits their strategy, from the alternatives. Many existing supply planning methods, however, create only one solution while most decision-makers do not like the black box system which proposes only one solution.

To produce an efficient supply plan, we introduced a Multi-objective Genetic Algorithm (GA) and Monte Carlo Simulation into the supply planning. GAs are considered as efficient ways of optimizing multi-objective problems[2,3,4]. In GAs, a number of individuals promote optimization in parallel and this characteristic is expedient to find Pareto optimal solutions all at once. We introduced Monte Carlo simulation[5] into the evaluation process of GA. Monte Carlo simulation simulates the uncertainty of demand, then profit and risk are evaluated from the simulation result. In order to search the solutions in an acceptable time, we proposed boundary initialization which initializes population on the boundary of constrained space and makes the search efficient.

We briefly describe the supply planning problems in section 2, and propose a GA with Monte Carlo simulation-based evaluation for supply planning and efficient population initialization method in section 3. Section 4 shows the result of computational experiments. Then we conclude in section 5.

2 Demand and Supply Planning

2.1 Supply Planning and Resource Constraints

Manufacturers and distributors deal with many kinds of products. A firm deals with I kinds of products and it wants to make a supply plan of a certain period consisting of T terms. d_{ti}, p_{ti}, and q_{ti} are the demand, the supply quantity, and the initial inventory quantity of product i in term t, respectively. Sales quantity s_{ti} and opportunity loss quantity l_{ti} are obtained by the following equations.

$$s_{ti} = \min\left(d_{ti}, p_{ti} + q_{ti}\right) \qquad (1)$$

$$l_{ti} = d_{ti} - s_{ti} \qquad (2)$$

The initial inventory quantity of the next term is

$$q_{(t+1)i} = p_{ti} + q_{ti} - s_{ti} . \tag{3}$$

The firm sells the product i for unit price u_{ti} in term t, the unit supply cost is v_{ti} and the unit inventory cost is w_{ti}. Then the gross profit of the firm through the planning period is obtained by equation 4.

$$G = \sum_{t=1}^{T} \sum_{i=1}^{I} (s_{ti} u_{ti} - p_{ti} v_{ti} - q_{ti} w_{ti}) \tag{4}$$

The first term represents the sales amount, the second and the third terms are the supply cost and inventory cost. Opportunity loss amount L is the sales amount which could be gained if the firm had enough supply and inventory, so it is defined as the following equation.

$$L = \sum_{t=1}^{T} \sum_{i=1}^{I} l_{ti} u_{ti} \tag{5}$$

Demand forecast d_{ti} is an uncertain estimate, that is a stochastic variable while supply quantity p_{ti} is a decision variable. Since d_{ti} is stochastic, as its dependent variables, gross profit G and opportunity loss L are naturally stochastic. The distributions of G and L depend on the decision variable p_{ti}. Thus, the objective of the supply planning problem is to decide the supply quantity so that the distributions are optimal.

Resources are consumed as products are supplied. For example, the machines to produce products and the trucks to transport them are the resources. The firm has J kinds of resources. r_{ij} of resource j is consumed to supply the unit quantity of product i. a_{tj} is the available quantity of resource j in term t. The amount of the consumption of the resource must not exceed the available quantity. On the other hand, supply quantity should not be negative. Therefore, the constraints of supply planning problems are defined as the following equations.

$$\sum_{i=1}^{I} r_{ij} p_{ti} \leq a_{tj} \quad (t = 1, ..., T, j = 1, ..., J) \tag{6}$$

$$p_{ti} \geq 0 \quad (t = 1, ..., T, i = 1, ..., I) \tag{7}$$

The constraints space is convex.

2.2 Optimization Criteria

Supply planning problems are multi-objective. They have a number of criteria of profit and risk. Statistics of the distributions of the gross profit and opportunity loss are used as the optimization criteria. Figure 1 summarizes the statistics. X axis indicates gross profit G or opportunity loss L, and y axis indicates the probability of it.

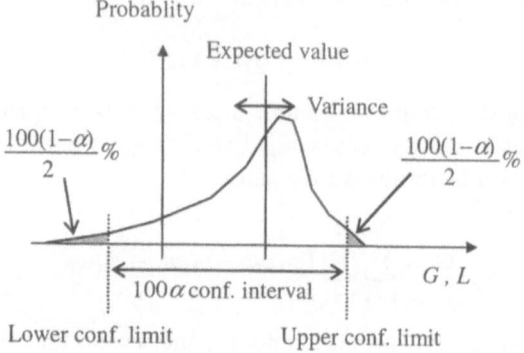

Fig. 1. Optimization Criterion

One of the most important objectives of business is maximizing profit. Thus, expected gross profit is naturally used as a criterion of profit, and it should be maximized. Variance of the gross profit reflects the volatility of outcome. Less volatility is preferable, thus variance is used as a risk criterion, and it should be minimized. Expected opportunity loss is also used as a risk criterion, and it should be minimized. The lower confidence limit of the 100α percent confidence interval is the lower $100(1-\alpha)/2$ percentile. For example, a lower confidence limit of 95 percent confidence interval of the gross profit is the lower 2.5 percentile of the profit. That means the probability of the gross profit falling below the limit is only 2.5 percent. Lower confidence limit is an inverse criterion of risk since it indicates the worst case of the profit, and it should be maximized. Likewise, upper confidence limit of the opportunity loss is a risk criterion since it indicates the worst case of the loss, and it should be minimized.

2.3 Demand Forecasting

Future demand can be forecasted according to the history of demand. There are several forecasting methods such as Moving Average Method, Exponential Smoothing Method, and Box-Jenkins Method [6].

In this paper, an existing commercial software was used to forecast future demand. The software analyzes the history of demand and automatically chooses the forecasting method which best fits to the historical data, then forecasts the expected value and the variance of the future demand. The demand forecast is supposed to follow Normal distribution though its left tail is truncated at zero.

3 Genetic Algorithms Approach

3.1 Genetic Representation

In order to optimize supply planning problems, we employed a real-coded genetic algorithm. Each individual is a vector of real value, and the vector

$x = (x_1, ..., x_{T \times I})$ corresponds to the set of supply quantities. Its element $x_{(t-1) \times I + i}$ corresponds to supply quantity p_{ti}, i.e.

$$x = (p_{11}, p_{12}, ..., p_{1I}, p_{21}, p_{22}, ..., p_{2I}, ..., p_{T1}, p_{T2}, ..., p_{TI}) . \qquad (8)$$

3.2 Fitness Function Using Monte Carlo Simulation

In order to evaluate the criteria of supply plan problems, we introduced Monte Carlo Simulation in the fitness function of GA. Monte Carlo simulation is a simulation method in which a large quantity of random numbers are used to calculate statistics. It calculates multiple scenarios of the gross profit and the opportunity loss by sampling demand quantities from the random number following their probability distributions.

Figure 2 describes the detail of the Monte Carlo simulation to evaluate supply planning. The future demand of each product is forecasted as the expected value and the variance. μ_{ti} and σ_{ti} denote the expected demand and its standard deviation of product i in term t. Demand is considered to follow normal distribution.

```
01 begin evaluation
02      m := 1
03      repeat                    *** Repeat simulation for an evaluation ***
04          Gₘ :=0,  Lₘ :=0      *** Initialize gross profit and opp. loss of m-th simulation ***
05          i :=1
06          repeat
07              q₁ᵢ :=0
08              t :=1
09              repeat
10                             *** Calculate the profit and the loss from product i in term t ***
11                             *** and sum up them into n-th profit and loss               ***
12                  dₜᵢ := Random number following N( μₜᵢ, σₜᵢ )
13                  sₜᵢ := min( dₜᵢ, pₜᵢ + qₜᵢ )
14                  lₜᵢ := dₜᵢ − sₜᵢ
15                  Gₘ := Gₘ + ( sₜᵢuₜᵢ − pₜᵢvₜᵢ − qₜᵢwₜᵢ )
16                  Lₘ := Lₘ + lₜᵢuₜᵢ
17                  q₍ₜ₊₁₎ᵢ := pₜᵢ + qₜᵢ − sₜᵢ
18                  t := t +1
19              until t <= T
20              i := i +1
21          until i <= I
22          m := m +1
23      until m <= M
24      Calculate optimization criteria
25 End evaluation
```

Fig. 2. Monte Carlo Simulation to Evaluate Supply Plan

The evaluation of one individual, or one supply plan, consists of M simulations of the gross profit and the opportunity loss. The block from line 4 to 22 in the figure corresponds to one simulation. In each simulation, the demand of each product in each term is simulated by the random number following normal distribution $N(\mu,\sigma)$(line 12 in the figure). After the iterations, we obtain M samples of gross profit G_m and opportunity loss L_m ($m = 1,...,M$). Statistics as the optimization criteria are calculated from the samples. For example, the expected gross profit and its variance are estimated by equation 9 and 10.

$$\hat{G} = \frac{1}{M} \sum_{m=1}^{M} G_m \qquad (9)$$

$$U(G) = \frac{1}{M-1} \sum_{m=1}^{M} \left(G_m - \hat{G}\right)^2 \qquad (10)$$

$\hat{G}$ should be maximized as a matter of course and $U(G)$ should be minimized since less volatility is preferable.

3.3 Genetic Operators and Selection

Supply planning problems are multi-objective. The outcome of multi-objective optimization is not a single solution, but a set of solutions known as Pareto optimal solutions. Among the solutions, each objective cannot be improved without the other objectives being degenerated. A vector $u = (u_1,...,u_n)$ is superior to $v = (v_1,...,v_n)$ when u is partially greater than v, i.e.,

$$\forall i, u_i \geq v_i \wedge \exists i, u_i > v_i . \qquad (11)$$

Any solution to which no other solution is superior is considered as optimal. Since supply plan optimization is multi-objective, it has several Pareto optimal solutions.

The supply plan optimization problem has convex constraints. The genetic operators proposed by Michalewicz [7][8] can handle convex constraints effectively. Thus we employed the operators and modified for multi-objective problems. The GA has two mutation operators, uniform and boundary, and two cross-over operators, arithmetic and heuristic.

The mutation operator selects one locus k from individual x randomly and changes the value of x_k in the range satisfying constraints. Uniform mutation changes x_k to a random number following uniform distribution, and boundary mutation changes x_k to the boundary of constrained space.

x and y denote parents, and z denotes offspring. Arithmetic crossover reproduces offspring as $z = \lambda x + (1-\lambda) y$ where λ is a random value following uniform distribution $[0,1]$. Since the constrained solution space is convex, whenever both x and y satisfy the constraints, arithmetic crossover guarantees the feasibility of z.

Heuristic crossover uses the evaluation of two parents to determine the search direction. The offspring reproduced by heuristic crossover is $z = x + \lambda(x - y)$ where the evaluation of individual x is superior to y and λ is a random number. The range of λ is $[0, 1]$, and in the case when the offspring is not feasible, the operator makes attempts to generate a feasible one.

The supply plan problems we deal with are multi-objective. Thus, in order to determine the search direction of heuristic crossover, we introduced a comparison procedure consisting of three steps. $E_i(x), (i = 1, ..., K)$ denotes the evaluation of i-th criterion of individual x. K is the number of criteria. In case of maximization, the procedure is as follows:

- Step 1

    Compare two parents according to Pareto optimality. Individual x is superior to y when

$$\forall i, E_i(x) \geq E_i(y) \land \exists i, E_i(x) > E_i(y) \ . \tag{12}$$

    Otherwise, go to step 2.
- Step 2

    Compare the number of individuals superior to each parent. The parent dominated by the smaller number of other individuals in the current population is considered to be superior. If the same number of individuals dominate both x and y, go to step 3.
- Step 3

    Randomly Select either of two parents as a superior.

We employed tournament selection [9]. It selects k individuals at random and selects the best one among these k individuals. k is a parameter called tournament size which determines selective pressure. Pareto optimality is also applied to determine which individual wins the tournament. Superior individuals have more chances to be selected. Consequently, Pareto optimal solution set is explored.

3.4 Boundary Initialization

In the GA we introduced, all the individuals in the initial population must satisfy the constraints. Thus, the population is generally initialized randomly in the constrained space.

In practice, it is more important to find good enough solutions in acceptable time than to find genuine optimal solutions. Thus, we propose a new initialization method, Boundary Initialization. Practically and empirically, most of optimal solutions of the real-world constraints problems are on the boundary of constraints. The method initializes the population randomly on the boundary of constrained space, and makes the search of solutions efficient. In figure 3, black dots are the individuals produced by Boundary Initialization, and white dots are produced by Random Initialization.

It is said that the diversity of initial population is very important because it ensures the exploration through the whole search space. Therefore, there is a

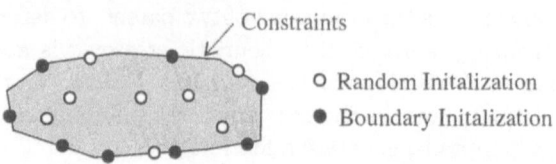

Fig. 3. Boundary Initialization and Random Initialization

possibility that search fails to reach optimal solutions when the initial population is biased on the boundary. However, we expect that the bias boosts the search efficiency in solving practical problems, and that Boundary Initialization should work well.

4 Computational Experiments

The supply planning method with GA was tested on the data provided by an electric appliances manufacturer. The data consist of ten product groups and four key resources. In our experiments, each product group is treated as a product.

In our experiments, population size was set to 100, termination was 50 generations, and tournament size was 4. We adopted the elitist policy [10] and elite size was 5. The iteration of simulations in one evaluation was 1000. That means gross profit and opportunity loss were calculated 5,000,000 times in one optimization run. It took about 255 seconds to run one optimization on Windows 2000 PC with Pentium IV 2.2GHz and 1GBytes RAM. The test program was implemented in C++. We carried out two experiments. In the first experiment, the objectives were maximizing expected gross profit and minimizing the standard deviation of the profit. The standard deviation was minimized since smaller volatility was preferable. In the second experiment, the objectives were maximizing expected gross profit and minimizing expected opportunity loss. Standard deviation of the profit and opportunity loss are the criteria of risk. We tested Random and Boundary Initialization in both experiments.

Figure 4 and Figure 5 show the Pareto optimal individuals, i.e. solutions at the last generation of five trials. Each marker corresponds to one solution and each line corresponds to the efficiency frontier of each trial. (a) is the result with Random Initialization and (b) is with Boundary Initialization. For comparison, we also tested conventional inventory management method.

Figure 4 is the result of the first experiment maximizing expected gross profit and minimizing the standard deviation. Obviously, the optimization with Boundary Initialization is better than that with Random Initialization. The standard deviation was minimized with both methods. However, Random Initialization could not obtain a better solution even than the conventional method.

Figure 5 is the result of the second experiment maximizing expected gross profit and minimizing opportunity loss. The optimization with Boundary Initialization is much better than that with Random Initialization.

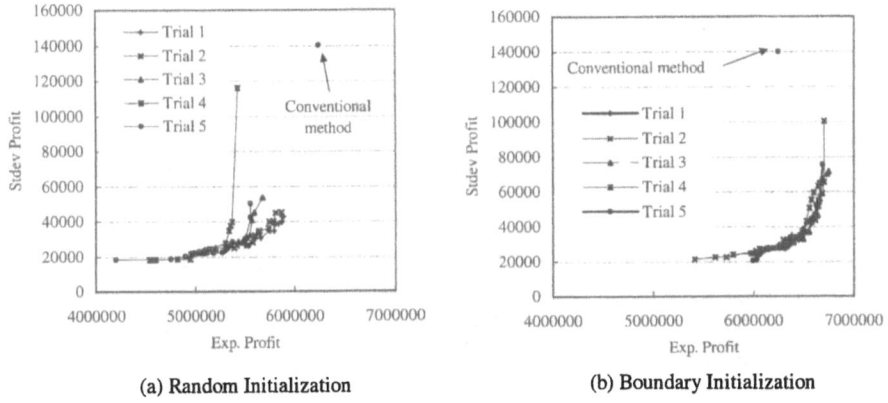

Fig. 4. Maximization of expected profit and minimization of standard deviation of profit

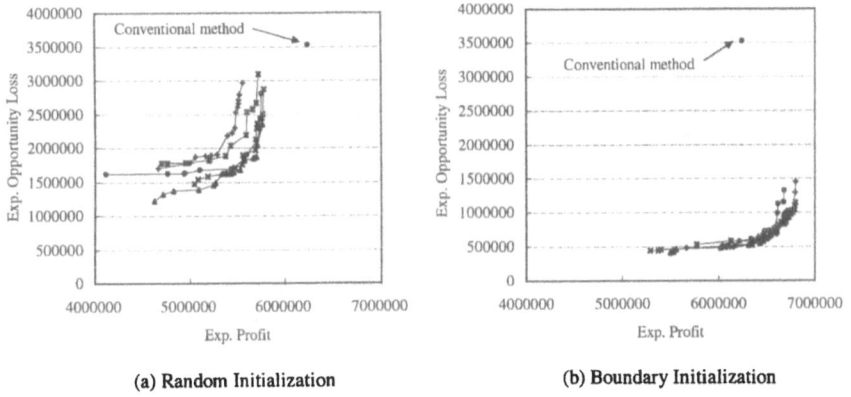

Fig. 5. Maximization of expected profit and minimization of opportunity loss

We can observe the trade-off between profit and risk from the figures. The solutions are appropriate as reasonable alternatives for decision-makers. We found that the supply planning GA with Boundary Initialization obtained extremely good solutions. The proposed approach could also provide several pareto-optimal solutions as alternatives, from high-risk and high-profit to low-risk and low-profit.

5 Conclusion

In this paper, we proposed a supply planning method employing a multi objective GA. The method uses Monte Carlo Simulation in the fitness function of GA. Monte Carlo simulation simulates uncertain demand, then profit and risk as fitness values are calculated from the simulation result. The GA searches a number of Pareto optimal solutions in one optimization run. We also proposed Boundary Initalization which initializes population on the boundary of constraints.

We tested our approach on the actual data from an electric appliances manufacturer. The proposed approach successfully optimized the supply planning problem. We also found that Boundary Initialization is more effective than Random Initialization.

The GA provided a number of Pareto optimal solutions covering from high-risk and high-profit to low-risk and low-profit. We believe this feature is very helpful to decision-makers. Since the Pareto optimal solutions can be the alternative choices, decision-makers can select one preferable solution from the alternatives according to their risk appetite and business strategies.

References

1. Hashimoto, F., Hoashi, T., Kurosawa, T. and Kato, K. : Production Management System. New edn., Kyoritsu Shuppan Co., Ltd., Tokyo (1993)
2. Schaffer, J. D. : Multiple Objective Optimization with Vector Evaluated Genetic Algorithms, Proc. First Intl. Conf. on Genetic Algorithms and their applications (1985) 93–100
3. Fonseca, C. M. and Fleming, P. J. : Genetic Algorithms for Multiobjective Optimization: Formulation, Discussion, and Genralization, Proc. 5th Intl. Conf. on Genetic Algorithms (1993) 416–423
4. Horn, J., Nafpliotis, N. and Goldberg, D. E. : Niche Pareto Genetic Algorithms for Multiobjective Optimization, Proc. First IEEE Conf. on Evolutionary Computation (1994) 82–87
5. Tsuda, T : Monte Carlo Method and Simulation. 3rd edn., Baifukan Co., Ltd., Tokyo (1995)
6. Goodrich, R. L. : Applied Statistical Forecasting, Business Forecast Systems, Inc., Belmont MA (1989)
7. Michalewicz, Z. and Janikow, C. Z. : Handling Constraints in Genetic Algorithms, Proc. 4th Intl. Conf. on Genetic Algorithms (1991) 151–157
8. Michalewicz, Z. : Genetic Algorithms + Data Structure = Evolution Programs. 3rd edn., Springer-Verlag (1996)
9. Goldberg, D. E., Deb, K., Korb, B. : Don't Worry, Be Messy, Proc. 4th Intl. Conf. on Genetic Algorithms (1991) 24–30
10. Liepins, G. E., and Potter, W. D. : A Genetic Algorithm Approach to Multiple-Fault Diagnosis, In: Davis, L. (ed.): Handbook of Genetic Algorithms, Van Nostrand Reinhold (1991)

Genetic Algorithms: A Fundamental Component of an Optimization Toolkit for Improved Engineering Designs

Siu Tong[1] and David J. Powell[2]

[1] Engineous Software, Cary, North Carolina
tong@engineous.com
[2] Elon University, Campus Box 2101, Elon, North Carolina 27244
dpowell2@elon.edu

Abstract. Optimization is being increasing applied to engineering design problems throughout the world. iSIGHT is a generic engineering design environment that provides engineers with an optimization toolkit of leading optimization algorithms and an optimization advisor to solve their optimization needs. This paper focuses on the key role played by the toolkit's genetic algorithm in providing a robust, general purpose solution to nonlinear continuous, mixed integer nonlinear and integer combinatorial problems. The robustness of the genetic algorithm is demonstrated on successful application to 30 engineering benchmark problems and the following three real world problems: a marine naval propeller, a heart pacemaker and a jet engine turbine airfoil.

1 Introduction

This paper describes a generic engineering design environment, iSIGHT, used by the Automotive, Aerospace, Industrial Manufacturing and Electronic industries in the United States, Japan, China, Korea and Europe [1]. Within these industries, the designer's knowledge of optimization varies from that of novice to expert. To meet the needs and challenges of a wide variety of industries and the wide disparity in designer's optimization expertise, iSIGHT provides an Optimization Toolkit. This paper focuses on the role of the Genetic Algorithm in iSIGHT's Optimization Toolkit and its application to three design problems.

The paper starts with a description of the optimization problem in engineering and presents a generic problem formulation. Thirteen well known packages that represent the major numerical and exploratory algorithms are benchmarked against a suite of thirty engineering problems that represent nonlinear continuous, mixed integer and integer combinatorial problems. The benchmark results are analyzed to demonstrate the strengths and weaknesses of each package and to demonstrate the critical role that genetic algorithms serve to the toolkit by providing a single algorithm to solve any nonlinear, constrained continuous or mixed integer problem. The benchmark conclusions are supported by describing the successful application of genetic algorithms to three customer applications for the design of a marine naval propeller, a heart pacemaker and a jet engine turbine airfoil. The paper concludes with lessons learned in the development and application of iSIGHT.

2 The Optimization Problem

During the past twenty years, product design and analysis has been routinely done with computer simulation programs developed either in house or by a commercial vendor. The designer supplies the design parameters for the product as input into one or more computer simulation programs, runs the program and then analyzes the results. If the results do not meet the design goals then the designer changes the design parameters and repeats the process. This process is typically called the design, evaluate and redesign process. The challenge to the designer is to find the best design in as short a time period as possible.

These design optimization problems are commonly found in manufacturing industries and can be represented by the following mathematically formulation.

Objective function: Minimize $y = f(X)$ where $X = \{x1, x2, ..., x_n\}$
Subject to:
Inequality constraints: $g_{kb}(X) <= g_k(X) <= g_{ku}(X)$
where $k = 0, 1, \cdots, K$
b is lower constraint boundary
u is upper constraint boundary
Equality constraints: $g_{le}(X) = h_l(X)$
where $l = 0, 1, ..., L$
e is the constraint boundary
Side constraints: $x_{ib} <= x_i <= x_{iu}$ where x_i is real or integer
or
x_i is a member of a discrete set of values

This formulation supports the specification of unconstrained and constrained problems with a single objective. (Note: Multiple objectives are automatically converted into a single weighted objective.) The constrained problems can have either or both inequality and equality constraints. The design parameters can be combinations of type real, integer or discrete. Depending on the mixture, the problem formulation can be considered a continuous nonlinear problem, a pure integer combinatorial problem or a mixed integer nonlinear problem (MINLP).

Non-linear optimization technologies are a natural fit to aid the designer but their successful application has been limited for many reasons to include:

1. The need to modify the simulation source code to interact with the optimization algorithm. If the simulation code is from a third party vendor then the source code may not be available.
2. Most simulation codes were not written with optimization in mind and would require significant reprogramming [2].

3. The difficulty in identifying which optimization algorithm(s) is appropriate for the product. For example, the domain modeled by the simulation code may be discontinuous, non convex, highly nonlinear, poorly scaled, have limited precision in output values, have parameters with different orders of magnitude, and have a mixture of discrete, integer and real parameter design variables.

To overcome these limitations and still solve real world problems, the authors decided to create a toolkit consisting of best-of-breed optimization algorithms that would work together in a coordinated manner. Accordingly, the first step was to identify the best algorithms for each class of industrial problems.

2.1 Optimization Algorithms

In order to identify the most suitable algorithms to be included in the optimization toolkit, a number of analytical and real world problems were set up as benchmarks. Each package was benchmarked to determine how often it achieved the known optimum (i.e. robustness) and the number of function evaluations needed to achieve an optimal result. Table 1 lists the packages, their underlying algorithms, whether they are gradient based or non gradient based, whether they support mixed integer, and their number of basic and advanced tuning parameters. A value of No for Mixed Integer indicates that the algorithm applies only for continuous problems. A value of Yes followed by (BB) indicates that the algorithm is based on a branch and bound algorithm and the underlying code must be able to support both real and integer values for integer parameters. A value of Yes not followed by (BB) for Mixed Integer indicates that the algorithm can work directly with integer values. The Tuning Parameters for each algorithm are set to default values recommended by either the implementer or expert users. The Basic Tuning Parameters are those most likely to be tuned by a user. The Advanced Tuning Parameters are those usually manipulated by only an expert in optimization.

Table 1 lists both numerical optimization algorithms and exploratory genetic algorithms used in this study. For certain algorithms (e.g., Genetic Algorithms and Sequential Quadratic Programming), multiple packages are listed. The rationale for testing multiple packages for an algorithm is due to the importance of the implementation on performance and robustness.

2.2 Benchmarks and Analysis

In 1977, Sandgren established a test set of single objective, nonlinear, continuous engineering problems with different numbers of design variables, inequality constraints and equality constraints [17]. Each problem had a given starting point and a known solution. Twenty eight of his thirty problems and two mixed integer nonlinear problems were used to benchmark the iSIGHT optimization algorithms. The two MINLP problems are a coil compression spring, San 31 [18], and a cantilevered beam, Van 32 [22]. Table 2 lists the benchmark data

Table 1. The optimization packages and their associated characteristics, technique and tuning parameters.

Package	Algorithm	Gradient Based	Mixed Integer	Basic Tuning Parameters	Advanced Tuning Parameters
1. GENEsYs [7]	Genetic Algorithm	No	Yes	2	11
2. Ga2k [8]	Genetic Algorithm	No	Yes	3	7
3. NSGA-II [6]	Genetic Algorithm	No	Yes	2	4
4. ASA [9]	Simulated Annealing	No	Yes	8	7
5. Hooke [11]	Hooke Jeeves	No	Yes	2	2
6. ADS-EP [10]	Exterior Penalty	Yes	No	6	22
7. ADS-SLP [10]	Sequential Linear Programming	Yes	No	6	25
8. Donlp [12]	Sequential Quadratic Programming	Yes	No	3	6
9. NLPQL [13]	Sequential QuadraticProgramming	Yes	No	4	3
10. MOST [14]	Sequential Quadratic Programming	Yes	Yes (BB)	3	5
11. LSGRG [15]	Generalized Reduced Gradient	Yes	No	3	5
12. ADS-MFD [10]	Method of Feasible Directions	Yes	No	5	11
13. ADS-MMFD [10]	Modified Method of Feasible Directions	Yes	No	5	25

for these thirty test cases. The relative error for the starting point, the optimization algorithm that achieved the published optimum in the fewest function evaluations, the relative error achieved by the ga2k genetic algorithm and the percentage of relative error reduced by the genetic algorithm from the starting point are shown. The calculation of relative error is shown in equation 1. The penalty is the sum of constraint violations.

$$\mathrm{RelativeError} = \frac{|\mathrm{current\ objective - published\ optimum}|}{|\mathrm{published\ optimum}|} + \mathrm{penalty} \quad (1)$$

The key results revealed by the benchmarks are:

1. No single algorithm worked the best in all test cases. In fact, 7 different algorithms proved to be the best for selective test cases.
2. If one had to pick a single algorithm to try for any optimization problem then the genetic algorithm would be a very robust and sound choice. In the benchmarks, the relative error of the optimum point found by ga2k averaged a reduction in relative error from the starting relative error by over 85%. This is a significant result. Keep in mind that the goal of the designer is to find the best design that meets customer constraints within the design deadline. The genetic algorithm found a feasible design on 29 of the thirty test cases. On only Sandgren 27 did the genetic algorithm fail to achieve a better design.
3. Genetic algorithms require one to two orders of magnitude more function evaluations than numerical algorithms.
4. Genetic algorithms are clearly superior to numerical optimization techniques on mixed integer problems when the simulation program does not accept non

Table 2. Benchmark results on test suite

Problem	Real Design Variables	Integer Design Variables	Inequality Constraints	Equality Constraints	Relative Error Starting Point	Feasible Starting Point	Name Best Algorithm	Function Evaluations Best Algorithm	Relative Error ga2k	Error Reduction ga2k
Sandgren 1	5	0	10	0	1.6192	YES	NLPQL	32	0.03096	0.98
Sandgren 2	3	0	2	0	0.697	YES	GRG	25	0.0003	0.99
Sandgren 3	5	0	6	0	0.0119	YES	DONLP	32	0.00799	0.32
Sandgren 4	4	0	0	0	19192	YES	ADS-EP	184	0.378	0.99
Sandgren 5	2	0	0	0	24.2	YES	GRG	166	0.026	0.99
Sandgren 6	6	0	0	4	1961.4	NO	DONLP	96	0.00528	0.99
Sandgren 7	2	0	1	0	53.554	NO	NLPQL	36	0.00031	0.99
Sandgren 8	3	0	2	0	0.6755	NO	NLPQL	21	0.00083	0.99
Sandgren 9	3	0	9	0	0.7892	YES	GRG	1692	0.27248	0.65
Sandgren 10	2	0	0	0	1468.7	YES	MMFD	32	5.7E-05	1.0
Sandgren 11	2	0	2	0	0.4236	YES	MMFD	18	0.00419	0.99
Sandgren 12	4	0	0	0	0.024	YES	NLPQL	61	0.00112	0.95
Sandgren 13	4	0	3	0	0.0456	YES	GA2K	10000	0.0	1.0
Sandgren 14	15	0	5	0	73.192	YES	DONLP	412	7.75308	0.89
Sandgren 15	16	0	0	8	18.212	NO	NLPQL	141	7.80919	0.57
Sandgren 16	3	0	14	0	0.253	YES	NLPQL	85	3.1E-05	0.99
Sandgren 17	12	0	3	0	2.4394	NO	GRG	132	0.29052	0.88
Sandgren 18	7	0	14	0	0.7315	YES	MOST	97	0.4954	0.32
Sandgren 19	8	0	4	0	0.6672	NO	MOST	136	0.17183	0.74
Sandgren 20	8	0	6	0	1.1829	NO	MOST	155	0.17048	0.85
Sandgren 21	13	0	13	0	4.0012	NO	MOST	1196	1.02186	0.74
Sandgren 22	7	0	4	0	647.91	NO	NLPQL	179	0.25652	0.99
Sandgren 24	4	0	5	0	5.6398	YES	NLPQL	56	0.79119	0.85
Sandgren 25	6	0	4	0	36.954	YES	NLPQL	180	0.71947	0.98
Sandgren 26	3	0	0	1	211.5	NO	GRG	121	0.47662	0.99
Sandgren 27	48	0	0	2	1.1545	YES	MOST	4708	1.1545	0.0
Sandgren 29	10	0	14	1	2.8149	YES	MOST	442	1.32457	0.52
Sandgren 30	19	0	1	11	6809.3	NO	NLPQL	132	5157.12	0.24
Sandgren 31	1	2	8	0	90.069	NO	GA2K	10000	0.48947	0.99
Van 32	4	6	11	0	2.5544	NO	GA2K	10000	0.11022	0.95

  integer values for integer parameters (e.g., problems Sandgren 31 and Van 32).
5. Genetic algorithms are least effective on problems with equality constraints (e.g. Sandgren problems 15, 27, 29 and 30).

This section describes the genetic algorithm in more detail as they are not as well known and not as well tested in the engineering community. Three genetic algorithm packages, GENEsYs, ga2k, and NSGA-II were benchmarked. All packages supported the parallel evaluation of each design in the population.

GENEsYs' default settings are for elitism, binary gray encoding, linear rank based selection with a maximum fitness of 1.1 and a minimum fitness of .9, two point crossover with a .6 crossover rate, standard mutation with a .01 mutation rate and seeding. The first population is seeded with the initial design point, 20% small creep, 20% large creep, 20% boundaries and random selection. With small creep, large creep and boundary each design parameter has a 50 percent chance of selection. If the parameter is selected with small creep, the parameter will be randomly varied by 0-3%. If selected by large creep the parameter will be varied by 0-10% and if selected with boundary, the parameter will be randomly set to either the upper or lower bound. The rationale for the small and large creep operators is to leverage the implicit knowledge provided in the initial user supplied design point. The boundary seeding leverages the heuristic that many design parameters are active at their boundaries at the optimal design point.

ga2k is a distributed genetic algorithm [16]. The package's default settings are for elitism, binary gray encoding, tournament based selection, 10 subpopulations of size 10 with a migration rate of .5 at every 5th generation, two point crossover with a 1.0 crossover rate, standard mutation with a .01 mutation rate and no seeding. This algorithm significantly outperformed GENEsYs in the set of benchmark test cases described in Section 2.2.

NSGA-II is a multiobjective genetic algorithm. The package's default settings are for elitism, real encoding, tournament selection, a population size of 100, 100 generations, SBX crossover at a rate of .9, real mutation crossover at a rate of 1/(number of design variables), a crossover distribution index of 20 and a mutation distribution index of 100. The NSGA-II algorithm was similar in performance to ga2k. In the 30 benchmarked test cases, it was better in 14, worse in 9 and equal in 7.

The benchmark tests are helpful but are not completely consistent with the types of applications encountered in industry. They are lacking in the following areas:

1. Their execution time is short compared to the times required of simulation programs which take anywhere from 1 minute to a few hours to execute a single evaluation. The simulation programs are typically thousands or hundreds of thousands of lines of code.
2. The continuous benchmark codes are relatively smooth, well behaved landscapes. In practice, the simulation programs are not smooth, have many discontinuities introduced by if statements, casts and parameter settings. In addition, the programs frequently provide less precision in their output parameters then desired by finite difference algorithms. For example, results may be provided to two digit decimal accuracy where numerical finite gradient algorithms typically are looking at changes in the 4th or 5th decimal digit.
3. The design parameters are typically not independent variables.

These differences make the results of the benchmark only a guideline rather than an absolute measurement of the usefulness of each algorithm in real world situations.

The iSIGHT design environment [1] was developed to provide an industrial solution to this design optimization problem. It has a process integration toolkit that allows simulation codes to be coupled into the environment without source code or reprogramming. The iSIGHT optimization toolkit, which is discussed in the next section, supplies a variety of algorithms which make different assumptions about the design space. An optimization advisor was developed to recommend the appropriate algorithms to novice users.

3 Optimization Toolkit

The iSIGHT Optimization Toolkit provides a number of industry leading implementations of optimization algorithms and a mechanism to provide an optimization plan from an interdigitation of individual algorithms to solve complex problems.

3.1 Strength and Weaknesses of Optimization Packages

More than a dozen packages from both public and private domains are provided in the toolkit and in a few cases multiple implementations of each. There are advantages and disadvantages to having this amount. The advantages are:

1. Each algorithm has one or more strengths that enables it to find a better design than another algorithm for certain design space conditions. For example, Exterior Penalty and Sequential Quadratic Programming work well when started from an infeasible starting point, GRG works well with equality constraints and Genetic Algorithm works well with non convex, multi-modal or mixed integer problems.
2. Some algorithms are extremely efficient for certain design space conditions. For example, gradient based techniques work well in smooth, unimodal, convex design spaces with twenty or fewer design variables.
3. An optimization expert has access to an assortment of algorithms, packages and tuning parameters to best match the characteristics of the algorithm to the characteristics of the product design space.
4. Each package has proven itself on at least one user application and is maintained for backward compatibility. However, certain packages such as GENEsYs have not been routinely updated by the developer and have fallen behind other packages such as ga2k in performance. If a designer selects one of these packages (e.g., GENEsYs) then the designer is notified that the selected package is not recommended and they should consider an alternative package (e.g., ga2k).

The disadvantages are:

1. For a novice user, the choices are confusing and intimidating. In fact, the choice of more than a dozen algorithms with close to 200 tuning parameters could be a larger optimization problem then the problem the designer is trying to solve. In this case, the problem of changing parameters has merely been transferred from the product parameters to the optimization parameters.
2. The diagnostic tools to determine how the package is performing are different for each package. A user experienced in one package cannot easily leverage

his experience to work with another package. As a result, designers will oftentimes use a single package regardless of the characteristics of the design space.

These disadvantages are overcome in two ways. First, iSIGHT has an optimization advisor to automatically pick packages for the user. (The advisor will be discussed in section 3.2). Second, the genetic algorithm is a general purpose algorithm that will work in any design space. A key lesson learned is that although the genetic algorithm cannot compete with numerical techniques in terms of function evaluations for smooth optimization problems, it will always find some improvement and with todays computing environment many of the function evaluations can be done in parallel.

iSIGHT actively supports only a handful of the best performing algorithms listed in Table 1 and deprecates the older or poorer performing packages. However, iSIGHT provides a set of application programming interfaces that allow the designers to couple their own package. In addition, iSIGHT supports Interdigitation which allows the user to create a hybrid technique from one or more existing techniques [21]. Interdigitation can be as simple as a sequential execution of an exploratory genetic algorithm followed by an exploitive numerical optimization or can have complicated scripting with conditional branches and loops among the optimization techniques.

3.2 Optimization Advisor

While the large number of algorithms with direct access to their tuning parameters provides a powerful tool for the expert in optimization, it is overwhelming to the majority of designers. In fact, less than a fraction of 1 percent of users using optimization are Operations Research professionals [19]. To address the needs of this vast majority of novice designers, iSIGHT provides an optimization advisor that automatically selects the best two optimization algorithms based on its analysis of the problem's characteristics. Alternatively, the user can view a priority listing of the recommended optimization packages. For example, on the Van32 benchmark which has a low number of design variables, a low number of design constraints, mixed types of parameters, small variable variance, no equality constraints, a discontinuous design space, non linear simulation codes, an initial infeasible design, low execution time, and no availability of simulation code gradients, the optimization advisor recommended a combination of a genetic algorithm, ga2k, followed by adaptive simulated annealing. For this example, the priority ranking of applicable optimization techniques was: genetic algorithm, adaptive simulated annealing, Hooke Jeeves, and MOST. The continuous numerical optimization techniques are not applicable and are not listed by the advisor for this mixed integer formulation. On the other hand, in a continuous design space with real parameters and high execution time of the simulation code, numerical optimization algorithms will have the highest priority.

During the past three years of iSIGHT application to a variety of real world problems, the genetic algorithm has been the optimization advisor's most frequently recommended technique. In many cases, it is not the highest recommended technique but its frequent recommendation recognizes the robustness of the technique.

4 Applications

This section demonstrates the robustness of the genetic algorithm in its successful application to three different types of real world problem formulations. The first application is a mixed integer design, the second is a pure integer combinatorial problem and the last is a continuous parameter problem in a multimodal design space.

4.1 Marine Propeller Design

Numerous competing design and performance parameters must be considered in propeller design. A properly designed propeller must balance the competing requirements characterized by cavitation, cost, efficiency, noise, strength, thrust, and vibration. The design space contains many local optimal designs.

Several codes are required for the preliminary design. The designers used the process integration toolkit to integrate six simulation codes whose function in the preliminary design process can be described as follows:

- Compute the mean and harmonic wake velocity components
- Calculate propulsive efficiency, tip vortex and cavitation inception speed
- Determine structural stress and weight
- Find surface cavitation inception speeds
- Compute vibratory forces and moments

The mixed integer design variables for this problem include the propeller diameter, speed (RPM), and several spline parameters that characterize the chord distribution, thickness distribution, and loading distribution. The design goal was to minimize the propeller weight, subject to limits on the minimum propulsive efficiency, minimum cavitation inception speed, minimum tip vortex speed and a maximum allowable stress.

Since the problem was mixed integer, the genetic algorithm was used with the default tuning parameter settings. A manual design was used as the baseline for comparison with the genetic algorithm solution. Figure 1 compares the chord distributions and propeller thickness distributions of the baseline and genetic algorithm optimized designs. The genetic algorithm design met all the design constraints and propulsion goal of 0.68 while reducing the weight 181 lbs below the baseline design weight. The net result was an improvement of 17% in the weight.

4.2 Heart Pacemaker Antenna Design

This case study details the design of an implantable patch antenna for a human heart pacemaker. Enabling two-way communication with pacemakers will make

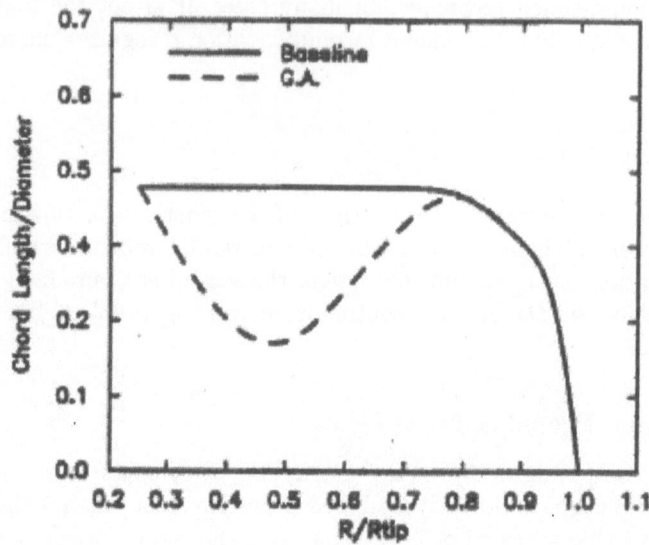

Fig. 1. Chord distribution of Baseline and genetic algorithm designs for marine propeller

it possible to download information about the condition of the pacemaker and the patients heart and upload improved instrument settings. Fitting pacemakers with antennas is challenging, because either the wavelength is too short to penetrate the body or the antenna is so large that it is impractical to implement. Antennas are normally 5 or 6 inches long to be used at the required frequency, long enough that they protrude into the body and risk infection or lung punctures. Researchers at Utah State University solved this problem by developing a two-inch-square 433 MHz patch antenna small enough to fit on a standard pacemaker battery pack [20]. The designers set out to design a 50-ohm antenna with the required performance characteristics that could fit onto the battery pack of the device, which would make it virtually noninvasive

Electromagnetic finite difference time domain software called XFDTD from Remcom Incorporated was used to evaluate the performance of specific designs. The design team worked on the problem and quickly realized that the design space was very sparsely populated with feasible designs that met all the stringent requirements. This made the discovery of an acceptable (much less optimal) patch antenna geometry a very difficult task. After 9 months of manual iteration they finally developed two optimized designs that met the requirements of the project - a u-shaped patch antenna and a spiral shaped antenna. But the design team recognized that the design would require numerous subsequent design iterations to meet additional requirements that arose as the project evolved.

iSIGHT was linked with the XFDTD simulation package. The design parameters included the length of the antenna and the locations of the feed and

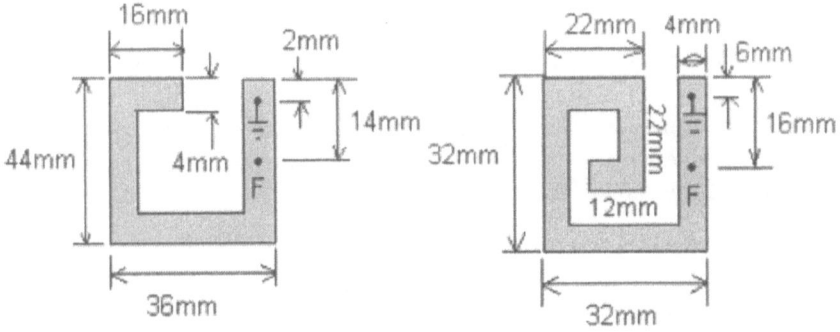

Fig. 2. Comparison of best antenna designs found by iSIGHT and manual design

ground pins. Design constraints on the maximum antenna length, pin locations, and impedance requirements were specified. The design objective was to minimize the antenna length and meet the target frequency of 433 MHz subject to the various design constraints. A genetic algorithm was used as this was a completely integer combinatorial problem. The genetic algorithm found a solution that met all the requirements and was better than the best manual designs. Figure 2 shows a comparison of the iSIGHT and manual trial-and-error best designs. The antenna length of the iSIGHT optimized design, shown on the left, was 128 mm compared with a length of 132 mm found by the 9 month trial-and-error method.

4.3 Jet Engine Airfoil Turbine Design

Researchers at Brigham Young University (BYU) recently completed a complex, multi-discipline design optimization of a cooled jet engine turbine airfoil. The optimization required running multiple software codes to generate a parametric solid model, and execute aerodynamic and structural analyses.

The parametric model was generated in Unigraphics (UG). This model was used as the foundation for the design and subsequent optimization. The structural analysis was performed in ANSYS while the GAMBIT and FLUENT software tools were used to compute the aerodynamic analysis.

The airfoil geometry was created from four sections, each section with seven parameters, for a total of 28 design variables. The design parameters or variables include the chord length, leading and trailing angles, leading and trailing radius, leading and trailing wall thickness and trailing radius thickness. The objectives of the optimization were to minimize the pressure loss and weight and to maximize the safety factor based on the stress in the airfoil subject to stress constraints and geometric limits.

The analysis for this problem was highly complex and required extremely long *computational times to solve*. The genetic algorithm available in iSIGHT was selected to complete the optimization study. The genetic algorithm was chosen for its abilities to search for a set of designs, explore the entire multi-modal design space, and run in parallel. A population size of 50 was used and 100 generations were run. The optimization was run automatically in parallel by iSIGHT on a 64 processor SGI and required approximately 20 hours to complete. 65 solutions were selected from the 5000 designs generated to form the pareto set. This set of 65 designs showed a decrease in the blade volume ranging from 2% to 19%, an increase in the safety factor of 2% to 8%, and pressure loss decreases of 2% to 38%.

5 Lessons Learned

The iSIGHT generic engineering design environment described in this paper has had a genetic algorithm as part of its optimization toolkit since 1990. The genetic algorithm has played a fundamental role in the robustness of the toolkit when applied individually as shown in the three real world applications discussed in this paper or when used with interdigitation to design an aircraft engine turbine [23]. It is the robustness of the genetic algorithm and its ease of use that makes it especially attractive to the majority of users who are not experts in optimization. The concept of the algorithm is easy to understand and there are no limiting design space assumptions of parameter independence or continuity to worry about. The genetic algorithms key advantage is it applies to almost every type of problem. This claim is supported by the genetic algorithm finding improved designs on 29 of 30 benchmarked engineering problems. This high success rate of 96% makes the genetic algorithm a reasonable choice when selecting an initial algorithm for a design when the problem domain is not very well known. The increased number of function evaluations required by the genetic algorithm is mitigated by the use of parallel computing and makes the use of the genetic algorithm feasible in real world problems.

References

1. Engineous Software Incorporated. www.engineous.com
2. Vanderplaats, G.: Numerical Optimization Techniques for Engineering Design. (1999).
3. Belegund, A. and Chandrupatla, T.: Optimization Concepts and Applications in Engineering. Prentice Hall (1999)
4. Onwubiko, C.: Introduction to Engineering Design Optimization. Prentice Hall (2000)
5. Gen, M. and Cheng, R: Genetic Algorithms and Engineering Optimization. John Wiley & Sons (2000).
6. Deb, K.: Multi-Objective Optimization Using Evolutionary Algorithms. John Wiley & Sons (2001).
7. Back, T.: A Users Guide to GENEsYs 1.0. University of Dortumnd. (1992)
8. Hiroyasu, T.: Spec Sheet: Distributed Genetic Algorithms ga2k (ver 1.1). Intelligent Systems Design Lab. Doushisya University (2002)
9. Inger, L.: Adaptive Simulated Annealing. http://www.ingber.com/. (2002)

10. Vanderplaats, G.: ADS – A Fortran Program for Automated Design Synthesis. Santa Barbara, CA: Engineering Design Optimization, Inc. (1988)
11. Johnson, M.: Hooke and Jeeves Algorithm. http://www.netlib.org/opt/hooke.c (1994)
12. Spellucci, P.: Donlp2 User Guide. http://www.netlib.org/opt/donlp2/donlp2doc.ps
13. Schittkowski, K.: NLPQL: A Fortran subroutine for solving constrained non linear programs. Annals of Operations Research, Vol.5, (1985–1986) 4850–500
14. Tseng, C.: MOST 1.1. Applied Optimal Design Laboratory, National Chiao Tung Univeristy, Technical Report AODL-9-01 (1996)
15. Smith, S. and Lasdon, L.: Solving large sparse nonlinear programs using GRG. ORSA J. Comput. 4, (1992) 1–15
16. Tanese, R.: Distributed genetic algorithms. Proceedings of the Third International Conference on Genetic Algorithms, (1989) 434–439.
17. Sandgren, E.: The utility of nonlinear programming algorithms. Purdue University Ph.D. Thesis, West Lafayette, IN (1977).
18. Sandgren, E.: Nonlinear integer and discrete programming in mechanical design optimization. Transactions of the ASME, Journal of Mechanical Design, 112(2), (1990) 223–229
19. Fylstra, D., Lasdon, L., Watson, J. and Waren, A.: Design and Use of the Microsoft Excel Solver. Interfaces 28 (1998) 29–55.
20. Furse, C.: Design an Antenna for Pacemaker Communication. Microwaves & RF, March (2000)
21. Powell, D.: Inter-GEN: A hybrid approach to engineering design optimization. Rensselaer Polytechnic Institute Ph.D. Thesis, Troy, NY (1990).
22. Vanderplaats, G.: Numerical Optimization Techniques for Engineering Design. (1999) 317–320.
23. Powell, D., Skolnick, M., and Tong, S.: Interdigitation: A Hybrid Technique for Engineering Design Optimization. Handbook of Genetic Algorithms. Van Nostrand Reinhold. (1991) 312–331.

Spatial Operators for Evolving Dynamic Bayesian Networks from Spatio-temporal Data

Allan Tucker[1], Xiaohui Liu[1], and David Garway-Heath[2]

[1] Brunel Univeristy, Middlesex, UK
{allan.tucker,xiaohui.liu}@brunel.ac.uk
[2] Glaucoma Unit, Moorfield's Eye Hospital, London, UK
david.garway-heath@moorfields.nhs.uk

Abstract. Learning Bayesian networks from data has been studied extensively in the evolutionary algorithm communities [Larranaga96, Wong99]. We have previously explored extending some of these search methods to temporal Bayesian networks [Tucker01]. A characteristic of many datasets from medical to geographical data is the spatial arrangement of variables. In this paper we investigate a set of operators that have been designed to exploit the spatial nature of such data in order to learn dynamic Bayesian networks more efficiently. We test these operators on synthetic data generated from a Gaussian network where the architecture is based upon a Cartesian coordinate system, and real-world medical data taken from visual field tests of patients suffering from ocular hypertension.

1 Introduction

Bayesian Networks (BNs) are probabilistic models that can be used to combine expert knowledge and data. They also facilitate the discovery of complex relationships in large datasets. A BN consists of a *directed acyclic graph* consisting of links between nodes that represent variables in the domain. The links are directed from a *parent* node to a *child* node, and with each node there is an associated set of *conditional probability distributions*. Learning the structure of a BN from data [Cooper92] is a non-trivial problem due to the large number of candidate network structures and as a result there has been substantial research in developing efficient algorithms within the optimisation and evolutionary communities. For example, Evolutionary Programs (EP) [Bäck93] and Genetic Algorithms (GA) [Holland95] have been used to search for candidate structures [Larranaga96, Wong99]. The Dynamic Bayesian Network (DBN) is an extension of the BN that can model time series [Friedman98]. See figure 1 for an example of a DBN with $N+1$ variables spanning two *time slices*. Note that links in a DBN can be between nodes in the same time slice or from nodes in previous time slices. We have developed algorithms (evolutionary and non-evolutionary) for efficiently learning DBN structures [Tucker01].

Spatial Data-mining [Ester00] involves the application of specialist algorithms and operators (e.g. neighbourhood distance, topology and direction) to data that has a spatial nature. That is, variables in the domain can be considered to have neighbourhood relations with other variables. This can vary from simple Cartesian coordinates to more complex spatial neighbourhoods. See [Roddick99] for a bibliography on spatial and temporal data mining. Spatial data is common in geographical research [Cofiño02].

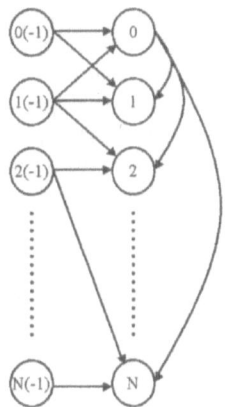

Fig. 1. A Typical DBN   **Fig. 2.** A Visual Field Test

To our knowledge, BN learning has not been investigated with respect to spatial data. We define a Spatial Bayesian Network (SBN) to be a BN that represents data of a spatial nature and a Spatial Dynamic Bayesian Network (SDBN) to be a BN that represents spatio-temporal data. We introduce and test an evolutionary algorithm and a set of spatial operators for learning SDBNs from spatio-temporal data. These operators are tested on synthetic data and visual field data of patients suffering from ocular hypertension. The algorithm and its operators are compared with a standard BN search technique. We validate resulting networks in several ways including inspection of the spatial links, calculating the structural difference of the original network used to generate the synthetic data, and using clinical expert knowledge on the visual field data.

2 Methods

2.1 Datasets

We have generated spatio-temporal data using DBNs with Gaussian probability distributions [Geiger94]. This data contains 1000 time points of 64 variables spatially located on an 8x8 grid, where each variable was influenced by its immediate neighbours at the previous time point.

Visual Field (VF) data can be recorded using the Humphrey Perimeter [Haley87]. This involves a patient fixating a point in the centre of a dimly illuminated bowl. The perimeter shines brighter stimuli onto the bowl at various points, corresponding to points in the visual field. The stimuli are varied in intensity and the patient presses a button when a stimulus has been observed. This technique determines the sensitivity to light of each point in the VF. The data that is used in this paper is from a study that contains tests recorded every few months, in patients with ocular hypertension, a major risk factor for glaucoma. The test measured 54 points on each eye (figure 2 shows an example of a patient's VF test). The dataset used in this paper involves 95 patients with 1809 measurements in all, concerning only the right eye. Two points in the VF correspond to the blind spot and should not contain any useful data. We have included these points to check for spurious relationships. We know of little research in using probabilistic models to understand VF data. Previously, a state space model has been used to classify glaucomatous patients [Anderssen98] and Bayesian statistics have been proposed to test VF data [Bengtsson97]. Both datasets were discretised into four states using equal bin sizes from the maximum value to the minimum value for each variable.

2.2 Algorithm

Candidate structures of a network, bn, given a dataset, D, are scored using the log-likelihood, calculated using the equation:

$$\log p(D \mid bn_D) = \log \prod_{i=1}^{N} \prod_{j=1}^{q_i} \frac{(r_i - 1)!}{(F_{ij} + r_i - 1)!} \prod_{k=1}^{r_i} F_{ijk}!,$$

where N is the number of variables in the domain, r_i denotes the number of states that a node x_i can take, q_i denotes the number of unique instantiations of the parents of node x_i, F_{ijk} is the number of cases in D, where x_i takes on its kth unique instantiation and, the parent set of i takes on its jth unique instantiation.

$$F_{ij} = \sum_{k=1}^{r_i} F_{ijk}.$$

In order to increase efficiency of our algorithms, we have developed an evolutionary approach without the necessity of storing a population of candidate solutions. Rather, we consider each point in the spatial dataset to be an individual within the population of points. Therefore, the population, itself, is the candidate solution. We have looked at a similar method before for grouping algorithms [Liu01]. The algorithm also makes use of a simulated annealing type of selection criteria [Kirkpatrick83], where good operations are always carried forward, but sometimes less good ones are also accepted dependant upon a temperature parameter. A form of elitism [Grefenstette86] is employed to ensure that the final structure is the best discovered. This is to prevent the simulated annealing process from moving away from a better solution when the temperature is still high. We formally define the algorithm below where *maxfc* is the maximum number of calls to the scoring function, c is the 'cooling parameter', t_0 is the initial temperature, b is the branching factor of a network, and $R(a, b)$ is a uniform random number generator with limits, a and b.

```
Input     t₀, b, maxfc, D
          fc = 0, t = t₀
          Initialise bn to a SDBN with no links
          result = bn
          While fc ≤ maxfc do
             score = L(bn)
             For each operator do
                Apply operator to bn
                If bn is valid given b Then
                   newscore =  L(bn)
                   fc = fc + 1
                   dscore=newscore-oldscore
                   If newscore > score Then
                      result = bn
                   Else If R(0,1)<e^{\frac{dscore}{t}}   Then
                      Undo the operator
                   End If
                End If
             End For
             t = t × c
          End While
Output    result
```

2.3 Operators

We now introduce three spatial and three non-spatial operators. All involve manipulating links within the SDBN. For this paper we have only investigated links that span one time slice, although we have developed equivalent operators for links within the same time slice (these require a more strict ordering assumption to ensure the structure is directed acyclic which is an essential property of a BN structure).

2.4 Non-spatial Operators

We have chosen three non-spatial operators as these represent common operators used in optimisation techniques such as hill climbing and simulated annealing.
    1) *Add* - A link with random parent and child is added to the network.
    2) *Take*– Randomly remove a single existing link.
    3) *Mutate* - Randomly change the parent of an existing link.

2.5 Spatial Operators

For the scope of this paper, we assume that the points in a spatial dataset are located according to Cartesian coordinates. Therefore, each point in a dataset with coordinates (x,y) has a first order *neighbourhood* which includes all nodes with coordinates (i,j) for $x-1 \leq i \leq x+1$ and $y-1 \leq j \leq y+1$. The spatial operators that we have developed exploit the Cartesian spatial nature of a dataset, whereby links are added to a node based upon the proximity of the parent to the child, and mutations are made from an existing parent's position to the positions of other nearby parents. A crossover operator has also been developed that swaps the relative positions of node's parents. Figure 3 shows examples of parent coordinates, relative to the child node, when applying the operators. Unfilled circles represent child nodes, filled circles represent parents of the child.

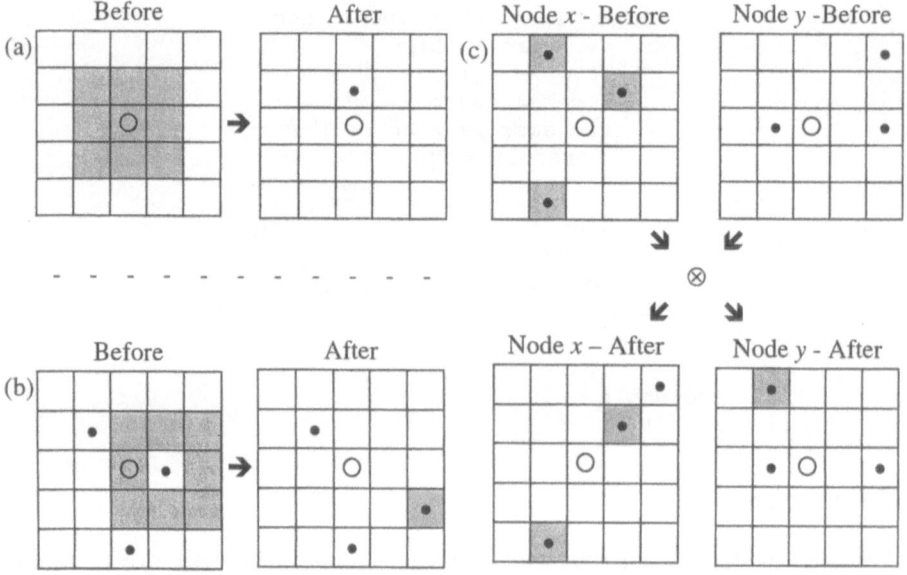

Fig. 3. Spatial Operator Examples: (a) Add (b) Mutate and (c) Crossover.

1) *Spatial Add* - Randomly add a link to a node, such that its parent is one of its first order neighbours (shaded in figure 3a). Note that the coordinates of a parent can be the same as its child because we are dealing with links that span a time-slice.

2) *Spatial Mutate* - Randomly replace the parent of an existing link with one of its first order neighbours (for example, the shaded region in figure 3b represents possible new positions for the parent immediately to the right of its child).

3) *Spatial Crossover* - Randomly select two nodes. For each parent of a node, x, there is a 50% chance of moving it to become a parent of the other node, y, so that its relative position to x becomes updated relative to y. See figure 3c where the shaded cells with filled circles show the parents of node x, before and after crossover, and the unshaded cells with filled circles show the parents of node y.

For all spatial operators, if a parent that is selected for a node is outside the boundary of the coordinate system, then it is deemed invalid.

3 Experiments

The experiments involved running the stochastic algorithm described earlier on the synthetic and VF datasets. One experiment involved using only non-spatial operators, one involved only spatial operators and one using all operators. All operators were applied in the order that they appear listed in the previous section. Due to the stochastic nature of the algorithm, we ran each experiment ten times and recorded the average learning curve. We also investigated Cooper's greedy search algorithm, K2, [Cooper92] when applied to the two datasets. To find out the performance of the individual operators, we also recorded the average success rate of each operator in the 'all operators' experiments. A success is recorded when an operator results in improved fitness.

In order to determine the quality of the final structures, we use the Structural Difference (SD) metric on the synthetic data. This is calculated from summing the missing links with the spurious links in the discovered structure, compared to the original network used to generate the data. Expert knowledge is used for the VF data, based upon the anatomical correspondence of VF locations to particular nerve fibre bundles and the proximity of the nerve fibre bundles to each other in terms of their position of entry (angular location) at the Optic Nerve (ON) margin [Garway00].

3.1 Parameters

In this experiments we set t_o=5, b=3, c=0.9999 and *maxfc*=35000. These values were chosen as they generated the most efficient results for all methods.

4 Results

4.1 Synthetic Data

Figure 4a shows the learning curves for each method when applied to the synthetic data. It can be seen that the simple K2 algorithm is not as efficient as any of the stochastic methods. What is more, the log likelihood of the final solution is not as high as the other methods. The spatial operators alone seem to perform the most efficiently. This could be due to the more limited type of relationships that exist. All are based upon the first order Cartesian neighbourhood. It should be noted that on the synthetic data that if all methods, including K2, are run for long enough they eventually reach solutions very close to one another (again perhaps due to the simplicity of the relationships within the network).

Figure 4b shows the mean success of individual operators during an experiment, figure 4b shows that the most successful were *Spatial Add* (SpatAdd), *Take*, and *Spatial Mutation* (SpatMut). The worst seems to be *Spatial Crossover* (SpatCross).

We have calculated the structural difference of the discovered structures from the original structures used to generate the data. Table 1 shows that on average the closest network to resemble the original structure that generated the data is surprisingly that from using K2. This is somewhat unexpected, but could be due to the simplistic nature of the synthetic data (where all dependencies are generated from a linear Gaussian process given each variable's first order neighbours). The next best structure is that from the spatial operators alone. The next closest is found when using spatial and non-spatial operators, and the worst when using only non-spatial operators.

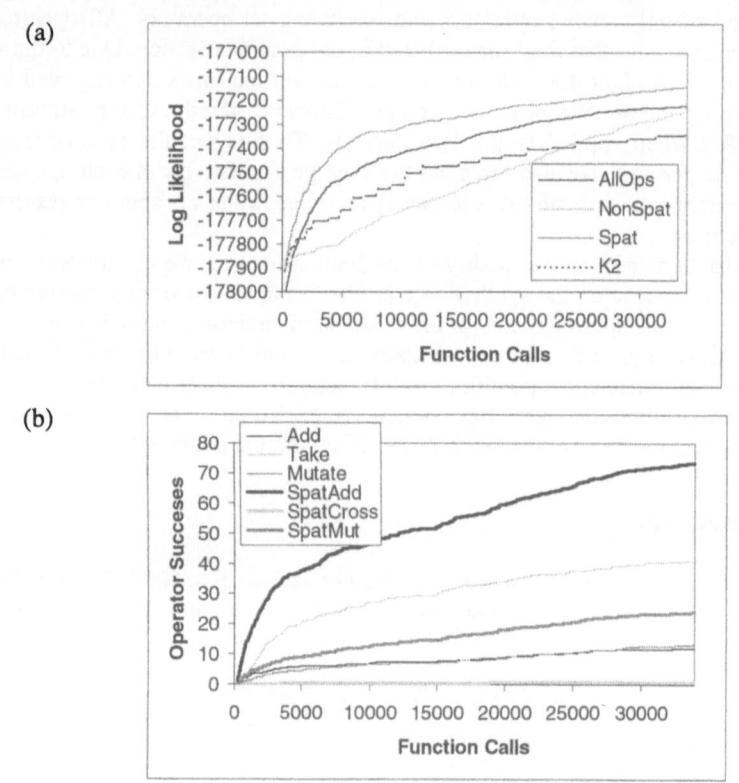

Fig. 4. (a) Mean Learning Curves and (b) Operator Successes for Synthetic Data

Table 1. Quality of Networks Learnt from Synthetic Data

	K2	Non-Spat	Spat	All
Structural Difference	119.0	142.0	122.3	129.2

Figure 5 shows the spatial nature of the final networks for the synthetic dataset. All networks appear to have strong spatial characteristics. The K2 result appears to have more links that lie on the border compared to the stochastic methods. This could be due to a bias in the search or an inability to remove links once they are discovered.

4.2 Visual Field Data

Figure 6a shows the learning curves for all methods on the VF data. It appears that the best operators are a combination of spatial and non-spatial. This results in more efficient learning curves compared to the experiments involving just spatial operators, which in turn generate more efficient learning curves than the non-spatial operators alone. However, figure 6a implies that if the non-spatial operator experiments are run for enough iterations then they eventually overtake the spatial operator curves (probably due to the over-restrictive nature of using these operators only). The VF data, unlike the synthetic data, appears to have a mixture of spatial and non-spatial relationships. Figure 6b shows the average success of individual operators during an experiment on the VF data. The most successful were *Spatial Add*, *Take*, and *Spatial Mutation*. The worst seems to be *Spatial Crossover*, though there is a steady success rate. In fact, successes involving *Spatial Crossover* generally result in higher fitness improvements than other operators.

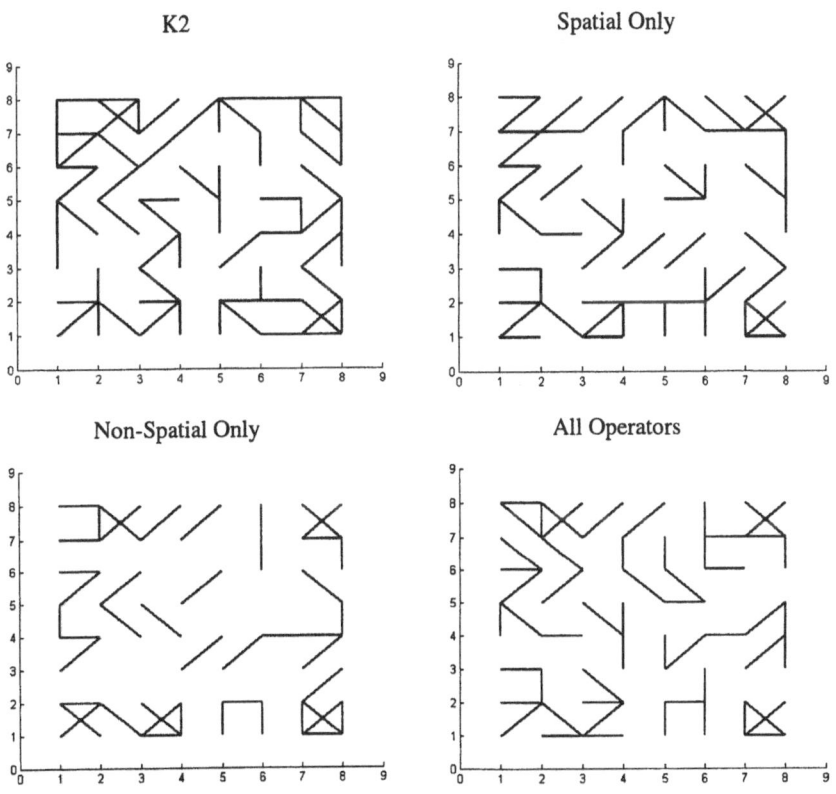

Fig. 5. Networks Learnt from Synthetic Data

We use the average Optic Nerve (ON) distance between each link's parent and child to validate the resulting networks. Table 2 shows that a similar ordering is found to that on the SD with the synthetic data but with K2 performing somewhat worse. The spatial operators generate networks that have least ON distance, followed by a combination of spatial and non-spatial, then just the non-spatial and finally the structure with largest ON distance is that generated using K2. This implies that for more complex real-world problems, a simple greedy search such as K2 is not suitable. We have also used knowledge of how *nerve fibre bundles* are arranged on the visual field [Garway00]. Figure 7 shows how these bundles are positioned with respect to the visual field. This means that it is likely that points in the visual field that share a nerve fibre bundle are likely to be related. Grey shaded points in figure 7 correspond to the blind spot. Table 2 shows the mean number of links that are contained within the same bundle as a percentage of all links, for each search method. A similar ordering to the mean ON distance is observed: a higher percentage of links in the same bundle implies a shorter average ON distance. The highest percentage of links contained within the same bundle are from networks generated from only spatial operators, the next best being all operators, followed by non-spatial operators and the worst being those generated using K2.

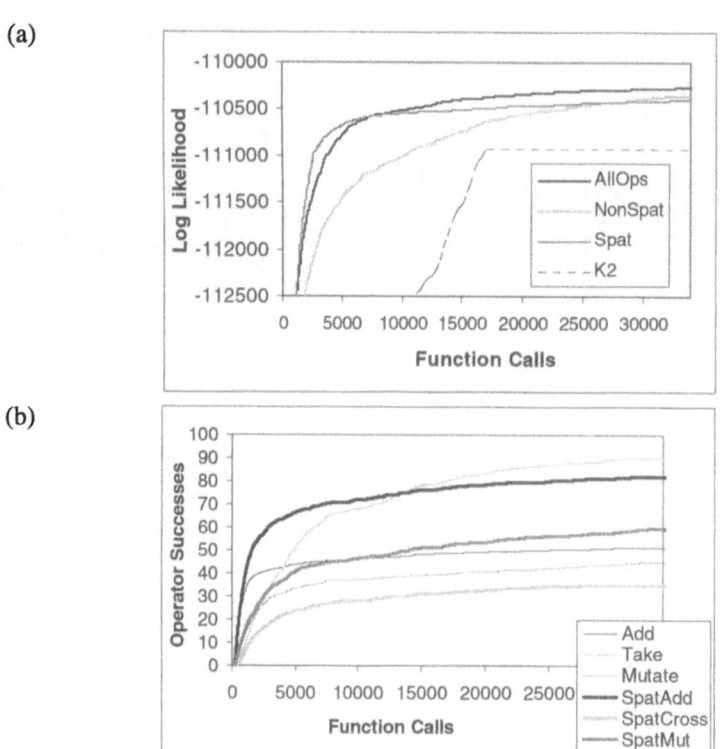

Fig. 6. (a) Mean Learning Curves and (b) Operator Successes for VF Data

Table 2. Quality of Networks learnt from VF Data

	% Links in same Bundle	Mean ON Distance
K2	62.963	41.056
Non-Spat	70.863	29.477
Spat	78.325	19.225
All	73.333	25.138

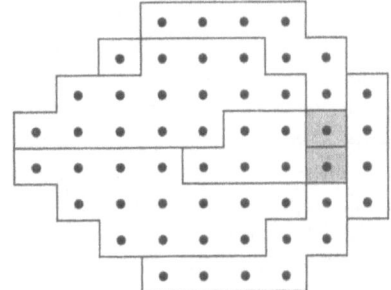

Fig. 7. Nerve Fibre Bundles

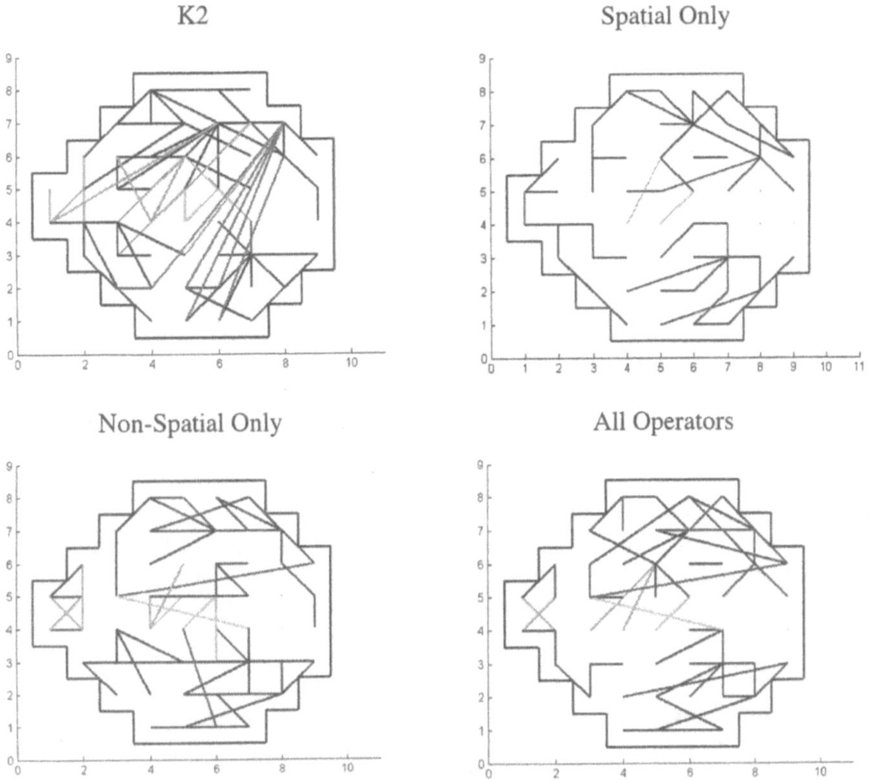

Fig. 8. Networks Learnt from VF Data

Figure 8 shows the spatial nature of the final networks for the different search methods on the VF data. ON distance is represented by the shading of the links (the larger the distance the lighter the link). Notice that the stochastic methods all have similar looking networks with strong spatial features. This is likely to be because these networks are the result of a large number of iterations and the methods have converged close to the optimum (figure 6a). The K2 result displays fewer spatial characteristics and many links with high ON distance (in lighter grey), including links across the horizontal midline that are anatomically unlikely. None of the networks have included links involving the blind spot, as was expected. For the stochastic methods, there are still some links that have high ON distance (links in lighter grey). Most of these links appear to be between VF points that are very close to one another and cross the central horizontal axis. Some nerve fibres interdigitate across the horizontal raphe in the nasal visual field, but this does not account for some of the more central links crossing the horizontal meridian. These could be a result of bias in the spatial operators, but they still exist in the non-spatial experiments. This implies errors in measurement of the visual field where the VF is shifted slightly or the disease process acts symmetrically in the visual field in anatomically unlinked locations. These biases and errors will be explored further.

5 Conclusions and Future Work

In this paper, we have developed evolutionary and non-evolutionary operators, for learning Bayesian network structures from spatial time series data. These operators have been tested on synthetic data and real-world visual field data. We have compared the efficiency of the operators as well as a well-known straw-man learning algorithm. Results using the proposed operators were very encouraging, particularly on the real-world data, where high quality networks were found more efficiently when spatial operators were used. Various methods were used to measure network quality, including structural difference and expert knowledge on the synthetic and visual field data, respectively. In the future, we will use spatial information such as optic nerve distance and geographical direction to guide search rather than simple Cartesian coordinates. It would be interesting to see how temporal lag search [Tucker01] interacts with spatial operators, and their performance on other datasets such as rainfall prediction of cities [Cofiño02]. We intend to explore how the operators compare to other methods for learning the network structures such as Estimation of Distribution Algorithms [Larranaga01] and methods based upon neural network optimisation [Kahng95]. We also plan to extend the modelling of the visual field to include data from both eyes, as well as clinical information.

Acknowledgements. We would like to thank Nick Strouthidis for his help with collating the visual field dataset and Stephen Swift for his general help and advice. In addition, we would like to thank the EPSRC and BBSRC for funding this research.

References

Anderssen, K.E., Jeppesen, V., Classifying Visual Field Data, Technical Report, MSc Thesis, Aalborg University, Denmark, (1998).
Bengtsson, B., Olsson, J., Heijl, A., Rootzen. H., A new generation of algorithms for computerized threshold perimetry, SITA. Acta Ophthalmologica Scandinavica 75, (1997), 368–375.

Cofiño, A.S., Cano, R., Sordo C., Gutiérrez, J.M., Bayesian Networks for Probabilistic Weather Prediction, Proc. of the 15th European Conference on Artificial Intelligence. IOS Press, (2002).

Cooper, G.F., Herskovitz, E., A Bayesian Method for the Induction of Probabilistic Networks from Data, Machine Learning 9, (1992), 309–347.

Ester, M., Frommelt, A., Kriegel, H.-P., Sander, J., Spatial Data mining: Database Primitives, Algorithms and Efficient DBMS Support, Special Issue on Integration of Data Mining with Database Technology, Data Mining and Knowledge Discovery, an International Journal, Kluwer Academic Publishers 4, Nos. 2/3, (2000).

Friedman, N., Learning the Structure of Dynamic Probabilistic Networks, Proc. of the 14th Annual Conference on Uncertainty in AI, (1998), 139–147.

Garway-Heath, D.F., Fitzke, F., Hitchings, R.A., Mapping the Visual Field to the Optic Disc, Opthalmology 2000, 107, (2000), 1809–1815.

Geiger, D., Heckerman, D., Learning Gaussian Networks, Proc. of the 10th Conference in Uncertainty in Artificial Intelligence, (1994), 235–243.

Grefenstette, J.J., Optimization of Control Parameters for Genetic Algorithms, IEEE Transactions on Systems, Man & Cybernetics 16, No. 1, (1986), 122–128.

Haley, M.J. (ed.), The Field Analyzer Primer, Allergan Humphrey, San Leandro, California, (1987).

Holland, J.H., Adaptation in Natural and Artificial Systems, University of Michigan Press, (1995).

Kahng, A.B., Moon, B.R., Toward More Powerful Recombinations, Proc. of the 6th International Conference on Genetic Algorithms, Morgan Kauffman, (1995), 96–103.

Kirkpatrick, S., Gelatt, C.D., Vecchi, M.P., Optimization by Simulated Annealing, Science 220, No. 4598, (1983), 671–80.

Larrañaga, P., Poza, M., Yurramendi, Y., Murga, R., Kuijpers, C., Structure Learning of Bayesian Networks using GAs, IEEE Transactions on Pattern Analysis and Machine Intelligence 18, No.9, (1996), 912–926.

Larrañaga, P., Lozano, J.A. (eds.), Estimation of Distribution Algorithms, Kluwer, (2001).

Liu, X., Swift, S., Tucker, A., Using Evolutionary Algorithms to Tackle Large Scale Grouping Problems, GECCO (2001), 454–460.

Roddick, J.F., Spiliopoulou, M., A Bibliography of Temporal, Spatial and Spatio-Temporal Data mining Research, ACM SIGKDD Explorations, Vol. 1, No. 1, (1999), 34–38.

Bäck T., Schwefel H.-P., An Overview of Evolutionary Algorithms for Parameter Optimization, Evolutionary Computation 1, No. 1, (1993) 1–23

Tucker, A., Liu, X., Ogden-Swift, A., Evolutionary Learning of Dynamic Probabilistic Models with Large Time Lags, International Journal of Intelligent Systems 16, No. 5, (2001), 621–646.

Wong, M., Lam, W., Leung, S., Using Evolutionary Programming and Minimum Description Length Principle for Data Mining of Bayesian Networks, IEEE Transactions on Pattern Analysis and Machine Intelligence, Vol. 21, No.2, (1999), 174–178.

An Evolutionary Approach for Molecular Docking

Jinn-Moon Yang

Department of Biological Science and Technology
National Chiao Tung University, Hsinchu, 30050, Taiwan
moon@csie.ntu.edu.tw

Abstract. We have developed an evolutionary approach for the flexible docking that is now an important component of a rational drug design. This automatic docking tool, referred to as the GEMDOCK (Generic Evolutionary Method for DOCKing molecules), combines both global and local search strategies search mechanisms. GEMDOCK used a simple scoring function to recognize compounds by minimizing the energy of molecular interactions. The interactive types of atoms between ligands and proteins of our linear scoring function consist only hydrogen-bonding and steric terms. GEMDOCK has been tested on a diverse dataset of 100 protein-ligand complexes from Protein Data Bank. In total 76% of these complexes, it obtained docked ligand conformations with root mean square derivations (RMSD) to the crystal ligand structures less than 2.0 Å when the ligand was docked back into the binding site. Experiments shows that the scoring function is simple and efficiently discriminates between native and non-native docked conformations. This study suggests that GEMDOCK is a useful tool for molecular recognition and is a potential docking tool for protein structure variations.

1 Introduction

The molecular docking problem is the prediction of a ligand conformation and orientation relative to the active site of a target protein. A computer-aided docking process, identifying the lead compounds by minimizing the energy of intermolecular interactions, is an important approach for structure-based drug designs [1]. Solving a molecular docking problem involves two critical elements: a good scoring function and an efficient algorithm for searching conformation and orientation spaces.

A good scoring function should be fast and simple for screening large potential solutions and effectively discriminating between correct binding states and non-native docked conformations. Various scoring functions have been developed for calculating binding free energy, including knowledge-based [2], physic-based [3], and solvent-based scoring functions [4]. In general the binding energy landscapes of these scoring functions are often complex and rugged funnel shapes [5].

Many automated docking approaches have been developed and can be roughly divided into rigid docking, flexible ligand docking, and protein flexible docking methods. The rigid-docking methods, such as DOCK program [6], treated both ligands and proteins as rigid. In contrast the ligand is flexible and the protein is rigid for flexible ligand docking methods including evolutionary algorithms [7,8,9,10], simulated annealing [11], fragment-based approach [12], and other algorithms. For reasonably addressing protein flexible problems, which both ligands and proteins are flexible, most of docking methods often allowed a limited model of protein variations, such as the side-chain flexible or small motions of loops in the binding site [13]. Most of these previous docking methods studied on a small test set (< 20 complexes), by contrast, the GOLD [8] and FlexX [12] were tested on a test set of over 100 complexes.

In this paper, we proposed an automatic program, GEMDOCK (Generic Evolutionary Method for DOCKing molecules), for docking flexible molecules. Our program used a simplified scoring function and a new evolutionary approach which is more robust than standard evolutionary approaches [14,15,16] on some specific domains [17,18,19,20]. Our energy function consisted only of steric and hydrogen-bonding terms with a linear model which was simple and fast enough to recognize potential complexes. In order to balance exploration and exploitation, the core idea of our evolutionary approach is to design multiple operators cooperating with each other by using the family competition which is similar to a local search procedure. We have successfully applied a similar idea to solve optimization problems in some differing fields [17,18,19,20].

In order to evaluate the performance and limitations of GEMDOCK on docking flexible ligands, we have tested it on a diverse dataset of 100 complexes from the Protein Data Bank. GEMDOCK achieved 76 ligands whose structures with RMSD values to the ligand crystal structures are less than 2.0Å. The rate increases to 86% when the structure water is considered. GOLD [8] achieved a 71% success rate in the same dataset and FlexX [12] achieved a 70% success rate on a dataset of 200 complexes extended from the data set of GOLD.

2 Method

The basic structure of the GEMDOCK (Figure 1) is as follows: Randomly generate a starting population with N solutions by initializing the orientation and conformation of the ligand relating to the center of the receptor. Each solution is represented as a set of four n-dimensional vectors $(x^i, \sigma^i, v^i, \psi^i)$, where n is the number of adjustable variables of a docking system and $i = 1, \ldots, N$ where N is the population size. The vector x represents the adjustable variables to be optimized in which x_1, x_2, and x_3 are the 3-dimensional location of the ligand; x_4, x_5, and x_6 are the rotational angles; and from x_7 to x_n are the twisting angles of the rotatable bonds inside the ligand. σ, v, and ψ are the step-size vectors of decreasing-based Gaussian mutation, self-adaptive Gaussian mutation, and self-adaptive Cauchy mutation. In other words, each solution x is associated with some parameters for step-size control. The initial values of x_1, x_2, and x_3 are randomly chosen from the feasible box, and the others, from x_4 to x_n, are randomly chosen from 0 to 2π in radians. The initial step sizes σ is 0.8 and v and ψ are 0.2. After GEMDOCK initializes the solutions, GEMDOCK enters the main evolutionary loop which consists of three main stages in every iteration: decreasing-based Gaussian mutation, self-adaptive Gaussian mutation, and self-adaptive Gaussian mutation. Each stage is realized by generating a new quasi-population (with N solutions) as the parent of the next stage. As shown in Figure 1, these stages apply a general procedure "FC_adaptive" with only different working population and the mutation operator.

The FC_adaptive procedure (Figure 1) employs two parameters, namely, the working population (P, with N solutions) and mutation operator (M), to generate a new quasi-population. The main work of FC_adaptive is to produce offspring and then conduct the

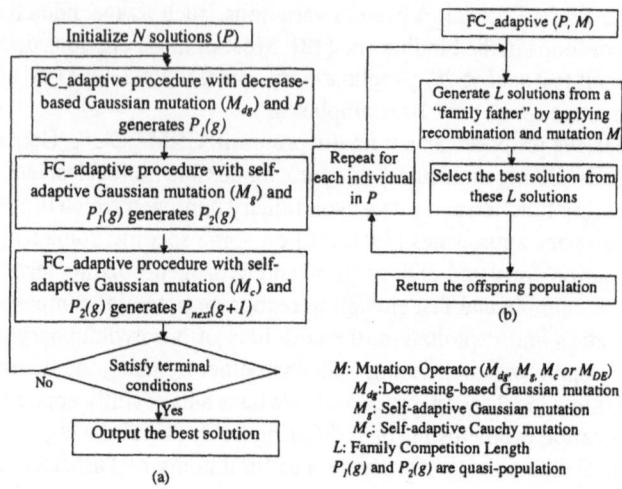

Fig. 1. Overview of GEMDOCK for molecular docking: (a) Main procedure (b) FC_adaptive procedure.

family competition. Each individual in the population sequentially becomes the "family father." With a probability p_c, this family father and another solution that is randomly chosen from the rest of the parent population are used as parents for a recombination operation. Then the new offspring or the family father (if the recombination is not conducted) is operated on by a mutation. For each family father, such a procedure is repeated L times called the family competition length. Among these L offspring and the family father, only the one with the lowest scoring function value survives. Since we create L children from one "family father" and perform a selection, this is a family competition strategy. This method avoids the population prematureness but also keeps the spirit of local searches. Finally, the FC_adaptive procedure generates N solutions because it forces each solution of the working population to have one final offspring.

In the following, genetic operators are briefly described. We use $a = (x^a, \sigma^a, v^a, \psi^a)$ to represent the "family father" and $b = (x^b, \sigma^b, v^b, \psi^b)$ as another parent. The offspring of each operation is represented as $c = (x^c, \sigma^c, v^c, \psi^c)$. The symbol x_j^s is used to denote the jth adjustable optimization variable of a solution s, $\forall j \in \{1, \ldots, n\}$.

2.1 Recombination Operators

GEMDOCK implemented modified discrete recombination and intermediate recombination [15]. A recombination operator selected the "family father (a)" and another solution (b) randomly selected from the working population. The former generates a child as follows:

$$x_j^c = \begin{cases} x_j^a & \text{with probability } 0.8 \\ x_j^b & \text{with probability } 0.2. \end{cases} \quad (1)$$

The generated child inherits genes from the "family father" with a higher probability 0.8. Intermediate recombination works as:

$$w_j^c = w_j^a + \beta(w_j^b - w_j^a)/2, \tag{2}$$

where w is σ, v, or ψ based on the mutation operator applied in the FC_adaptive procedure. The intermediate recombination only operated on step-size vectors and the modified discrete recombination was used for adjustable vectors (x).

2.2 Mutation Operators

After the recombination, a mutation operator, the main operator of GEMDOCK, is applied to mutate adjustable variables (x).

Gaussian and Cauchy Mutations: Gaussian and Cauchy Mutations are accomplished by first mutating the step size (w) and then mutating the adjustable variable x:

$$w_j' = w_j' A(\cdot), \tag{3}$$
$$x_j' = x_j + w_j' D(\cdot), \tag{4}$$

where w_j and x_j are the ith component of w and x, respectively, and w_j is the respective step size of the x_j where w is σ, v, or ψ. If the mutation is a self-adaptive mutation, $A(\cdot)$ is evaluated as $\exp[\tau' N(0,1) + \tau N_j(0,1)]$ where $N(0,1)$ is the standard normal distribution, $N_j(0,1)$ is a new value with distribution $N(0,1)$ that must be regenerated for each index j. When the mutation is a decreasing-based mutation $A(\cdot)$ is defined as a fixed decreasing rate $\gamma = 0.95$. $D(\cdot)$ is evaluated as $N(0,1)$ or $C(1)$ if the mutation is, respectively, Gaussian mutation or Cauchy mutation. For example, the self-adaptive Cauchy mutation is defined as

$$\psi_j^c = \psi_j^a \exp[\tau' N(0,1) + \tau N_j(0,1)], \tag{5}$$
$$x_j^c = x_j^a + \psi_j^c C_j(t). \tag{6}$$

We set τ and τ' to $(\sqrt{2n})^{-1}$ and $(\sqrt{2\sqrt{n}})^{-1}$, respectively, according to the suggestion of evolution strategies [15]. A random variable is said to have the Cauchy distribution $(C(t))$ if it has the density function: $f(y;t) = \frac{t/\pi}{t^2+y^2}$, $-\infty < y < \infty$. In this paper t is set to 1. The formulation of the self-adaptive Gaussian mutation is similar to the self-adaptive Cauchy mutation and is given

$$v_j^c = v_j^a \exp[\tau' N(0,1) + \tau N_j(0,1)], \tag{7}$$
$$x_j^c = x_j^a + v_j^c N_j(0,1). \tag{8}$$

Our decreasing-based Gaussian mutation uses the step-size vector σ with a fixed decreasing rate $\gamma = 0.95$ and works as

$$\sigma^c = \gamma \sigma^a, \tag{9}$$
$$x_j^c = x_j^a + \sigma^c N_j(0,1). \tag{10}$$

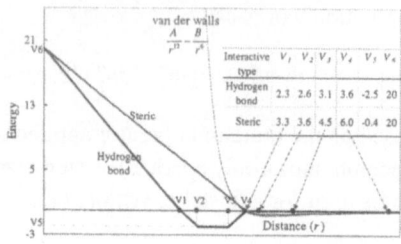

Fig. 2. The linear energy function of the pair-wise atoms for steric and hydrogen bonds in GEM-DOCK (bold line) with a standard Lennard-Jones potential (light line).

Rotamer-Mutation: This operator is only used for x_7 to x_n to find the conformations of the rotatable bonds inside the ligand. For each ligand, this operator mutates all of the rotatable angles according to the rotamer distribution and works as:

$$x_j = r_{ki} \text{ with probability } p_{ki}, \qquad (11)$$

where r_{ki} and p_{ki} are the angle value and the probability, respectively, of ith rotamer of kth bond type including $sp^3 - sp^3$ and $sp^3 - sp^2$ bond. The values of r_{ki} and p_{ki} are based on the energy distributions of these two bond types.

2.3 Scoring Function

In this work, we used a simple scoring function given as

$$E_{tot} = E_{inter} + E_{intra} + E_{penal}, \qquad (12)$$

where E_{inter} and E_{intra} are the intermolecular and intramolecular energy, respectively, E_{penal} is a large penalty value if the ligand is out of range of the search box. In this paper, E_{penal} is set to 10000.

The intermolecular energy is defined as

$$E_{inter} = \sum_{i=1}^{lig} \sum_{j=1}^{pro} \left[F(r_{ij}^{B_{ij}}) + 332.0 \frac{q_i q_j}{4 r_{ij}} \right], \qquad (13)$$

where r_{ij} is the distance between the atoms i and j, q_i and q_j are the formal charges and 332.0 is a factor that converts the electrostatic energy into kilocalories per mole. The lig and pro denote the numbers of the heavy atoms in the ligand and receptor, respectively. $F(r_{ij}^{B_{ij}})$ is a simple atomic pair-wise potential function (Figure 2) modified from previous works [7,21] and given as

Table 1. Atom types of GEMDOCK

Atom type	Heavy atom name
Donor	primary and secondary amines, sulfur, and metal atoms
Acceptor	oxygen and nitrogen with no bound hydrogen
Both	structural water and hydroxyl groups
Nonpolar	other atoms (such as carbon and phosphorus)

$$F(r_{ij}^{B_{ij}}) = \begin{cases} V_6 - \frac{V_6 r_{ij}^{B_{ij}}}{V_1} & \text{if } r_{ij}^{B_{ij}} \leq V_1 \\ \frac{V_5(r_{ij}^{B_{ij}} - V_1)}{V_2 - V_1} & \text{if } V_1 < r_{ij}^{B_{ij}} \leq V_2 \\ V_5 & \text{if } V_2 < r_{ij}^{B_{ij}} \leq V_3 \\ V_5 - \frac{V_5(r_{ij}^{B_{ij}} - V_3)}{V_4 - V_3} & \text{if } V_3 < r_{ij}^{B_{ij}} \leq V_4 \\ 0 & \text{if } r_{ij}^{B_{ij}} > V_4 \end{cases} \quad (14)$$

$r_{ij}^{B_{ij}}$ is the distance between the atoms i and j with bond type B_{ij} which is the interaction bonding type forming by the pair-wise heavy atoms of a ligand and a protein. B_{ij} is either hydrogen binding or steric state. The values of parameters, $V_1, \ldots, V_6$, are given in Figure 2. In this atomic pair-wise model, the interactive types are only hydrogen binding and steric potential which have the same function form but with different parameters, $V_1, \ldots, V_6$. The energy value of hydrogen binding should be larger than the one of steric potential. In this model, the atom is divided into four different atom types (Table 1) : donor, acceptor, both, and nonplar. The hydrogen binding can be formed by the following pair atom types: donor-acceptor (or acceptor-donor), donor-both (or both-donor), acceptor-both (or both-acceptor), and both-both. Other pair-atom combinations are to form the steric state.

The intramolecular energy of a ligand is

$$E_{intra} = \sum_{i=1}^{lig} \sum_{j=i+2}^{lig} F(r_{ij}^{B_{ij}}) + \sum_{k=1}^{dihed} A[1 - \cos(m\theta_k - \theta_0)], \quad (15)$$

where $F(r_{ij}^{B_{ij}})$ is defined as Equation 14 except the value is set to 1000 when $r_{ij}^{B_{ij}} <$ 2.0 Å and $dihed$ is the number of rotatable bonds. We followed the work of Gehlhaar et al. (1995) to set the values of A, m, and θ_0. For the $sp^3 - sp^3$ bond A, m, and θ_0 are set to 3.0, 3, and π; and $A = 1.5$, $m = 6$, and $\theta_0 = 0$ for the $sp^3 - sp^2$ bond.

3 Results

3.1 Parameters of GEMDOCK

Table 2 indicates the setting of GEMDOCK parameters, such as initial step sizes, family competition length ($L = 3$), population size ($N = 200$), and recombination probability ($p_c = 0.3$) in this work. The GEMDOCK optimization stops when either the convergence is below certain threshold value or the iterations exceed a maximal preset value which

Table 2. Parameters of GEMDOCK

Parameter	Value of parameters
Initial step sizes	$\sigma = 0.8, v = \psi = 0.2$ (in radius)
Family competition length	$L = 3$
Population size	$N = 200$
Recombination rate	$p_c = 0.3$
# of the maximum generation	100

was set to 100. Therefore, GEMDOCK generated 2400 solutions in one generation and terminated after it exhausted 240000 solutions in the worse case. These parameters were decided after experiments conducted to recognize complexes of test docking systems with various values.

3.2 Test Data Set

In order to evaluate the strength and limitation of GEMDOCK, we tested it on a highly diverse dataset of 100 protein-ligand complexes proposed by Jones et al. [8] (Table 3). The ligand input files were generated by GENLIG which assigned the formal charge and atom type (donor, acceptor, both, or nonplar) of each atom and the bond type ($sp^3 - sp^3$, $sp^3 - sp^2$, or others) of a rotatable bond inside a ligand. These materials were used in Equation 12 to calculate the scoring value of a solution. Table 3 shows the ligand summary, including the minimum, average, and maximum values of the number of rotatable bonds and the number of heavy atoms.

When preparing the proteins, we removed all structural water molecules and metal atoms except we discussed the influence of considering these hetero atoms. In order to decide the size of active site, GEMDOCK was tested on four different sizes (d Å): 6Å, 8Å, 10Å, and 12Å. The size with d Å means that all protein atoms in the active site are selected if they are located less than d Å apart from each ligand atom. GEMDOCK automatically decide the cube of a binding site based on the maximum and minimum of coordinates of these selected protein atoms. Experiments shows that GEMDOCK had little influence on the different sizes. In this paper, the distance d is set to 10Å when a ligand is docked back into the active site. Among these 100 test systems, the minimum cube is 23Å×24Å×20Å (2mcp) and the maximum cube is 41Å×40Å×30Å (2r07).

3.3 Results on the Dataset of 100 Complexes

GEMDOCK executed 10 independent runs for each complex. The solution with lowest scoring function was then compared with the observed ligand crystal structure. Table 3 shows the summary information and performance. We based the results on root mean square deviation (RMSD) error in ligand heavy atoms between the docked conformation and the crystal structure. By contrast, Jones et al [8] used four subjective categories (good, close, error, and wrong) to evaluate the performance. Because they found all of good and close solutions with RMSD were below 2.0 Å, we considered that a docking result is acceptable if the RMSD value is less than 2.0 Å [8,12].

Table 3. GEMDOCK results and summary of ligands in the dataset of 100 complexes

RMSD(Å)	rank 1	any rank	PDB code with rank 1
≤ 0.5	24	41	1abe 1acm 1aco 1did 1epb 1fki 1hdy 1hsl 1ida 1lst 1pbd 1pha 1rob 1stp 1tpp 2ada 2cgr 2cht 2dbl 3aah 3tpi 4phv 6abp 6rsa
> 0.5, ≤ 1.0	39	32	1acj 1ack 1acl 1aha 1dbb 1dbj 1die 1dr1 1dwd 1eap 1etr 1fkg 1ghb 1hri 1hyt 1ldm 1lic 1mrk 1nis 1phd 1phg 1rds 1slt 1srj 1tka 1tmn 1ulb 1glq 2ak3 2ctc 2mcp 2pk4 2r07 2sim 3cpa 3hvt 4cts 4dfr 4est
> 1.0, ≤ 1.5	10	11	1aaq 1apt 1cbx 1cps 1hdc 1icn 1poc 4fab 5p2p 8gch
> 1.5, ≤ 2.0	3	6	1aec 1tdb 3ptb
> 2.0, ≤ 2.5	1	0	1mdr
> 2.5, ≤ 3.0	3	6	1ase 1blh 2yhx
> 3.0	20	10	1eta 1eed 1azm 1rne 6rnt 1mcr 2mth 1mup 2plv 1baf 1ive 2phh 3gch 1hef 1igj 1coy 1xid 1xie 3cla 7tim

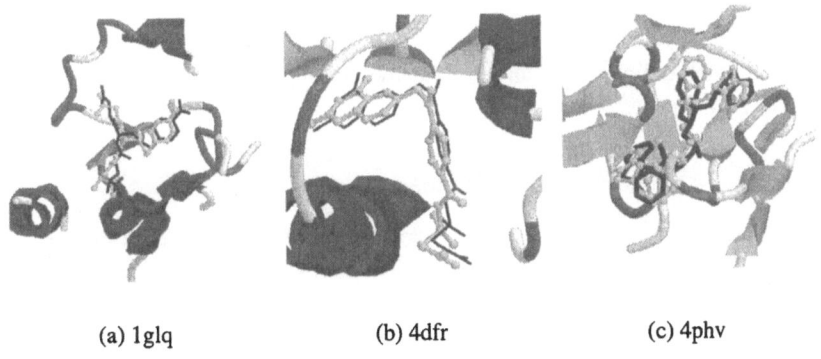

(a) 1glq (b) 4dfr (c) 4phv

Fig. 3. Acceptable docking examples: The docked ligand conformation (red) is similar to the crystal ligand structure (yellow). The RMSD values are 0.78 Å for 1glq, 0.56 Å for 4dfr, and 0.42 Å for 4phv.

Table 3 shows that GEMDOCK achieved a 76% success rate in identifying the experimental binding model if the solutions at the first rank are considered. The RMSD values of 63 complexes are less than 1.0 Å. This rate further rises to 84% based on the solutions with any rank. The performance of GEMDOCK was little influenced by the number of rotatable bonds and the number of heavy atoms of a ligand. When the structural water and metal atoms are considered, the success rate is improved to 86% with the first rank. On average GEMDOCK took 305 seconds for a docking run on Pentium 1.4 GHz personal computer with single processor. The maximum time was 883 seconds for the complex, 1rne, and the shortest time was 102 seconds for 2pk4. In the following, we discussed some acceptable examples and unacceptable examples.

Figure 3 shows three typical acceptable solutions in which GEMDOCK predicted correct positions of all ligand groups. The predicted ligand is red and crystal ligand is yellow. All of these three examples are identical with the crystal structures and the RMSD values are 0.78 Å (1glq: nitrophenyl ligand for glutathione S-transferase), 0.56 Å

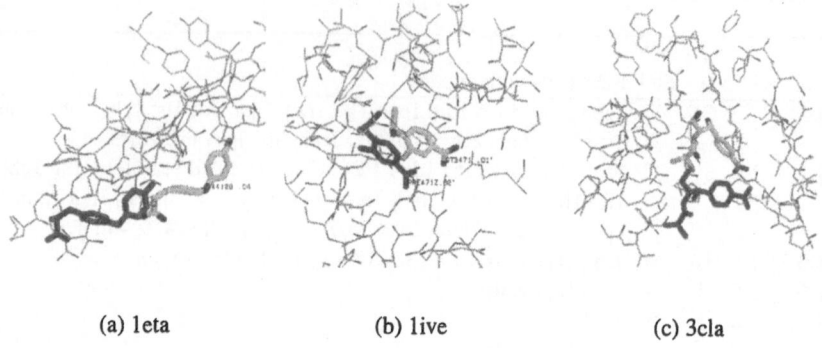

(a) 1eta   (b) 1ive   (c) 3cla

Fig. 4. Results of three poor docking examples. The docked ligand conformation is red and the crystal ligand structure is cpk. The RMSD values are 3.62 Å for 1eta, 6.32 Å for 1ive, and 7.56 Å for 3cla.

(4dfr: methotrexate ligand for dihydrofolate reductase), and 0.42 Å (4phv: peptide-like ligand for HIV-1 protease).

3.4 Examples of Unacceptable Solutions

Table 3 shows 24 unacceptable docking complexes with RMSD values more than 2.0 Å. Three poor examples are shown in Figure 4 in which the structures of predicted ligands and crystal ligands are displayed with red and cpk color, respectively. The RMSD values are 3.62 Å for 1eta (tetraiodo L-thyronine ligand for transthyretin), 6.32 Å for 1ive (acetylamino ligand for influenza), and 7.56 Å for 3cla (chloramphenicol ligand for Type III chloramphenicol acetyltransferase).

We have analyzed these poor examples to understand why GEMDOCK failed to recognize the binding models by using numerical experiments. These experiments were based on three main factors: the scoring functions, the docking materials, and the search methods. For the scoring functions we tested various uses and parameter values (Equation 12) on the dataset of 100 complexes. According to our experimental results, the E_{inter} was the main factor in our system, the E_{intra} and E_{penal} were minor factors that influenced some specific docking cases. The element, $F(r_{ij}^{B_{ij}})$, of the E_{inter} (Equation 13) dominated the performance. Figure 5 shows the relationship between the RMSD values and scoring values with 100 independent runs. For the good docking example (4dfr) 95 solutions with RMSD values are less than 1.0 Å and the scoring value is similar in each run (Figure 5(a)). By contrast, for a poor example the RMSD value is more than 3.0 Å and the score is diverse (Figure 5(b)). These experiments indicate that our scoring function seems to be simple and fast to discriminate native binding state and non-native docked conformations for 90% testing complexes.

For the docking materials we have discussed the influences of the sizes of the binding site (see Subsection Test data set) and of the hetero atoms in the binding site. Among these 100 complexes there are 17 proteins with metal atoms, and 84 proteins with structural water atoms. When the metal atoms are included, GEMDOCK can consistently im-

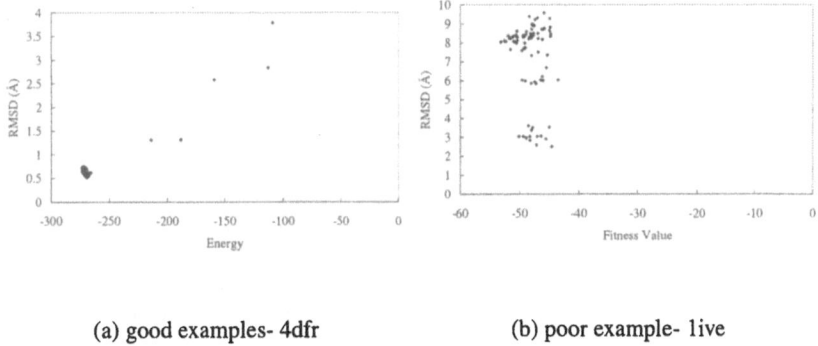

(a) good examples- 4dfr (b) poor example- 1ive

Fig. 5. Typical relationship between the values of the scoring function and the RMSD on 100 independent runs. (a) For a good docking example the 95 solutions with RMSD values are less than 0.5 Å and the scoring values are similar. (b) For a poor docking example the RMSD values are often more than 3.0 Å and the scoring values are diverse.

prove docking accuracy, such as the complexes 1xid and 1xie. In general GEMDOCK is able to improve the docking accuracy when both structural water and metal atoms are considered.

3.5 Comparison with Other Approaches

According to our best survey, most of previous docking works studied on a small dataset (< 20 complexes) except the GOLD [8] and FlexX [12] used. Here we compared GEMDOCK with GOLD and FlexX on the dataset of 100 complexes. Table 4 shows the summary of these three docking tools. The rates of FlexX based on a test set of 200 complexes enlarged the GOLD test set. GOLD was a steady-state genetic algorithm and FlexX was an incremental approach. GEMDOCK obtained a 76% success rate based on the condition of an RMSD value less than 2 Å. In contrast GOLD [8] achieved a 71% success rate in identifying the experimental binding model based on their assessment categories, and the rate was 66% if based on the RMSD condition. FlexX [12] achieved a 70% success rate based on solutions with any rank and the RMSD condition. The rate was 46.5% if the solutions at the fist rank were considered. The results of FlexX were often sensitive to the choice of the base fragment and its placement and the number of the fragments.

4 Conclusions

In this work, we have developed a robust evolutionary approach with a simple fitness function for docking flexible molecules. Experiments on 100 test systems verify that the proposed approach achieved a 76% success rate in recognizing the binding models.

Table 4. Comparsion GEMDOCK with GOLD and FlexX on the dataset of 100 complexes

RMSD(Å)	GEMDOCK	GOLD[a]	FlexX[b]
≤ 0.5	24%	8%	12.5%
$> 0.5, \leq 1.0$	39%	27%	38.5%
$> 1.0, \leq 1.5$	10%	20%	12.5%
$> 1.5, \leq 2.0$	3%	11%	5.5%
$> 2.0, \leq 2.5$	1%	2%	7.5%
$> 2.5, \leq 3.0$	3%	4%	2%
> 3.0	20%	28%	21.5%

[a]: GOLD [8] is a steady-state genetic algorithm.
[b]: The rate of FlexX [12], a fragment-based approach, is based on any rank with a dataset of 200 complexes extended from the GOLD data set.

GEMDOCK seamlessly blends local search and global search to work cooperatively by the integration of a number of genetic operators, each having unique search mechanism. In summary, we have demonstrated the robustness and adaptability of GEMDOCK for exploring the conformational space of a molecular docking problem and efficiently finding the solution under the constraint of the fitness function used. Our scoring function seems to be simple and fast to discriminate native binding states and non-native docked conformations. We believe that GEMDOCK is an effective tool for docking flexible molecules.

Acknowledgements. This work was supported in part by grant 91-2320-B-009-001 from the National Science Council of Taiwan and in part by grand DOH92-TD-1132 from Department of Health, Taiwan.

References

1. I. D. Kuntz. Structure-based strategies for drug design and discovery. *Science*, 257:1078–1082, 1992.
2. H. Gohlke, M. Hendlich, and G. Klebe. Knowledge-based scoring function to predict protein-ligand interactions. *Journal of Molecular Biology*, 295:337–356, 2000.
3. S. J. Weiner, P. A. Kollman, D. A. Case, U. C. Singh, C. Ghio, G. Alagona, S. Profeta, Jr., and P. Weiner. A new force field for molecular mechanical simulation of nucleic acids and proteins. *Journal of the American Chemical Society*, 106:765–784, 1984.
4. B. K. Shoichet, A. R. Leach, and I. D. Kuntz. Ligand solvation in molecular docking. *Proteins: Structure, Function, and Genetics*, 34:4–16, 1999.
5. D. W. Miller and K. A. Dill. Ligand binding to proteins: the binding landscape model. *Protein Science*, 6:2166¡V2179, 1997.
6. I. D. Kuntz, J. M. Blaney, S. J. Oatley, R. Langridge, and T. E. Ferrin. A geometric approach to macromolecular-ligand interactions. *Journal of Molecular Biology*, 161:269–288, 1982.
7. D. K. Gehlhaar, G. M. Verkhivker, P. Rejto, C. J. Sherman, D. B. Fogel, L. J. Fogel, and S. T. Freer. Molecular recognition of the inhibitor AG-1343 by HIV-1 protease: conformationally flexible docking by evolutionary programming. *Chemistry and Biology*, 2(5):317–324, 1995.

8. G. Jones, P. Willett, R. C. Glen, A. R. Leach, and R. Taylor. Development and validation of a genetic algorithm for flexible docking. *Journal of Molecular Biology*, 267:727–748, 1997.
9. J. S. Taylor and R. M. Burnett. Darwin: A program for docking flexible molecules. *Proteins: Structure, Function, and Genetics*, 41:173V191, 2000.
10. G. M. Morris, D. S. Goodsell, R. S. Halliday, R. Huey, W. E. Hart, R. K. Belew, and A. J. Olson. Automated docking using a lamarckian genetic algorithm and empirical binding free energy function. *Journal of Computational Chemistry*, 19:1639–1662, 1998.
11. C. J. Sherman, R. C. Ogden, and S. T. Freer. *De Novo* design of enzyme inhibitors by monte carlo ligand generation. *Journal of Medicinal Chemistry*, 38(3):466–472, 1995.
12. B. Kramer, M. Rarey, and T. Lengauer. Evaluation of the flexX incremental construction algorithm for protein-ligand docking. *Proteins: Structure, Function, and Genetics*, 37:228–241, 1999.
13. F. Österberg, G. M. Morris, M. F. Sanner, A. J. Olson, and D. S. Goodsell. Automated docking to multiple target structures: Incorporation of protein mobility and structural water heterogeneity in autodock. *Proteins: Structure, Function, and Genetics*, 46:34–40, 2002.
14. D. E. Goldberg. *Genetic Algorithms in Search, Optimization and Machine Learning*. Addison-Wesley Publishing Company, Inc., Reading, MA, USA, 1989.
15. T. Bäck. *Evolutionary Algorithms in Theory and Practice*. Oxford University Press, New York, USA, 1996.
16. D. B. Fogel. *Evolutionary Computation: Toward a New Philosophy of Machine Intelligent*. IEEE Press, New York, 1995.
17. J.-M. Yang and C.-Y. Kao. Flexible ligand docking using a robust evolutionary algorithm. *Journal of Computational Chemistry*, 21(11):988–998, 2000.
18. J.-M. Yang, C.-H. Tsai, M.-J. Hwang, H.-K. Tsai, J.-K. Hwang, and C.-Y. Kao. GEM: A gaussian evolutionary method for predicting protein side-chain conformations. *Protein Science*, 11:1897–1907, 2002.
19. J.-M. Yang and C.-Y. Kao. A robust evolutionary algorithm for training neural networks. *Neural Computing and Application*, 10(3):214–230, 2001.
20. J.-M. Yang, J.-T. Horng, C.-J. Lin, and C.-Y. Kao. Optical coating designs using an evolutionary algorithm. *Evolutionary Computation*, 9(4):421–443, 2001.
21. R. M. A. Knegtel, J. Antoon, C. Rullmann, R. Boelens, and R. Kaptein. Monty: a monte carlo approach to protein-dna recongnition. *Journal of Molecular Biology*, 235:318–324, 1994.

Evolving Sensor Suites for Enemy Radar Detection

Ayse S. Yilmaz[1], Brian N. McQuay[1], Han Yu[1], Annie S. Wu[1], and John C. Sciortino, Jr.[2]

[1] School of Electrical Engineering and Computer Science
University of Central Florida, Orlando, FL 32816, USA
{selen,bmcquay,hyu,aswu}@cs.ucf.edu
[2] Naval Research Laboratory, Washington, DC 20375, USA
john.sciortino@nrl.navy.mil

Abstract. Designing optimal teams of sensors to detect the enemy radars for military operations is a challenging design problem. Many applications require the need to manage sensor resources. There is a tradeoff between the need to decrease the cost and to increase the capabilities of a sensor suite. In this paper, we address this design problem using genetic algorithms. We attempt to evolve the characteristics, size, and arrangement of a team of sensors, focusing on minimizing the size of sensor suite while maximizing its detection capabilities. The genetic algorithm we have developed has produced promising results for different environmental configurations as well as varying sensor resources.

1 Introduction

The problem of determining an optimal team of cooperating sensors for military operations is a challenging design problem. Tactical improvements along with increased sensor types and abilities have driven the need for automated sensor allocation and management systems. Such systems can provide valuable input to human operators in terms of exploring and offering candidate solutions, providing a testbed on which to evaluate candidate solutions, and providing a penalty-free environment on which to test for severe failure conditions.

Given a possible configuration of enemy radars, our goal is to find a team of sensors that can sense as many enemy radars as possible. Sensors are available in a wide range of capabilities and costs and the number of sensors in a team can vary. The importance of a mission may restrict the types and numbers of available sensors as well as the maximum cost. There is a tradeoff between the need to maximize the sensing capability of a sensor suite and the desire to minimize cost.

The open-endedness of this problem makes it an interesting design problem. The number of potential components (sensor type, characteristics, and placement) that make up a solution is extremely large. There are very few restrictions as well as very little guidance as to how many and what types of these components would make up a good solution.

We use a genetic algorithm (GA) to tackle this problem of designing optimal sensor suites. We have developed a GA that will generate candidate sensor suites that are evaluated using a simulator based on military specification. In this paper, we describe experiments which explore the performance of our GA in different types of enemy environments using different types of sensor resources. Our experimental results indicate that such a GA approach is able to design reasonable and effective sensor suites.

2 Related Work

Evolutionary Algorithms have been used to generate optimal designs effectively. Peter Bentley's [1] pioneering work in evolutionary design incorporated evolutionary computation techniques into the evolution of design components that would build wide range of solid objects such as tables, prisms and cars. The system takes the input specifications for the object to be designed and evolves the shape of the design that performs the required function. Funes et al. [2] apply evolutionary techniques to the design of structures in a simulation environment that are then buildable using Lego parts. Instead of using expert engineering knowledge, they use a fitness function to evaluate the feasibility and functionality of evolved structures. Husbands et al. [3] report the results of a comparative study of ten different optimization techniques in structural design optimization. The distributed Genetic Algorithm (DGA) and various hybrid methods (DGA with gradient descent, Simulated Annealing with gradient descent) appear to have significant advantages over the other techniques tested. Lee et. al. [4] develop a hybrid approach to evolve both controllers and robot bodies for performing specific tasks. They evolve physical body parameters such as the number and location of sensors and show that the design of robot body considerably can affect the behavior of the system.

Bugajska et al. [5] investigate the co-evolution of form and function for autonomous agents. The form component in [5] focuses on the characteristics of a sensor suite. All sensors within a sensor suite are assumed to have the same characteristics. Characteristics that can be evolved are the number of the sensors in a sensor suite and the sensor coverage area of each sensor. The maximum number of sensors is limited to 32 and the detection angle varies from 5 to 30 degrees. The location of the sensors are fixed and the coverage degradation by the distance is not taken into consideration. In [6], Bugajska et al. added the detection angle and the placement of the sensor to the list of evolved characteristics. They assume that power efficiency degrades with increased sensor coverage area. They evolve the number of the sensors implicitly by allowing a zero detection angle to indicate the existence of no sensor. The maximum number of sensors is limited to nine. The sensor detection angle ranges from 0 to 45 degrees. By contrast, in our system, detection angle can vary from 0 to 359 degrees and the placement of the sensors is not limited to fixed locations. The evolvable characteristics of our sensors are location, detection coverage, power and the frequency band at which each sensor operates.

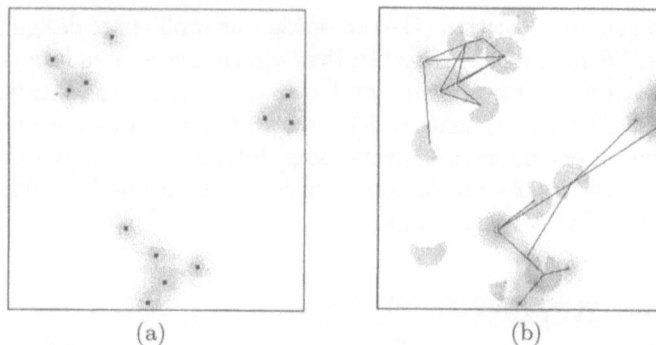

Fig. 1. (a) Example problem environment; (b) Problem environment with candidate solution.

The sensor management problem has been investigated by researchers using non-evolutionary techniques as well. Gaskell *et al.* [7] use a decision theoretic approach to develop a sensor management system for mobile robots. Popoli [8] proposes a sensor management scheme that uses fuzzy set theory, knowledge-based reasoning and expert systems. Schmaedeke *et al.* [9] uses an information theoretic approach for sensor to target assignment optimization. We believe the evolutionary technique and the advances in our work will have an advantage in terms of being robust and easily transferable between different problem scenarios with little or no modification to the system.

3 Problem Environment

Before looking at the details of our GA implementation, it is necessary to understand the problem to which we apply the GA. The problem environment consists of a collection of stationary enemy radars located in a two-dimensional plane. Figure 1(a) shows an example environment consisting of twelve randomly placed enemy radars. Radars are represented as points surrounded by gradually fading circles. The environment is restricted to two dimensions in our current work but can be extended to three dimensions in the future.

The location, power, and frequencies of the enemy radar are configured beforehand and remain static throughout a run. Each radar can operate on one of four different frequencies: A, B, C, or D. A radar can only be detected by sensors that are configured to sense on the same frequency. A radar must be detected by at least three difference sensors to be *fully detected*. (Three measurements are necessary for accurate triangulation of position.) Radars that are detected by two sensors are *partially detected* and radars that are detected by one sensor are *minimally detected*. Partially and minimally detected radars contribute less to the fitness evaluation than fully detected radars. In the experiments described here, we set all radars to operate on a single frequency in order to test the GA's ability to evolve minimal cost teams.

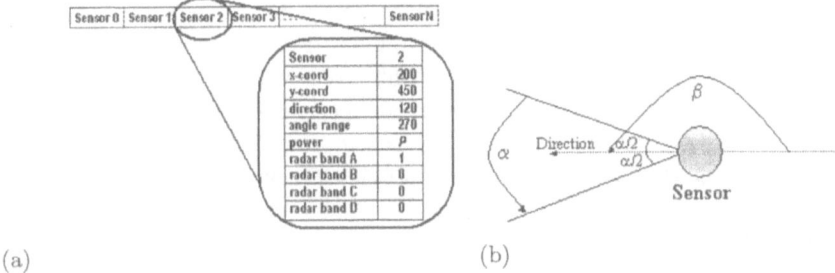

Fig. 2. (a) Problem representation for the evolvable team of sensors; (b) Sensor characteristics, α = Detection angle range and β= Direction angle

Figure 1(b) shows the same environment along with a candidate solution. The pie shaped elements indicate sensors and their detection angle and range. Lines indicate detection of a radar by a sensor.

4 GA Design

4.1 Problem Representation

Each individual in a GA population specifies the composition and arrangement of a team of sensors encoded as a vector of genes. Each gene encodes the evolvable characteristics for a single sensor. Figure 2(a) shows an example individual which represents a team of N sensors. Example parameter values for Sensor 2 are shown in detail. As the optimal number of sensors may not be known in advance, we allow the GA to evolve variable length individuals. Initially, each individual contains 20 randomly configured sensors. The maximum possible length of an individual is 100, indicating a maximum team size of 100 sensors.

The location of each sensor is encoded as x and y coordinate values. Sensors can be active or inactive. As we have no restriction on representation, multiple sensors in an individual may have the same location in the environment. If that's the case, only the sensor that appears first in the individual (the leftmost sensor) is active. The others are inactive and are unable to detect any radars; however, they are still included in the cost component of the fitness function.

Figure 2(b) illustrates the direction angle and the detection angle range. The direction angle indicates the "front" of a sensor and is a counter-clockwise angle relative to the 3 o'clock position. The detection angle range is centered around the direction angle and indicates the range within which a sensor can detect signals. Suppose that we have a square environment with a radar in each corner and a sensor located in the center of the square. If the sensor has a direction angle of 45 degrees and a detection angle range of 90 degrees. This particular sensor is only capable of detecting the radars located in the upper right quadrant of the environment, as long as other requirements such as frequency and power are satisfied.

The minimum power threshold for each sensor indicates the minimum power value that a sensor must receive from an enemy radar in order for detection to occur. A sensor is initially randomly configured with a minimum threshold power for radar detection. The power a sensor receives from a radar is inversely proportional to the distance between the sensor and the radar. The received power is also proportional to the base power of each enemy radar. Thus, $p = r/d$, where p is the power received by a sensor, r is the enemy radar power, and d is the distance between the sensor and the enemy radar. A sensor can detect a radar only if the power received by the sensor is greater than the minimum threshold for that sensor. The minimum threshold value that can be evolved is configurable as a system parameter which enables us to vary the coverage capabilities of the sensors that can be generated within a GA run.

Each sensor can detect radar frequencies in zero to four bands (as represented by A, B, C, and D in section 3). Detection is represented by a four bit binary string in each gene. Each bit refers to one radar band. A 1 denotes the ability to detect the radar frequency represented by that bit and a 0 indicates inability to detect the corresponding radar frequency. Sensors are capable of detecting on more than one band. A sensor can detect a radar only if it operates on the same radar frequency as the radar.

4.2 Fitness Evaluation

The fitness function consists of two parts, the detection capability and the total cost of a solution. The fitness function is:

$$f = \rho/\sigma, \tag{1}$$

where f is the raw fitness, ρ is the detection fitness that indicates the detection capability, and σ is the total cost of a solution. To calculate ρ, the detection fitness, we count the number of radars that are fully, partially, and minimally detected. The detection fitness is:

$$\rho = \frac{3*F + 2*P + M}{R}, \tag{2}$$

where R is the number of enemy radars and F, P, M are the numbers of fully, partially and minimally detected radars, respectively. Detection angle affects how much the environment a sensor can see, which implicitly affects the number of detectable radars.

The raw fitness is inversely proportional to the total solution cost. The total cost of a solution is its basic cost plus the total cost of all of the sensors:

$$\sigma = b + \sum_{i=1}^{n}(p_i + q_i), \tag{3}$$

where b is the fixed basic cost of deployment, n is the total number of sensors, p_i is the fixed basic cost of each sensor and q_i is the cost due to the band frequencies that are equipped in each sensor.

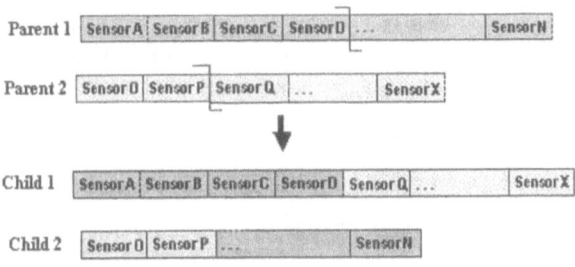

Fig. 3. Inter-gene level crossover operation

$$q_i = 1 + 2 * r_i, \qquad (4)$$

where r_i is the number of bands a sensor has. The cost of radar bands favors multi-band sensors such that it is cheaper to have a sensor with four bands than to have two sensors with two bands each. The costs given to a sensor according to the number of bands it has are 1, 3, 5, 7 and 9 for zero, one, two, three and four bands respectively. Two 2-band sensors would cost 10 while one 4-band sensor would cost 9, thus, favoring multi-band sensors.

4.3 Selection and Genetic Operators

The selection method we use is deterministic tournament selection with size two. One-point crossover operator is applied to variable length individuals where the crossover point on each parent is chosen independently. The length of an offspring may be different from its parents. Crossover points always fall in between the genes as shown in Figure 3.

We use a mutation scheme that is different from traditional mutation. Mutation is done on the intra-gene level. Each characteristic of each gene is subject to the mutation rate. If frequency band is to be mutated, we randomly generate a value (0 or 1) for the mutated bit. If the location, direction angle, detection angle range, or minimum threshold power are to be mutated, a Poisson distribution function is used to generate a value as an offset from the original value. That is, the new value is not randomly selected from a specified range, but rather follows a distribution probability that favors smaller changes. As a result, mutation is more likely to generate values that are similar to the original value instead of simply mutating randomly to any new value. We expect this mutation scheme to encourage accurate adjustment of the location, direction angle, detection angle range and power of the sensors.

5 Experiments

The goal of our experiments is to demonstrate that a GA is capable of designing teams of sensors for the detection of the enemy radars in a reasonable amount of time. We test our GA on a variety of environmental scenarios and sensor configurations to study the robustness of the system with respect to the inputs for the environmental conditions.

5.1 Test Sets

The two environmental scenarios we test are :

- Radars organized in a 5 × 5 grid pattern totaling up to 25 radars. We expect a GA to evolve a relatively distributed placement of sensors in this environment. Placing sensors within the area covered by radars is expected to be most efficient; however, the rightmost column and the bottom row of the environment are left empty to see if the GA places sensors outside the perimeter of the radars.
- Radars clustered at random locations, 12 radars in three clusters. We expect sensors to be clustered near radar clusters. The locations that have no enemy radars are expected to have no sensors.

The two general types of sensors we test are:

- Long range sensors that are able to cover the entire environment. As any sensor should be able to detect all radars, we expect the GA to evolve solutions that are close to three sensors, the minimum number of sensors required to fully detect a radar.
- Short range sensors that cover less than a quarter of the environment. We expect solutions to consist of more sensors as short range sensors cannot sense the entire environment.

Using the test cases explained above, we test four different scenarios for in our experiments:

1. Long range sensors with radars placed evenly in a grid pattern.
2. Longe range sensors with radars placed randomly in three clusters
3. Short range sensors with radars placed evenly in a grid pattern.
4. Short range sensors with radars placed randomly in three clusters

All enemy radars are set to operate on a single band, A. We selected the following GA parameter settings based on the performance of previous experiments:

Population size	: 200, initialized randomly
Initial length	: 20
Parent Selection	: Tournament, size:2
Crossover type	: one-point
Crossover rate	: 0.7
Mutation rate	: 0.01 (per gene)
Max number of generations	: 400
Number of runs	: 100

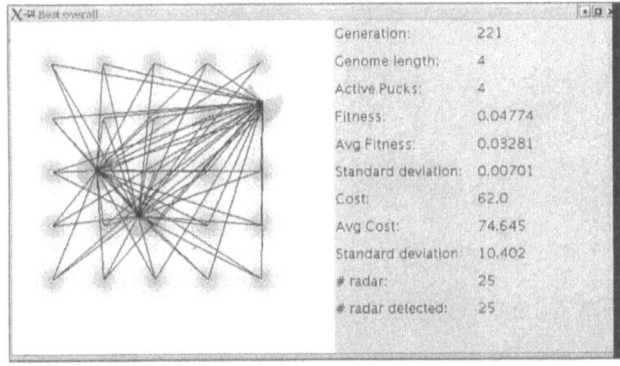

Fig. 4. Long range sensors with enemy radars located in a grid for a single run

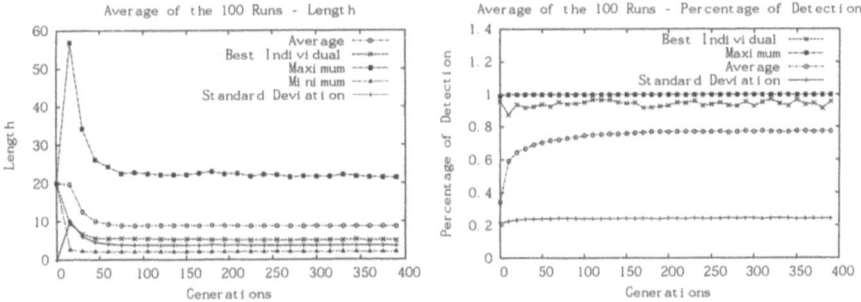

Fig. 5. Length (Number of Sensors) and percentage of detection averaged over 100 runs for long range sensors and grid placement scheme for enemy radars

5.2 Test Results

Figure 4 shows an example solution from our first scenario. In this experiment, sensors are able to extend their coverage area to include the whole environment. We observe that the GA is able to evolve a team of four sensors which is close the minimum requirement of three to fully detect all enemy radars. There are no undetected radars. 24 out of 25 enemy radars are *fully* detected . Only one enemy radar located at the top right corner is detected by two sensors. The results we obtain over 100 runs are consistent with the example solution. Figure 5 shows the evolution of individual length, i.e. number of sensors evolved, and percentage of radar detection averaged over 100 runs. The number of sensors evolved in the best individuals levels off around four, and the percentage of radar detection in best individual is typically 90% or higher. The second scenario uses long range sensors in a clustered environment. Our best solutions are again teams of four sensors as shown in Figure 6. All radars except for one are fully detected; the rest of the radars are partially detected. The average behavior of the GA over 100 runs as shown in Figure 7 is similar to that shown in Figure 5 suggesting that the behavior of this GA, in terms of the number of evolved sensors and the

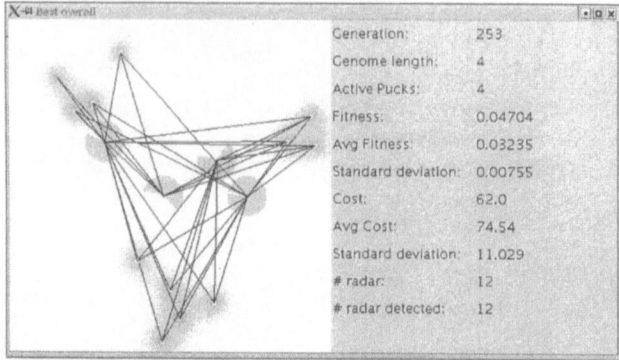

Fig. 6. Long range sensors with enemy radars located randomly in 3 clusters for a single run

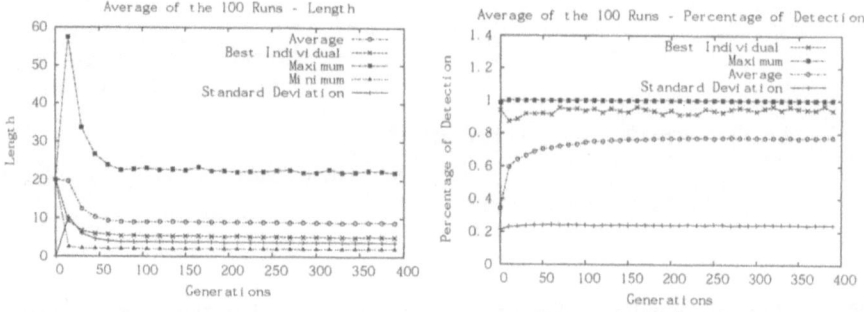

Fig. 7. Length (Number of sensors) and percentage of detection averaged over 100 runs for long range sensors and cluster placement scheme for enemy radars

percentage of detection, is independent of the placement of the enemy radars. This result is expected as long range sensors should have the same capability regardless of the enemy radar configuration. In general, sensors tend to cluster around the center of the environment in order to detect all radars most efficiently.

The third scenario employs sensors that can sense a maximum of one quarter of the environment. Given the distributed radars and assuming the need for at least three sensors for each quarter of the environment, we anticipate solutions to contain a minimum of twelve sensors. The GA is able to evolve a team of fourteen sensors as shown in the example solution in Figure 8. All enemy radars are detected by at least one sensor. The fourth scenario employs short range sensors in a clustered environment. We expect each cluster to require at least three sensors for full detection. Figure 10 shows that the GA is successful at evolving a minimum number of sensors of nine. There are no undetected radars. The number of radars detected by one, two and three sensors are one, seven and four respectively. Because of their decreased range capacity, all of the sensors can only detect radars within the one of the clusters.

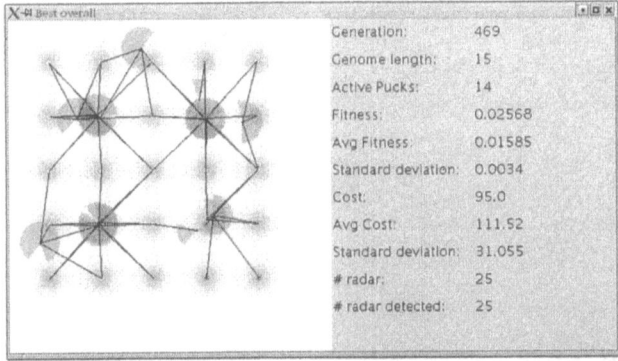

Fig. 8. Short range sensors with enemy radars located in a grid for a single run

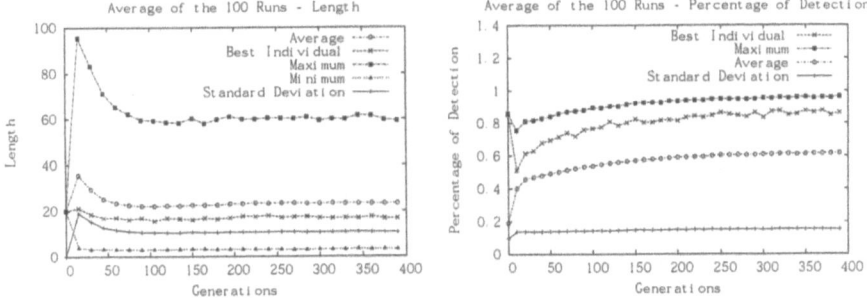

Fig. 9. Length (Number of sensors) and percentage of detection averaged over 100 runs for short range sensors and grid placement scheme for enemy radars

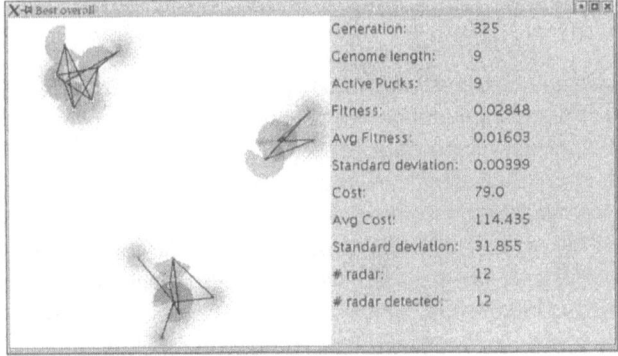

Fig. 10. Short range sensors with enemy radars located randomly in 3 clusters

The results we obtain over 100 runs are also consistent with the single run results. The individual length, i.e. the number of sensors evolved and the percentage of detection averaged over 100 runs are reported in Figure 9 and Figure

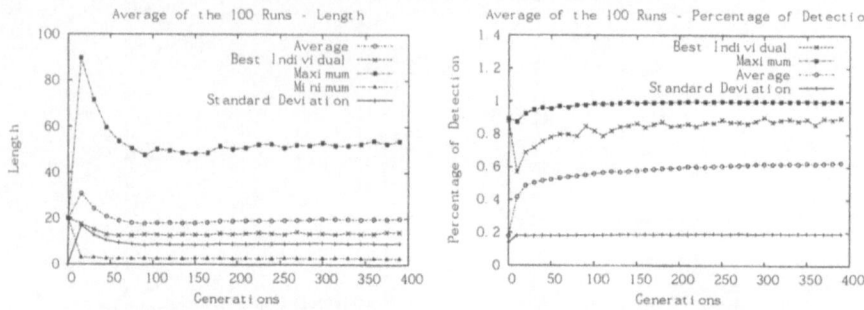

Fig. 11. Length and percentage of detection averaged over 100 runs for short range sensors and cluster placement scheme for enemy radars

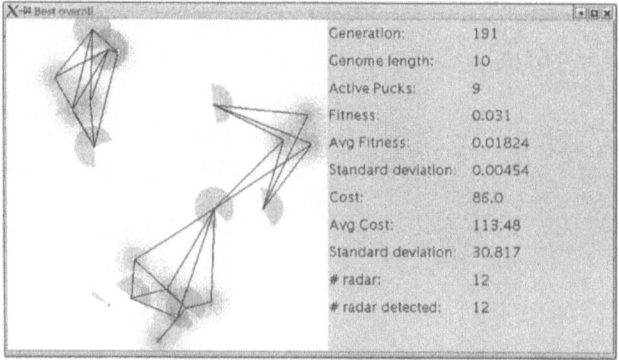

Fig. 12. Medium range sensors with enemy radars located randomly in 3 clusters

11, for grid and the clustered placement of the enemy radars respectively. GA is able to detect the enemy radars with same performance for both placement scheme with a detection percentage around 90%. However the number of sensors evolved in best individuals levels off around sixteen for grid placement while it's near fourteen for the clustered placement which makes short range sensors more sensitive to the changes in the environment than the long range sensors. Due to limited capability of the sensors, we observe a considerable increase in the number of sensors evolved compared to the scenarios with long range sensors.

We explored additional varying sensor ranges and include an example solution using medium range sensors which can potentially cover approximately 75% of the working environment. This solution, shown in Figure 12, includes sensors that are mainly focused within clusters, with a few that are able to reach multiple clusters.

6 Conclusion

In this study, we examine the problem of designing teams of sensors to detect enemy radars located in different configurations. We extend the basic GA design to enable evolving as much sensor characteristics as possible. Based on the detection criteria we have, the results we obtain are promising. Our GA is able to evolve close to the optimal number of sensors based on the capabilities of the sensors. Our system is demonstrated to be robust to the changes in the placement of the enemy radars. The detection percentage is around 90% with the limited sensing capability of the sensors and around 100% with the sensors that have full coverage capability independent from the enemy radar placement. Future goals include testing the robustness and the performance of our GA on more realistic scenarios such as ones with moving radars. In addition we plan to investigate the adaptibility of our system to changing environments, to introduce more problem specific genetic operators and to investigate the identification of the building blocks that form effective subteams.

Acknowledgements. The study we reported in this paper is sponsored by the Naval Research Laboratory, ITT Industries Inc., and partially supported by NSF. We would like to thank the reviewers for their valuable comments.

References

1. Bentley, P.J.: Generic evolutionary design of solid objects using a genetic algorithm. In: Ph.D. thesis, Division of Computing and Control Systems, School of Engineering, University of Huddersfield. (1996)
2. Funes, P., Pollack, J.: Computer evolution of buildable objects. In: Procs. of fourth European Conference on AI. (1997)
3. Husbands, P., Jermy, G., McIlhagga, M., Ives, R.: Two applications of genetic algorithms to component design. In: Workshop on EC. (1996)
4. Lee, W., Hallam, J., Lund, H.: A hybrid gp/ga approach for co-evolving controllers and robot bodies to achieve fitness-specified tasks. In: Procs of IEEE third International Conference on Evolutionary Computation. (1996)
5. Bugajska, M., Schultz, A.: Co-evolution of form and function in the design of autonomous agents: Micro air vehicle project. In: GECCO-2000 Workshop on Evolution of Sensors in Nature, Hardware and Simulation, Las Vegas, NV (2000)
6. Bugajska, M., Schultz, A.: Co-evolution of form and function in the design of micro air vehicles. In: NASA/DoD Conference on Evolvable HW. (2002)
7. Gaskell, A., Probert, P.: Sensor models and a framework for sensor management. In: Sensor Fusion VI. Proceedings of SPIE. Volume 2059. (1993)
8. Popoli, R.: The sensor management imperative. In: Chapter in Multitarget-Multisensor Tracking: Applications and Advances. Volume II. (1992)
9. Schmaedeke, W., Kastella, K.: Information based sensor management and immkf. In: Signal and data processing of small targets 1998: proceedings of the SPIE. Volume 3373., Orlando, FL (1998)

Optimization of Spare Capacity in Survivable WDM Networks[1]

H.W. Chong and Sam Kwong

Department of Computer Science
City University of Hong Kong, 83 Tatchee Avenue, Hong Kong
Chonghw@cs.city.edu.hk, Cssamk@cityu.edu.hk

Abstract. A network with restoration capability requires spare capacity to be used in the case of failure. Optimization of spare capacity is to find the minimum amount of spare capacity for the network to survive from network component failures. In this paper, this problem is investigated for wavelength division multiplexing (WDM) mesh networks without wavelength conversion. We propose a hybrid genetic algorithm approach (GA) for the problem. Simulated Annealing (SA) and Tabu Search (TS) are also applied to this problem for comparison purpose. Simulation results show very favorable results for the Genetic Algorithm approach.

1 The Proposed Algorithm

In this paper, we are primarily concerned with the restoration at the optical layer. The discussion is limited to wavelength-continuity optical WDM mesh network with static traffic. Given a network topology, a traffic demand consisting of the connections to be established, and set of alternate routes for each connection request, Our objective is to find sets of primary lightpaths and backup lightpaths such that the sum of working and backup capacity usage can be minimized while a set of customer traffic demands can still be satisfied and the traffic is 100% restorable under the constraint to be immune against a single link failure. Our algorithms will use path-based restoration method and backup multiplexing to improve the channel utilization.

To minimize the spare capacity, we will optimize both routing and wavelength assignment, this combinatorial problem is usually called Routing and Wavelength Assignment (*RWA*) problem and it is known to be NP-complete. For the routing problem, we employ the alternate routing approach. For each connection request, a set of candidate routes is precomputed off-line, our algorithms will select the primary routes and backup routes that can optimize the total capacity usage. For the wavelength assignment problem, we reduce it to a graph-coloring problem and the nodes are assigned colors in the order found by our algorithms. The *RWA* problem is divided into two sub-problems in this paper but our proposed GA will solve these two these Sub-problems simultaneously.

Figure 1 shows the structure of a chromosome, the chromosome consists of three parts.

---

[1] This work is supported by City University Strategic Grant 7001337.

Part one is a set of working routes for connection requests, part two is a set of backup routes for connection requests and part three is the wavelength assignment order for each route. Since the structure of the chromosome is divided into three parts and each part codes different sub-problem, we design different type of crossover operation for different part of chromosome. One-point crossover and two-point crossover operations are applied to part 1 and part 2 of the chromosome, after that, uniform order-based crossover operation is applied to part 3 of the chromosome. Similar to crossover operation, we design different mutation operator for different part of chromosome structure. For the mutation on part 1 and 2 of chromosome, we need to select a working route and its corresponding backup route before mutation is implemented, and then a new working route and backup route are found to replace the current routes in the selected chromosome. After performing mutation on part 1 and 2 of chromosome structure, we will apply scramble sublist mutation [1] to the third part of chromosome structure.

2 Experimental Results

We conduct experiment to study the performance of proposed GA. For benchmarking purpose, SA and TS are also applied to the problem as well. The three approaches are applied on the network shown in figure 2. A performance comparison between GA, SA and TS is shown in the chart. From the result, we can find that GA approach clearly outperforms SA and TS approaches. Also, TS approach is the poorest of the three. Therefore, GA approach is recommended to be the best choice for solving the spare capacity allocation problem arising in the design of survivable WDM mesh network with static traffic.

Fig. 1. Structure of a chromosome

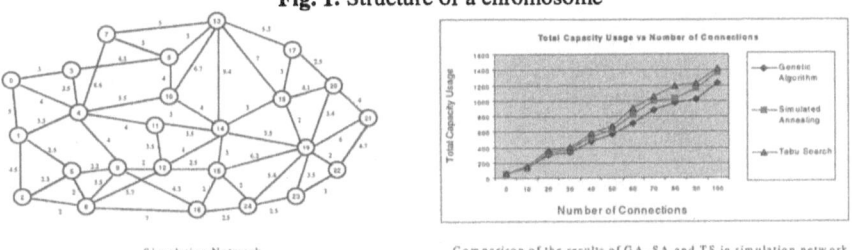

Fig. 2. Simulation network and results

Reference

1. Davis, L.: Handbook of genetic algorithms. 1991, New York: Van Nostrand Reinhold

Partner Selection in Virtual Enterprises by Using Ant Colony Optimization in Combination with the Analytical Hierarchy Process

Marco Fischer[1], Hendrik Jähn[1], and Tobias Teich[2]

[1] Chemnitz University of Technology, 09107 Chemnitz, Germany
marco.fischer@wirtschaft.tu-chemnitz.de
http://www.tu-chemnitz.de/wirtschaft/bwl7/
[2] The University of West Saxony at Zwickau, PSF 201037
08012 Zwickau, Germany

1 Summary

Within the scope of the collaborative research centre "Non-hierarchical regional production networks" at Chemnitz University of Technology, a virtual enterprise model is developed. This is based on very small performance units - the competence cells (CCs). The precondition of the efficient and thereby competitive operation of those networks is a network controlling for objectively selecting the best suitable CCs for every order. This problem gains complexity because the manufacture of a product can be carried through in different ways.

The central task within the network consists in guaranteeing the best possible realization of an order. First of all, a customer offer has to be generated to the competence cell network and the necessary CCs for the processing of the several partial performances within an order have to be selected. Thereby, a difficult economical decision problem arises for the network management. An *elemen-*

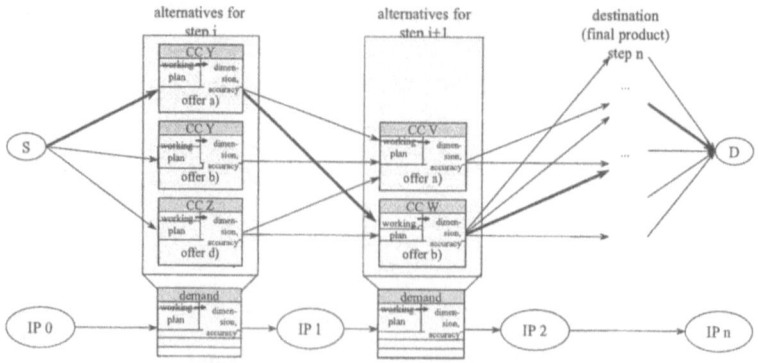

Fig. 1. Illustration of the Problem

tarized process variant plan are described exactly by the demand vectors (DV). Those define the necessary process plans in order to complete an intermediate product. According to the according DVs, the corresponding CCs are searched for all elements, which are potentially able to complete the performances. This means, the offer vectors (OVs) of the CCs have to be equal to the DVs to a special degree. The OV has to express the possibilities of a CC as exactly as possible and to make those comparable in order to make possible a machine evaluation. For the optimization all manufacturing variants within the CC-offer network

```
1  begin
2    initialization(problem − structure);
3    i = source;
4    while not(exit − condition) do
5        for k := 0 to m step 1 do
6            while (N_i^k ≠ ∅) ∩ (i ≠ destination) do
7                random(z);
8                if z ≤ q then p_{ij}^k(t) = [τ_{ij}(t)]^α · [η_{ij}]^β;
9                else p_{ij}(t) = [τ_{ij}(t)]^α · [η_{ij}]^β / Σ_{l∈N_i^k} [τ_{il}(t)]^α · [η_{il}]^β;
10               fi
11               decide(j); Ψ_k = Ψ_k ∪ CC(max(AHP(j)));
12               /* localPheromonupdate */
13               τ_{ij}(t+1) = (1 − ρ') · τ_{ij}(t) + ρ' · Δτ_{ij};  i := j
14           od
15       od
16       /* globalPheromonupdate */
17       τ_{ij}(t+1) ← (1 − ρ) · τ_{ij}(t) + Δτ_{ij}(t) ∀τ_{ij};
18       max − min − rule τ_{ij};
19       decide(L_k);
20   od
21   Ψ_k ∈ M :  ∀ Ψ_k mit L_k > κ · L_k*   0 ≤ κ ≤ 1;
22   compute(E_{Ψ_k,CC}  ∀ CC ∈ Ψ_k,  K_{Ψ_k} ∀Ψ_k ∈ M);
23   decide(Ψ_k^{max} : max(aggregation(MK_k, SK_k)))
24 end
```

Fig. 2. Procedure of the Search for an optimal Manufacturing Variant

are illustrated in a *digraph*. Each manufacturing alternative includes a source, subsequently it disposes of CCs for carrying through the partial performances and finally it ends in a drain. The objective function of the algorithm is the maximization of the combined AHP-values of the CCs. The larger the value of a manufacturing variant, the more advantageous it is to realize that variant. Figure 1 illustrates the modelling as a digraph. For every step of processing along the way to the final product, several manufacturing variants exist, out of whom the best has to be selected by the help of an algorithm illustrated in Figure 2.

References

1. Teich, T., Fischer, M., Vogel, A., Fischer, J.: A new Ant Colony Algorithm for the Job Shop Scheduling Problem. In: Proceedings of the Genetic and Evolutionary Computation Conference, San Francisco, California (2001) 803

Quadrilateral Mesh Smoothing Using a Steady State Genetic Algorithm

Mike Holder[1] and Charles L. Karr[2]

[1] American Buildings Company, 1150 State Docks Road, Eufaula, AL 36027-3344
[2] Aerospace Engineering and Mechanics, The University of Alabama
Tuscaloosa, AL 35487-0280

Abstract. This paper investigates the use of a steady state genetic algorithm (GA) to perform quadrilateral finite element mesh smoothing. GAS short for genetic algorithm smoother moves one to 64 nodes at the same time. GAs smooth as well as untangle (removing twisted and inverted elements), which has been a separate operation in the past.

1 Introduction

A common use of finite elements analysis (FEA) is to determine displacements and stresses of a structure. Automeshers fill the model with finite elements establishing connectivity between the element nodes; it is the objective of mesh smoothers to insure the quality of the mesh once connectivity has been established. Poorly shaped finite elements result in incorrect results. This paper compares a GA smoothing tool with the Laplace smoothing method.

2 Finite Element Smoothing with a GA

FlexGA, the genetic algorithm engine used in this paper, is written in MATLAB and was developed by Dr. K. K. Kumar, The Flexible Intelligence Group, Tuscaloosa, AL. The objective function created by Alan Oddy, at Carleton University in Ottawa, Canada was written in C and interfaced with MATLAB using CMEX, the MATLAB/C interface code [1].

The GA works on a family of chromosomes that in this problem represents the X/Y coordinates of each element node. The distortion metric was augmented with penalty functions. There are penalty functions that check for quadrilateral interior angles being equal to 90 degrees; aspect ratios equal to one and twisted or inverted element. The angular penalty function is

$$pf = \sum_{i=1}^{4} \frac{(\varphi_i - 90)^2}{SF}$$

GAS-N moves all of the element nodes at a time. An upper limit of about 64 nodes was run using GAS-N. A rectangle with a cutout was created to fool the Laplace smoother to see how the GA would perform using movable boundary nodes. (See Figure (a)).

3 Results

For simple meshes, the Laplace smoother outperformed the GAS-N smoother when both accuracy and computational time were considered. However, GAS-N usually outperformed the Laplace method in terms of accuracy when complex geometries were encountered. An example can be seen in the rectangle shown in Figure (a) with a semicircular cutout. Figure (b) presents this mesh smoothed with the Laplace method (maximum distortion of 84.3) with Figure 1 giving the mesh smoothed with GAS-N using movable boundary nodes (maximum distortion of 1.9).

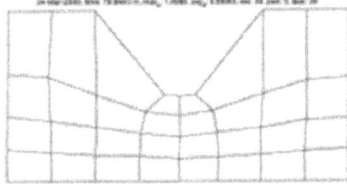
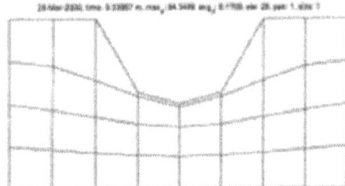

(a) GAS-smoothed with movable boundary nodes

(b) Laplace-smoothed

Using a series of square "checker board" models ranging in size from and 8x8 mesh to a 3x3 mesh, it was determined that 64 X/Y node coordinates can be successfully manipulated at a time. A more reasonable upper limit is 35 active nodes. It was also determined that 16 movable nodes is the controllable limit of GAS-N without the use of feasible circles to limit the individual nodal search areas.

In conclusion, GAS-N's ability to move small groups of nodes simultaneously during the smoothing process makes it potentially an effective tool for smoothing large meshes.

References

1. Holder, E. M., (2001). Quadrilateral Mesh Smoothing Using a Genetic Algorithm, Doctoral Dissertation, University of Alabama.

Evolutionary Algorithms for Two Problems from the Calculus of Variations

Bryant A. Julstrom

Department of Computer Science
St. Cloud State University, St. Cloud, MN, 56301 USA
julstrom@eeyore.stcloudstate.edu

Abstract. A brachistochrone is the path along which a weighted particle falls most quickly from one point to another, and a catenary is the smooth curve connecting two points whose surface of revolution has minimum area. Two evolutionary algorithms find piecewise linear curves that closely approximate brachistochrones and catenaries.

Two classic problems in the calculus of variations seek a brachistochrone, the path along which a weighted particle falls most quickly from one point to another, and a catenary, the smooth curve of arc length l between two points whose surface of revolution has minimum area. Analytical solutions to these problems have long been known. In a uniform gravitational field and without friction, a brachistochrone is an arc of a cycloid, the curve traced by a point on a rolling circle. The curve of specified length whose surface of revolution has minimum area is a catenary, an arc of a hyperbolic cosine.

Two evolutionary algorithms seek approximate solutions to these problems. They search spaces of piecewise linear functions, which they represent as sequences of y-coordinates associated with evenly-spaced x's. The algorithms apply mutation and $(\mu+\mu)$ reproduction to the chromosomes in their populations. The mutation operator perturbs elements of parent chromosomes with random values from a normal distribution with mean zero, and the standard deviation of this distribution diminishes as each algorithm runs.

Previously, Zitar and Homaifar [2] described a genetic algorithm for the brachistochrone problem, and Erickson, Killmer, and Lechner [1] applied genetic programming to it.

In the EA for the brachistochrone problem, a chromosome's fitness, which the EA seeks to minimize, is the time required for the particle to fall along the path that the chromosome represents. Consider the particle as it passes through a point $P = (x, y)$ on a curve. The particle was initially at rest, and its potential energy relative to P was mgy, where m is the particle's mass and g is the acceleration due to gravity. At P, the particle is moving with velocity v and its kinetic energy is $\frac{1}{2}mv^2$. Energy is conserved, so $mgy = \frac{1}{2}mv^2$. Thus the particle's velocity at P is $v = \sqrt{2gy}$.

The length of the segment from (x_i, y_i) to (x_{i+1}, y_{i+1}) is $d_i = \sqrt{(x_{i+1} - x_i)^2 + (y_{i+1} - y_i)^2}$, and the particle's average velocity over the segment is the average of its velocities at the segment's endpoints: $v_i = (\sqrt{2gy_i} +$

$\sqrt{2gy_{i+1}}/2$. The time the particle spends traversing the segment is then $t_i = \frac{d_i}{v_i}$, and the time that the particle requires to fall the entire distance on the path a chromosome specifies is the sum of these values over all the subintervals.

In the EA for the catenary problem, we seek to minimize both the area of the surface generated by rotating the represented curve about the x-axis and the difference between the length of that curve and a target length l. The segment connecting (x_i, y_i) and (x_{i+1}, y_{i+1}) sweeps out a frustum of a cone whose surface area is $A_i = 2\pi r_i d_i$, where $r_i = (y_i + y_{i+1})/2$ is the average radius of the frustum and d_i is again the length of the segment. The area of the surface is the sum of the areas of these frusta, and the length of the curve is the sum of the lengths of its segments. A chromosome's fitness is a weighted sum of the former and the deviation of the latter from l:

$$\text{fitness} = w_A \cdot \text{area} + w_d \cdot |\text{length} - l| \ .$$

In tests of the two algorithms, the number of linear segments was $n = 30$. The EAs' populations contained $\mu = 50$ chromosomes, each of $n+1 = 31$ values. The initial standard deviation of the mutating distribution was 1.0, and it was multiplied by 0.998 after each generation. Each algorithm ran through 2 500 generations. In the EA for the catenary problem, $w_A = 1$ and $w_d = 25$.

Both EAs were effective on a variety of test instances. Figure 1(a) shows a curve of rapid descent generated by the first algorithm; it is indistinguishable from the optimum arc of a cycloid. Figure 1(b) shows a curve of small area and specified length generated by the second algorithm; it is indistinguishable from the optimum arc of a hyperbolic cosine.

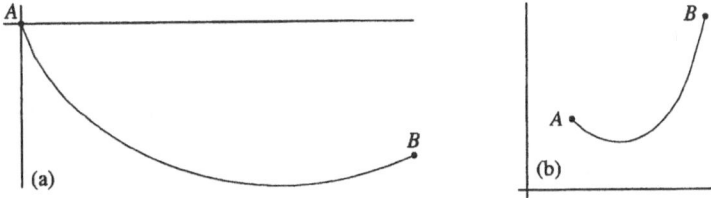

Fig. 1. (a) An approximate brachistochrone generated by the first EA and (b) an approximate catenary generated by the second EA

References

1. Damon Erickson, Charles Killmer, and Alicia Lechner. Solving the brachistochrone problem with genetic programming. In H. R. Arabnia and Youngsong Mun, editors, *Proceedings of the International Conference on Artificial Intelligence: IC-AI'02*, volume III, pages 860–864. CSREA Press, 2002.
2. R. A. Abu Zitar and Abdullah Homaifar. The genetic algorithm as an alternative method for optimizing the brachistochrone problem. In *Proceedings of the International Conference on Control and Modeling*, Rome, 1990.

Genetic Algorithm Frequency Domain Optimization of an Anti-Resonant Electromechanical Controller

Charles L. Karr[1] and Douglas A. Scott[2]

[1] Associate Professor and Head, Aerospace Engineering and Mechanics
The University of Alabama, Box 870280, Tuscaloosa, AL, 35487-0280
[2] Graduate Student, Mechanical Engineering, Purdue University, 1288 ME
West Lafayett, IN, 47907-1288

Abstract. The new standard for actuators in the aerospace industry is electromechanical actuators. This paper describes an approach whereby PID and DFF controllers are used in unison to effectively manipulate a thrust vectoring system.

1 Introduction and Background

Thrust vector control (TVC) is a prime example of an aerospace system in which electromechanical actuators (EMAs) can potentially be effective. Researchers at NASA's Marshall Space Flight Center are currently developing EMAs to replace the hydraulic servo actuators for use on space vehicles with TVC. The current paper focuses on the control of EMAs in TVC applications and on the optimization of such control systems with genetic algorithms (GAs).

2 Problem Environment

Thrust vector control involves the manipulation of engine thrust direction to achieve attitude adjustment of a vehicle. Figure (a) illustrates the placement and function of the actuators for TVC. The objective of the current effort is to develop an EMA controller that can provide accurate engine position control (high bandwidth) while at the same time rejecting the disturbance forces encountered during startup and shut down.

Scott and Karr [1] presented a control system for manipulating the TVC that incorporated the PID and DFF controller architectures. This control architecture is presented in Figure (b).

3 Results

In this problem, a GA is asked to determine values for five control gains used in the control system. Each parameter was represented with an eight-bit substring. A fitness function was defined that considered (1) the magnitude of the

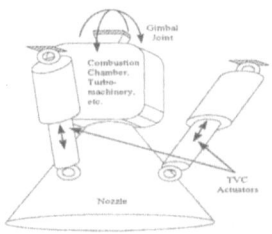
(a) TVC operation

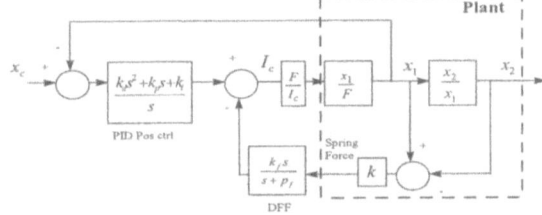
(b) Block Diagram representation of total system

frequency response, (2) the number of peaks in the frequency response, and (3) the rate at which the frequency response decreases.

Figures (c) and (d) compare the performance of the GA-designed controller to that of the previously best-known Bode controller. This figure shows that the GA controller produces a smaller phase lag in the engine closed loop frequency response for higher frequency ranges when compared closed loop system incorporating the Bode controller. This accentuates the tracking ability of the GA controller. The GA design produces a higher bandwidth than the Bode controller. Additionally, by reducing the resonant peak in the Bode controlled system, the GA controller provides a higher overall phase margin.

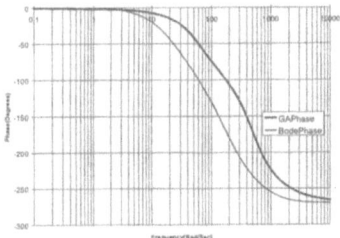
(c) Phase of the Engine frequency response

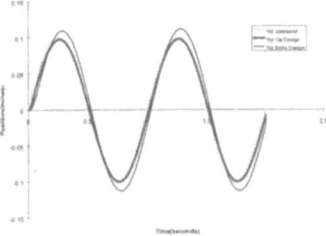
(d) Engine response due to a .1in 1hz sine input

References

1. Scott, D. A., Karr, C. L., & Schinstock, D. E. (2003). Genetic algorithm frequency domain optimization of an anti-resonant electromechanical controller. *Engineering Applications of Artificial Intelligence*, **12**, 201–211.

Genetic Algorithm Optimization of a Filament Winding Process

Charles L. Karr[1], Eric Wilson[2], and Sherri Messimer[3]

[1] Associate Professor and Head, Aerospace Engineering and Mechanics
The University of Alabama, Box 870280, Tuscaloosa, AL, 35487-0280
[2] Research Engineer, Delta Research, 315 Wynn Dr, Suite 1, Huntsville, AL 35805
[3] Associate Professor, Industrial Engineering Dept.
University of Alabama in Huntsville, Huntsville, AL 35899

Abstract. This paper describes a research effort to improve the efficiency of a filament winding process by combining the simulation capabilities of the WITNESS modeling program with the search capabilities of a genetic algorithm. Results show that the genetic algorithm is able to reduce the cost of producing filament wound mandrels used in the defense industry.

1 Introduction

This paper describes an effort to use a genetic algorithm to enhance the modeling and optimization capabilities in a specific simulation environment, WITNESS – a well-known simulation environment designed by AT&T and Istel that is generally geared toward industrial engineering applications.

The current work represents an expansion of an earlier work completed by the authors [2]. This previous effort focused on a very simple optimization problem in which only the production costs of a filament winding operation were considered. In the current work, the total cost (production cost, materials cost, labor costs, etc.) are considered.

2 Filament Winding Model

The filament winding process is an assembly process used by the advanced composites industry to make composite material filaments. It involves taking composite fibers, winding them around mandrels, hardening them, and then cutting and assembling them into a finished product. Products produced using the filament winding process include rocket-motor cases, helicopter blades, piping, tubing, and drive shafts. For the purpose of this research, the same filament winding model formerly used by the authors [2] is once again addressed. The objective of the current effort is to use a genetic algorithm to minimize the cost of producing a composite mandrel. The production process is simulated in the WITNESS environment. Figure (a) shows a schematic of the filament winding model, along with the means by which the genetic algorithm interacts with the filament winding model.

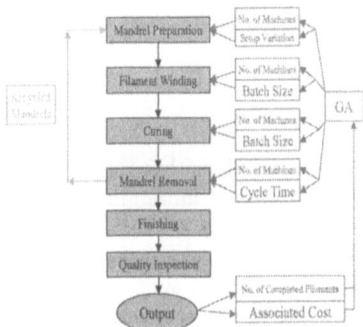

(a) Filament Winding Model

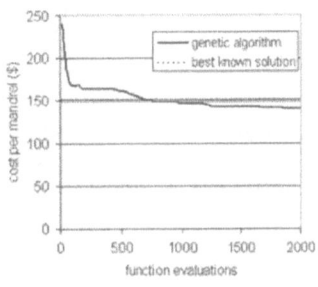

(b) Genetic Algorithm Performance

3 Results

The optimization problem involved determining two parameters for each workstation. Thus, the problem solved is actually a 10 parameter optimization problem. Minimizing the total cost of mandrel production involved material costs, labor costs, operation costs, and market value of the product. In addition, penalties were incurred if the products were not delivered on time. The exact objective function used here can be found in the thesis by Wilson [1].

For the purpose of the current study, the best solution determined by the genetic algorithm ($140.70) was compared to the best know solution as determined by process engineers addressing the problem ($152.10). Figure (b) indicates that the genetic algorithm clearly located a more cost-effective layout for the filament winding process.

References

1. E. Wilson (2000). "Genetic Algorithm Optimization of Assembly Lines Modeled in the WITNESS Simulation Environment." MS thesis, Tuscaloosa, AL: The University of Alabama.
2. E. Wilson, C.L. Karr, C.L., & S. Messimer (2001). Genetic algorithm optimization of a filament winding process modeled in WITNESS. In A. De Wilde and L. C. Jain (Eds.), *Practical applications of soft computing techniques* (pp. 223–240). New York: Kluwer.

Circuit Bipartitioning Using Genetic Algorithm

Jong-Pil Kim and Byung-Ro Moon

School of Computer Science and Engineering
Seoul National University
Shilim-dong, Kwanak-gu, Seoul, 151-742 Korea
{jpkim,moon}@soar.snu.ac.kr

Abstract. In this paper, we propose a hybrid genetic algorithm for partitioning a VLSI circuit graph into two disjoint graphs of minimum cut size. The algorithm includes a local optimization heuristic which is a modification of Fiduccia-Matheses algorithm. Using well-known benchmarks (including ACM/SIGDA benchmarks), the combination of genetic algorithm and the local heuristic outperformed hMetis [3], a representative circuit partitioning algorithm.

The Fiduccian-Matheyses algorithm (FM) [2] is a representative iterative improvement algorithm for a hypergraph partitioning problem. FM improves an initial solution through short-sighted moves based on the gain. Thus, the quality of the FM is not stable. Kim and Moon [4] introduced lock gain as a primary measure for choosing the node to move. It uses the history of search more efficiently. Lock gain showed excellent performance for general graphs. We adapt the lock gain for hypergraphs within the framework of FM.

To apply the lock gain to the hypergraph bisection, considerable modification is necessary for the lock gain calculation method, because lock gain was originally designed for the general graphs [4]. We propose a new lock gain calculation method for the hypergraph bisection. Let's define $l_e(v)$ to be the lock gain of a node v due to the net e. $l_e(v)$ is obtained as the following. We assume that the node v is on the left side without loss of generality.

Positive lock gain　　　　　　　　Negative lock gain

If all nodes on the right side are locked and there is no locked node on the left side (Case 1) or if there exists locked nodes on the right side and there is no free node on the left side (Case 2) then $l_e(v) = 1$. On the other hand, if all nodes are on the left side and at least one node is locked (Case 3) or if all nodes on the left side except v are locked and there is no locked node on the right side

Table 1. Bipartition Cut Sizes of GA and hMetis

Circuits	GA			hMetis200		
	Average[1]	CPU[2]	Best	Average[3]	CPU[2]	Best
Test02	**88.16**	17.20	88	89.81	12.12	88
Test03	**58.00**	5.59	58	59.90	11.74	58
Test04	**51.15**	8.90	51	54.77	10.02	54
Test05	72.05	27.85	71	**71.33**	19.22	71
Test06	**63.12**	4.89	63	64.05	12.30	64
Prim1	53.87	1.14	53	**53.00**	6.33	53
Prim2	**149.02**	19.45	146	177.76	34.39	146
19ks	**110.22**	21.44	110	110.81	25.89	110
Industry2	186.97	211.25	183	**181.15**	233.35	180

1. The average cut size of 100 runs
2. CPU seconds on Pentium III 1GHz
3. The average cut size of 100 runs, each of which is the best of 200 runs of hMetis

(Case 4) then $l_e(v) = -1$. Then, the lock gain $l(v)$ of the node v is defined to be $l(v) = \sum_{e \in N(v)} l_e(v)$ where $N(v)$ is a set of nets to which the node v is connected.

We tested the proposed algorithm on 9 benchmarks including ACM/SIGDA benchmarks [1]. We compare the performance of the hybrid GA which uses the lock-gain based FM as a local optimization engine against a well-known partitioner hMetis [3]. Because the GA took roughly 200 times more than a single run of hMetis, it is not clear how critical the genetic search is to the performance improvement. Thus, hMetis200, that is a multi-start version of hMetis with 200 runs, was compared.

Table 1 shows the performance of the GA. On the average, the proposed GA performed best in six graphs among nine.

Acknowledgments. This work was partly supported by Optus Inc. and Brain Korea 21 Project. The RIACT at Seoul National University provided research facilities for this study.

References

1. BENCHMARK. http://vlsicad.cs.ucla.edu/~cheese/benchmarks.html.
2. C. Fiduccia and R. Mattheyses. A linear time heuristics for improving network partitions. In *19th IEEE/ACM Design Automation Conference*, pages 175–181, 1982.
3. G. Karypis, R. Aggarwal, V. Kumar, and S. Shekhar. Multilevel hypergraph partitioning: Application in VLSI domain. In *Proc. Design Automation Conference*, pages 526–529, 1997.
4. Y. H. Kim and B. R. Moon. A hybrid genetic search for graph partitioning based on lock gain. In *Genetic and Evolutionary Computation Conference*, pages 167–174, 2000.

Multi-campaign Assignment Problem and Optimizing Lagrange Multipliers*

Yong-Hyuk Kim and Byung-Ro Moon

School of Computer Science & Engineering, Seoul National University
Shillim-dong, Kwanak-gu, Seoul, 151-742 Korea
{yhdfly,moon}@soar.snu.ac.kr

Customer relationship management is crucial in acquiring and maintaining royal customers. To maximize revenue and customer satisfaction, companies try to provide personalized services for customers. A representative effort is one-to-one marketing. The fast development of Internet and mobile communication boosts up the market of one-to-one marketing. A personalized campaign targets the most attractive customers with respect to the subject of the campaign. So it is important to expect customer preferences for campaigns. Collaborative Filtering (CF) and various data mining techniques are used to expect customer preferences for campaigns. Especially, since CF is fast and simple, it is widely used for personalization in e-commerce. There have been a number of customer-preference estimation methods based on CF. As personalized campaigns are frequently performed, several campaigns often happen to run simultaneously. It is often the case that an attractive customer for a specific campaign tends to be attractive for other campaigns. If we perform separate campaigns without considering this problem, some customers may be bombarded by a considerable number of campaigns. We call this overlapped recommendation problem. The larger the number of recommendations for a customer, the lower the customer interest for campaigns. In the long run, the customer response for campaigns drops. It lowers the marketing efficiency as well as customer satisfaction. Unfortunately, traditional methods only focused on the effectiveness of a single campaign and did not consider the problem with respect to the overlapped recommendations. In this paper, we define the multi-campaign assignment problem (MCAP) considering the overlapped recommendation problem and propose a number of methods for the issue including a genetic approach. We also verify the effectiveness of the proposed methods with field data.

Let N be the number of customers and K be the number of campaigns. The MCAP is to find customer-campaign assignments that maximizes the effects of campaigns. The main difference with independent campaigns lies in that the customer response for campaigns is influenced by overlapped recommendations. More detailed description is omitted by space limit. In case of overlapped campaign recommendations, the customer response rate drops as the number of recommendations grows. We introduce the response suppression function for

---

* This work was partly supported by Optus Inc. and Brain Korea 21 Project. The RIACT at Seoul National University provided research facilities for this study.

Table 1. Comparison of Algorithms

Method	Independent	Random	CAA[†]	LM-GA
Fitness	26452.75	25951.19	66473.63	**66980.76**

[†] Starting at the situation that no customer is recommended any campaigns, we iteratively assign campaigns to customers by a greedy method. We call this algorithm Constructive Assignment Algorithm (CAA). We use an AVL tree for the efficient management of real-valued gains. The time complexity of the algorithm is $O(NK^2 log N)$.

the response-rate degradation with overlapped recommendations. We used the response suppression function derived from Gaussian function.

Since the MCAP is a constraint optimization problem, it can be solved using Lagrange Multipliers (LMs). But, since it is a discrete problem which is not differentiable, it is formulated to a restricted form. The LM method guarantees optimal solutions. The suboptimality of heuristic algorithms can be measured by using the optimality of the LM method. By using LM method, the problem of finding the optimum campaign assignment matrix becomes that of finding a K-dimensional real vector with LMs. The LM method takes $O(NK2^K)$ time. It is more tractable than the original problem. Roughly, for a fixed number K, the problem size is lowered from $O(N^{K+1})$ to $O(N)$. We also propose a genetic algorithm (GA) for optimizing LMs. Our GA provides an alternative search method to find a good campaign assignment matrix by optimizing K LMs instead of directly dealing with the campaign assignment matrix. A typical steady-state genetic algorithm is used in our GA. In the following, we describe each part of the GA. A real encoding is used for representing a solution. A gene corresponding to a LM has a real value. The GA first creates 100 real vectors at random. We use a proportional selection scheme and the uniform crossover. After the crossover, mutation operator is applied to the offspring. We use a variant of Gaussian mutation that we devised. Our evaluation function is to find a LM vector that has high fitness satisfying the constraints as *much* as possible.

We used the preference values estimated by CF from field data with 48,559 customers and 10 campaigns. We examined the Pearson correlation coefficient of preferences for each pair of campaigns. Thirty three pairs (about 73%) out of the totally 45 pairs showed higher correlation coefficient than 0.5. This property of field data provides a good reason for the need of MCAP modeling. Table 1 shows the performance of the independent campaign and various multi-campaign algorithms in the multi-campaign formulation. The figures in the table represent the fitness values. The result of "Independent" campaign is from 10 independent campaigns without considering their relationships with others. Although the independent campaign was better than the "Random" assignment in multi-campaign formulation, it was not comparable to the other multi-campaign algorithms. The solution fitness of CAA heuristic was more than 2.5 times higher than that of the independent campaign. When LMs are optimized by a genetic algorithm (LM-GA), we found the best-quality solution satisfying all the constraints. Our LM method is fast and outputs optimal solutions. But, it is not easy to find LMs satisfying all constraints. When combined with the genetic algorithm, we could find high-quality LMs.

Grammatical Evolution for the Discovery of Petri Net Models of Complex Genetic Systems

Jason H. Moore and Lance W. Hahn

Program in Human Genetics, Vanderbilt University, Nashville, TN, USA 37232-0700
{Moore,Hahn}@phg.mc.Vanderbilt.edu

Abstract. We propose here a grammatical evolution approach for the automatic discovery of Petri net models of biochemical systems that are consistent with population level genetic models of disease susceptibility. We demonstrate the grammatical evolution approach routinely identifies interesting and useful Petri net models in a human-competitive manner. This study opens the door for hierarchical systems modeling of the relationship between genes, biochemistry, and measures of health.

Petri nets are a type of directed graph that can be used to model discrete dynamic systems. The goal of this study was to develop a grammatical evolution (GE) strategy for the automatic discovery of Petri net (PN) models of biochemical systems that are consistent with gene-gene interactions that increase susceptibility to human disease. Understanding the relationship between genes, biochemistry, and measures of health is an important endeavor in the domain of human genetics. We first summarize the PN modeling strategy and then briefly present the grammar used and the results.

The goal of identifying PN models of biochemical systems that are consistent with observed population-level gene-gene interactions is accomplished by developing PN that are dependent on specific genotypes from two (or more) genetic variations. Here, we make firing rates of transitions and/or arc weights genotype-dependent yielding different PN behavior. Each PN model is related to the genetic model using a threshold model. With the threshold model, it is the concentration of a substance that is related to the risk of disease. For each model, the number of tokens at a particular place is recorded and if they exceed a certain threshold, the appropriate risk assignment is made. If the number of tokens does not exceed the threshold, the alternative risk assignment is made. The high-risk and low-risk assignments made by the discrete threshold from the output of the PN can then be compared to the high-risk and low-risk genotypes from the genetic model. A perfect match indicates the PN model is consistent with the gene-gene interactions observed in the genetic model. Here, the fitness function of the GE algorithm is proportional to the number of high-risk and low-risk assignments incorrectly made.

We developed a grammar for PN in Backus-Naur Form (BNF). Nonterminals form the left-hand side of production rules while both terminals and nonterminals can form the right-hand side. A terminal is essentially a model element while a nonterminal is the name of a production rule. For the PN models, the terminal set includes, for example, the basic building blocks of a PN places, arcs, and transitions. The nonterminal set includes the names of production rules that construct the PN. For example, a nonterminal might name a production rule for determining whether an arc has weights that are fixed or genotype-dependent. We show below in (1) the production rule that is executed to begin the model building process.

(1) <root> ::= <pick_a_gene> <pick_a_gene> <net_iterations> <expr> <transition> <place_noarc>

When the initial <root> production rule is executed, a single PN place with no entering or exiting arc (i.e. <place_noarc>) is selected and a transition leading into or out of that place is selected. The arc connecting the transition and place can be dependent on the genotypes of the genes selected by <pick_a_gene>. The nonterminal <expr> is a function that allows the PN to grow. The production rule for <expr> is shown below in (2). Here, the nonterminal (0, 1, 2, or 3) selected in the right-hand side of the production rule is determined by a combination of bits in the genetic algorithm chromosome.

(2) <expr> ::= <expr> <expr>   0
             | <arc>   1
             | <transition>   2
             | <place>   3

The base or minimum PN that is constructed using the <root> production rule consists of a single place, a single transition, and an arc that connects them. Multiple calls to the production rule <expr> by the genetic algorithm chromosome can build any connected PN. In addition, the number of times the PN is to be iterated is selected with the nonterminal <net_iterations>. Many other production rules control the arc weights, the genotype-dependent arcs and transitions, the number of initial tokens in a place, the place capacity, etc. All decisions made in the building of the PN model are made by each subsequent bit or combination of bits in the genetic algorithm chromosome.

The GE algorithm was run a total of 100 times for each of two hypothetical nonlinear gene-gene interaction models. For each genetic model, GE always yielded a PN model that was consistent with the high-risk and low-risk assignments for each combination of genotypes with no classification error. Most PN models consisted of one place, two arcs, and two transitions. We found that there was a clear preference for making arcs, rather than transitions, genotype-dependent. Understanding how interactions at the biochemical level manifest themselves as interactions among genes at the population level, will provide a basis for understanding the role of genes in diseases susceptibility.

This work was supported by National Institutes of Health grants HL65234, HL65962, GM31304, AG19085, and AG20135.

Evaluation of Parameter Sensitivity for Portable Embedded Systems through Evolutionary Techniques

James Northern III and Michael Shanblatt

Department of Electrical and Computer Engineering
Michigan State University, East Lansing, MI 48910
{norther1,mas}@msu.edu

Abstract. Power consumption and portability issues are becoming increasingly significant in embedded system architectures. Therefore, it is important that chip architects and integrated circuit designers focus on power and portability, they must also be conscious of choosing the right set of parameters for an embedded processor. An evolutionary approach to configure the parameters of an embedded processor can be used to address these issues. From a given set of parameters the design space is explored using a simple genetic algorithm. This paper investigates the sensitivity of these parameters as they change during the process of the genetic algorithm.

1 Introduction

Embedded systems have been motivated by the demands of the rapidly growing market in portable electronic devices. The development of lighter products, thermal efficient systems and longer battery life are effects of reduced power consumption in mobile embedded systems (*e.g.*, medical equipment, cellular phones, pagers and video game consoles). Embedded systems usually employ one specific application. To design such a system, various hardware units are configured based on a set of parameters for a particular application. As portable computing systems continue to grow, the focus of development will be the integration of these processors for communication and computation into a single unit (*i.e.*, system-on-a-chip). In this paper, the population size of a genetic algorithm is adapted to find an "ideal" configuration and evaluate the sensitivity of processor parameters. An expansion of Goldberg's Simple Genetic Algorithm is selected as the method for optimizing these configurations based on power consumption.

2 Methodology

For the purpose of this study, the set of experiments includes 16 different parameters in the configuration of an embedded processor, with at least four options per parameter. If searched exhaustively, it would take 8,153,726,970 different configurations to search the entire design space, which could take hundreds of years. Also, if different parameter sets were randomly selected, the impact of each characteristic could possibly have a dynamic effect on power consumption (*e.g.*, cache size and pipeline depth). Based on these parameterized characteristics and a particular embedded application, the search for an "ideal" configuration of an embedded processor is NP-complete. Evolutionary algorithms offer the ability to traverse the design space to find quality solutions from large search spaces. The simple genetic algorithm (SGA) seeks an "ideal" solution for minimum power consumed by a configuration based on natural selection methods. The SGA calls an architectural power estimation tool to evaluate each configuration. Two tools, Genetic Algorithm Optimized for Portability and Parallelism System (GALOPPS) and Wattch, are selected to help complete this task [1, 2]. Each parameter is evaluated for sensitivity by counting the number of mutations of the "best fit" individual in a given run, and all values are sorted based on a total average per parameter.

3 Summary

The experiments that are reported used estimated values without a targeted workload for maximum power consumption per configuration. The population size was varied from 50 to 100. A population size of 70 yielded the best result with a minimal power consumption of 39.97 watts. The GA in its initial form (with a population size of 70) achieved a power reduction of 25% from the initial SoC configuration. Moreover, when population size was varied, additional gains of up to 5% additional power reduction were attained. The comparison of population sizes showed that there is a tradeoff between population size and number of evaluations. Smaller population sizes yielded higher power consumption per number of evaluations, therefore inferring that the population was not large enough. When the population size was greater than 70, the final power consumption values were also higher, concluding that the larger population size may increase the time needed for the SGA to converge to a configuration that consumes minimum power. From our sensitivity evaluation, certain parameters that were less sensitive maintained the same characteristic for the final configuration (*e.g.*, data cache size, integer and floating point ALUs and rate of instructions implemented). Ongoing work involving additional runs will be used to show the statistical significance of power reduced and sensitivity of parameters from using a simple genetic algorithm.

References

1. Goodman, E.,"An Introduction to GALOPPS--The 'Genetic Algorithm Optimized for Portability and Parallelism' System," Technical Report GARAGe 96-07-01, Michigan State University, 1996.
2. Brooks, D., Tiwari, V., Martonosi, M., "Wattch: A Framework for Architectural-Level Power Analysis and Optimizations," Proc. of the 27th International Symposium on Computer Architecture, June 2000, Vancouver, BC, pp. 83–94.

An Evolutionary Algorithm for the Joint Replenishment of Inventory with Interdependent Ordering Costs

Anne Olsen

Winthrop University, Rock Hill, SC 29733, USA
olsen@winthrop.edu

Abstract. The joint replenishment of inventory problem (JRP) requires independence of minor ordering costs. In this paper we propose an evolutionary algorithm (EA) for a modification of the JRP. The modified JRP allows for interdependence of minor ordering costs. Our proposed EA (EARP) is a nested EA that searches for a solution to minimize the total cost of inventory replenishment. It combines an EA which uses a direct grouping method with an EA that uses an indirect grouping approach (EA_ind) by nesting EA_ind inside EARP. We test EARP against partial enumeration and show that it provides close to optimal results for some problems. We know of no other algorithm to solve this problem.

1 Introduction

Replenishment of multiple items from a single supplier is called joint replenishment. The savings realized by joint replenishment can be significant. The joint replenishment problem (JRP) is to determine an inventory replenishment policy that minimizes the total cost, TC, of replenishing multiple items from a single supplier. The TC depends on the cost of holding items in inventory (h_i), the cost of placing an order, and the demand, D_i. The cost of placing an order includes a fixed cost of preparing an order (S) and a handling cost associated with each item in the order (s_i) [4]. The problem is to find the close to optimal grouping of items for each order [4].

The JRP assumes independence of the s_i's which means s_i of item i is not affected by other items included in the same order. We consider problems for which the s_i's are interdependent. Interdependence occurs two ways. First the value of s_i may change depending on which other items are in the same order. Second, item i may be constrained from being in the same order as item j.

Two methods covered in the literature to find the best grouping of items in an order are indirect grouping and direct grouping. Indirect grouping replenishes inventory using a basic cycle time. Not all items are necessarily ordered every cycle. An integer, k_i, indicates in which cycles item will be ordered. Direct grouping divides items into disjoint groups and a fixed cycle time is determined for each group. All items in a group are ordered every cycle time for that group [3].

E. Cantú-Paz et al. (Eds.): GECCO 2003, LNCS 2724, pp. 2416–2417, 2003.
© Springer-Verlag Berlin Heidelberg 2003

In this paper we propose a nested EA (EARP) to find a near optimal solution to the JRP under conditions of interdependence. The next section describes EARP, the testing procedure, and ends with a brief description of results.

2 The Evolutionary Algorithm

EARP uses a direct grouping method to place n items into groups, G_j, where $j = 1, ..., J$, which satisfy the constraints. It then calls EA_ind which uses an indirect grouping method to find the basic cycle time, T_j, and the integer multiples, k_i, for the items in G_j. At the end of N generations, the individual with the lowest TC represents the best ordering policy.

Each chromosome represents a replenishment policy and contains group ids (integers) and bits representing the k_i's. See Olsen [2] for details. Each group, G_j, in the chromosome is evaluated by the following equation:

$$eval_j = \left[2 * \left(S + \sum_i (s_i/k_i) + \sum penalty\right) * \left(\sum_i h_i k_i D_i\right)\right]^{1/2} \quad (1)$$

where *penalty* is an amount added due to dependence. This is a modification of the equation used by RAND [2], an indirect grouping method. is found by where . indicates the "fitness" of each chromosome and guides the search. Parameters used follow guidelines in [1] and are: pop_size = 20, p_c, p_m = 0.01, and the number of generations is 1000 for EARP with 500 for EA_ind, the nested EA.

We tested EARP with 810 problems randomly generated within selected ranges. Results were compared with a partial enumeration process. EARP gave results equal to partial enumeration 5.4% of the time, was worse 87.3% of the time, and showed improvement 7.3% of the time. The average percent difference between EARP and partial enumeration shows EARP is 1.2% worse than partial enumeration overall for the 810 problems. These results suggest an EA solution may produce a close to optimal replenishment policy. Additional work is scheduled for publication.

References

1. Michalewicz, Z.: Genetic Algorithms + Data Structures = Evolution Programs, Springer-Verlag, New York (1996)
2. Olsen, A.: An Evolutionary Algorithm for the Joint Replenishment Problem. Ph.D. thesis, Charlotte, NC: University of North Carolina at Charlotte, USA (2002)
3. Rosenblatt, M.J.: Fixed Cycle, Basic Cycle and EOQ Approaches to the Multi-item Single-supplier Inventory System. International Journal of Production Research, vol. 23:6, Taylor & Francis, Ltd. (1985) 1131–1139
4. Silver, E.A.: A Simple Method of Determining Order Quantities in Joint Replenishments Under Deterministic Demand. Management Science, vol. 22:12. The Institute of Management Sciences, U.S.A. (1976) 1351–1361.

Benefits of Implicit Redundant Genetic Algorithms for Structural Damage Detection in Noisy Environments

Anne Raich and Tamás Liszkai

Texas A&M University, Dept. of Civil Engineering, College Station, Texas, 77843, USA
{araich,tamas}@tamu.edu

Abstract. A robust structural damage detection method that can handle noisy frequency response function information is discussed. The inherent unstructured nature of damage detection problems is exploited by applying an implicit redundant representation (IRR) genetic algorithm. The unbraced frame structure results obtained show that the IRR GA is less sensitive to noise than a SGA.

1 Unstructured Problem Domain of FRF-Based Damage Detection

The goal of structural damage identification methods (SDIM) is to accurately assess the condition of structures. Most SDIMs assume that vibration signatures are sensitive indicators of structural integrity. In this research, FRF data was used to identify the location and severity of damage. An optimization problem was defined using an error function between the measured data and the discrete analytical model. Although the total number of structural elements typically is large, the number actually damaged is smaller. This unique situation defines an unstructured problem, in which the number of damages is unknown. The optimization problem is solved using genetic algorithms (GA) by altering member properties. A damage vector is obtained that identifies the location and severity of damage(s) in the structure. This formulation requires minimal measurement information. A comprehensive review of model parameter updating methods is provided in [1]. Two GA representations were investigated (Fig. 1). A fixed number of variables were encoded using a SGA representation to represent a complete solution by defining a damage indicator for each element. The IRR representation [2] considered the unstructured nature of damage detection by allowing the number of damaged elements to change during optimization, which is beneficial when the number and location of damages are unknown. A complete solution is encoded using only the damages for a small subset of the elements, instead of all elements.

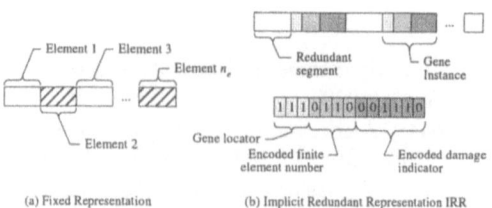

Fig. 1. Comparison of design variable encoding for SGA and IRR GA representations

2 Results of Cases Studies with Added Measurement Noise

Case studies were performed on a three-story, three-bay frame with 10% damage imposed on a first floor beam (Element 21). Three measurement locations and an excitation at the upper story were used to generate the simulated FRF data. Four noise levels were investigated by adding normally distributed random noise to the FRF data. A more detailed discussion of the case study and results is provided in [3]. Results obtained for the IRR GA trials after 300 generations are shown in Fig. 2. Without noise, the IRR GA found the global optimum in 241 generations. For all noise levels, the correct damaged element was identified with close to the exact 10 % damage value. The number of falsely identified elements with observable damage magnitude increased as the noise level increased.

The IRR GA outperformed the SGA in all trials. The SGA encoded all of the damage indicators in the finite element model. Therefore, the number of falsely identified damaged elements was large. In comparison, the adaptive characteristic of the IRR GA was beneficial since the number of damage indicators was not explicitly encoded. The IRR GA determined the number of damaged elements while minimizing the error function. For the noise free case, the best IRR individual initially encoded 15 gene instances, while the best IRR individual in the final population encoded 3 gene instances, One instance identified the correct damaged element and two others instances identified an element with zero damage. When noise was added, the number of gene instances changed from 15 in the initial population to 8-10 in the final population. In a noise free environment, the global optimum was always found. Overall, the IRR GA was considerably less sensitive to measurement noise compared with the SGA. Even on larger problems, the IRR GA was able to identify the damaged elements. Seeding the initial population with the zero damage individual was not necessary for the IRR GA to find the optimal solution, but was beneficial in many trials.

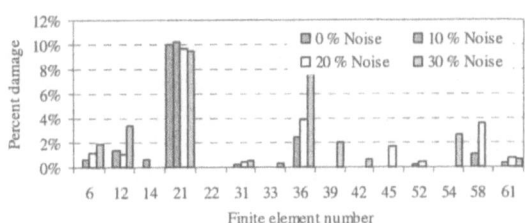

Fig. 2. Damage detection results for the frame problem (Element 21 - 10% damage)

References

1. Mottershead, J.E., Friswell, M.I.: Model updating in structural dynamics: A survey. Journal of Sound and Vibration 167(2) (1993) 347–375
2. Raich, A.M., Ghaboussi, J.: Implicit redundant representation in genetic algorithms. Evolutionary Computation 5(3) (1997) 277–302
3. Liszkai, Tamas R.. Modern heuristics in structural damage detection using frequency response functions. Ph.D. Dissertation, Texas A&M University (2003).

Multi-objective Traffic Signal Timing Optimization Using Non-dominated Sorting Genetic Algorithm II

Dazhi Sun, Rahim F. Benekohal, and S. Travis Waller

University of Illinois at Urbana-Champaign
Urbana, IL 61801, USA
{dazhisun,rbenekoh,stw}@uiuc.edu

Abstract. This paper presents the application of Non- dominated Sorting Genetic Algorithm II (NSGA II) in solving multiple-objective signal timing optimization problem (MOSTOP). Some recent researches on intersection signal timing design optimization and multi-objective evolutionary algorithms are summarized. NSGA II, which can find more of the Pareto Frontiers and maintain the diversity of the population, is applied to solve three signal timing optimization problems with 2-objective and 3-constraint, which account for both deterministic and stochastic traffic patterns. Mathematical approximation of the resulting Pareto Frontiers are developed to provide more insight into the trade-off between different objectives. GAs experimental design and result analysis are presented with some recommendations for prospective applications.

1 Multi-objective Traffic Signal Timing Optimization Problem

Minimizing the average delay and minimizing the number of stops per unit of time are important objectives for traffic signal timing design. However, none of feasible solutions could achieve the simultaneous optimality of these two objectives for an intersection with asymmetric traffic demand. A generic multi-objective traffic signal timing optimization problem at an isolated intersection with two-phase control strategy and permissive left turn can be formulated as:

$$\text{minimize } F(G) = [f_1(G), f_2(G)]$$

Where: G- vector of effective green time for each phase i; $f_1(G)$- the first objective function with respect to delay; $f_2(G)$- the second objective function with respect to stops.

Webster delay formulation and Akçelik stops function, which are widely used for calculating the corresponding performance index of delay and number of stops, are modified to be the objective functions mentioned in the above equation. Due to the impact of cycle length on the intersection overall effective capacity, an additional constraint on minimum cycle length was introduced to the above optimization problem.

2 GAs Experiment Design and Result Analysis

Three signal design problems are defined to minimize average delay and the average number of stops, using the effective green time at each signal phase as the design variable. NSGA II, first introduced by Deb et. al (2002), is modified to solve these problems. The used GA parameters are - maximum generation: 500; population size: 100; binary string length: 30; probability of crossover: 0.9; probability of mutation: 0.01; minimum green: 10; maximum green: 100. The designed scenario is a two-phase isolated intersection with permissive left turn. The critical flow ratios are 0.47 and 0.39 and saturation flow for each approach is 1800 pcphpl. It was observed that a clear frame of actual Pareto Frontiers can be located within 20 generations. As the generation number grows, more Pareto Frontiers were discovered and a well-fitted third degree polynomial function can be constructed to evaluate the tradeoff between the conflicting objectives. In the meanwhile, it was observed that the Pareto-optimal design variables locate along a certain straight line in the feasible space and the corresponding regression functions were developed in this study as well.

The following figures show the population and objective values at generation 1, 10 and 20 for the multi-objective signal optimization problem under stochastic traffic pattern with Webster cycle length constraint.

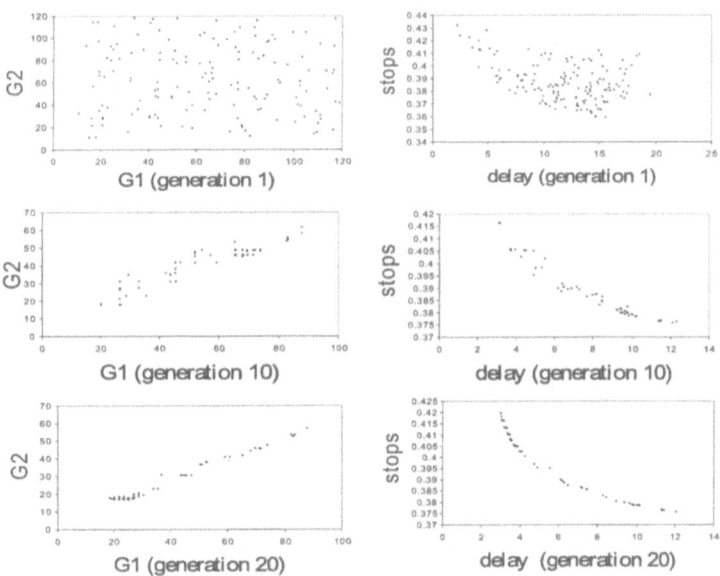

This study solved multi-objective traffic signal optimization problem by NSGA II and analyzed its effectiveness. Results showed that NSGA II can find a much better spread of optimal signal design plans on the true Pareto-optimal frontier at a high convergence speed.

Exploration of a Two Sided Rendezvous Search Problem Using Genetic Algorithms

T.Q.S. Truong[1] and A. Stacey[2]

[1] Air Operations Division, Defence Science and Technology Organisation
Melbourne, Australia
tracy.truong@dsto.defence.gov.au
[2] Mathematics Department, RMIT University, Melbourne, Australia
stacey@rmit.edu.au

Abstract. The problem of searching for a walker that wants to be found, when the walker moves toward the helicopter when it can hear it, is an example of a two sided search problem which is intrinsically difficult to solve. Thomas et al [1] considered the effectiveness of three standard NATO search paths [2] for this type of problem. In this paper a genetic algorithm is used to show that more effective search paths exist. In addition it is shown that genetic algorithms can be effective in finding a near optimal path of length 196 when searching a 14 × 14 cell area, that is a search space of 10^{100}. ...

1 Problem Description

Rendezvous search, or looking for someone who wants to be found, such as a walker lost in the desert was recently considered by Thomas and Hulme [1]. The basic problem is as follows, the lost walker is searched for by a helicopter. The speed of the walker is 3 m.p.h. and that of the helicopter 60 m.p.h.. If the walker hears the helicopter it moves towards it, if not it stays still. The walker can detect the helicopter to a radius of 1 mile, and the helicopter can detect the walker to a radius of 0.11 miles. The search region is broken up into 196 cells in a 14 × 14 grid. The helicopter takes 9 seconds to transit between cells and spends 10 seconds searching each cell. The walker could move freely in the search space but the helicopter was constrained to move up and down but not diagonally between adjacent cells. The cell size was the largest square that could be inscribed within the circle of detection of the helicopter, and it was assumed that detection could only occur if the walker and helicopter were in the same cell at the same time. This problem is hard to solve analytically.

If detection in the simulation (occurring with probability 0.78) is determined by the throw of a dice, this causes fluctuations in the values of the fitness function. The GA was found to not converge unless up to $1,000$ trials were averaged for each fitness value. This then makes the problem computationally intractable.

To overcome this, for each search path and starting location of the walker, a simulation was done to determine at what times the walker and helicopter were in the same cell. This information was used to construct a theoretical cumulative probability curve, as a function of time. These results were then averaged over all possible walker starting locations to give a cumulative probability distribution which characterized the effectiveness of the path. These distributions have no random fluctuations and can be used to construct a noise free fitness function for the GA. The fitness function was a weighted sum of the maximum probability at the end of the search and the area under the curve. Including the area under the curve ensured that higher fitness values were given to paths that had higher probabilities of early detection.

2 Results

The three NATO search paths are the decreasing square (DESQ) starting in cell one (at one corner) and spiraling in to the centre, the expansion square (EXSQ) covering the same path in reverse, and the scan search (SCAN) a path which zig zags across the area with strips parallel to the boundaries. The path found by the GA was compared with the three NATO paths, results for which are shown in Fig. 1.

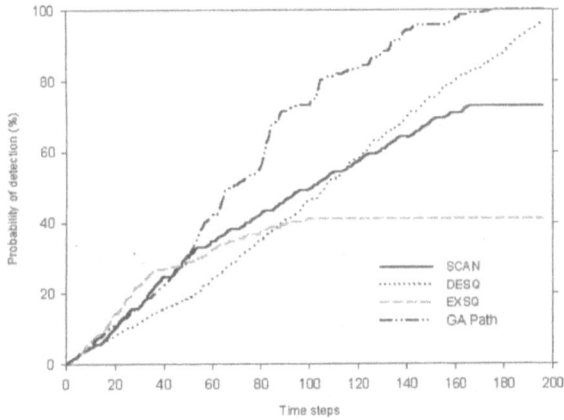

Fig. 1. 14 × 14 region, velocity of walker 3 m.p.h.

References

1. Thomas, L.C.,Hulme, P.B.: Searching for targets who want to be found. Journal of the Operational Research Society **48** (1997) 44–52
2. N.A.T.O.: Manual on Search and Rescue, ATP-10(C), Annex H of Chapter 6. NATO: Brussels (1988)

Taming a Flood with a T-CUP – Designing Flood-Control Structures with a Genetic Algorithm

Jeff Wallace and Sushil J. Louis

Department of Computer Science
University of Nevada, Reno, USA
sushil@cs.unr.edu
http://www.cs.unr.edu/~sushil/

Abstract. This paper describes the use of a genetic algorithm to solve a hydrology design problem - determining an optimal or near-optimal prescription of Best Management Practices in a flood-prone watershed. The model has proved capable of discovering design prescriptions that are more cost-effective than existing designs, promising significant financial benefits in a shorter time period. The approach is flexible enough to be applied to any watershed with basic precipitation and soil data.

1 Introduction and Problem Description

We use the T-CUP (Total Capture of Urban Precipitation) hydrology model to drive a genetic algorithm that finds cost-effective prescriptions of Best Management Practices (BMPs) to address a severe flooding problem in the Sun Valley Watershed of northern Los Angeles County, California. Due in part to the T-CUP model, the Los Angeles County Department of Public Works recently diverted funds from a proposed $42 million storm drain system in order to retrofit the entire watershed with BMPs, a decentralized method of capturing runoff locally. BMP examples include cisterns, drywells, pavement removal, and planting trees, and are designed to retain most runoff on-site. The Sun Valley watershed is divided into 38 sub-basins that flow into each other.

Each chromosome represents a spatially-explicit prescription of BMP installations. We know a little about the search space and were able to dramatically improve our results through selective initialization - we initialize the first chromosome in the population to be all zeros - a prescription of zero installations of each BMP in each sub-basin. We use a modified version of Eshelman's Adaptive CHC genetic algorithm [1] and added a "tweak" option that allows a user to manually initiate a new epoch through cataclysmic mutation.

We found that the best solutions were not neccessarily a function of compute time or number of evaluations. As seen in Table 1, When large populations run for relatively few generations, the single null chromosome does not have time to propogate through the population before time runs out. Figure 1 plots objective function values (to be minimized) versus number of generations and shows the

Table 1. Summary table of performance for different population sizes and generations

# of generations	Population size				
	25	50	100	200	400
20	69,524	43,300	31,971	33,046	29,824
50	159,322	62,306	19,456	14,529	16,231
100	246,582	136,185	26,636	6796	5725

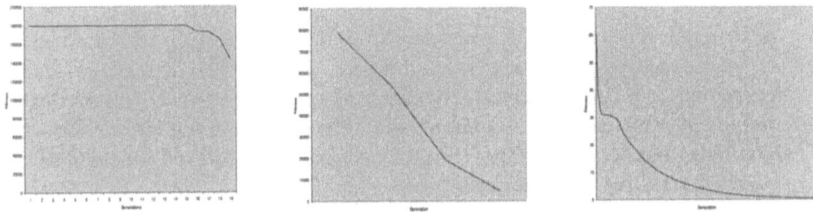

Fig. 1. System performance over three sequential phases.

three phases, geographic viability, meeting runoff targets, and cost minimization that our system goes through. The system quickly finds high quality solutions and we find that our users like to run with small population sizes to more quickly obtain feedback on the structure of promising prescriptions - even if larger population sizes lead to prescriptions with lower costs. The single all-zero individual also has much larger effect on a small population size.

T-CUP provides a useful tool for hyrdologists analysing and designing flood control structures. Our initialization procedure allows T-CUP to provide good flood control prescriptions in reasonable time. With large population sizes, the system can be used to provide a set of promising design solutions. We also find that users like to run the system with small populations to get real-time feedback on the structure of promising prescriptions and so getting a "feel" for the kinds of solutions that would suit a particular watershed. We have a number of improvements in mind and we encourage you to try the T-CUP demo available at www.parconline.com/ga/ – any feedback will be appreciated. The full paper is available at www.cs.unr.edu/~sushil.

References

1. Larry Eshelman: The CHC Adaptive Search Algorithm, How to have safe search when engaging in nontraditional genetic recombination. in Rawlins G.J.E., editor. Foundations of Genetic Algorithms – 1, Morgan KauffMan (1991) 265–283.

Assignment Copy Detection Using Neuro-genetic Hybrids

Seung-Jin Yang, Yong-Geon Kim, Yung-Keun Kwon, and Byung-Ro Moon

School of Computer Science and Engineering
Seoul National University
Shilim-dong, Kwanak-gu, Seoul, 151-742 Korea
{sjyang,dvp,kwon,moon}@soar.snu.ac.kr

Abstract. It is difficult for teachers to find out copies in their program assignments. We propose a method which detects the similarity of assignments. We first transform a program into a sequence of tokens and provide it to an artificial neural network. The neural network is optimized by a hybrid genetic algorithm. Experimental results showed considerable improvement on artificial neural networks.

It is one of the most time-consuming problems for teachers or grading assistants to find out students who copied other program. It is easy to copy other text with various editors. But it is not an easy job to examine the similarity of the programs manually. There are thus many efforts to develop automated copy detection methods.

We want to develop a system that takes a number of programs as input and returns suspicious pairs of programs. In the system the core engine takes two programs as input and returns the degree of suspicion for plagiarism.

We use token counting method on detecting plagiarism. The frequency of each token is determined by the object and programer of the program. We designed the system as follows. We define 40 tokens for the Java language. And we count the frequencies of these tokens and express the frequencies as a vector. We obtain two vectors $(a_1, a_2, ... , a_{40})$ and $(b_1, b_2, ... , b_{40})$ from the two programs. Here, a_i and b_i match the 40 attributes each corresponding to a token. These two vectors are used for the input of the neural network. There are two versions in this paper. One uses the vector $(a_1, b_1, a_2, b_2, ... , a_{40}, b_{40})$ as the input; the other uses $(|a_1 - b_1|, |a_2 - b_2|, ... , |a_{40} - b_{40}|)$ as the input. To test the proposed approaches, we prepared two different program assignment sets submitted in past classes. The program sets are used for the training and the test, respectively. Each set was expanded in the following way. Some base programs were selected from each set, intensively checked to see that they are not copies of one another, and distributed to our laboratory fellows. Then, they generated copy programs from the base programs in their own ways. We added the copy programs into the original data set.

Table 1 describes the input data and Table 2 shows the experimental results. In the table, the "candidates" means the number of program pairs that

Table 1. Input Data

Assignment 1 (training data)		Assignment 2 (test data)	
total pairs	copy pairs	total pairs	copy pairs
2000	159	1475	42

Table 2. Results of copy detection (average of 20 trials)

	output			
	candidates	copies	no copies	accuracy
40-input ANN	44.3	19.3	25	43.6
80-input ANN	51.4	16.3	35.1	31.7
40-input hybrid GA	44.7	20.8	23.9	46.5
80-input hybrid GA	48.7	24.2	24.5	49.7

each algorithm outputs as suspicious candidates. And the columns "copies" and "no copies" represent the number of copy pairs and non-copy pairs, respectively, among candidates. The column "accuracy" means the rate of copies over candidates. The numbers 40 and 80 in the row titles correspond to the number of input neurons.

The 80-input hybrid GA (Genetic Algorithm) performed best among them. We can observe an interesting contrast in the results of ANN (Artificial Neural Networks) and hybrid GAs. In case of ANNs, 40-input ANN performed better than the 80-input ANN. In case of hybrid GAs, on the other hand, 80-input hybrid GA performed better than the 40-input hybrid GA. This is strong evidence that shows the gap between the powers of ANNs and hybrid GAs. Although 80 inputs provide more information than 40 inputs, the problem space of ANN optimization might perhaps be too large for the backpropagatoin algorithm. On the contrary, the hybrid GA took advantage of the full information. The 80-input hybrid GA catched on average 24.2 copies among the 42 copies, although the absolute value would vary depending on the degrees of copies.

To the best of our knowledge, this work is the first for the copy detection problem using genetic algorithms combined with neural networks. The result showed considerable performance improvement over ANNs. It also showed that the model of token counting is attractive, and that the GA plays a crucial role in high accuracy detection of copies.

Acknowledgments. This work was partly supported by Optus Inc. and Brain Korea 21 Project. The RIACT at Seoul National University provided research facilities for this study.

Structural and Functional Sequence Test of Dynamic and State-Based Software with Evolutionary Algorithms

André Baresel[1], Hartmut Pohlheim[1], and Sadegh Sadeghipour[2]

[1] DaimlerChrysler AG, Research and Technology, Methods and Tools
Alt-Moabit 96a, 10559 Berlin, Germany
{andre.baresel,hartmut.pohlheim}@daimlerchrysler.com
[2] IT-Power Consultants, Jasmunder Str. 9,
13355 Berlin, Germany
sadegh@itpower.de

Abstract. Evolutionary Testing (ET) has been shown to be very successful for testing real world applications [10]. The original ET approach focuses on searching for a high coverage of the test object by generating separate inputs for single function calls.
We have identified a large set of real world application for which this approach does not perform well because only sequential calls of the tested function can reach a high structural coverage (white box test) or can check functional behavior (black box tests). Especially, control software which is responsible for controlling and constraining a system cannot be tested successfully with ET. Such software is characterized by storing internal data during a sequence of calls.
In this paper we present the Evolutionary Sequence Testing approach for white box and black box tests. For automatic sequence testing, a fitness function for the application of ET will be introduced, which allows the optimization of input sequences that reach a high coverage of the software under test. The authors also present a new compact description for the generation of real-world input sequences for functional testing. A set of objective functions to evaluate the test output of systems under test have been developed. These approaches are currently used for the structural and safety testing of car control systems.

1 Introduction to Sequence Testing

When analyzing the code of real world applications, a large set of implementations can be identified which use a functional model whereby a controller procedure is called periodically. In the world of car control units it is very common for a software function to be based on an initialization and step function. Whereas the initialization function is only called once at the beginning, the step function is executed at regular time intervals e.g. every 10 milliseconds. A cruise control and vehicle-to-vehicle distance controller program (Distronic) which checks the velocity and distance to a leading vehicle at regular intervals to find out if a safe distance is maintained is an example of control software (structure shown in figure 7).

This kind of software seems to be very specific. However, an equivalent situation can be found within object oriented software. OO software systems implement objects which often create their own internal storage and methods. The methods initialize and alter the objects data during its life-time. It is very common for an object to provide an interface to initialize and change its data. The implemented functions behave differently on the basis of the object's inputs and internal settings.

Depending on the complexity of the system under test automatic testing can be performed in three different ways. The approach commonly used is to generate pairs of initial states and input situations shown in Figure 1. This approach works well for simple software models but raises several problems. At first, the direct setting of internal states is not possible in all cases and needs changes to be made to program under test. By generating the internal states the test system has to make sure that the states produced are valid according to the specification. Forcing the system into a generated state is, in most cases, not useful for the tester. This is because the tester has to use the test data generated to analyze software bugs. He needs first to ascertain how to produce the initial state which causes the problem. This information is not provided by this test automation.

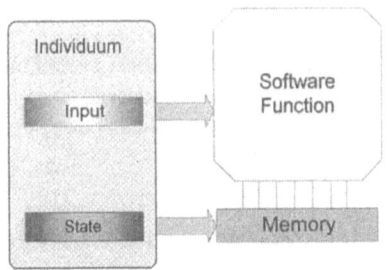

Fig. 1. Automatic testing by generating (state, input) pairs (first approach)

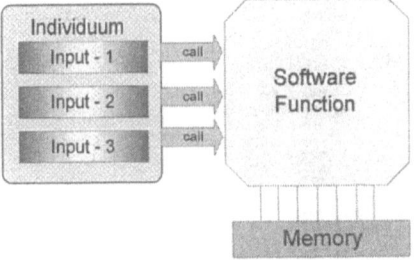

Fig. 2. Automatic testing by generating list of inputs (second approach)

For complex systems it is necessary to search for an input sequence. One approach is to create lists of inputs for sequential calls of the function under test. This is often sufficient for software systems based on state models, see Fig. 2.

For control systems requiring long 'real world' input sequences, encoded input functions must be generated. Figure 3 presents this third approach.

Monitoring the sequence test of a system differs from the original ET approach. For white box tests, a list of execution paths has to be analyzed. Black box tests are performed on sequences of output values instead of single values. This allows the assessment of dynamic functional and safety criteria.

The two approaches for generating input sequences are, from the tester's point of view, the best solution since they guarantee that the system is tested in the same way as it will later be used.

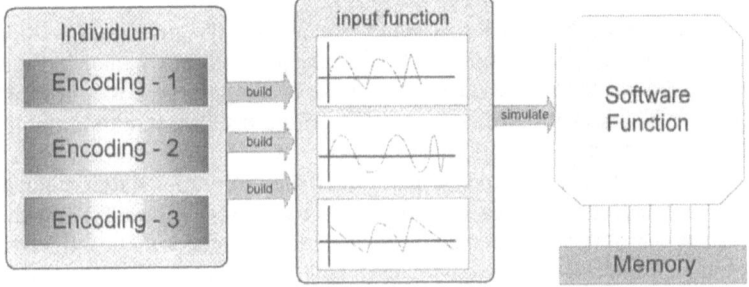

Fig. 3. Automatic testing by generating input sequences from encoded input functions (third approach)

Section 2 describes how evolutionary algorithms (EA) have been applied to structural and functional sequence testing. Both applications have been implemented by the authors and results of real world experiments will be presented. Structural Sequence Testing has been tested with the generation of input sequences, see section 3. Functional Sequence Testing, see section 4, has been applied to Safety Tests by creating encoded input functions.

2 Overview of Evolutionary Testing (ET)

Evolutionary algorithms (EA) have been used to search for data for a wide range of applications. EA is an iterative search procedure using different operators to copy the behavior of the biologic evolution. Using EA for a search problem it is necessary to define the search space and the objective function (fitness). The algorithms are implemented in a widely used tool box [7]. It consists of a large set of operators e.g. real and integer parameter, migration and competition strategies.

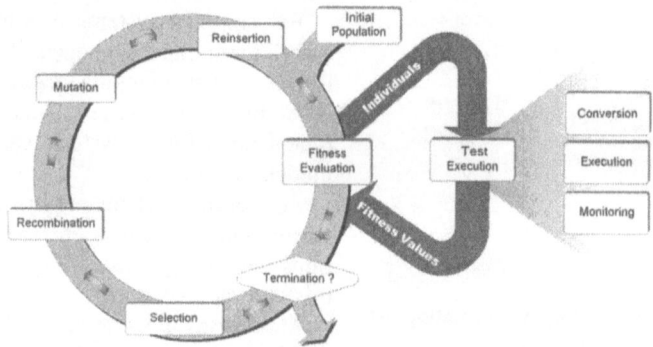

Fig. 4. The structure of Evolutionary Testing

Evolutionary Testing uses EA to test software automatically. The different software test criteria formulate requirements for a test case set to be generated. Until now, the generation of such a set of test data usually had to be carried out manually. Automatic software testing generates a test data set automatically, aiming to fulfill the requirements in order to increase efficiency and resulting in an enormous cost reduction.

Evolutionary Structural Testing. Structural Testing has the goal of automating the test case design for white box testing criteria [2]. Taking a test object, namely the software under test, the goal is to find a test case set (selection of inputs) which achieves full structural coverage. The general idea is a separation of the test into test aims and the use of EA to search for test data fulfilling the test aims.

The separation of the test into partial aims and the definition of fitness functions for partial aims are performed in the same manner for each test criterion. Each partial aim represents a program structure that requires execution in order to achieve full coverage of the structural test criterion selected, i.e. each single statement represents a partial aim when using statement coverage.

The definition of a fitness function, that represents the test aim accurately and supports the guidance of the search, is a prerequisite for the successful application of Evolutionary

Test. In order to define the fitness function, this research builds upon previous work dealing with branching conditions (among others [8], [5], and [9]). These are developed in [10] by introducing the idea of an approximation level.

Evolutionary Functional Testing. Complex dynamic systems must be evaluated over a long time period (longer than the highest internal dead time or time constant). This means that the systems must not be stimulated for only few simulation steps, but rather the input signals must be up to hundreds of time steps long. Long input sequences are therefore necessary in order to simulate these systems.

Several disadvantages result from the length of the sequences necessary: the number of free variables during optimization is very high and the correlation between the variables is great. For this reason, one of the most important aims is the development of a very compact description for the input sequences. This must contain as few elements as possible but, at the same time, offer a sufficient amount of variety to stimulate the system under test as much as necessary.

Moreover, possibilities for automatically evaluating the system answers must be developed, which allow differentiation between the quality of the individual input sequences. These requirements and the solutions developed will be presented in section 4.

3 Evolutionary Algorithms for Structural Sequence Testing

The evolutionary structural test was designed originally to find a set of test data for single function calls which results in a high structural coverage of the software under test. This approach has been defined for the most common test criteria in [10]. The fundamental idea is the transformation of the test data generation into a search problem which is then solved by evolutionary algorithms. The optimization function designed for this calculates so-called fitness values. The evolutionary algorithm uses these values to optimize the solution to meet the test aim currently selected.

Structural Sequence Testing targets the structural coverage of a given test object. For this reason, test aims are defined in the same manner as introduced for ET. However, in order to apply evolutionary algorithms it is necessary to redefine the search space and the fitness function because test cases are now input sequences each executing several paths (see figure 5).

3.1 Search Space Structure

In the original approach the search space is formed by the input parameters of the function under test. This is different for sequence tests. The authors decided to implement the generation of input sequences for the first experiments in the area of structural tests. With this approach one test datum is a list of inputs. Monitoring a test case will return a list of execution paths (one per input). The length of this list is not defined in the test object and is, in general, not limited for control systems.

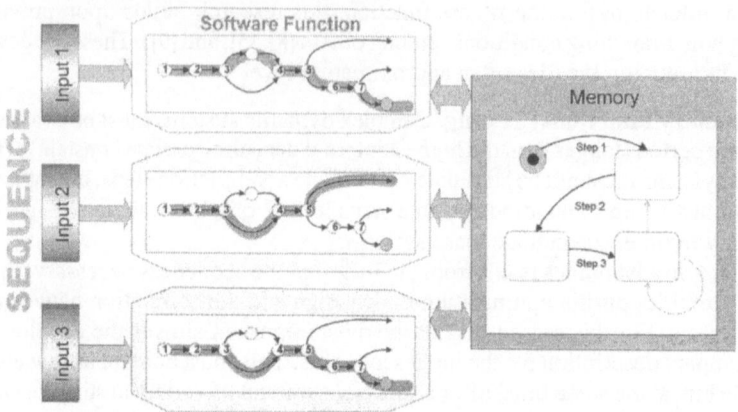

Fig. 5. Test case (sequence) and different execution paths for the inputs; On the right hand side an example memory model is shown, where each step results in a state change

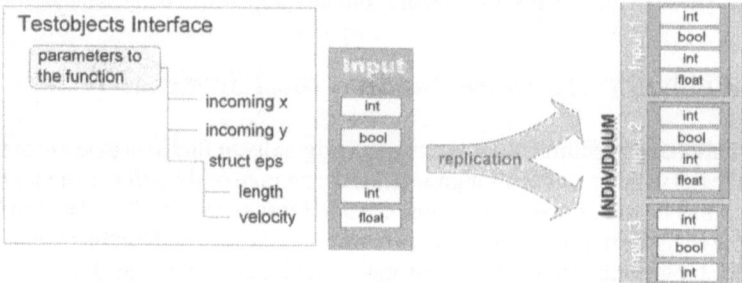

Fig. 6. The encoding of the test case data to different runs

For evolutionary algorithms, an encoding of the individuals generated (*mapped to test data*) has to be defined. A straightforward approach is to let the user specify the length of the input sequences manually. With the information on the sequence length a search space is formed by replicating all input parameters in such a way that a separate value will be provided for each call and each parameter. Every individual generated by the EA represents the data of one sequence of calls (here called test data). Figure 6 shows the mapping of the data.

3.2 Sequence Evaluation

As mentioned previously, the test criteria for structural coverage are defined on the basis of the test object source code. An automatic test system has to monitor and analyze all runs of the test object to find out which test aims have been reached (structures that have been covered) and to provide a fitness value for the test aim currently selected. This fitness is determined by monitoring the test execution.

In the original ET approach, fitness is calculated by analyzing one execution path through the test object. This is different in sequence testing, since each test case run results in a list of execution paths. This list is the basis upon which a fitness value has to be calculated. The sequence testing fitness function introduced here uses the original idea of the fitness function and creates a sequence fitness value from that.

For structural coverage it is not important at which step of a sequence a structural element is covered. For this reason, the authors decided to design a fitness function which analyzes all the execution paths of a test case sequence and uses the closest path to the test aim as the fitness of the sequence. With this approach it does not matter whether a close execution path is traversed earlier or later in the sequence and whether or not the EA is able to create better performing solutions by ordering the sequence in different ways (see [1] for details of sequence optimization).

The original ET approach uses decisive branches to ascertain the fitness of an execution path (see [10]). This idea can be reused by analyzing each path of the sequence and determining an overall fitness. Due to the fitness function's structure, it has to be minimized by the EA. For this reason, the overall fitness is the minimum of the path fitness values. Using this approach the best path will define the fitness of the sequence.

3.3 Experiments

The experiments on Structural Sequence Testing were performed with an extended version of the automatic structural test system. The GEATbx [7] is the optimization component used by the system. The authors applied the standard settings for this class of problem (namely, 6 subpopulations with 50 individuals each, linear ranking, a selection pressure of 1.7, migration and competition between subpopulations). A real valued representation is used. The subpopulations employ different search strategies by using different settings for the recombination (discrete and line recombination) and mutation operator (real valued mutation with differently sized mutation steps – large, medium and small).

Table 1. Information on the complexity of the test objects

Name	Size	LOC	nodes	param.	if–then	conditions	nesting level
Ctrl_S	16kB	220	20	28	5	6	2
Enable_S	54kB	800	140	15	51	70	10
Enable_A	44kB	520	86	11	39	56	8

Three software functions which are part of a large real-world functional model were tested using the extended ET system. Table 1 provides an overview of the complexity measures for the test objects. The test object *Ctrl_S* is relatively small but is code generated from a state diagram containing flags in all conditions.

Table 2. Experimetal results, showing the numbers of test aims, number of individuals generated and reached coverage

	Coverage criteria	test aims	num. individuals	coverage	not reached
Ctrl_S	StatementCover	17	550 000	95 %	1
	BranchCover	23	650 000	91 %	2
	ConditionCover	12	370 000	92 %	1
Enable_S	StatementCover	135	225 000	99 %	1
	BranchCover	196	550 000	98 %	5
	ConditionCover	168	650 000	97 %	5
Enable_A	StatementCover	87	390 000	95 %	5
	BranchCover	126	495 000	92 %	11
	ConditionCover	116	425 000	94 %	8

The three standard test criteria were employed. Statement cover requires a test case set that executes all statements of the software under test. The branch cover criterion requires a test case set traversing all branches of a test object. Last but not least, condition cover necessitates a collection of test data evaluating all the atomic conditions of a program with the values *true* and *false*.

The high coverage reached for all the criteria can be seen in table 2. Only a few of the test aims not reached are a sequence test specific problem. Some of them refer to the problem of unreachable code created by the code generator and some to the problem of performing an evolutionary test with flag conditions. However, in the next paragraphs we would like to give an short overview of the reasons why certain test goals were not reached.

Test result details for test object 'Ctrl_S'. This test object was difficult to test with the original ET approach due to the source code structure. The code was generated from a state diagram. All the program's conditional statements use flags that have been assigned previously. Some conditions access flags that have been assigned in previous calls of the function which result in the test object requiring sequence tests. The test runs have shown that this kind of software does not pose a problem for the approach, since all sequence test specific test aims have been covered. For one statement, the corresponding two branches and the flag condition of a sequence could not be found because the search was not guided to a flag assignment within a function call.

Test results for test object 'Enable_S'. The testing of this module demonstrates the potential of the evolutionary testing approach. The test object consists of 800 lines of code leading to 135 control flow nodes, 196 branches and 168 test aims for simple condition cover.

During the statement coverage test only one statement was not executed. The non-executable statements, branches or conditions appear when using current versions of code generators. This occurs if the developer uses template library functions and configures them with constants. This leads to a comprehensible model with reuse of components but the code generated contains lines of code that are not executable.

The performance of the branch coverage test was more challenging, resulting in coverage of 98%. Only five branches were not traversed. Upon checking the code we found that two branches belong to the non-executable statement. One branch is not executable because the associated condition cannot be evaluated as *false*. The evolutionary test system was not able to find an input for two conditions which depend on sequential calls. This is due to the high nesting level of dependent variable assignments and uses. At the moment it is not possible to guide the search to a solution that executes an assignment in one sequence step and traverses the corresponding condition in later steps. This would require further research.

The results of simple condition cover were good in that the evolutionary test found a test case set covering 97% of the test aims. For the reason why the five conditions not covered were missed, see the paragraph on branch coverage above.

Test results for test object 'Enable_A'. This module is not particularly complex with regard to metrics, however, it contains some program structures that are difficult to test. First of all, the code is state oriented and many conditions depend on the settings of state variables. The function has a high nesting level of *if-then-else* statements and employs a state encoded using a set of flags. Again, the test object contained unfeasible code because of the use of library templates (3 statements). Two other statements were not covered because of the flags used in the conditions. The results of branch coverage are similar to those of statement coverage. The test case set found did not cover 4 branches starting at flag conditions and 6 branches are placed at unreachable code. One branch could not be traversed because a precondition had not been satisfied. This branch requires an input sequence assigning two variables in previous calls of the function. The approach does not find a solution for this. The condition cover test returned equivalent results. Four of the conditions are not feasible because of the generation of code. The test system did not find an input sequence covering three flag conditions. Again, the condition which requires variable settings in previous calls was not fulfilled.

4 Evolutionary Algorithms for Functional Sequence Testing

The generation of test sequences for dynamic real-world systems pose a number of challenges:
- The input sequences must be long enough to stimulate the system realistically.
- The input sequences must possess the right qualities in order to stimulate the system adequately. This concerns aspects such as the type of signal, speed and rate of signal change.
- The output sequences must be evaluated regarding a number of different conditions. These conditions are often contradictory and several of them can be active simultaneously.

In order to develop a test environment for the functional test of dynamic systems, which can be applied in practice, the following main tasks must therefore be completed.
- Creation of a compact description for the input sequences,
- Evaluation of output sequences regarding defined requirements,
- Inclusion of input sequence generation and evaluation of system requirements into an optimization process.

Throughout the whole section examples are given using a real-world system. The structure of the integrated cruise control and vehicle-to-vehicle distance control (Distronic) is shown in figure 7. An extensive description is contained in [3].

The test environment presented in this section is implemented in Matlab and used for the testing of Simulink/Stateflow models. However, we concentrate on the presentation of the main concepts and the demonstration of results.

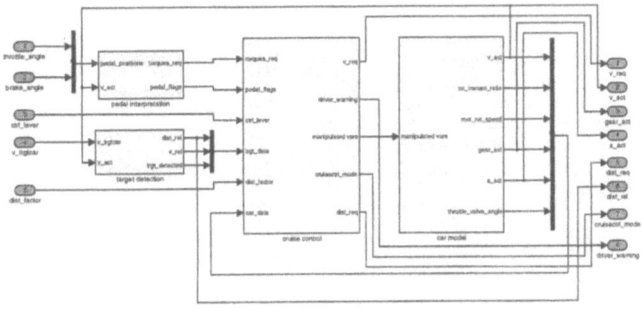

Fig. 7. Model of an integrated cruise control and vehicle-to-vehicle distance control (Distronic)

4.1 Description of Input Sequences

In order to describe the input sequences, several descriptions must be differentiated. On the one hand we have the description of the signal, which is used as input for the simulation of the dynamic system (simulation sequence). On the other hand we have the compact description which the user stipulates (compact user description). Between these is the description of the boundaries of the variables for the optimization (description of sequence bounds) as well as the instantiation of the individual sequences (optimization sequence). These different sequence description levels are illustrated in figure 8.

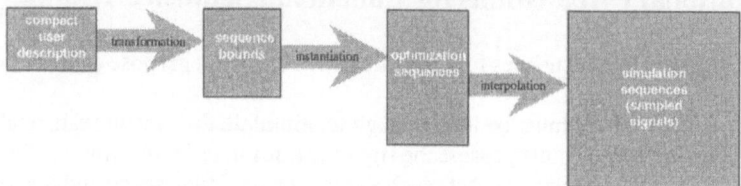

Fig. 8. Different levels of input sequence description

For a compact description the long simulation sequence is subdivided into individual sections. Each section is represented by a base signal, which is parameterized by the variables: signal type, amplitude and length of section. We use the following base signal types: step, ramp (linear), impulse, sine, spline. These base signal types are sufficient to generate any signal curve.

Only the possible areas for these parameters are specified for the optimization. In this way, the bounds are defined, within which the optimization generates solutions, which are subsequently evaluated by the objective function with regard to their quality.

The amplitude of each section can be defined absolutely or relatively. The length of the section is always defined relatively (to ensure the monotony of the signal). The base types are given as an enumeration of possible types. These boundaries must be defined for each input of the system. For nearly every real-world system these boundaries can be derived from the specification of the system under test.

An example of an input sequence description is provided in figure 10. The textual description as defined by the tester is given. Figure 10 provides examples of three different input signals generated by the optimization.

```
Input.Names = {'throttle_angle', 'brake_angle', 'ctrl_lever', ...
               'v_trgtcar', 'dist_factor'};
Input.BasePoints = [10, 10, 10, 10, 2]
Input.Amplitude.Bounds = [  0,   0, 0,  0, 1;...
                          100, 100, 3, 30, 1]
Input.Amplitude.Interpret = {'abs', 'abs', 'abs', 'abs', 'abs'};
Input.Basis.Bounds = [1, 1, 1, 1, 1; ...
                      3; 3; 5; 3; 3]
Input.Transition.Pool = { {'linear', 'spline'}; {'linear'}; {'impulse'}; ...
                          {'spline'}; {'step'} };
Sim.Length = 200; Sim.SamplingRate = 2.5;
```

Fig. 9. Textual description of input sequences

The system under test (Distronic) has 5 inputs. We use 10 sections for each sequence. The amplitude for the throttle pedal can change between 0 and 100, the control lever can only have the values 0, 1, 2 or 3.

In real-world applications the variety of an input signal is often constrained with regard to possible base signal types. An example of this is the control lever in figure 9. This input contains only impulses as the base signal type.

To generate a further bounded description it is possible to define identical lower and upper bounds for some of the other parameters. In this case, the optimization has an empty search space for this variable – this input parameter is a constant. An example is the distance factor (a measure for the relative distance to the preceding vehicle) in figure 10. This input signal is always set to a constant value of 1. Thus, it is not part of the optimization (but used as a constant value for the generation of the internal simulation sequence).

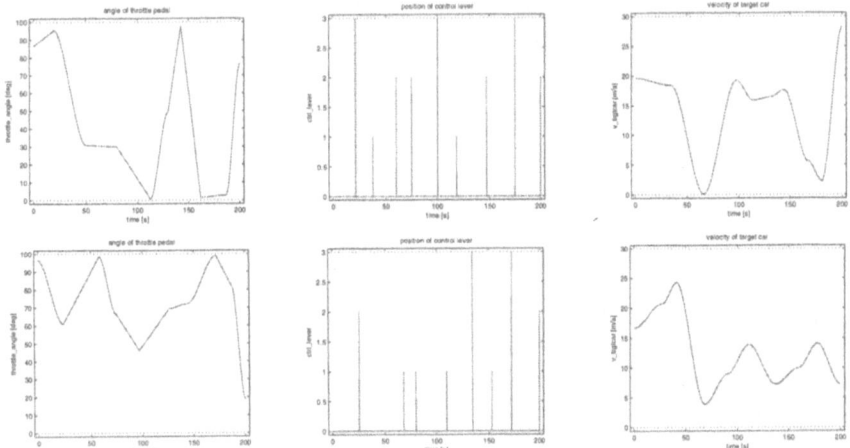

Fig. 10. Instances of simulation sequences generated by the optimization based an the textual description in figure 9; left: throttle pedal, middle: control lever, right: velocity of target car

All these different levels of the description of the input sequences ensure that the requirements for an adequate simulation of the system and a very compact and comprehensible description by the tester are fulfilled. The compact description is used for the optimization ensuring a small number of variables. When comparing the size of both descriptions for the example dynamic system used (5 inputs – one of the inputs is constant, 10 signal sections, 3 variable parameters for each section, 200 seconds simulation time, sampling rate 2.5 Hz) the differences are enormous:

$$SizeSimulationSignal = 5 \cdot 200 \cdot 2.5 = 2500$$
$$SizeCompactDescription = (5-1) \cdot 10 \cdot 3 = 120$$
$$CompressionRatio = \frac{2500}{120} = 20.8 \quad (1)$$

Only this compact description opens up the opportunity to optimize and test real-world dynamic systems within a realistic time frame.

4.2 Evaluation of Output Sequences and Objective Function

The test environment for the functional test of dynamic systems must perform an evaluation of the output sequences generated by the simulation of the dynamic system. These output sequences must be evaluated regarding the optimization aims. During the test we always search for violations of the defined requirements. Possible aims of the test are to check for violations of:

- signal amplitude boundaries,
- signal dependencies,
- maximal overshoot and maximal settlement time.

Each of these checks must be evaluated over the whole or a part of the signal lengths and an objective value generated. Due to space constraints we describe only the first two requirements in detail. However, our test environment can assess all of the checks (and more will be added).

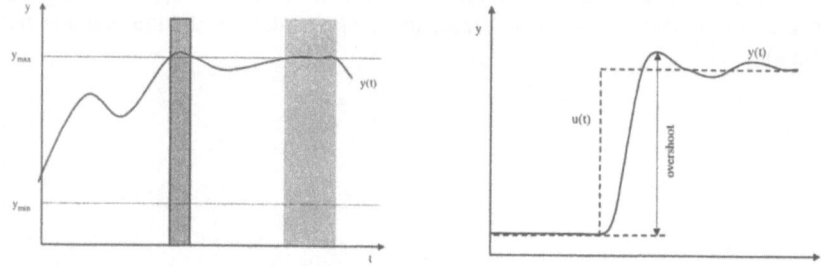

Fig. 11. Left: violation of maximum amplitude; right: assessment of signal overshoot

An example of the violation of signal amplitude boundaries is given in figure 11, left. A minimal and maximal amplitude value is defined for the output signal y. In this example the output signal violates the upper boundary. The first violation is a serious violation as the signal transgresses the bound by more than a critical value y_c (parameter for this requirement). In this case, a special value indicating a severe violation is returned as the objective value (value -1), see equation (2). The second violation is less severe, as the extension over the bound is not critical. At this point an objective value indicating the closeness of the maximal value to the defined boundary is calculated. This two-level concept allows a differentiation in quality between multiple output signals violating the bounds defined. The direct calculation of the objective value is provided in equation (2).

$$signal_{max} = \max(y(t)) \qquad ObjVal_{max} = \begin{cases} -1 & signal_{max} \geq y_{max} + y_c \\ \left(\dfrac{signal_{max}}{y_{max}}\right)^6 & signal_{max} < y_{max} + y_c \end{cases} \qquad (2)$$

A similar assessment is used for calculating the objective value of the overshoot of an output signal after a jump in the respective reference signal. First, the maximal overshoot value is calculated. Next, the relative height of the overshoot is assessed. A severe overshoot outside the specification returns a special value (again −1). This special value is used to terminate the current optimization. The test was successful, as we were able to find a violation of the specification and thus reach the aim of the optimization. In all other cases, an objective value equivalent to the value of the overshoot is calculated (similar to equation (2)).

Each of the requirements tested produces one objective value. For nearly every realistic system test we receive multiple objective values. In order to assess the quality of all objectives tested we employ multi-objective ranking as supported by the GEATbx [7]. This includes Pareto-ranking, goal attainment, fitness sharing and an archive of previously found solutions.

4.3 Experiments

The test environment was used for the functional testing of a number of real-world systems. One of these is the Distronic model described earlier ([3], structure shown in figure 7), for which the results of one test will be presented in this subsection.

For the car system with an activated Distronic the maximum speed was specified at 44 m/s (the critical value y_c was set to 0). This means that a higher speed is not permitted under any circumstances. Thus, a test was specified to search for inputs which produce a speed greater than this boundary.

With an active Distronic the car can be accelerated only by pushing the control lever upwards (the respective input value is 1). The car is decelerated by pushing the control lever downwards (input value: 2). Beside the amplitude of the input control lever, the relative length of the signal sections could be changed between 1 and 5. The results of one successful optimization are shown in figures 12 and 13.

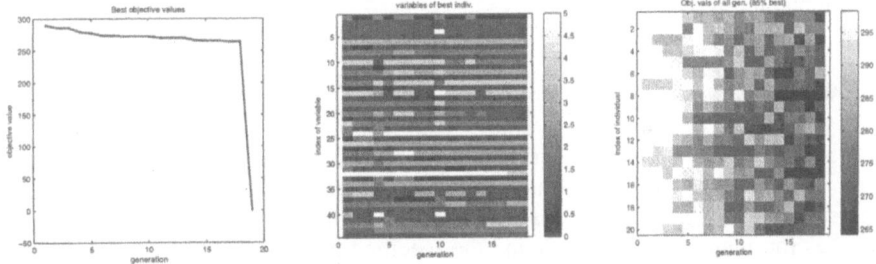

Fig. 12. Visualization of the optimization process, left: best objective value; middle: variables of the best individual; right: objective value of all individuals

The optimization process is visualized in figure 12. The left graphic presents the progress of the best objective value over all the generations. The optimization continually finds better values. In the 19th generation a serious violation is detected and an objective value of -1 returned. The middle graphic presents the variables of the best individual in a (gray) color quilt. Each line represents the value of one variable over the optimization. The graphic on the right visualizes the objective values of all the individuals during the optimization (generations 1 to 18).

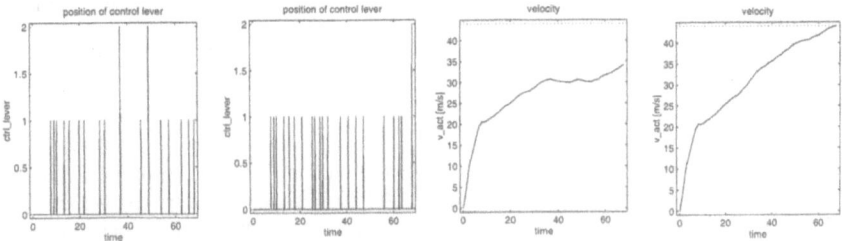

Fig. 13. Problem-specific visualization of the best individual during the optimization, left: input of the control lever - begin and end of optimization (2nd and 19th generation), right: vehicle velocity - begin and end of optimization (2nd and 19th generation);

The graphics in figure 13 provide a much better insight into the quality of the results, visualizing the problem-specific results of the best individual of each respective generation. The input of the control lever is shown in the two left side graphics. The resulting velocity of the car is presented at the right. The graphics are taken from the 2nd (left) and 19th generation (right). At the beginning the maximal velocity is far below the critical boundary. During the optimization the velocity is increased (the control lever is pushed up more often and at an earlier stage as well as being pushed down less frequently). At the end an input situation is found, in which the velocity is higher than the bound specified. By looking at the respective input signals the developer can check and change the implementation of the system.

During optimization we employed the following evolutionary parameters: 20 individuals in 1 population, discrete recombination and real valued mutation with medium sized mutation steps, linear ranking with a selection pressure of 1.7 as well as a generation gap of 0.9. Other tests with a higher number of relevant input signals and thus more optimization variables employ 4-10 subpopulations with 20-50 individuals each. In this case, we use migration and competition between subpopulations. Each subpopulation uses a different strategy by employing different parameters (most of the time differently sized mutation steps).

5 Conclusion

In this paper we have presented different approaches for applying Evolutionary Testing to sequence testing. The first method aims at automating test case generation to achieve high structural coverage of the system under test (white box test). The other approach searches for input sequences violating the specified functional behavior of the system (black box test).

Both test methods were implemented. Experiments with software modules and dynamic systems with varying complexity were conducted. We have presented a small selection of the results. The results show the new test methods to be promising. It was possible to find test sequences, without the need for user interaction, for problems which could previously not be solved automatically.

During the experiments a number of issues were identified which could further improve the efficiency of the test methods presented. For the structural sequence test, a fitness function which expresses the dependency between successive sequence steps would significantly aid the search process. In the case of the functional sequence test, it is necessary to include as much as possible of the existing problem-specific knowledge into the optimization process.

References

[1] *Baresel, A., Sthamer, H., Schmidt, M.*: Fitness Function Design to improve Evolutionary Structural Testing. Proceedings of GECCO2002, New York, USA, pp. 1329–1336, 2002.
[2] *Beizer, B.*: Software Testing Techniques. New York: Van Nostrand Reinhold, 1983.
[3] *Conrad, M., Hötzer, D.*: Selective Integration of Formal Methods in the Development of Electronic Control Units. Proceedings of Second IEEE International Conference on Formal Engineering Methods ICFEM'98, IEEE Computer Society, pp. 144–155, 1998.
[4] *Harman, M., Hu, L., Munro, M., Zhang, X.*: Side-Effect Removal Transformation. IEEE International Workshop on Program Comprehension (IWPC) Toronto, Canada, 2001.
[5] *Jones, B.-F., Sthamer, H., Eyres, D.*: Automatic structural testing using genetic algorithms. Software Engineering Journal, vol. 11, no. 5, pp. 299–306, 1996.

[6] *Korel, B.*: Automated Test Data Generation. IEEE Transactions on Software Engineering, vol. 16 no. 8, pp. 870–879, 1990.
[7] *Pohlheim, H.*: *GEATbx* - Genetic and Evolutionary Algorithm Toolbox for *Matlab*. http://www.geatbx.com/, 1994-2003.
[8] *Sthamer, H.*: The Automatic Generation of Software Test Data Using Genetic Algorithms. *PhD* Thesis, University of *Glamorgan, Pontyprid*, Wales, Great Britain, 1996.
[9] *Tracey, N., Clark, J., Mander, K., McDermid, J.*: An Automated Framework for Structural Test-Data Generation. Proceedings of the 13th IEEE Conference on Automated SE, Hawaii, USA, 1998.
[10] *Wegener, J., Sthamer, H., Baresel, A.*: Evolutionary Test Environment for Automatic Structural Testing. Special Issue of Information and Software Technology, vol. 43, pp. 851–854, 2001.
[11] *Wegener, J., Sthamer, H., Jones, B., Eyres, D.*: Testing Real-time Systems using Genetic Algorithms. Software Quality Journal, vol. 6, no. 2, pp. 127–135, 1997.

Evolutionary Testing of Flag Conditions

André Baresel and Harmen Sthamer

DaimlerChrysler AG, RIC/SM Methods and Tools
Alt-Moabit 96a, 10559 Berlin, Germany
{andre.baresel,harmen.sthamer}@daimlerchrysler.com

Abstract. Evolutionary Testing (ET) has been shown to be very successful in testing real world applications [16]. However, it has been pointed out [11], that further research is necessary if flag variables appear in program expressions. The problems increase when ET is used to test state-based applications where the encoding of states hinders successful evolutionary tests. This is because the ET performance is reduced to a random test in case of the use of flag variables or variables that encode an enumeration type.
The authors have developed an ET System to provide easy access to automatic testing. An extensive set of programs has been tested using this system [4], [16]. This system is extended for new areas of software testing and research has been carried out to improve its performance. This paper introduces a new approach for solving ET problems with flag conditions. The problematic constructs are explained with the help of code examples originally found in large real world applications.

1 Introduction to the Flag Problem

Evolutionary Structural Testing has to generate a set of test data for a given test object in order to obtain a high coverage of the program structures. The automatic process creates several test aims, which are tried to be executed in separate search processes. Test aims for statement coverage are the statements of the test object. Each search to solve a test aim is guided by a fitness function. This function defines numerically the proximity of a test datum to the current test aim.

The function uses the values examined at the condition statements during the program execution. For instance, the instrumentation of an equivalence condition on a and b will report the distance of a and b as fitness. A very small distance results in a very good fitness. This function guides the search to a equals b.

In the case of flag values, the fitness function is simply a Boolean function returning only one poor fitness value for all test data resulting in the undesired flag value. The Boolean fitness function does not guide at all the search to a desired solution and results in an evolutionary test behaving like a random test.

Fig. 1 shows two fitness landscapes close to the solution of executing the test aim. On the left hand side a Boolean function created by flag variables is shown and, in contrast to that, the right hand side shows a local distance function formed by equivalence operators. The minimum of both functions is searched for. Whereas a solution for one value of the Boolean function is easily found, the solution leading to the other Boolean value is very hard to find. This is different for the function created for the equivalence operators. Values for this function can be easily optimized.

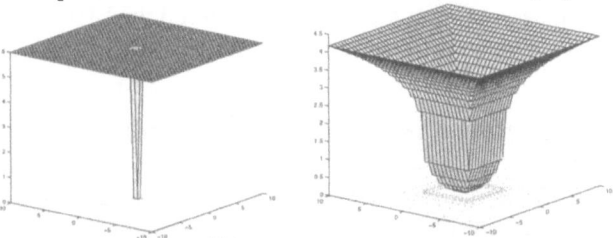

Fig. 1. The diagram on the left shows a Boolean function and the diagram on the right shows a fitness landscape created by an equivalence operator of two double variables.

It is very common to use flag variables in real world applications. Even code generators e.g. for matlab / simulink / stateflow [7] create code containing flags. This creates the problem of a Boolean-like function, which does not guide the search for test data. It also occurs when using variables that encode only a small subset of the integer type. This often emerges for variables defined as enumeration types. The measured distance between enumeration type elements does not help to find the right input to fulfill a condition that compares elements of this type. In this case the fitness function returns values that do not direct the search to the solution because the function only expresses that right or wrong values have been monitored.

The authors would like to show that random search behavior can be avoided by using a new fitness function for flag conditions. This also works for enumeration type values. The improvement is using additional information about the flag assignments in a program. These assignments can be identified by static data flow analysis.

2 Short Overview on Evolutionary Testing

Evolutionary algorithms (EA) have been used for searching data for a wide range of applications. EA is an iterative search procedure using different operators to copy the behavior of biologic evolution. When using EA for a search problem it is necessary to define the search space and the objective function (fitness). The algorithms are implemented in the widely used tool box GEAtbx [12]. It consists of a large set of operators e.g. real and integer parameter, migration and competition strategies.

ET uses EA for automatic software testing. The different software test criterions formulate requirements for a test case set to be generated. The creation of such a test data set usually has to be carried out manually. Automatic software testing generates test data sets automatically, trying to fulfill the requirements in order to increase efficiency and achieve considerable cost reduction [15].

This paper covers the automation of structural testing which is a white box test defining the quality of a test case set on the basis of the structural coverage of the test object. Evolutionary structural testing defines test aims and fitness functions to transform the problem of generating a test case set into searches solved by EA [16]. The quality of the fitness function plays an important role for the success of this approach. Some research has carried out to improve the quality (e.g. [2], [9]).

ET works very well for many real world applications. However, *'interesting test objects'*, where even an experienced software tester will have problems finding a full covering test data set, are still a challenge for the original ET. This is often due to certain constructs in the programs, which are not taken into account by the fitness.

3 The Approach to Testing Flag Conditions

This section introduces the idea of improving the fitness function by means of additional information in case of flag variables appearing in the code. The goal is to design a fitness function with a better guidance of the search, so that test data generation for flag conditions is improved and does not result in a random search.

The section is divided into subsections explaining the solution step by step. The paper ends with the analysis of a real world example and shows that the approach introduced helps gain full coverage of this test object.

3.1 Direct Assignment of a Boolean Value

The basic idea of the new approach will be explained first by using a simple example. The source code in Example 1 shows the assignment and the use of a flag variable within a nested if-then structure.

Example 1: Usage of a flag assignment
```
1:  flag = false;                                  [ENTRY-NODE]
2:  if (a==0) flag=true;   /* flag assign */      [TARGET-NODE-1]
3:  ...                    /* no other assignments to this flag*/
4:  if (c == 4) {
5:      if (flag && b> c)  /* contains a flag condition */
6:          /* test aim */                         [TARGET-NODE-2]
```

The original ET approach will immediately reach a high coverage without, however, fully covering the structure. The last search is performed for the test aim at node 6. The fitness function for this test aim is based on the control dependencies, which are created by the if-statements of line 4 and 5. This provides the search with good guidance for reaching the if-statement in line 5. However, the local distance there is a Boolean function and does not direct the search.

The authors suggest a better fitness function, which makes use of the static analysis of the test object. The *use-definition-analysis* returns the assignments in the code, which have an influence on the flag condition. The estimation of the *use-definition-chain* ([1]) will return, that the flag assignment of line 2 is the only assignment influencing the flag use of line 5. This information is the basis for a new fitness function which first guides the search to test data, which execute the flag assignment in line 2, and goes to the flag condition of line 5.

A fitness function targeting two locations in the source code has been defined as a node-node-oriented fitness function in [16]. The test aim is split into two target nodes. The first target node is the assignment, the second one the original test aim.

After applying the new fitness function, the test data searched for has to execute the assignment as well as the test aim. It is now explained how this changes the fitness function applied to Example 1.

- The original approach searches for " c==4" and "flag=true && b>c".

  This is created by the control dependency of the if-statements of line 4 and 5. Because of the last condition containing a flag an ET behaves like a random search.

- The improvement searches for "a==0 " and "c==4" and "flag=true && b>c"

  The first part is created by the control dependencies for the flag assignment of line 2 and the second part by the control dependencies for the test aim. This new fitness function directs the search to "a==0" in the first instance, automatically resulting in fulfilling the flag condition.

The following shows a short overview of the evaluation for a node-node-oriented fitness. Evolutionary Structural Testing monitors the execution of the program under test. The monitoring result for each test data contains information on the execution path and the evaluation of the conditions during runtime. On the basis of this information a fitness value has to be calculated. The original ET approach has introduced the idea of decisive branches and approximation levels assigned to the control flow graph of the test object. The approximation level at a decisive branch expresses the global distance to the test aim, whereas a local distance can be used to compare solutions with the same level reached. A detailed description on this can be found in [16].

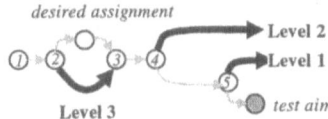

Fig. 2. Control flow graph of Example 1 with highlighted decisive branches and annotated approximation levels

Approximation levels are assigned to all branches that create a control dependency for the test aim. The dependencies can be calculated by standard algorithms (see [13]). For the *node-node-oriented* fitness function, it is necessary to estimate the dependencies for progressing from *entry node* to *target node 1* (for target nodes see Example 1, right hand side) and the dependencies progressing from *target node 1* to *target node 2*.

The assignment of the levels is performed by analyzing the possible execution orders of the identified nodes. Nodes that come later in the execution path will achieve better approximation levels than nodes at the beginning of the paths.

control dependencies for "entry-node" to "target node 1":
  node 2
control dependencies for "target node 1" to "target node 2":
  node 4, node 5
execution order of the identified nodes and assigned approximation levels:
  Node 2   Level 3
  Node 4   Level 2
  Node 5   Level 1

Fig. 3 shows a graphical interpretation of the approximation levels comparing the original (left hand side) and the new fitness evaluation approach (right hand side).

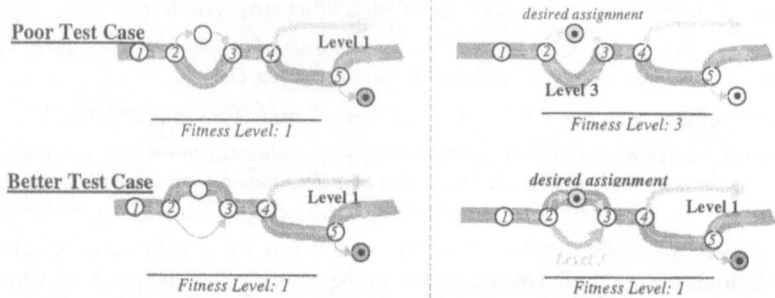

Fig. 3. The original ET (left) estimates the same fitness for both test cases; the new approach (right) assigns a poor fitness for the path not executing the flag assignment (top)

Whereas in the original approach only one decisive branch can be executed per execution path and test aim (because of the definition of the control dependency), it is necessary in the new approach to make sure that the first decisive branch (e.g. top right path) defines the fitness. This is due to the concatenation of the decisive branches of the two target nodes. A complete definition of the fitness calculation rules will be provided later in Definition 1 after all details have been explained. Fig. 4 shows the experiment results. Two scales are used in the diagram, since the fitness values are of different value ranges. The original fitness for the selected test aim ranges from 0 to 2, and the new fitness has an increased range of [0, 3].

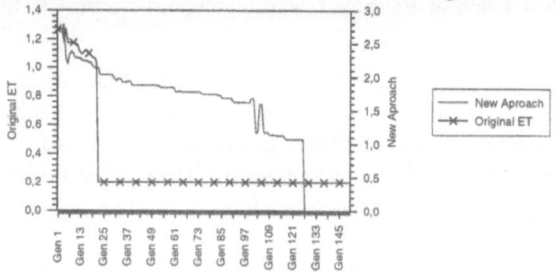

Fig. 4. The diagram shows the fitness of the best individual created by the EAs over all generations; the fitness function of the new approach guided a search so that the solution was found in Generation 130, whereas the original ET stagnates at fitness 0,4.

With the new fitness function targeting the flag assignments the evolutionary test can be improved to reach a higher coverage for applications that use single flags with just one assignment. In the next steps the authors would like to show how this idea can be extended to solve more complicated flag uses and to provide a universal solution for any kind of flag definition and use.

3.2 Dealing with Undesired Flag Assignments

The previous subsection gave an example where the execution of the test aim depended on a desired assignment. It is certain that in real world applications program code can include assignments that should not be executed to fulfill a test aim.

A very simple source code showing this can be seen in Example 2. In order to execute the test aim a flag value "true" is necessary. For this reason, a solution that reaches the test aim cannot execute the flag assignment in line 2.

Example 2: Source code containing an undesired flag assignment

```
1: flag = true;
2: if ( a!=0 || c>5 ) flag=false;/* avoid this assignment execution */
3: switch (er) {
4: case 5:
5:    if (flag)
6:       /* test aim */
```

The original ET approach does not use information about the undesired assignment. For the original ET a high probability of the flag assignment execution results in a random search, because the fitness function does not take into account the condition at line 2 (it does not create a control dependency for the test aim). A static analysis of the program under test can be helpful in this situation. By using the information about undesired flag assignments it is possible to create a better fitness function. This function has to guide the search to *not execute* the identified flag assignment. The control dependencies of the undesired assignment will be used to define the new function, but, in contrast to the solution explained for desired flag assignments, the fitness is now based on the branches not leading to the assignment.

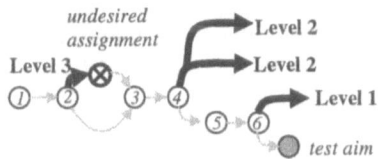

Fig. 5. Control flow graph of the example with highlighted decisive branches. The branch leading to the undesired assignment is decisive and gets a poor approx. level.

After applying the new fitness function, the test data that is searched for must *not execute* the assignment and then traverse the test aim. It is explained next how this changes the fitness function applied to Example 2.

- The original approach searches for "er==5" and "flag=true".

  This is because of the control dependency created by the switch- and if-statements of line 3 and 5. The flag condition creates a Boolean fitness function.

- The new approach is searching for "not (a!=0 || c > 5)" and "er==5" and "flag=true"

  The first part is created by the control dependencies for the flag assignment of line 2 and the second part by the control dependencies for the test aim at line 6. This new fitness function allows only solutions with "not (a!=0 || c > 5)" in the first instance (this will automatically result in fulfilling the flag condition of the second term).

An experiment with the new fitness function is presented in Fig. 6. Again, the fitness curves have been made comparable by using two scales. The optimization using the original approach stagnates at value 0,4 because it does not find a solution which results in a flag value "true".

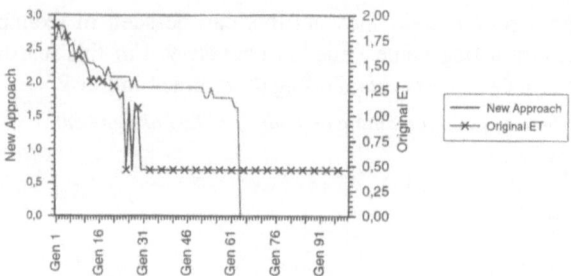

Fig. 6. Diagram demonstrating the improvement of the search process; the original approach does not find a solution for the test aim (remains at fitness about 0,4)

As shown by means of this example it is also possible to improve the original ET approach for test objects containing undesired flag assignment.

3.3 Multiple Assignments to Flag Variables

The examples shown in the previous sections only use single flag assignments. The approach is now extended for test objects with multiple flag assignments. To achieve this, the idea of *include* and *exclude* lists for flag assignments is introduced. The explanations refer to Example 3. The test aim has been commented on and the desired and undesired flag assignments for this test aim have been marked in the source code.

Applying static analysis for the example will return three flag definitions where the execution of one assignment is not desired for covering the test aim. All desired assignments will be collected in a so-called *include-assignment* list and the undesired ones will be placed on the corresponding *exclude-assignment* list.

Example 3: Source code containing multiple flag assignments

```
1: flag = false;
2: If (a==0) flag = true;      /* execute this */
3: If (b==0) flag = true;      /* or execute this */
4: ...
5: If (z > y) flag = false;    /* do not execute ! */
6: ...
7: If (c==0) {
8:    If (flag) /* test aim: go here */
```

In order to obtain a solution with a higher structural coverage it is necessary for the fitness function to return good values for any test data that:

- are close to or execute any of the *include-assignments* and
- do not execute any of the *exclude-assignments*.

All *include-assignments* have to be treated equally since it cannot be decided which of the assignments creates an executable solution for the test aim. Even with these new requirements for the fitness calculation is it possible to reuse the idea of approximation levels. Assigning the right levels to the nodes in the control flow graph and some additional calculation rules will result in fitness values that meet the requirements.

To enable a fitness calculation for the suggested improvements, branches are additionally defined as *'decisive'* (execution has an effect on fitness calculation). These are branches *avoiding the execution of a desired flag assignment* (Fig. 7 - highlighted branches of nodes 2 and 3) or branches *leading to an undesired flag assignment* (highlighted branch of node 3). If a path traverses one of those identified branches fitness has to be calculated because of the depending flag assignments.

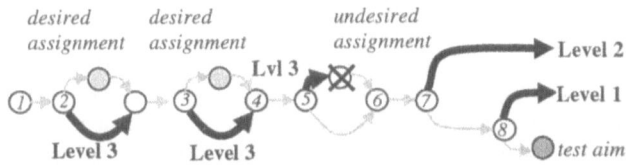

Fig. 7. Decisive branches assigned with approximation levels

The introduction of additional decisive branches makes new fitness calculation rules necessary, because some execution paths might run through several decisive branches. This cannot happen in the original ET.

DEFINITION 1: FITNESS CALCULATION RULES
1. *first decisive branch temporarily determines the fitness value, fitness can only be changed again when rules 2 and 3 are enabled.*
2. *within loops the best level over all iterations establishes the fitness value*
3. *whenever a desired flag assignment is executed it fixes the temporary fitness value; only rule 4 can make this fitness invalid*
4. *whenever an undesired flag assignment is executed it reactivates rule 1 and resets the fitness calculation.*

Some examples illustrating the usage of the rules will be described next (all referring to Example 3). If a test case misses the assignment of line 2 (rule 1 is enabled) and then executes the assignment of line 3, it will achieve a good fitness because of rule 3 (level 2). However, if the assignment of line 5 is executed afterwards, rule 4 is activated and fitness is set back to a poor value (level 3). A path that does not execute a decisive branch (highlighted in Fig. 7) and does not fulfill the condition of node 7 is assigned a fitness value of level 2. Last but not least, a test reaching line 8 and not having executed any decisive branch before will meet the test aim.

It will be described next how the *include-* and *exclude lists* change the fitness function and how this affects the search process. Example 3 is used for descriptions.

- The original ET searches for data fulfilling the condition "c==0" and "flag=true"

  This is because of the control dependency created by the if-statements of line 7 and 8. As described previously, this fitness function will result in a random search.

- The suggested improvement searches for a solution of the following term: "(a=0 || b=0) and not (z>y) and (c=0 && flag=true)"

  The first part is created by the include-assignments of the example (line 2 & 3); the second part comes from the exclude-assignment of line 5, and the last part is created by the control dependencies of the test aim. The new fitness function only allows solutions with "(a==0 or b==0) and not (z>y)" in the first instance, which automatically fulfill the flag condition.

The new fitness function targets the flag assignments, in this way enabling the evolutionary test to find a high covering test data set for test objects that uses flags. The authors will show next that even *flag uses within loops* and *expressions in flag assignments* will no longer be a problem with this approach.

3.4 Flag Assignments within Loops

As mentioned previously, applications can also assign flag variables within loops. ET generally has no problems with loops appearing in the test object source code. The method can search for test data fulfilling test aims outside and inside any kind of loops. For this reason the new approach also works if the flag assignment is placed within a loop. The short Example 4 demonstrates this case. The flag assignment of line 3 is placed within a *while-loop* and the flag use is at line 4.

Example 4: Flag assignment appearance within a loop
```
1:  flag = false;
2:  while (i<10)
3:     if (a[i]==0) flag = true;
4:  if (flag)
5:     /* test aim */
```

In this situation, an evolutionary search for a solution of the test aim at line 5 requires an improved fitness function because of the flag condition at line 3. The calculation of the *include-assignments* and *exclude-assignments* will return just one *include-assignment* in line 3 (line 1 is an initialization that is always executed).

ET has no problem guiding the search to a test datum that executes the assignment of line 3. The new fitness function is the following logical term:

- The suggested improvement is searching for a solution of the following term: "{*in-any-iteration* a[i]==0}" and "flag==true"

   The operator "in-any-iteration" is created by the include assignments which is placed within the loop. The assignment needs to be executed only once.

This new fitness function improves the evolutionary test. However, using the new fitness function implemented via approximation levels does not solve the problem of undesired flag assignment appearing in loops. This is because of the incomplete use of the monitored information. With the appearance of *undesired-assignments* inside loops it is necessary to define a fitness function which numerically evaluates the *"in-all-iterations"* operator. In Example 4 it is the fulfillment of the condition:

"{*in-all-iteration* not a[i]==0 }"

But an implementation using approximation levels with the rules of Definition 1 results in a fitness calculated on the basis of a transformed version of this term:

"not {in-any-iteration a[i]==0 }.

This is logically equivalent to the first but results in a fitness function which is calculated on the basis of only one iteration that fulfills the condition: *"a[i] = 0"*.
Analyzing the results of ET has shown that this will lead to a random search. The reason is that the fitness values do not take into account that a test datum with just one iteration meeting the condition "a[i] equals zero" is better than a test datum executing more iterations meeting this condition. A different implementation is needed here.

Using the new approach the ET can also be improved in the case of loops appearing in the code. The new fitness function guides the search to execute desired flag assignments appearing within any kind of loop statements even if the loop is created by 'goto' statements. Only the appearance of *undesired-assignments* within loops still causes the flag problem for our implementation.

3.5 Boolean Expressions Assigned to Flags

The previous subsections have shown how to improve the fitness function if the flag is directly assigned to just one value. Real world applications often make use of assigning expressions to flags which are evaluated at runtime. Static analysis cannot distinguish which value these assignments return. Nevertheless, the approach can also handle this very common case. Example 5 shows one function code containing an assignment of an expression in line 1 and the use of the flag in line 3.

Example 5: Flag assigned by a Boolean expression
```
1: flag = (a==0) || (b>0 && b<5);
2: ...
3: if (flag) /* test aim */
```

Running the original ET approach will perform poorly with a low coverage. There is only a slim chance that it will find a solution. An ET inserting the Boolean expression at line 3 will work without any problems. Unfortunately the transformation is sometimes a difficult task (more details in [6]). A more simple transformation can help obtain a test object which can be handled with the approach explained in the previous subsections. This transformation changes the assignment expression into an if-then-else construct which behaves completely equivalent to the original code. The transformed version is shown in Example 6. This code has just two direct value assignments to the flag variable.

Example 6: Transformed expression
```
1: if ( ( a == 0 ) || ( b > 0 && b < 5 ) )
       flag = true;   else   flag = false;
```

Applying the new approach does not actually require such a transformation as seen in Example 6. Simple function calls inserted into the Boolean expression will obtain the information necessary for the fitness calculation,

e.g. `flag=DistEqual(a,0) || DistGreater(b,0) && DistLess(b,5))`.

A fitness function using the data collected during the execution of the instrumented expression guides the search to a solution assigning the required flag value.

The approach in [5] describes a similar idea for guiding the search by data dependencies. All paths containing assignments relevant for the flag condition under test are generated and optimized in a sequence. However, our approach uses a single optimization considering all these paths simultaneously. [5] can solve the previously described flag problems, but cannot guide the search in case of Boolean expressions.

Up until now a general approach for guiding the search in the case of just one flag variable has been presented. The next subsection extends the applicability to a wider range of test objects. The use of more than one flag takes place in many code generated examples. However, it becomes more difficult to guide the search in such cases.

3.6 Using More than One Flag Variable

The idea is extended for test problems with more than one flag variable in the condition that needs to be fulfilled. The next example shows an instance of this case.

Example 7: Source code showing the use of more than one flag variable

```
1 If (a==0) initialized = true;          /* execute this*/
2 If (b==0) has_been_fired = true;       /* and execute this */
3 ....
4 if (initialized)
5 {
6     If (b==3) has_been_fired = true;   /* or execute this */
7 ..
8     If (has_been_fired )
9         /* test aim */
```

The idea of this example is that the test aim is only executed if the two flags are assigned in the right way before reaching the test aim. In real world applications any combination of flag assignments within nested conditions and loops are conceivable. The fitness evaluation has to check in parallel for each flag appearing in the code whether or not it is assigned in the desired way. That is why the original approach using approximation levels does not work properly.

With the use of approximation levels and local distances the instrumentation can decide how close the test datum is to a flag assignment. This value (*temporary flag fitness*) can be stored together with the flag value at runtime. If the flag is assigned as necessary for the test aim this information is not used, but, if it is not, the temporary flag fitness will be used to calculate the overall fitness of the test datum. The fitness calculation is prepared by checking the control dependencies of all flag uses; estimating the *exclude-assignment* and *include-assignment* lists for each identified flag, calculating the decisive branches on the basis of the *include* and *exclude* lists, and assigning the *flag-approximation-levels* to the branches independently for each flag. The assigned *flag-approximation levels* are used to calculate temporary *flag fitness*. This fitness forms the basis for the improved fitness function for multiple flag uses.

DEFINITION 2: FITNESS CALCULATION RULES FOR MULTIPLE FLAGS
1. *temporary flag fitness is estimated as defined in the rules of Definition 1 for each flag appearing in the code*
2. *upon reaching a control-dependent condition containing flags it is checked whether or not the flags are assigned in the desired way. If the flags are assigned correctly the usual fitness calculation proceeds. Whenever one or more flags are not assigned correctly, fitness is calculated using the corresponding temporary flag fitness.*

Annotation. *In the case of multiple flags occurring in a condition we suggest calculating an overall fitness using all temporary flag fitness values. See [2] for the reasons for parallel condition optimization.*

The implementation of this approach is quite complex and the calculation of the approximation levels requires some extra processing time. However, when applying the original approach on examples using multiple flags, the coverage reached with the original ET is worse. A related paper to this idea is [3]. A complete solution needs further research.

3.7 Real World Example

The authors have tested the approach using a real world application extracted from software for a car-controlling unit. The function under test is responsible for regulating the internal states of the energy control of a small sub-function within body and comfort electronics of a car. The function has 100 Loc and an *if-then* nesting level of 5. It has 4 input parameters and internally uses two flags. For testing reasons the initial state and the input situation have been generated.

The original approach did not find a solution for the condition statements using the two flags. Within a test of about 320.000 individuals it has reached a branch coverage of about 90%. This happens with the standard settings of the ET-system using 300 individuals in 6 populations with competition enabled. A maximum of 200 generations per test aim were allowed. The authors ran the same test with a bigger population (700 individuals) and a later stopping criterion (max. 400 generations), but no improvements were noticeable.

Using the new approach no problems were distinguished. One of the flags was assigned by an expression which had a very low probability evaluating to 'true'. This was the reason why even more tests did not perform better using the original approach. The second flag was assigned within a nested if-then structure and tested later in a different nesting level. In the original ET approach there was only a very low chance of executing the flag assignment and the condition within one execution. The fitness function did not guide the search to this solution.

The new approach used the standard EA settings and covered the test object fully after 60 generations and testing 14500 individuals. This is 5 percent of the workload of the original approach and leads to 100 % branch coverage.

4 Conclusion

Evolutionary Testing uses metaheuristic search methods to automate software testing aspects. The occurrence of flag variables has been pointed out to be problematic because of a poor guidance of the search at conditions containing flags. The authors introduce a new fitness function, which improves evolutionary structural testing in case of flag conditions. This function uses additional information on the flag assignments occurring in the test object. The solution is explained by using short code examples extracted from real world applications.

The introduced improvements cannot only solve the problem of flag variables, the authors argue that the test of sources containing enumeration type conditions, also known to be problematic, can be improved using the introduced approach.

By using a fitness function that guides the search to variable assignments as well as variable uses it has been shown that the flag problem can be solved. We believe that future research on sequence testing can reuse this idea. This is because in sequence testing state variables are assigned and used in different areas of a program and sometimes also in the different steps of a sequence. A fitness function guiding the search to the execution of state variable assignments and to the condition testing the state variable will perform better than the original ET.

References

[1] Appel, A. W.: Modern Compiler Implementation in C, Cambridge, New York: Cambridge University Press, 1998
[2] Baresel, A., Sthamer, H. and Schmidt, M.: Fitness Function Design to improve Evolutionary Structural Testing, Proceedings of the GECCO, New York, USA, July 2002
[3] Bottaci, L.: Instrumenting Programs With Flag Variables For Test Data Search By Genetic Algorithms, GECCO 2002: Proceedings of the Genetic and Evolutionary Computation Conference, p. 1337–1342, July 2003
[4] Buhr, K.: Complexity Measures for the Assessment of Evolutionary Testability (only german version). Diploma Thesis, Technical University Clausthal, 2001
[5] Ferguson, R.; Korel, B.: The Chaining Approach for Software Test Data Generation, Transactions on Software Engineering and Methodology, Vol. 5 No.1, pp.63–86, 1996
[6] Harman, M., Hu, L., Hierons, R., Baresel, A. and Sthamer,H.: Improving Evolutionary Testing by Flag Removal, Proceedings of GECCO, New York, USA, 9–13th July 2002
[7] http://www.mathworks.com/
[8] Harman, Hu, Hierons, Fox, Danicic, Baresel, Sthamer; Wegener: Evolutionary Testing Supported by Slicing and Transformation, 18th IEEE International Conference on Software Maintenance (ICSM 2002), 3–6 Oct. 2002. Montreal, Canada. Page 285.
[9] Jones, B.-F.; Sthamer: H.-H.; Eyres, D.: Automatic structural testing using genetic algorithms. Software Engineering Journal, vol. 11, no. 5, pp. 299–306, 1996
[10] Korel, B.: Automated Test Data Generation. IEEE Transactions on Software Engineering, vol. 16 no. 8 pp.870–879; August 1990
[11] Michael, C.C,. Mcgraw, G, Schatz, M.A.: Generating Software Test Data by Evolution, IEEE Transactions on Software Engineering, vol. 27, No. 12 pp. 1085–1110; Dec. 2001
[12] Pohlheim, H.: GEATbx – Genetic and Evolutionary Algorithm Toolbox for Matlab. http://www.geatbx.com/, 1994–2001
[13] Schaeffer: A mathematical theory of global program optimization, Prentice-Hall Inc. 1973
[14] Tracey, N., Clark, J., Mander, K. and McDermid, J.: An Automated Framework for Structural Test-Data Generation, Proceed. of the 13th IEEE Conf. on Automated SE, 1998
[15] Wegener, J., Sthamer, H., Baresel, A.: Application Fields for Evolutionary Testing, Eurostar 2001 Stockholm, Sweden, November 2001
[16] Wegener, J., Sthamer, H., Baresel, A.: Evolutionary Test Environment for Automatic Structural Testing, Special Issue of Information and Software Technology, vol. 43, pp. 851–854, 2001

Predicate Expression Cost Functions to Guide Evolutionary Search for Test Data

Leonardo Bottaci

Hull University, Hull, HU6 7RX, UK
l.bottaci@dcs.hull.ac.uk

Abstract. Several researchers are using evolutionary search methods to search for test data with which to test a program. The fitness or cost function depends on the test goal but almost invariably an important component of the cost function is an estimate of the cost of satisfying a predicate expression as might occur in branches, exception conditions, etc. This paper reviews the commonly used cost functions and points out some deficiencies. Alternative cost functions are proposed to overcome these deficiencies. The evidence from an experiment is that they are more reliable.

1 Introduction

Several researchers are using evolutionary search methods to search for test data with which to test a program. The fitness or cost function depends on the test goal but almost invariably an important component of the cost function is an estimate of the cost of satisfying a predicate expression as might occur in a branch condition, an exception condition, etc.

As an example of the most basic instrumentation, consider a search for test data that will execute a given sequence of branches in a program. A record may be kept of the values of all branch predicate expressions executed. A cost for the given input is computed by counting the number of branches that have not been satisfied. This is the method used by Pargas et al. [5].

Although a count of undesired branch decisions provides some guidance to the search; all the test cases that fail to satisfy the same branch are given the same cost. At this point, the cost surface over the search space has become flat and the search becomes random. As an example, consider the program fragment below.

```
...
if (a <= b)
    ...            // EXECUTION REQUIRED TO ENTER THIS BRANCH
```

Suppose a test case is required to cause execution of the true branch of the conditional shown above. If the required branch is difficult to enter, many test cases will cause a <= b to be false. To discriminate between these tests, the program is instrumented to calculate a cost measure that penalises those tests that may be considered to be "far from" satisfying a <= b. As an example of a possible cost function for the condition a <= b, the value of $a - b$ increases as a becomes larger than b and a zero or negative cost indicates that a solution has been found. Through instrumentation, the subject program has in effect been converted into another program that computes a function that is to be minimised to zero. This approach has been used by Korel [2], Tracey et al. [7], [8], Wegener et al. [9], Jones [1] and Michael [4] [3].

Clearly, the effectiveness of the search depends on the reliability of the cost functions used for the relational and logical expressions. This paper reviews the commonly used cost functions and points out some cases where they are deficient. Alternative cost functions are proposed to overcome these deficiencies. A small experiment provides some evidence that the alternative cost functions are more reliable, and more so for relatively simple programs.

2 Cost Functions for Relational Predicates

Table 1 shows the commonly used cost functions for the relational predicates. a, b are numbers and ϵ is a positive number.

Since, logically equivalent expressions express the same condition, ideally they should have equal costs. In purely integer domains, $a + 1 \leq b \Leftrightarrow a < b$. Since $cost(a + 1 \leq b) = a + 1 - b$ and $cost(a < b) = a + \epsilon - b$, this entails $\epsilon = 1$. In real domains, ϵ is the smallest positive real.

It should be emphasised that $a - b$ as the cost of satisfying $a \leq b$ is at best a heuristic. Usually, the input test case determines the values of the operands a, b only indirectly and any intervening statements have the potential to produce a cost surface as a function of the inputs that is far more complex than $a - b$. Nonetheless, many arithmetic operations do not destroy the reliability of the heuristic and it remains effective where inputs are modified in the manner illustrated in the example below

```
a := a * a - k;
if (a <= 0)
```

It is, of course, possible for the heuristic to be deceptive. Consider the program fragment

```
a := (a * a + 1) mod 65;
if (a <= 0)
```

Table 1. Relational predicate cost table using conventional cost functions

Predicate expression	Cost of not satisfying predicate expression
$a \leq b$	$a - b$
$a < b$	$a - b + \epsilon$
$a = b$	$abs(a - b)$
$a \neq b$	$\epsilon - abs(a - b)$

In this fragment, the cost of a <= 0 decreases as a (input) approaches 0. At $a = 0$ the cost function attains a local minimum value of 1 but the solution is $a = 8$. It is also possible for the heuristic to be completely uninformative as in

```
a := random(a);
if (a <= 0)
```

3   Cost Functions for Logical Operators

Cost functions may be defined for the logical operators *not*, *or* and *and* in order to define a cost for compound predicate expressions such as a <= b and not(a > 0). Consider first the logical negation operator.

Some researchers avoid an explicit cost function for logical negation and instead rewrite expressions that contain negation into a form that is negation free. So for example, not (a <= b) is rewritten as a > b. Introducing a cost function for logical negation avoids the need to rewrite expressions; indeed by introducing a cost function for each relational and logical predicate, cost functions may be isomorphic to predicate expressions which allows cost functions to be constructed in a simple syntax directed manner. This is an important consideration when building tools.

The cost function for negation can be derived from the requirement that $cost(a \leq b) = cost(\neg(a > b))$, i.e. $a - b$ should equal $-(b - a + \epsilon) + \epsilon$ and hence $cost(\neg a) = -cost(a) + \epsilon$.

The use of logical negation also introduces the need for a cost to be assigned to a true predicate expression. In particular, the logical constant, *true* may be given a cost of $-maxcost$ and the logical constant, *false* has a cost of $maxcost$. The use of these large absolute values reflects the fact that the logical constant *false* can never be satisfied, and *true* can never be falsified. In this scheme, all cost functions are bounded by $-maxcost$ and $maxcost$.

Consider next the cost function for a disjunction. Clearly, when the operands have different truth values, the cost of the disjunction should be the cost of the true operand. This leaves the cases where the operands have the same truth value. In this case, a popular choice for the cost of a disjunction is the cost of the operand with the lowest cost i.e. the cost function is the *min* function. The common corresponding cost function for the conjunction is the *max* function.

The cost tables for logical negation, *or* and *and* are summarised in Table 2, where c_a is the cost of a boolean expression a. In Table 2, c_a and c_b are positive (*false*) and $c_{a'}$ and $c_{b'}$ are non-positive (*true*).

The cost functions of Table 2 have been used by a number of researchers. Tracey et al. [7] use essentially the same cost functions although their's are restricted to nonnegative values, they measure only the cost of not satisfying an expression and negation is removed by rewriting expressions. A notable difference, however, is the use of $+$ rather than max for conjunction so that $cost(a \wedge b) = cost(a) + cost(b)$.

Table 2. Logical operator cost table using conventional cost functions

a	b	$\neg a$	$a \vee b$	$a \wedge b$
c_a	c_b	$-c_a + \epsilon$	$min(c_a, c_b)$	$max(c_a, c_b)$
c_a	c'_b		c'_b	c_a
c'_a	c_b	$-c'_a + \epsilon$	c'_a	c_b
c'_a	c'_b		$min(c'_a, c'_b)$	$max(c'_a, c'_b)$

4 Cost Function Reliability

4.1 Analytical Considerations

The functions min and max have become popular for various forms of logical reasoning under uncertainty [6] since Zadeh proposed them for fuzzy logic [11]. In fuzzy logic, *true* is represented by 1 and *false* by 0 and intermediate values lie in between. The truth value of a fuzzy disjunction is the maximum of the truth values of the operands, the value of a fuzzy conjunction is the minimum of the operand truth values. The common use of the functions min and max[1] should not obscure the fact that the intended interpretation in test data search is quite different.

Fuzzy logic is concerned with formalising reasoning with vague concepts where the vagueness is formalised as a fuzzy set. The justification for the use of min and max in fuzzy logic comes from the corresponding union and intersection operations for fuzzy sets. The costs associated with logical expressions to guide search are not intended to measure vagueness.

Given that the costs of predicate expressions are intended to estimate search effort, some properties which could be considered essential are:

1. The cost of a disjunction should be no more than the cost of either disjunct, i.e. $cost(a) \geq cost(a \vee b)$ and $cost(b) \geq cost(a \vee b)$.
2. The cost of a conjunction should be no less than the cost of either conjunct, i.e. $cost(a) \leq cost(a \wedge b)$ and $cost(b) \leq cost(a \wedge b)$.
3. The cost of logically equivalent expressions should be equal.

Property 1 can be justified on the grounds that adding an alternative means of satisfying a condition cannot make that condition more difficult to satisfy and so cannot increase the cost. The min function satisfies property 1 but makes the assumption that the cost of a disjunction is the maximum cost consistent with property 1. The argument for property 2 is analogous to that for the disjunction. The max function satisfies property 2 but makes the assumption that the cost of a conjunction is the lowest cost consistent with property 2.

Property 3 requires the cost functions to be consistent with the associative, commutative and distributive laws. Examples of other laws include: $cost(a)$

---
[1] In fuzzy logic truth increases with numerical value so min corresponds to max and vice versa.

should equal $cost(a \vee a)$ and $cost(a \vee b)$ should equal $cost(\neg(\neg a \wedge \neg b))$. The cost functions of Table 2 satisfy all three of these properties.

Although all three properties would appear necessary, it might be advantageous to trade some violation of the third property for a more reliable or informative function. Recall the use by Tracey [7] of $+$ instead of max as the cost of a conjunction. To use $+$ instead of max for the cost of a conjunction, while retaining min for the cost of a disjunction is to give up the property that logically equivalent expressions have the same cost. In particular, the distributive law of disjunction over conjunction is not satisfied because $cost(a \vee (b \wedge c)) = min(a, b+c)$[2] but $cost((a \vee b) \wedge (a \vee c)) = min(a, b) + min(a, c)$. De Morgan's law is not satisfied either.

A possible consequence of this is illustrated in the code fragment below where two logically equivalent expressions appear in distinct subgoals. Assume that the test goal is to find a test that will execute either of the statements z := 0;, i.e. to satisfy either (x < 4 or y < 4) or not (x >= 4 and y >= 4).

```
if (x + y >= 16)
    if (x < 4 or y < 4)
        z := 0;
else
    if (not (x >= 4 and y >= 4))
        z := 0;
```

To exaggerate the inconsistency and also for the sake of clarity, variables are taken to be integer so that $\epsilon = 1$. When x = 8, y = 8, the cost is $min(5,5) = 5$ but when x = 7, y = 8, which is a better test, the cost is higher at $-(-3+-4)+1 = 8$.

The possible bias this might introduce in examples such as that shown above might well be compensated for by a more general reliability advantage of $+$ over max. $+$ rewards a decrease in the costs of both conjuncts more than it rewards a corresponding decrease in the cost of just one. For example, consider the problem of satisfying the condition x = 0 and y = 0. A move by a search algorithm from the (x, y) point $(4, 6)$ to the point $(3, 5)$ is rewarded by $+$ in a cost decrease of 2 but max produces only a cost decrease of 1. Similarly, a detrimental move from $(4, 6)$ to $(5, 7)$ is penalised more heavily by $+$ than by max. The function $+$ would seem to be more discriminating.

Continuing in this direction, the use of min as the cost function for disjunction may be reconsidered in the hope of finding some different function, analogous to $+$, that is more discriminating than min. Consider, for example, the following program fragment.

```
x := 1;
while (x <= 0 or y = 5)    // EXECUTION REQUIRED TO ENTER LOOP
```

---

[2] Note that the notation of c_a as the numeric cost of the boolean expression a is being dropped from here on. A symbol a may denote either a boolean expression or its numeric cost, as determined by the context.

Table 3. As y approaches 5, the cost decreases towards 0

	y	9	8	7	6	5
x <= 0		1	1	1	1	1
y = 5		4	3	2	1	0
x <= 0 or y = 5		0.8	0.75	0.67	0.5	0

Table 4. Proposed logical *or* and logical *and* cost table

a	b	$a \vee b$	$a \wedge b$
a	b	$\frac{ab}{a+b}$	$a+b$
a	b'	b'	a
a'	b	a'	b
a'	b'	$a'+b'$	$\frac{a'b'}{a'+b'}$

When searching for values for the variables x and y to enter the while loop, the value of x is 1 when the conditional is first evaluated and so the cost of x <= 0 is 1 and this will in fact be the cost of the predicate expression for all input values unless $y = 5$. A flat surface in the cost function provides no guidance to the search.

One way to interpret this problem is to consider that the cost of 1 for x <= 0 should, initially at least, be much higher given the impossibility of changing the value of x from this value until entry into the loop. Determining such facts about the values of arbitrary variables in a program is of course just as hard a problem as that of finding test data.

Since the *min* function ignores improvement in the cost of the more costly operand, a more discriminating function might be constructed that takes account of a cost improvement in either operand. Such an alternative is the ratio of the product of the costs to the sum of the costs, i.e.

$$cost(a \vee b) = \frac{ab}{a+b}.$$

The table of costs below shows how this cost function solves the problem in the previous example.

The proposed cost functions for *or* and *and* are shown in Table 4. Note that when $a' = b' = 0$ then $cost(a \wedge b) = 0$. The above cost functions satisfy properties 1 and 2 but do not satisfy the property of equal costs for logically equivalent expressions. The error in satisfying De'Morgan's law, i.e. $cost(a \vee b) \neq cost(\neg(\neg a \wedge \neg b))$ is due to the presence of ϵ, the positive offset from zero for the costs of all false predicates that is absent from the costs of true expressions.

This anomaly can be removed by modifying the relational predicate cost functions to produce values symmetrical about zero in the range $[-maxcost, -\epsilon] \cup [\epsilon, maxcost]$ (as shown in Table 5) and to define the cost of logical negation as $cost(\neg p) = -cost(p)$.

Other inconsistencies remain, however. For example, $cost(a) \neq cost(a \vee a)$. The difference between $cost(a)$ and $cost(a \vee a \vee a)$ is greater still with the dif-

Table 5. Relational predicates with costs symmetrical about zero

Predicate expression	Cost of predicate expression	
$a \leq b$	$a - b,$	$a > b$
	$a - b - \epsilon,$	$a \leq b$
$a < b$	$a - b + \epsilon,$	$a \geq b$
	$a - b,$	$a < b$
$a = b$	$abs(a - b),$	$a \neq b$
	$-\epsilon,$	$a = b$

Table 6. A flat cost function when x = 0.0

i	6.0	7.0	8.0	9.0	10.0
i <= 9.0	$-3.0 - \epsilon$	$-2.0 - \epsilon$	$-1.0 - \epsilon$	$-\epsilon$	$1.0 + \epsilon$
x = 0.0	$-\epsilon$	$-\epsilon$	$-\epsilon$	$-\epsilon$	$-\epsilon$
i <= 9.0 and x = 0.0	$-\epsilon$	$-\epsilon$	$-\epsilon$	$-\epsilon$	$1.0 + \epsilon$
not(i <= 9.0 and x = 0.0)	ϵ	ϵ	ϵ	ϵ	$-1.0 - \epsilon$

ference bounded by $cost(a)$. The difference between $cost(a)$ and $cost(a \wedge a)$ is $cost(a)$ and is unbounded as the number of conjunctions of a increases. In practice, such expressions are likely to be relatively rare because programmers tend to avoid writing expressions that are clearly inefficient.

One of the problems with the cost function $\frac{ab}{a+b}$ is that when one operand is very small, changes in the other are not significant. When the cost of $b = \epsilon$ then $\frac{ab}{a+b} = \frac{a\epsilon}{a+\epsilon}$ which because of rounding error can only safely be taken to be ϵ. As an example of this problem, consider the condition not((i <= 9.0) and (x = 0.0)) where x and i are real and so ϵ is the smallest positive real. Table 6 shows the cost calculations for different values of i when $x = 0.0$. The cost remains the same at ϵ for all values of i up to 9 and so provides no guidance for the search.

To assign a cost of ϵ (the least positive value is the relevant number domain and the lowest possible cost for a false predicate expression) to a single relational expression such as $a < b$ leaves no room to give a lower cost to disjunctive expressions that include $a < b$ as one of the disjuncts. Recall the requirement that $cost(a) \geq cost(a \vee b)$.

Cost functions for relational predicates can be modified to overcome this problem by setting the absolute minimum cost for any single relational predicate expression to be some value significantly larger than ϵ, say $R \geq 1$. The costs of relational predicates in this scheme is given in Table 7.

In Table 8 the cost calculations of the previous example are repeated but with the use of the modified relational predicate cost functions with $R = 1$. The cost of not(i <= 9.0 and x = 0.0) now decreases as i approaches 9.0.

4.2 Experiment to Compare Cost Function Reliability

Ultimately, the validity of cost functions for test data search is an empirical issue. Ideally, cost functions should be compared over a large sample of programs from

Table 7. Modified relational predicate cost functions

Predicate expression	Cost of predicate expression	
$a \leq b$	$a - b + R - \epsilon,$	$a > b$
	$a - b - R,$	$a \leq b$
$a < b$	$a - b + R,$	$a \geq b$
	$a - b - R + \epsilon,$	$a < b$
$a = b$	$abs(a - b) + R - \epsilon,$	$a \neq b$
	$-R,$	$a = b$

Table 8. A decreasing cost function

i	6.0	7.0	8.0	9.0	10.0
i <= 9.0	−4.0	−3.0	−2.0	−1.0	2.0
x = 0.0	−1.0	−1.0	−1.0	−1.0	−1.0
i <= 9.0 and x = 0.0	−0.8	−0.75	−0.67	−0.5	2.0
not(i <= 9.0 and x = 0.0)	0.8	0.75	0.67	0.5	−2.0

various application areas. A less time consuming experiment can be done with synthetic programs generated automatically. Consequently, a sample of programs was generated by modifying the following program.

```
x := a + b + c;
y := a + b + c;
z := a + b + c;
if (x = a or y = b or z = c)
```

The modification rules, allowed: all occurrences of + to be replaced by any arithmetic operator, all occurrences of = to be replaced by any relational predicate, all occurrences of or to be replaced by any binary logical operator with the possible insertion of a negation operator, in addition two binary operators leftoperand and rightoperand were allowed throughout (each returning the value of one operand) as a way of reducing the length of expressions. The modification rules, allowed variables (other than x, y and z) to be replaced with any other variable or a constant drawn from a small set of integers.

For each program generated, at random, three copies were produced, the first was instrumented with the *min* and *max* cost functions of (Table 2) (herein called the min-max functions), the second with the $\frac{ab}{a+b}$ and $a + b$ functions of (Table 4) (herein called the ratio-sum functions) and the third with constant functions so that a uniform random search is done. The relational predicate expression cost functions of Table 7 were used throughout with $R = 1$. For all programs, the test goal was to find values of a, b and c, each from the domain $[-4999, 5000]$ to cause entry into the conditional.

The genetic algorithm used for the search was of the steady-state variety and similar to Genitor [10]. Test inputs were coded not as binary strings but as strings of integer. Reproduction takes place between two individuals who produce one or two offspring (depending on the choice of reproduction operator). These

Table 9. Performance of cost functions

program type	min-max		ratio-sum		random		trivial	failed
	solns	evals	solns	evals	solns	evals		
simple	1116	504	1309	456	97	436	0	0
complex	1853	308	2059	313	789	301	16992	6237

offspring are then immediately inserted into the population (of size 50) expelling the one or two least fit. The population is kept sorted according to cost and the probability of selection for reproduction is based on rank in this ordering.

4.3 Results

For each set of cost functions, Table 9 shows the number of trials (solns column) in which a solution was found. Three attempts were made at each problem and the table shows the total number of solutions found. The column headed 'evals' shows the mean number of fitness function evaluations used per solution, counting only cases where a solution was found. A number of the programs generated were not counted against any cost function because they were either too easy to solve (column headed 'trivial'), namely a solution was found in less than 10 random attempts, or too difficult (column headed 'failed'), because a solution was not found within the limit on fitness function evaluations, set at 1000.

A crude attempt was made to distinguish between simple and more complex programs according to the syntactic complexity of the arithmetic expressions. All programs of the following form (where only or, and may be replaced by or, and) were classed as simple.

```
x := a;
y := b;
z := c;
if (x = 1 or y = 1 and z = 1)
```

Simple programs all have a smooth cost surface (smoothest in the case of ratio-sum) but the solution set is small. In Table 9, the set of programs classed as complex is simply the entire sample of synthetic programs generated.

It can be seen that for simple programs, the performance of the ratio-sum cost functions is better, the data shows a 17% outperformance. For all programs, the ratio-sum cost functions outperform the min-max cost functions by about 10%. A possible explanation for this difference is that there is less opportunity for ratio-sum to take advantage of an improvement in more than one predicate expression cost at once since such moves are less likely to occur as the jaggedness of the cost surface increases.

5 Conclusion

Several researchers are using evolutionary search methods to search for test data with which to test a program. The fitness or cost function depends on the test goal but almost invariably an important component of the cost function is an estimate of the cost of satisfying a predicate expression as might occur in a branch condition, an exception condition, etc.

It has been shown that the set of commonly used cost functions for the satisfaction of logical predicates (the min-max functions) perform poorly in certain cases. An alternative set of cost functions (the ratio-sum functions) has been proposed which overcome these specific problem cases. To determine the effectiveness of the ratio-sum functions on a wider range of problems, an experiment was done to compare cost functions for a sample of synthetic programs. It has been shown the ratio-sum cost functions modestly outperform the min-max cost functions but more so for relatively simple programs. A possible explanation for this is that the ability of ratio-sum to take advantage of an improvement in more than one predicate expression cost at once can be better exploited in simple programs.

Synthetic programs are an inexpensive way of subjecting cost functions to a relatively large sample of programs but they can at best provide only an insight into the behaviour of different cost functions. In terms of assessing the reliability of different cost functions in practice, they can be no more than a prelude to an experiment with a large sample of real programs. Work is underway to do this.

References

1. B. F. Jones, H. Sthamer, and D.E. Eyres. Automatic structural testing using genetic algorithms. *Software Engineering Journal*, 11(5):299–306, 1996.
2. B. Korel. Automated software test data generation. *IEEE Transactions on Software Engineering*, 16(8):870–879, August 1990.
3. G. McGraw, C. Michael, and M Schatz. Generating software test data by evolution. Technical Report RSTR-018-97-01, RST Corporation, Suite 250, 21515 Ridgetop Circle, Sterling VA 20166, 1998.
4. C. Michael, G. McGraw, M. Schatz, and C. Walton. Genetic algorithms for dynamic test data generation. Technical Report RSTR-003-97-11, RST Corporation, Suite 250, 21515 Ridgetop Circle, Sterling VA 20166, 1997.
5. R. P. Pargas, M. J. Harrold, and R. P. Peck. Test-data generation using genetic algorithms. *Software Testing, Verification and Reliability*, 9:263–282, 1999.
6. Judea Pearl. *Probabilistic reasoning in intelligent systems*. Morgan Kaufmann, 1988.
7. N. Tracey, J. Clark, and K. Mander. Automated program flaw finding using simulated annealing. *Software Engineering Notes*, 23(2):73–81, March 1998.
8. N. Tracey, J. Clark, K. Mander, and J. McDermid. Automated test data generation for exception conditions. *Software – Practice and Experience*, 30:61–79, 2000.
9. J Wegener, A. Baresel, and H. Sthamer. Evolutionary test environment for automatic structural testing. *Information and Software Technology*, 43:841–854, 2001.
10. D. Whitley. The genitor algorithm and selective pressure: why rank based allocation of reproductive trials is best. *Proceedings of the Third International Conference GAs.*, pages 116–121, 1989.
11. L. A. Zadeh. Fuzzy logic and approximate reasoning. *Synthese*, 30:407–428, 1975.

Extracting Test Sequences from a Markov Software Usage Model by ACO

Karl Doerner[1] and Walter J. Gutjahr[2]

[1] Department of Management Science, University of Vienna
Bruenner Strasse 72, A-1210 Vienna, Austria
[2] Department of Statistics and Decision Support Systems, University of Vienna
Universitaetsstrasse 5/3, A-1010 Vienna, Austria
{Karl.Doerner,Walter.Gutjahr}@univie.ac.at

Abstract. The aim of the paper is to investigate methods for deriving a suitable set of test paths for a software system. The design and the possible uses of the software system are modelled by a Markov Usage Model which reflects the operational distribution of the software system and is enriched by estimates of failure probabilities, losses in case of failure and testing costs. Exploiting this information, we consider the tradeoff between coverage and testing costs and try to find an optimal compromise between both. For that purpose, we use a heuristic optimization procedure inspired by nature, Ant Colony Optimization, which seems to fit very well to the problem structure under consideration. A real world software system is studied to demonstrate the applicability of our approach and to obtain first experimental results.

1 Introduction

Markov usage models, as introduced by Whittaker, Poore, Walton and other authors (see [27]–[28], [26], [21], [24]) are versatile model descriptions for the use of a given software product in the application field. It is common knowledge that users do not exploit all functions described in the specification of a product with the same intensity; some features of a software application get "everyday routine", others are very rarely touched – or never. A software developer can schedule her/his resource allocations during different phases of the development cycle more cost-effectively, if she/he takes account of these differences. For example, it does not make sense to perform extensive design reviews, code inspections and tests for specific usage constellations that will probably never occur in practice, at least if a failure of the product on these constellations will not cause much harm.

A Markov usage model (MUM) is based on a *Markov chain* representation of the software usage. For putting it up, it is necessary to identify (1) *states* of the software, (2) possible *transitions* between these states (during these transitions, processing takes place), and (3) *probabilities* (or: relative frequencies) for the possible transitions. It is convenient to use a directed graph for the representation of the MUM; therein, the nodes are the states, the arcs are the transitions, and the transition probabilities are indicated as labels assigned to the arcs.

Related, but not quite identical to Markov usage models in the sense above is another class of probabilistic, graph-based processing models for software the first of which have already been developed three decades ago (see Littlewood [18], [19], [20], Cheung [5]); more recent articles concerning this other type of models are Siegrist [25] or Rajgopal and Mazumdar [22]: In these models, the nodes of the graph do not represent states of the processing, but components (modules etc.) of the software product, and the arcs do not represent processing steps, but control transfer between the components. In so far as they assume a Markovian (history-independent) probability structure, these models are often simply called *Markov models*; to distinguish them from Markov usage models in the sense defined above, we call them *component-based Markov models* in this paper. On a non-formal level, the main difference is that component-based Markov models already exploit "whitebox" knowledge of the software developer for putting up the model graph, whereas MUMs rather start with the "blackbox" viewpoint of the software user: The developer knows the module structure of the given program and can estimate transfer probabilities between modules (knowledge on the actual use of the program is nevertheless necessary for that); the user, on the other hand, identifies states, e.g., certain input windows, menus, intermediate or final outputs etc., and can estimate probabilities of transitions between such states in her/his application environment. The present paper is based on the MUM modelling type, but we shall enrich the MUMs by some additional information accessible only to the developer.

The aim of the paper is to automatically derive a suitable set of *test paths* through the MUM graph, which may be elaborated in a second step to complete test cases for the software product under consideration. Similar problems have already been investigated in the literature, mainly in the context of protocol conformance testing or user interface testing (see. e.g., Belli and Grosspietsch [2], Aho et al. [1], Csöndes et al. [6], Zhang and Cheung [29]). While these approaches strive for a satisfaction of pre-defined coverage criteria (usually: arc coverage)[1], we do not force complete coverage (or coverage on a pre-defined level) in the present paper. Instead, we consider the tradeoff between coverage and testing costs and try to find an optimal compromise between both. For determining this solution, the uncertain knowledge the developer has on possible failures and losses in case of failure needs to be quantified, and the MUM must be enriched by this information.

We do not deal here with the question of how the selected test paths are extended to full test cases. We must treat this aspect as beyond the scope of the present paper.

A Markov usage model intends to represent the *operational distribution* of program inputs. For purposes of reliability estimation, it would be desirable to use this distribution also for testing, or to test with a distribution changed in a controlled way, as suggested in [11], [12], [7] and [8]. However, as argued by

---

[1] An exception is [6], where, among others, the problem of maximizing coverage subject to cost restrictions is treated. Contrary to our approach, however, information on program usage does not enter into the problem formulation.

Rivers and Vouk in [23], resource limits often force the developer to give up this aim and to resort to non-operational testing. Our approach proposes a kind of compromise for such situations: while dropping the goal of making the derivation of reliability estimates from the test outcomes possible, we nevertheless try to exploit as much information about the operational use of the program when deriving test cases, including information on processing steps that are "critical" with respect to failure probability or to damage in case of failure.

2 The Model

As input we take a MUM represented by a strongly connected directed graph $G = (V, A)$ without multiple arcs, where a specific node (labeled by the number 1) is distinguished as the start node. A MUM with unique start node and unique end node, as considered in [11]–[12], can be reduced to this model by adding an arc from the end node to the start node, representing a new call of the program. The transition probabilities $p(e)$ assigned to the arcs $e \in A$ are assumed as strictly positive. Furthermore, we suppose that to each arc e, the following estimates are assigned: the failure probability $f(e)$, the loss $l(e)$ caused in case of failure, and the cost $c(e)$ for testing e. Note that since there are no multiple arcs, each arc e can be identified with the ordered pair (i, j) of its incident nodes. If $e = (i, j)$, we shall also write $p(i,j)$, $f(i,j)$, $l(i,j)$ and $c(i,j)$ instead of $p(e)$, $f(e)$, $l(e)$ and $c(e)$, respectively.

Remark. Let us emphasize that in our model, processing steps of the program are assigned to *arcs* and not to nodes of the graph. This convention is different from that used in component-based Markov models (see our remarks in the Introduction). For example, in our formalism, node $i = 2$ can refer to the state that a specific menu, from which the user is to select one of the program functions fun1, fun2 or fun3, is presented to the user. According to the actual choice by the user, the program starts processing the selected function and gives an output to the user; node $i = 3$, $i = 4$ and $i = 5$ refer to the state that the output of function fun1, fun2 and fun3, respectively, is presented to the user. In this way, the processing of function fun1, fun2 and fun3 corresponds to arc $(2, 3)$, $(2, 4)$ and $(2, 5)$, respectively. In particular, the chosen representation implies that estimates on failure probabilities and losses in case of failure have to be assigned, as indicated above, to the arcs $e \in A$ and not to the nodes $i \in V$. Since in our model, the states do not refer to processing activities, also sojourn times etc. are not essential in our case.

For modelling failures, we use the *static* model described in [12], i.e., we assume that a failing arc always fails, and that the resulting loss occurs only one time, no matter how often the arc is (or would be) traversed. Our technique can also be extended to the *dynamic* model of [12]. A discussion of the two models from the viewpoint of application can be found in [12].

We define a test sequence as a closed walk $w = (e_1, \ldots, e_m)$, $e_k \in A$ ($k = 1, \ldots, m$), containing the start node 1. Arcs e may occur more than once in w.

The cost for testing $w = (e_1, \ldots, e_m)$ is

$$C(w) = \sum_{k=1}^{m} c(e_k). \tag{1}$$

The assumption that w is closed allows a decomposition of w in a series of single test paths from the start node 1 to an "end" node, i.e., a predecessor node of 1 (a node where a new program call is necessary to continue the test).

After test, tested arcs are considered as correct (not failing). Of course, this is a simplification, since other program inputs than those actually chosen for the test of a given arc could have caused this arc to fail. The simplification can be justified by the consideration that not testing an arc at all leaves a failure probability of a higher order of magnitude than testing it with just one specific input combination. It should be mentioned that the well-known coverage measures for whitebox testing use the same simplification.

Whether an arc is failing or not is uncertain before test. For modelling this uncertainty, we use the probabilistic approach.

As an indicator variable for the stochastic event that arc e fails in case it is traversed, we introduce $I_f(e) = 1$ if e is failing, given that e is traversed, and $I_f(e) = 0$ otherwise. Evidently, the expected value of the random variable $I_f(e)$ is just $f(e)$.

In a similar way, we introduce an indicator variable for the traversal of arc e. When defining this variable as a random variable, it seems appropriate to refer to the *steady state* of the operational usage of the program, i.e., the distribution of traversals of arcs as it is reached after some time of use of the program, when "starting effects" have already become insignificant. Therefore, we set $I_r(e) = 1$ if a random transition in steady state passes through e, and $I_r(e) = 0$ otherwise.

The random variables $I_r(e)$ and $I_f(e)$ are independent by definition, since we have defined the event $I_f(e) = 1$ as the event that e fails, conditional on the event that e is traversed, so whether e is actually traversed or not has no influence on whether $I_f(e) = 1$ or not.

Let $\bar{A} \subseteq A$ be the set of *not* tested arcs. According to our assumption, only the arcs in $\bar{A}$ can cause losses by failure. Denoting the mathematical expectation by E, the expected loss by failure for one steady-state transition can then be calculated as follows:

$$L(\bar{A}) = E\left(\sum_{e \in \bar{A}} I_r(e) \cdot I_f(e) \cdot l(e)\right) = \sum_{e \in \bar{A}} E(I_r(e) \cdot I_f(e)) \cdot l(e).$$

Using independence of $I_f(e)$ and $I_r(e)$ and the fact that the expectation of a product of independent random variables is the product of the expectations, we further obtain

$$L(\bar{A}) = \sum_{e \in \bar{A}} E(I_r(e)) \cdot f(e) \cdot l(e). \tag{2}$$

For abbreviation, let us set $q(e) = q(i,j) = E(I_r(e))$ for the probability that in steady state, arc $e = (i,j)$ is traversed. These probabilities can be computed as follows: Let x_i denote the probability with which state (node) i occurs in the steady-state distribution. Evidently, $q(i,j) = x_i \cdot p(i,j)$. The probabilities x_i result by the solution of the system

$$(P^t - I)(x_1,\ldots,x_n)^t = (0,\ldots,0)^t, \quad (1,\ldots,1)(x_1,\ldots,x_n)^t = 1 \qquad (3)$$

of linear equations, where $P = (p(i,j))$ is the matrix of transition probabilities, I is the identity matrix, and n is the number of states (nodes).

By using a sufficiently long test sequence w, it is of course possible to test *all* arcs, such that $\bar{A} = \emptyset$ and hence $L(\bar{A}) = 0$. However, there is a tradeoff between the achieved coverage and the testing costs. Let $\bar{A}(w)$ denote the set of arcs not covered by the walk w. Then we may determine the best point of this tradeoff by solving the following optimization problem:

Problem 1. Find a test sequence w, such that $\gamma L(\bar{A}(w)) + C(w) = \min$. Therein, γ is a factor indicating how high the program developer values an average loss of one currency unit per transition in operational usage as his/her own loss. γ depends on several factors, e.g., contracts, number of customers, market reputation effects etc. An alternate problem formulation where γ needs not to be estimated is the following:

Problem 2. Find a test sequence w, such that $L(\bar{A}(w)) = \min$ under the constraint $C(w) \leq B$. Here, B is a fixed given budget limit for testing.

Proposition. Both Problem 1 and 2 are NP-hard combinatorial optimization problems.

Proof. By choosing specific values for $f(e)$ or $l(e)$, the *Rural Postman Problem* (RPP), which is known to be NP-hard, can be reduced to Problem 1 or Problem 2. This can be seen as follows: The RPP consists in finding a shortest closed walk w in a graph $G = (V, A)$ with the property that w contains each arc $e \in A_0$, where A_0 is a given subset of A, at least once. Now define $f(e) = 1$ for all $e \in A_0$ and $f(e) = 0$ for all $e \notin A_0$, and choose γ very high. Then a solution of Problem 1 solves the given RPP. In a similar way, the RPP can be reduced to Problem 2. Therefore, Problems 1 and 2 are NP-hard as well. □

The property stated in Proposition is an indicator that for larger problem instances, *heuristic* solution approaches will be required.

Remark. We want to emphasize that the problem of obtaining *complete* coverage, as investigated in [17] and in a series of related articles, is contained in our problem 1 as the special case where the constant γ is very large and the values $f(e)$ and l strictly positive, such that an optimal solution must be a solution where all arcs are covered.

3 Heuristic Solution by ACO

One can easily design a straightforward greedy heuristic for solving Problem 1 or 2. Nevertheless, our experiments indicate that greedy-type heuristics can produce very poor solutions at our problems (see Section 4). So we apply instead a metaheuristic technique, *Ant Colony Optimization* (ACO), for which some types of convergence behavior towards the optimal solution have been shown ([13]–[14]), to our problems. The basic ideas of ACO have been developed by Dorigo, Maniezzo and Colorni [10] at the beginning of the 90ies; a formulation as a flexible metaheuristic applicable to the overall range of combinatorial optimization is given by Dorigo and Di Caro [9]. Compared with other possible metaheuristics like genetic algorithms, simulated annealing or tabu search, the application of ACO to our problems has the advantage that the graph structure of the problems fits very well to the graph-oriented "philosophy" of ACO (see [13]).

Roughly speaking, we can consider an ACO implementation as a learning system with a fixed number of cooperative agents, implemented on the same or on different processors. Each agent constructs solutions for the given problem, governed (1) by randomness, (2) by a pre-evaluation of possible construction steps according to simple heuristic considerations, yielding so-called *attractiveness values*, and (3) by a blackboard mechanism representing the "collective memory" of the agent community. Learning takes place via this blackboard which contains *trail levels* used by all agents for the computation of the probabilities for performing possible partial construction steps. Agents that have found the best solution up to now, or the best solution generated within a certain construction period, or, in the so-called rank-based variant (see [4]), one of the r best solutions generated within the period, are allowed to "write" on the blackboard, which means that the trail levels for the partial construction steps they applied in their solution are increased. For details of the general algorithm, we refer the reader to the literature cited above.

In our computational experiments, we applied three different variants; in each of them, trail levels are assigned to the arcs of G, and a construction step is an extension of the current partial walk $w^{(k)} = (e_1, \ldots, e_k)$ to the partial walk $w^{(k+1)} = (e_1, \ldots, e_k, e_{k+1})$, where e_{k+1} is selected from the set of arcs starting at the end node of e_k. (Construction begins by choosing an arc e_1 starting at the start node 1.) In the first variant ("Standard"), the probability of an arc of being appended is chosen proportional to $\tau(e) \cdot \eta(e)$, where $\tau(e)$ is the trail level and $\eta(e)$ is the attractiveness value; we decided to compute $\eta(e)$ as $q(e) \cdot f(e) \cdot l(e)$. In the second variant ("Variant A"), the probability that a particular arc e is appended is chosen proportional to $\frac{\tau(e)}{n(e)+1} \cdot \eta(e)$, where $n(e)$ counts the total number how often e has been traversed in all previous construction rounds in any walk constructed by any of the agents. Thus, arcs, that have already been used frequently get a lower probability of being considered in the future, which favors diversification of the walks. In the third variant ("Variant B"), we apply the same selection mechanism as in the first (Standard) variant, but the trail level update mechanism is changed: Only an arc that lies on the best walk found so far with a minimum number of traversals in this best walk gets a trail level increment. Also this variant favors diversification.

4 Experimental Results

In this section we apply our approach to a medium-size real world system representing an outbound call center which is used by the Austrian Red Cross. We give a brief overview of the system and present numerical results of our analysis. The considered MUM consists of 47 nodes and 85 arcs and reflects three main menues and submenues including the functions for the *customer processing*, the *supervisor functions* and the functions for the *campaign management*. To increase transparency, we have partitioned the entire MUM description into a set of 4 diagrams–the *main section* and three subsections. Entry node and exit node of a subsection are represented in the *main section* and are repeated in the subsection.

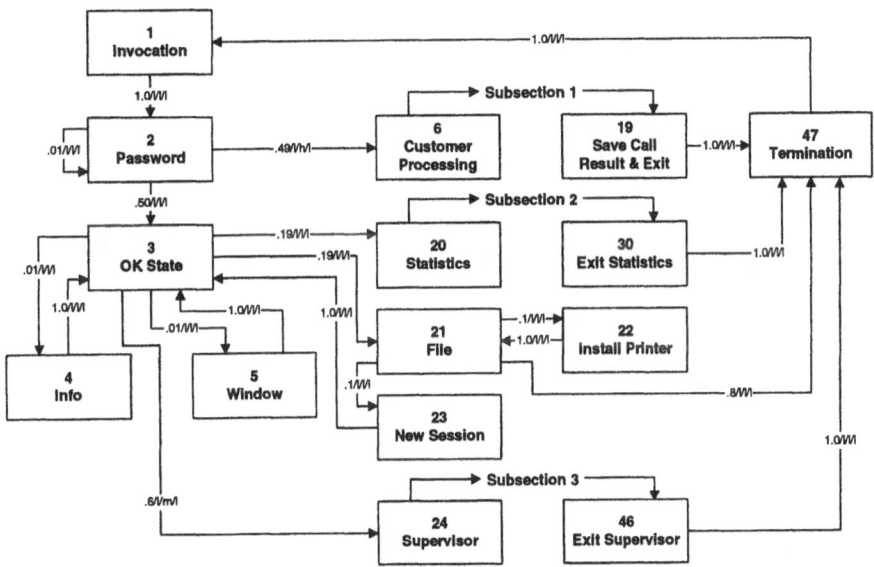

Fig. 1. Call Center Application – Main Section

The main functionality is *customer processing* (node 6 and the subsequent part of the graph given in Fig. 2). By means of customer processing, the operator can select a potential new or an already registered customer, and the customer's data can be seen on the screen (node 7). The system provides also further information about the customer: information on the last phone calls with the customer (node 8) and the complete chronology of interactions with him/her (node 9). As soon as the operator knows the important facts about the potential customer he/she will, with a certain probability, proceed to perform a phone call (node 10). After the call the operator can modify customer data, e.g., register a new e-mail address. If there are no changes the operator can go directly to a function where he can store the call results (node 15).

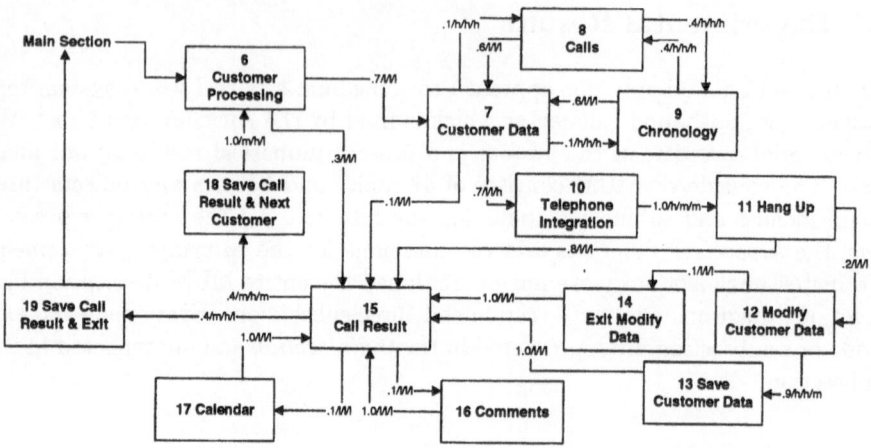

Fig. 2. Call Center Application – Customer Processing

In the subsection *statistics*, several performance statistics can be computed and printed or exported to an accounting program.

In the subsection *supervisor* (see Fig. 3), the three main functionalities *check campaign results* (node 31), *distribute campaigns* (node 36) and *maintain campaigns* (node 39) are implemented. One important task of the supervisor is to control the work of the operators. For this purpose he/she uses the functionality *check campaign results*. By using the functionality *send e-mail* the checked results can be distributed (node 34). Another task of the supervisor is to distribute the work to the available operators (node 36). Certain customers are grouped to so-called *campaigns*, and each campaign is assigned to an operator.

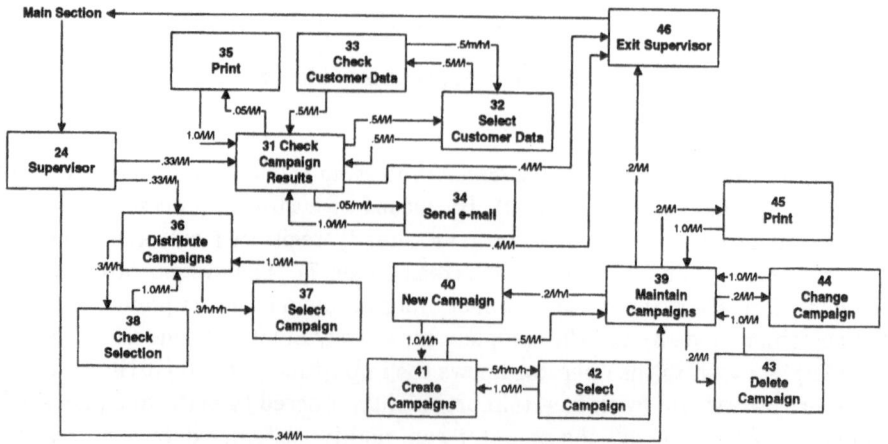

Fig. 3. Call Center Application – Supervisor Functions

Our description in Fig. 1–Fig. 4 also includes the operational probability $p(i, j)$, an estimation of the failure probability $f(i, j)$ and of the loss in case of failure $l(i, j)$, and also the cost $c(i, j)$ for testing the function (i, j). These values are integrated into the MUM description. Each arc (i, j) is labelled with the following estimates: $p(i, j) / f(i, j) / l(i, j) / c(i, j)$. Therein, the constants l (low), m (medium), and h (high) have the following values: For $f(i,j)$: $l = 0.001, m = 0.01, h = 0.1$. For $l(i,j)$: $l = 1, m = 10, h = 100$. For $c(i,j)$: $l = 1, m = 10, h = 1000$. Failure probabilities, losses in case of failure and testing costs have been estimated by "educated guesses". A high failure probability is expected for the functions (7,8), (7,9), (8,9) and (9,8). Also the loss in case of failure is high for these functions, since it is not possible to perform a phone call if the operator has wrong or missing information about the (potential) customer. The costs for testing these functions are high as well, because the data are extracted from a legacy system. To test this system an additional expert has to join the test team. Also the testing costs for testing function (7,10) are high: To verify the test results of function (7,10), a call center application expert is required.

We tested Problem 1 with $\gamma = 1$ at the problem instance described above for the three different variants of our ACO algorithm. To evaluate the performance of our novel approach, we compared the results with the results of a greedy heuristic. Our greedy heuristic works as follows: we choose the next not visited arc that produces the maximum value of $(q(e) \cdot f(e) \cdot l(e))/(d(e)+1))$, where $d(e)$ denotes the distance from the current position to arc e. For using this heuristic one has to specify the coverage rate of the arcs to be tested. In order to make a fair comparison, we use about the same coverage rate as it has been recognized as favorable in good solutions found by the ACO algorithm (about 25 %). The solution quality obtained by the greedy heuristic turned out as 1045.45 cost units. To evaluate the performance of the ACO algorithm, we will also plot the performance of the variant where the trail levels are not used, i.e., construction is only guided by the attractiveness values $\eta(e)$. In Fig. 4, we call this variant "Attractiveness Heuristic".

We performed six runs with each of the algorithms. The plots presented in Fig. 4–5 show best solutions values found after a certain number of iterations, averaged over these six runs. In Fig. 4, the results of the attractiveness heuristic and our first (Standard) ACO variant are shown. It is immediately seen that the solution quality achieved by the attractiveness heuristic is about the same as that of the greedy heuristic, and that both are not able to find low-cost solutions: the best found solution value is higher than 1000 cost units. Standard ACO finds good solutions, but only after about 35 iterations. After 10 – 20 iterations, the values are still rather poor.

Fig. 5 shows the results of the ACO modifications, Variant A and B. We see that, compared to Standard ACO, relatively good solutions are obtained already after a smaller number of iterations. This makes Variants A and B particularly interesting for large applications, where it cannot be expected that computation time suffices for driving the optimization process to the point where optimal solutions are reached.

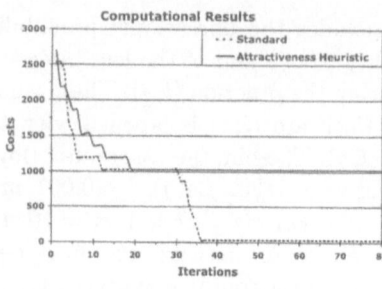
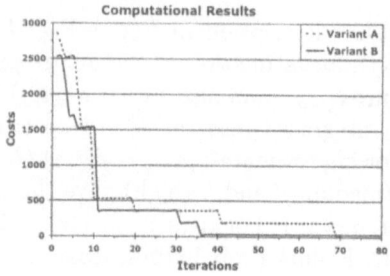

Fig. 4. Computational Results – Attractiveness Heuristic, Standard ACO

Fig. 5. Computational Results – Variant A, Variant B

All the three ACO algorithms find very reasonable solutions after 70 iterations. These iterations took about three seconds. Of course, the trends of the presented results will have to be verified by more extensive experiments in the future.

5 Conclusion

We have proposed a technique for obtaining appropriate software test sequences from a Markov Usage Model description of the expected use of a software product. Our approach addresses the tradeoff between testing costs and failure risk caused by untested processing steps (state transitions). For the search for good solutions, we have applied a metaheuristic, Ant Colony Optimization. In this respect, our work is in the spirit of the currently emerging paradigm of "Search-Based Software Engineering", as it has been formulated by Harman and Jones [15] and propagated in a series of recent initiatives. As an additional input information completing the Markov Usage Model description, we use estimates of failure probabilities and of losses in case of failure. This gives the considered problem the character of a decision problem under uncertainty. Our technique has been applied to a medium-size real-world problem. The experimental results indicate that our ACO approach outperforms a greedy-type solution heuristic. Two variants of the standard ACO technique have been specifically designed for the problem under consideration; they produce still more promising results than standard ACO. Nevertheless, much more experimental work will be required in the future as a prerequisite for the development of suitable toolkits supporting software testing along the lines of the described approach on an industrial level.

Practical application of our technique requires that the parameters values $p(e)$, $f(e)$, $l(e)$ and $c(e)$ are estimated. Sensitivity of the optimization results with respect to these estimates is an important topic that deserves further investigation. Our first impression is that the results are relatively robust with respect to all parameter types except the transition probabilities $p(e)$. On the other hand, just these probabilities can be estimated in a rather precise, objective way by using an instrumentation of the program code and measuring the transfer frequencies in a test with similar usage profile as that expected in the application field.

Our approach takes Markov usage models of any type as inputs and can therefore be used in a broad range of applications. Some recent articles indicate that the application of graph-based models for GUI (graphical user interface) testing is an area of growing importance, see, e.g., Kallepalli [16] or Belli [3]. Taking account of the specific requirements of GUI testing in the context of our problem formulation seems to be an interesting topic for future research. Another topic of research would be an extension to semi-Markov usage models for applications where the Markov assumption (claiming that state transition probabilities are history-independent) seems not adequate. Of course, an important issue for further research is how to test paths are to be extended (semi-)automatically to test cases.

Acknowledgement. The authors are indebted to Fevzi Belli for drawing their attention to the optimization approaches in protocol conformance testing and for some helpful disussions.

References

1. Aho, A. V., Dahbura, A. T., Lee, D., Uyar, M. Ü, "An optimization technique for protocol conformance test generation based on UIO sequences and rural chinese postman tours", *IEEE Trans. on Communications* 39 (1991), pp. 1604–1615.
2. Belli, F., Grosspietsch, K.-E., "Specification of fault-tolerant system issues by predicate/transition nets and regular expressions – approach and case study", *IEEE Trans. Software Eng.* 17 (no. 6) (1991), pp. 513–526.
3. Belli, F., "Finite-state testing and analysis of graphical user interfaces", *Proc. 12th Int. Symp. on Software Reliability Engineering (ISSRE)*, IEEE CS Press (2001), pp. 34–43.
4. Bullnheimer, B., Hartl, R. F., Strauss, C., "A new rank-based version of the Ant System: A computational study", *Central European Journal for Operations Research* 7(1) (1999), pp. 25–38.
5. Cheung, R. C., "A user-oriented software reliability model", *IEEE Trans. Software Eng.*, vol. 6 (1980), pp. 118–125.
6. Csöndes, T., Kotnyek, B., Szabó, J. Z., "Application of heuristice methods for conformance test selection", *European J. of Operational Research*, vol. 142 (2002), pp. 203–218.
7. Doerner, K., Gutjahr, W. J., "Representation and Optimization of Software Usage Models with Non-Markovian State Transitions", *Information and Software Technology* vol. 47 (2000), pp. 873–884.
8. Doerner, K., Laure, E., "High performance computing in the optimization of test plans", *Optimization and Engineering*, vol. 3 (2002), pp. 67–87.
9. Dorigo, M., Di Caro, G., "The Ant Colony Optimization metaheuristic", in: *New Ideas in Optimization*, D. Corne, M. Dorigo, F. Glover (eds.), pp. 11–32, McGraw-Hill (1999).

10. Dorigo, M., Maniezzo, V., Colorni, A., "The Ant System: An autocatalytic optimization process", Technical Report 91–016, Dept. of Electronics, Politecnico di Milano, Italy (1991).
11. Gutjahr, W. J., "Importance sampling of test cases in Markovian software usage models", *Probability in the Engineering and Informational Sciences*, vol. 11, (1997), pp. 19–26.
12. Gutjahr, W. J., "Software Dependability Evaluation Based on Markov Usage Models", *Performance Evaluation*, vol. 40 (2000), pp. 199–222.
13. Gutjahr, W. J., "A graph–based Ant System and its convergence", *Future Generation Computing Systems* 16 (2000), pp. 873–888.
14. Gutjahr, W. J., "ACO Algorithms with Guaranteed Convergence to the Optimal Solution", accepted for publication in: *Information Processing Letters*.
15. Harman, M., Jones, B. F., "Search-based software engineering", *Information and Software Technology*, vol. 43 (2001), pp. 833–839.
16. Kallepalli, Ch., "Measuring and modeling usage and reliability for statistical Web testing", *IEEE Trans. Software Eng.* vol. 27 (no. 11) (2001), pp. 1023–1036. *IEE Proc.-Softw.* 146, no. 4 (1999), pp. 187–192.
17. Kumar, G. P., Venkataram, P., "Protocol test sequence generation using MUIOS based on TSP Problem", *Proc. IFIP TC6 Conf.* (1994), pp. 165–191.
18. Littlewood, B., "A reliability model for systems with Markov structure", *J. Royal Statistical Soc., Series C (Applied Statistics)*, vol. 24 (1975), pp. 172–177.
19. Littlewood, B., "A Semi-Markov model for software reliability fith failure costs". In: *MRI Symp. Computer Software Engineering*, Polytechnic Press, Polytechnic of New York, New York (1976), pp. 281–300.
20. Littlewood, B., "Software Reliability Models for modular program structure", *IEEE Trans. Reliab.*, vol. 28 (no. 3) (1979), pp. 241–246.
21. Poore, J. H., Trammell, C. J., "Application of statistical science to testing and evaluating software intensive systems". In: *Statistics, Testing, and Defense Acquisition*, Washington: National Academy Press (1998).
22. Rajgopal, J., Mazumdar, M., "Modular operational test plans for inferences on software reliability based on a Markov model", *IEEE Trans. Software Eng.*, vol. 28 (no. 4) (2002), pp. 358–363.
23. Rivers, A. T., Vouk, A. M., "Resource-constrained non-operational testing of software", *Proc. 9th Int. Symp. on Software Reliability Engineering (ISSRE)*, pp. 154–163 (1998).
24. Sayre, K., Poore, J. H., "Partition testing with usage models", *Information and software technology*, vol. 42 (2000), pp. 845–850.
25. Siegrist, K., "Reliability of systems with Markov transfers of control", *IEEE Trans. Software Eng.*, vol. 14 (no. 7) (1988), pp. 1049–1053.
26. Walton, G. H., Poore, J. H., Trammell, J., "Statistical testing of software based on a usage model", *Software - Practice and Experience*, vol. 25 (1) (1995), pp. 97–108.
27. Whittaker, J. A., Poore, J. H., "Markov analysis of software specifications", *ACM Trans. on Software Eng. and Method.*, vol. 2 (1) (1993), pp. 93–106.
28. Whittaker, J. A., Thomason, M. G., "A Markov chain model for statistical software testing", *IEEE Trans. Software Eng.*, vol. SE-20 (1994), pp. 812–824.
29. Zhang, F., Cheung, T. Y., "Optimal Transfer Trees and Distinguishing Trees for Testing Observable Nondeterministic Finite-State Machines", *IEEE Trans. Software Eng.*, vol. SE-29 (2003), pp. 1–14.

Using Genetic Programming to Improve Software Effort Estimation Based on General Data Sets

Martin Lefley and Martin J. Shepperd

School of Design Engineering and Computing, University of Bournemouth, Talbot Campus,
Poole, BH12 5BB, UK
mlefley@bournemouth.ac.uk

Abstract. This paper investigates the use of various techniques including genetic programming, with public data sets, to attempt to model and hence estimate software project effort. The main research question is whether genetic programs can offer 'better' solution search using public domain metrics rather than company specific ones. Unlike most previous research, a realistic approach is taken, whereby predictions are made on the basis of the data available at a given date. Experiments are reported, designed to assess the accuracy of estimates made using data within and beyond a specific company. This research also offers insights into genetic programming's performance, relative to alternative methods, as a problem solver in this domain. The results do not find a clear winner but, for this data, GP performs consistently well, but is harder to configure and produces more complex models. The evidence here agrees with other researchers that companies would do well to base estimates on in house data rather than incorporating public data sets. The complexity of the GP must be weighed against the small increases in accuracy to decide whether to use it as part of any effort prediction estimation.

1 Introduction

Reliable predictions of project costs — primarily effort — are greatly needed for better planning of software projects, but unfortunately size and costs are seldom, if ever, proportionally related [8]. There has been extensive research into software project estimation, with researchers assessing a number of approaches to improving prediction accuracy. One of the first methods to estimate software effort automatically was COCOMO [4], where in its simplest form effort is expressed as a function of anticipated size as:

$$E = aS^b \quad (1)$$

where E is the effort required, S is the anticipated size and a and b are domain specific parameters. Independent studies, for example, Kitchenham and Taylor [14] and Kemerer [12] found many errors considerably in excess of 100% even after model calibration.

Subsequently attempts have been made to model data automatically based on local collection, for example, Kok et al. [15] published encouraging results using stepwise regression. Such linear modelling can cover only a small part of the possible solution space, potentially missing the complex set of relationships evident in many complex domains such as software development environments.

A variety of machine learning (ML) methods have been used to predict software development effort. Good examples include artificial neural nets (ANNs) [3, 28], case based reasoning (CBR) [9, 23] and rule induction (RI) [20]. Hybrids are also possible, e.g. Shukla [24] reports better results from using an evolving ANN compared to a standard back propagation ANN. Dolado and others [6, 7] analyse many aspects of the software effort estimation problem and present encouraging results for a genetic programming (GP) based estimation using a single input variable. Burgess and Lefley [5] also had some success using GP based estimation.

One characteristic to all ML methods is the need for training data. However, recent software engineering research has found that shared or public data sets are much less effective than restricting the prediction system to potentially very few local cases [11, 21]. The issue at stake is whether a company can improve prediction accuracy by incorporating results from other companies. Generally the more data available to a learner, the better it can model behaviour. However, no matter how good the control, some metrics are likely to be measured differently across companies and the working environments may have a marked difference. Thus there will be some distortion of the models accuracy. The results reported by [11, 21] for non-evolutionary models found the larger data sets to provide less accurate estimates. This suggests that companies should only use their own data, assuming they have sufficient examples close to a new case to make an estimate. Their research used models based on regression, CART, robust regression, stepwise ANOVA and analogy based estimation. Genetic programming offers an evolutionary solution to estimation problems that may better take into account the source of data. For example a variable indicating in house or external could be used as a multiplier to ignore data from outside sources and so has the potential to build a prediction system at least as accurate as one based on only internal data.

To summarise, the aim of this paper is to consider the question: can the use of evolutionary models overcome the problems of using non-local data to build project effort prediction systems. The ability to do this is of particular interest to environments that have collected little local data, a seemingly common occurrence within software engineering. The remainder of the paper describes the data set used for our analysis, considers how prediction accuracy might be assessed and then reviews the application of ANNs, nearest neighbour (NN) and GP methods. Next we provide more detailed information on our GP method. This is followed by results, conclusions and suggestions for further work.

2 The Data Set

The data used for the case study is often referred to as the 'Finnish data set' and is made up of project data collected by the software project management consultancy organisation SSTF. This data set contains 407 cases described by more than 90 features including metrics such as function points. This data set used in this paper is quite large for this type of application due to the fact that it comprises project data from many organisations interested in benchmarking their software productivity. The features are a mixture of continuous, discrete and categorical. However, there are missing data values and features not be known at prediction time and so are not included in the development of a prediction system. Removing features with missing values or values that cannot be known until after the prediction is required, leaves a subset of 83 features used for this case study. The data set exhibits significant multicollinearity, in other words there are strong relationships between features as well as with the dependent variable. More background information on this can be found in [21] which describes an earlier version of the same data set.

The total project effort variable was removed and used as the output or dependent variable. This is used to model the training data and to evaluate the test data. Where data was in the form of a string it was changed to an index, so that it could be processed numerically. This included the company number; to assist the models the chosen company was allocated the special code index 0. All values were scaled by range to fall between 0 and 1. In order to provide a realistic assessment of the hypothesis, the data was split chronologically into training and test sets. The data was sorted by start dates in such a way that a reasonable number of the projects from one company were available for training and testing. Thus we model the situation on 15th October 1991 when the selected company had completed 48 projects, with another 15 yet to commence. At this cut off date, there was data available from another 101 projects from other companies apart the one chosen for modelling. Data from the other companies beyond the cut off date was discarded.

3 Comparing Prediction Techniques

Our chief requirement is to obtain a quantitative assessment of the accuracy of the different prediction techniques. This achieved by comparing software project effort estimates made over a number of cases with known actual effort. Typically these cases are not made available during the configuration and training stages in order to avoid bias. There are many ways for calculating and combining the accuracy measures obtained, with no simple way of ranking these techniques, even for a given domain [13]. Most learners use an evaluation or fitness function to determine error to decide on future search directions. The choice of method to make this evaluation is crucial for accurate modelling. All of the learners considered here use root average mean square error. This is important also when considering accuracy estimates as the evaluation function will bias search towards similar evaluation metrics.

A number of accuracy — strictly speaking residual — summary statistics can be used to make comparisons between models. Each captures different aspects of the notion of model accuracy. Ultimately the decision should be made by the user, based on estimates of the costs of under or over estimating project effort, how risk averse they are and so forth. All are based on the calculated error or residual, that is the difference between the predicted and observed output values for a project. Most use an average performance over the whole set so we also include the worst case as an illustration of a less generalised metric. We also assess results using AMSE, adjusted mean square error (see Equation 2). This is the sum of squared errors, divided by the product of the means of the predicted and observed outputs. For further details on these metrics, see [5].

$$\text{AMSE} = \sum_{i=1}^{i=n} \frac{(E_i - \hat{E}_i)^2}{\left(\overline{E_i} * \overline{\hat{E}_i}\right)} \quad (2)$$

In summary the accuracy metrics used for this research are:
1. Correlation coefficient (Pearson) of actual and predicted
2. Adjusted mean square error - AMSE
3. Percentage of predictions within 25% - Pred(25)%
4. Mean magnitude of relative error - MMRE
5. Balanced mean magnitude of relative error - BMMRE
6. Worst case error as %

We also consider qualitative factors adapted from Mair *et al.* [20]: accuracy, explanatory value and ease of configuration. The ease of set up and quality of information provided for the estimator is of great importance. Empirical research has indicated that end-users coupled with prediction systems can outperform either prediction systems or end-users alone [25]. The more explanations given for how a prediction was obtained, the greater the power given to the estimator. If predictions can be explained, estimators may experiment with "what if" scenarios and meaningful explanations can increase confidence in the prediction.

4 Machine Learning and Software Effort Estimation

Many machine learning methods have been used for effort estimation, though there are few comparisons. Work by Burgess and Lefley [5] found that GPs performed similarly to other advanced models. As part of this investigation we consider different learners to benchmark the achievements of the GP system. Results are presented for models based on

- Random
- Least squares regression
- Nearest neighbour
- Artificial neural network
- Genetic programming
- Average of all non-random estimators

The random method uses the substitution of an output variable, that is effort, randomly selected from the training set and is included as a benchmark.

4.1 Artificial Neural Networks (ANNs)

ANNs are parallel systems inspired by the architecture of biological neural networks, comprising simple interconnected units, called artificial neurons. The neuron computes a weighted sum of its inputs and generates an output activated by a stepped function. This output is excitatory (positive) or inhibitory (negative) input fed to other neurons in the network. Feed-forward multi-layer perceptrons are the most commonly used form of ANN, although many more neural network architectures have been proposed. The ANN is initialised with random weights. The network then 'learns' the relationships implicit in a data set by adjusting the weighs when presented with a combination of inputs and outputs that are known as the training set.

Studies concerned with the use of ANNs to predict software development effort focus on comparative accuracy with algorithmic models, rather than on the suitability of the approach for building software effort prediction systems. An example is the investigation by Wittig and Finnie [28]. The results from two validation sets are aggregated and yield a high level of accuracy viz. Desharnais set MMRE=27% and ASMA set MMRE=17%, although some outlier values are excluded. Mair et al. [20] show neural networks offer accurate effort prediction, though not as good as Wittig and Finnie, concluding that they are relatively difficult to configure and interpret.

A number of experiments were needed to determine an appropriate number of hidden nodes, with 20 being used for this investigation. Other parameters such as learning rate were set to values chosen by experience and experimentation. It was found that nearby values for parameters only slightly affected convergence performance.

4.2 Nearest Neighbour Techniques

Examples of successful case based reasoning (CBR) tools for software project estimation include: cEstor [26], a CBR system dedicated to the selection of similar software projects for the purpose of estimating effort, and more recently, FACE [2] and ANGEL [22]. We use a simple NN technique and a weighted (by the reciprocal of the Euclidean distance) average of three neighbours as simple form of CBR. The Euclidean distance is determined by the root sum of squares of the difference between the scaled input variables for any two projects.

5 Background to Genetic Programming

Genetic Algorithms were developed as an alternative technique for tackling general optimisation problems with large search spaces. They have the advantage that they do not need any prior knowledge, expertise or logic related to the particular problem being solved. Occasionally, they can yield the optimum solution, but for most problems with a large search space, a good approximation to the optimum is a more likely outcome. The basic ideas used are based on the Darwinian theory of evolution, which in essence says that genetic operations between chromosomes eventually leads to fitter individuals which are more likely to survive. Thus, over time, the population of the species as a whole improves. However, not all of the operations used in the computer analogy of this process necessarily have a biological equivalent.

Genetic Programming is an extension of Genetic Algorithms, which removes the restriction that the chromosome representing the individual has to be a fixed length binary string. In general for GP, the chromosome is some type of program, which is then executed to obtain the required results. One of the simplest forms of program, used for this application, is a binary tree containing operators and operands. This means that each solution is an algebraic expression that can be easily evaluated. Koza "offers a large amount of empirical evidence to support the conclusion that GP can be used to solve a large number of seemingly different problems from many different fields" [16].

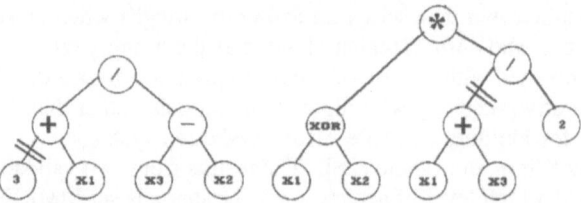

Fig. 1. Illustration of crossover operator before operation. The double line illustrates where the trees are cut

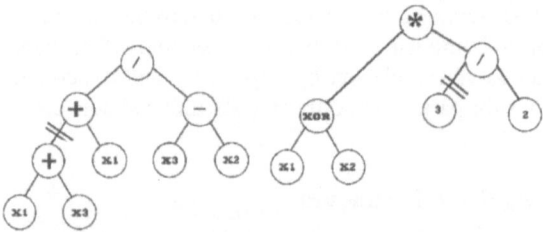

Fig. 2. Illustration of crossover operator after operation. The double line illustrates where the sub-trees have been swapped.

The main genetic operations used are crossover and mutation. The crossover operator chooses a node at random in each chromosome and the branches to those nodes are cut. The sub trees produced below the cuts are then swapped. The method of performing crossover is easily illustrated using an example (see figures 1 and 2). Although this example includes a variety of operations, only the set { + , - , * , / } were made available for our experiments. More complex operators can, of course, be developed by combining these simple operations. Simple operators eliminate the bounding problems associated with more complex operations such as XOR, which is not defined for negative or non-integer values. The multiply is a protected multiply which prevents the return of any values greater than 10^{20}. Similarly there is a protected divide, division by 0 returning 1000. This is to minimise the likelihood of real overflow occurring during the evaluation of the solutions.

The initial solutions were generated using "ramped half and half". This means that half the trees are constructed as full trees, i.e. operands only occur at the maximum depth of the tree, and half are trees with a random shape. Often elitism, where the top $x\%$ of the solutions, as measured by fitness, is copied straight into the next generation, is used to maintain the best solutions, where x can be any value but is typically 1 to 10.

Since trees tend to grow as the algorithm progresses, a maximum depth is normally imposed (or maximum number of nodes) in order that the trees do not get too large. This is often larger than the maximum size allowed in the initial population. Any trees produced by crossover that are too large are discarded. The reasons for imposing the size limit is to save on both the storage space required and the execution time needed to evaluate the trees. There is also no known evidence that allowing very large trees will lead to better results and smaller trees are more likely to generalise. Key parameters that have to be determined, in order to get good solutions without using too much time or space, are: the best genetic operators, the population size, maximum tree size, number of generations, etc. More information related to Genetic Programming can be found in [1, 17, 18].

6 Applying GP to Software Effort Estimation

The software effort estimation problem may be modeled as a symbolic regression problem, which means, given a number of sets of data values for a set of input parameters and one output parameter, construct an expression of the input parameters which best predicts the value of the output parameter for any set of values of the input parameters. In this paper, GP is applied to the data set already described in section 2, with the same 149 projects in the Training Set and the remaining 15 projects in the Test Set in a similar way to the other ML methods. Ease of configuration depends mainly on the number of parameters required to set up the learner. For example, a nearest neighbour(NN) system needs parameters to decide the weight of the variables and the method for determining closeness and combining neighbours to form a solution.

Koza [16] lists the number of control parameters for genetic programming as being 20 as compared to 10 for neural networks, but it is not always easy to count what constitutes a control parameter. However, it is not so much the number of the parameters as the sensitivity of the accuracy of the solutions to their variation. It is possible to search using different parameters but this will depend on their range and the granularity of search. In order to determine the ease of configuration for a genetic program, we test empirically whether parameter values suggested by Koza offer sufficient guidance to locate suitably accurate estimators. Similarly, all of our solutions use either limited experimentation or commonly used values for control parameters. The parameters chosen after a certain amount of experimentation with the Genetic Programming system are shown in Table 1.

Table 1. Main parameters used for the GP system

Parameter	Value
Size of population	1000
Number of generations	500
Number of runs	10
Maximum initial full tree depth	5
Maximum no. of nodes in a tree	64
Percentage mutation rate	5%
Percentage of elitism	5%

No function 'seeding' was used and only a simple mutation operator was available based on substitution of parameters. The initial solutions were generated using "ramped half and half". This means that half the trees are constructed as full trees, i.e.

operands only occur at the maximum depth of the tree, and half are trees with a random shape.

The GP system, as with all the systems considered here, is written in power basic running on a 1GHz Pentium PC with 512Mbytes of RAM. However, the run-time memory usage is less than 1Mbytes to store the trees and population data required for each generation. Careful garbage management ensures this does not grow with successive generations. One run of 500 generations takes about 40 minutes processor time for the full data set, 25 minutes using the smaller, company specific set.

The In order to appreciate the performance of the GP the test is also performed on some other suitable techniques.

7 Results of the Comparison

This study evaluates various statistical and machine learning techniques, including a genetic programming tool, to make software project effort predictions. The main hypothesis to be tested is whether GP can significantly produce a better estimate of effort using a public data set rather than a company specific database. We make comparisons mostly based on the accuracy of the results but we also consider the ease of configuration and the transparency of the solutions. Note that in the comparison, learners that contain a stochastic element are tested over a population of solutions, independently generated with the best modeller (smallest average squared error) used.

7.1 Accuracy of the Predictions

The ANN and GP solutions required some experimentation to find good control parameters. All of the training sessions were successful and no problems with local minima were encountered. Table 2 lists the main results used for using general, company wide data. Table 3 shows the same results for the company specific data set.

Table 2. Comparing different prediction systems for the full data set. Best performing estimators (within 5% of best) are highlighted in bold type. Best performers across both data sets are shown in italics

	Estimated effort						
	Random	LSR	1-NN	3-NN	ANN	GP	Average
correlation	-0.161	0.846	0.390	0.497	0.806	**0.937**	*0.890*
AMSE	28.091	4.601	13.970	14.720	6.584	**1.981**	3.596
Pred(25)%	13.333	*40.000*	33.333	20.000	26.667	*40.000*	33.333
MMRE	166.39	46.925	85.401	59.192	68.856	*37.670*	43.467
BMMRE	238.35	73.629	110.022	79.049	85.478	**62.797**	*47.343*
Worst case MRE	609	150	332	185	195	**125**	167

Table 3. Comparing different prediction systems for the company specific data set. Best performing estimators (within 5% of best) are highlighted in bold type. Best performers across both data sets are shown in italics.

	Estimated effort						
	Random	LSR	1-NN	3-NN	ANN	GP	Average
correlation	-0.167	*0.927*	0.104	0.793	**0.951**	**0.933**	*0.939*
AMSE	77.762	2.523	19.347	5.868	2.015	12.672	*1.796*
Pred(25)%	6.667	26.667	26.667	20.000	**40.000**	33.333	*40.000*
MMRE	381.422	45.648	129.045	114.709	69.228	**37.959**	65.694
BMMRE	448.695	63.995	217.353	128.525	71.166	**54.015**	71.888
Worst case MRE	1865	174	436	495	254	**79**	198

The three strongest techniques overall appear to be LSR, ANN and GP with GP achieving the best (or within 5%) level of accuracy the most often. The NN techniques are consistently less accurate. Using an average of each technique performs well for the company specific data set but is less successful for the full data set.

In terms of our main hypothesis, that it is better to restrict data to locally relevant projects, the evidence is somewhat mixed. For the poorest techniques, most notably 1-NN, this is clearly not the case as in all cases the results from the full data set are to be preferred to the company specific data set. With ANN and LSR the results for the company specific data set are generally better than those for the full data set. This agrees with other researchers that companies would do well to base estimates on their own data rather than incorporating public data sets. The GP results are almost identical in terms of the correlation coefficients and pred(25) indicator, substantially improved for BMMRE and the worst case, yet worse for MMRE and AMSE. It may be that the latter two indicators are affected by a different distribution of under and overestimates since they are both in asymmetric in their treatment of such estimates. Overall the results are weakly suggestive of the benefits of using preferring local data particularly for LSR, ANN and too a lesser extent GP. However, where local data is very limited, it is possible that the public data set could improve estimation accuracy.

7.2 Qualitative Assessment

It was fairly easy to configure the ANN but we have a fair degree of past experience in this area. Generally, different architectures, algorithms and parameters made little difference to results, but some expertise is required to choose reasonable values. Although heuristics have been published on this topic [10, 27], we found the process largely to be one of trial and error, guided by experience. ANN techniques would not be easy to use for project estimation by end-users as some expertise is required. We have found that investigation of using methods such as those reported in [19] can improve understanding of results from the ANN but that such information is limited.

GP has the most degrees of freedom in design, for example the choice of functions, mutation rate and strategy, percentage elitism, dealing with out of bounds conditions, creation strategy and setting maximum tree sizes and depths and so on. We found by using suggestions by Koza [16] and experimentation, we obtained convergence but again, at present, expertise is necessary. Many experiments were tried with various approaches and a number are still pending. Such effort suggests that GP only be used if it provides greater accuracy, not supported by our quantitative results for this data. The resulting program, like most of the techniques here, may provide information on the relative importance of input variables. Typical GP solutions for this problem are provided below.

$$((X13 + X17 + (((((((X62 + X56) * (X20 + X33)) * (((0.25 * X34) * (X32 + X70 - X19)) + (X20 + X80))) - (X40 - X9)) - (((X4 * X69) + X39 + X47) + ((2.1 + X28 + X62) * X4 * X56))) - X75 + X2) + 2.1 + X28 + X62) + (X20 + X33))) + (3 * X56 + X9 - X21 - X19 + X62)) * 0.25 * X83 \qquad (3)$$

$$X83 / (X45 * X4) \qquad (4)$$

Equation 3 is not easy to comprehend, unlike equation 4, which is of forced, limited length. The latter type of solution was found to be marginally less accurate but might provide the software manager with clearer insight to the problem.

8 Conclusions and Future Work

There is no clear winner on accuracy but GP performed well for this problem. This is in good agreement with other research [5, 7] using different approaches and data sets. The company specific ANN also performed well and the average over all methods for this data is also particularly good. Generally the better techniques (LSR, ANN and GP) are slightly more accurate with the company specific database.

The results, however, highlight that measuring accuracy is not simple and researchers should consider a range of measures. Perhaps a useful conclusion is that based on the evidence from this experiment, the more elaborate estimation techniques such as ANNs and GPs provide better fitting models but this slight advantage must be weighed against the extra effort in design, and a simple LSR can still provide useful estimates.

Future work will center on a greater variety of GP models, though we need to ensure that by trying so many models leads to one model performing very well by chance over fit. The stochastic population nature of GP can be used to reduce this effect. Models that will be used include better mutation (by tree generation rather than parameter swapping), seeding (for example with the LSR) and parameter tuning.

Acknowledgements. The authors are indebted to STTF Ltd for making the Finnish data set available.

References

[1] W. Banzhaf, P. Nordin, R. E. Keller, and F. D. Francome, *Genetic Programming: An introduction*. San Mateo, CA: Morgan Kaufmann, 1998.
[2] R. Bisio and F. Malabocchia, "Cost estimation of software projects through case base reasoning," presented at 1st Intl. Conf. on Case-Based Reasoning Research & Development, 1995.
[3] J. Bode, "Neural networks for cost estimation," *Cost Engineering*, vol. 40, pp. 25–30, 1998.
[4] B. W. Boehm, *Software Engineering Economics*. Englewood Cliffs, N.J.: Prentice-Hall, 1981.

[5] C. J. Burgess and M. Lefley, "Can genetic programming improve software effort estimation? A comparative evaluation," *Information & Software Technology*, vol. 43, pp. 863–873, 2001.
[6] J. J. Dolado, "Limits to methods in software cost estimation," presented at 1st Intl. Workshop on Soft Computing Applied to Software Engineering, Limerick, Ireland, 1999.
[7] J. J. Dolado, "On the problem of the software cost function," *Information & Software Technology*, vol. 43, pp. 61–72, 2001.
[8] S. Drummond, "Measuring applications development performance," in *Datamation*, vol. 31, 1985, pp. 102–8.
[9] G. R. Finnie, G. E. Wittig, and J.-M. Desharnais, "Estimating software development effort with case-based reasoning," presented at 2nd Intl. Conf. on Case-Based Reasoning, 1997.
[10] S. Huang and Y. Huang, "Bounds on the number of hidden neurons," *IEEE Trans. on Neural Networks*, vol. 2, pp. 47–55, 1991.
[11] R. Jeffery, M. Ruhe, and I. Wieczorek, "Using public domain metrics to estimate software development effort," presented at 7th IEEE Intl. Metrics Symp., London, 2001.
[12] C. F. Kemerer, "An empirical validation of software cost estimation models," *Communications of the ACM*, vol. 30, pp. 416–429, 1987.
[13] B. A. Kitchenham, S. G. MacDonell, L. Pickard, and M. J. Shepperd, "What accuracy statistics really measure," *IEE Proceedings - Software Engineering*, vol. 148, pp. 81–85, 2001.
[14] B. A. Kitchenham and N. R. Taylor, "Software cost models," *ICL Technical Journal*, vol. 4, pp. 73–102, 1984.
[15] P. Kok, B. A. Kitchenham, and J. Kirakowski, "The MERMAID approach to software cost estimation," presented at Esprit Technical Week, 1990.
[16] J. R. Koza, *Genetic programming: On the programming of computers by means of natural selection*. Cambridge, MA: MIT Press, 1992.
[17] J. R. Koza, *Genetic Programming II: Automatic discovery of reusable programs*: MIT Press, 1994.
[18] J. R. Koza, F. H. Bennett, D. Andre, M. A. Keane, and (). *Genetic Programming III: Darwinian Invention and Problem Solving*. San Mateo, CA: Morgan Kaufmann, 1999.
[19] M. Lefley and T. Kinsella, "Investigating neural network efficiency and structure by weight investigation," presented at European Symp. on Intelligent Technologies, Germany, 2000.
[20] C. Mair, G. Kadoda, M. Lefley, K. Phalp, C. Schofield, M. Shepperd, and S. Webster, "An investigation of machine learning based prediction systems," *J. of Systems Software*, vol. 53, pp. 23–29, 2000.
[21] K. Maxwell, L. Van Wassenhove, and S. Dutta, "Performance evaluation of general and company specific models in software development effort estimation," *Management Science*, vol. 45, pp. 787–803, 1999.
[22] M. J. Shepperd and C. Schofield, "Estimating software project effort using analogies," *IEEE Transactions on Software Engineering*, vol. 23, pp. 736–743, 1997.
[23] M. J. Shepperd, C. Schofield, and B. A. Kitchenham, "Effort estimation using analogy," presented at 18th Intl. Conf. on Softw. Eng., Berlin, 1996.
[24] K. K. Shukla, "Neuro-genetic prediction of software development effort," *Information & Software Technology*, vol. 42, pp. 701–713, 2000.
[25] E. Stensrud and I. Myrtveit, "Human performance estimating with analogy and regression models: an empirical validation," presented at 5th Intl. Metrics Symp., Bethesda, MD, 1998.
[26] S. Vicinanza, M. J. Prietula, and T. Mukhopadhyay, "Case-based reasoning in effort estimation," presented at 11th Intl. Conf. on Info. Syst., 1990.
[27] S. Walczak and N. Cerpa, "Heuristic principles for the design of artificial neural networks," *Information & Software Technology*, vol. 41, pp. 107–117, 1999.
[28] G. Wittig and G. Finnie, "Estimating software development effort with connectionist models," *Information & Software Technology*, vol. 39, pp. 469–476, 1997.

The State Problem for Evolutionary Testing

Phil McMinn and Mike Holcombe

Department of Computer Science, The University of Sheffield,
Regent Court, 211 Portobello Street, Sheffield, S1 4DP, UK.
{p.mcminn,m.holcombe}@dcs.shef.ac.uk

Abstract. This paper shows how the presence of states in test objects can hinder or render impossible the search for test data using evolutionary testing. Additional guidance is required to find sequences of inputs that put the test object into some necessary state for certain test goals to become feasible. It is shown that data dependency analysis can be used to identify program statements responsible for state transitions, and then argued that an additional search is needed to find required transition sequences. In order to be able to deal with complex examples, the use of ant colony optimization is proposed. The results of a simple initial experiment are reported.

1 Introduction

Evolutionary testing (ET) is a technique by which test data can be generated automatically through the use of optimizing search techniques. The search space is the input domain of the software under test. ET has been shown to be successful for generating test data for many forms of testing, namely specification testing [9], extreme execution time testing [12] and structural testing [10].

It has also been shown that certain features of programs can inhibit the search for test data, for example flag variables [3, 6]. This paper introduces another such feature: states in test objects. States can cause a variety of problems for ET, since test goals involving states can be dependent on the entire history of input to the test object, as well as just the current input. In addition, guidance must be provided so that statements responsible for state transitions are executed, so as to put the test object into the required states for certain test goals to become feasible. Internal states have hindered test data generation for automotive components used at DaimlerChrysler. The aim of this work is to extend the DaimlerChrysler ET system [14] to enable it to generate test data when presented with such troublesome test objects.

This paper is organized as follows. Section 2 reviews evolutionary testing. Section 3 introduces the state problem with examples. Section 4 discusses the use of data dependency analysis to identify program statements that are responsible for state transitions. Section 5 discusses how this could be applied to the state problem, and argues that an additional search is needed to find required sequences of transitional statements. In the case of simple examples an exhaustive search may be all that is required, however for more complex cases an optimization technique may be needed, and the use of ant colony optimization is proposed. Results of a simple initial experiment are reported. Section 6 then closes with conclusions and outlines future work.

2 Evolutionary Testing (ET)

Evolutionary Testing (ET) uses optimizing search techniques such as evolutionary algorithms to generate test data. The search space is the input domain of the test object, with each individual, or potential solution, being an encoded set of inputs to that test object. The fitness function is tailored to find test data for the type of test that is being undertaken.

This paper discusses the state problem in the context of structural testing. Here the aim is to find test data to execute every structural component of some coverage type, for example all branches of the program's control flow graph, or the execution of every definition-use pair for every variable. In order to retrieve fitness information, the test object must be instrumented.

Previous work [10] has argued that higher levels of coverage are obtained when each structural element of the chosen coverage type is targeted individually as a partial aim. For each partial aim, the minimizing fitness function is made up of two components, namely the approximation level and a branch distance calculation [10, 11]. The approximation level supplies a value indicating how close in structural terms an individual is to reaching the target. For node-oriented coverage types, for example statement coverage, this value is calculated as the number of branching statements lying between branches covered by an individual and the target branch. At the point where the individual diverged away from the target node, a normalized branch distance calculation is computed. This value indicates how close the individual was to evaluating the branch predicate in the desired way. For example if a condition (x == y) needs to be executed as true, the branch distance is calculated using |x-y|. For the thermostat function in figure 1, and the partial aim where the node 6 must be executed, the fitness values are computed as follows. Individuals reaching node 4 and evaluating the branching condition heater_on as false receive an approximation level of 1 and a branch distance of 1 (1 - heater_on). On the other hand, individuals reaching node 5 but evaluating the condition as false receive an approximation level of zero, and a branch distance computed using the formula POWER_THRESHOLD - power. The value of the fitness function when the target is finally reached is therefore zero.

With path-oriented coverage types, the approximation level is formed from the length of all identical path sections. Branch distances are taken at each point in which the flow of execution diverged from the intended path. These are then accumulated and normalized.

3 The State Problem

The state problem occurs with functions at higher system levels that exhibit state-like qualities by storing information in internal variables, which retain their values from one execution of the function under test to the next. Such variables are hidden from the optimization process because they are not available to external manipulation. The only way to change the values of these variables is through execution of statements that perform assignments to them. If such variables can be described as state variables, then these assignments, or definitions, are the transitions of the underlying state machine.

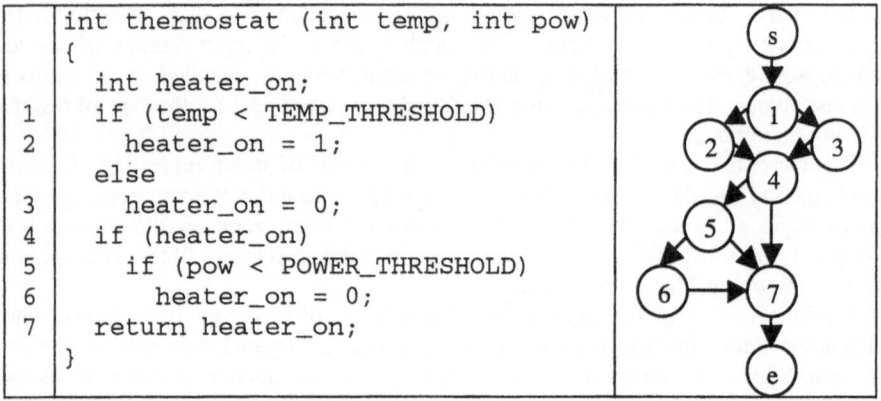

Fig. 1. C Code fragment of the simple thermostat example, with control flow graph (right column) and control flow graph node numbers corresponding to program statements in the left column

Figure 2 illustrates a function that is a controller for a smoke detector, the style of which mimics real-life code witnessed at our industrial partner. It takes three arguments - the first being the current room smoke level, followed by two arguments used as outputs to signal that the alarm should be switched on or off (the alarm works on a latch whereby after an "on" signal is received; the alarm stays on until it receives an "off" signal). The function is designed to be cycled once every second by the hardware. When the room smoke level becomes higher than a given threshold for a certain period of time, the alarm is raised (lines 13-14). When the room smoke level returns to safe levels for a given time, a special `waiting` flag becomes true (lines 17-18). The alarm then stays on for another 20 seconds (lines 21-22), unless the smoke levels breach acceptable limits again (lines 19 -20). The static storage class is used to declare several local variables. This allows them to maintain their values after the function is executed, however, as they are internal to the function, they cannot be directly optimized by the evolutionary search process.

Of course, states can also exist in system components. This can occur again through the use of the static storage class in C, or in object-oriented languages, through class variables that are protected from external manipulation using access modifiers. In our work however, the focus is only on test objects written using the C language.

The next sections discuss the problems that states in test objects can cause for ET.

```
1    const double LEVEL = 0.3;
2    const int DANGER = 5, WAIT_TIME = 20;
3    void smoke_detector(double level,
4                       int* signal_on, int* signal_off)
5    {
6      static int time = 0, off_time = 0;
7      static int detected = 0, alarm_on = 0, waiting = 0;
8      time ++;
9      if (level > LEVEL && detected < DANGER)
10        detected ++;
11     else if (detected > 0)
12        detected --;
13     if (!alarm_on && detected == DANGER)
14     { alarm_on = 1; *signal_on = 1; }
15     if (alarm_on)
16     {
17        if (!waiting && detected == 0)
18        { waiting = 1; off_time = time + WAIT_TIME; }
19        if (waiting && detected == DANGER)
20           waiting = 0;
21        if (waiting && time > off_time)
22        { waiting = 0; alarm_on = 0; *signal_off = 1;}
23     }
24   }
```

Fig. 2. C code for the smoke detector example, line numbers appear in the left column

3.1 Input Sequences

Partial aims relying on the values of state variables often require input sequences in order for them to be possible. This is because the execution of a series of transitions is required in order to set state variables to desired values. For the smoke detector example, the true branch from line 15 requires the detected variable to be equal to the DANGER threshold. This requires five calls to the function, each of which must execute the transitional statement on line 10, which increments the detected variable. Not until this has been done does the target become feasible.

A problem with the generation of sequences is that it is generally impossible to automatically determine beforehand roughly how long the required sequence is going to be, as different test objects can require radically different sequence lengths.

3.2 Disseminated Functionality

Where system functionality is dispersed across a series of components, it is possible that the function under test is not the function responsible for manipulating the state in the desired way. For these more complicated state problems, input sequences must be found to call different functions in a certain order. This is illustrated by the exam marks example in figure 3. The target is the true branch from the if statement in the compare function (part (a)). This depends on the state of the module portrayed in part (b). Only until the function has_max in this module has been executed a certain number of times will the branch in compare become feasible.

3.3 Guidance to Transitional Statements

A sequence of function calls is not always enough to ensure that transitions will be invoked for test goals to become feasible. Extra guidance must be provided to find inputs that ensure that transitional statements are actually executed, and executed in the correct order. In the exam marks example, the incremental statement involving the num_top variable in the dependent module must be executed for the test goal in the compare function to become feasible (assuming last_year_top is greater than zero). This statement is unlikely to be executed at random since the actual input value that will cause the statement to be reached (where mark is equal to 100) occupies a very small portion of the input domain.

The use of control variables can further complicate matters. Control variables are used to model the fact that a system's behavior should change given the fact that certain events have occurred, as with the flags used in the smoke detector (alarm_on and waiting). Such variables are often implemented as Boolean variables, to indicate the system is in a state or not in a state; or as enumerations, to indicate the system is in one of a collection of states. ET has further trouble with such variables, for the same reason that it has trouble with the closely related flag problem [3, 6]. The evolutionary search receives little guidance from branch distance calculations using these types of variables, due to their "all or nothing" nature. As state problems are not only dependent on the current input but also the history of inputs, the problem is accentuated.

| ```
// ...
void compare()
{
   if (get_num_top() >
       last_year_top)
   {/* target branch */}
}
// ...
``` | ```
const int MAX_MARK = 100;
static int num_top = 0;
void has_max(int mark)
{
   if (mark == MAX_MARK)
     num_top ++;
}
int get_num_top ()
{ return num_top; }
``` |
|---|---|
| (a) Function under test | (b) Dependent module |

Fig. 3. C code for the exam marks example

4 Possible Solutions

Solving state problems for ET would be seemingly straightforward if some state machine specification of the test object existed. Unfortunately such a model might not always be available, and even if it were, the required information may not be present.

Such representations tend to model control only (due to state explosion problems), whereas partial aims in ET may depend also on data states.

An alternative is to identify transitional statements using data dependency analysis [1]. The chaining approach of Ferguson and Korel [5] is relevant here. The chaining approach was not specifically designed for state problems but rather for the structural test data generation for "problem" program nodes, whose executing inputs could not be found using their local search algorithms (hence many nodes were declared problematic in their work, since these techniques often became stuck in local optima). As a secondary means of trying to change the flow of execution at the problem node, data dependency analysis was used to find previous program nodes that could somehow affect the outcome at the problem node. Through trial and error execution of sequences of these nodes, it was hoped that the outcome at the problem node would be changed.

In figure 1, execution of the if statement at node 4 as true may have been declared as problematic. The use of a flag variable produces fitness values of zero or one, which provides little guidance to the search for identifying input values for the true case in the event that current test data had led to the false case, or vice versa. In this situation the chaining approach would look for the last reachable definitions of variables used at the problem node. In this instance the variable heater_on is the only variable used at node 4, and its last reachable definitions can be found at nodes 2 and 3. Therefore two node sequences, known as *event sequences*, are generated - one requiring execution of node 2 before node 4, and the other requiring the execution of node 3 before node 4. In order to execute node 2 in the sequence <$s, 2, 4$>, the true branch from node 1 must be taken, requiring the search for input data to execute the condition temp < TEMP_THRESHOLD as true. This becomes the new goal of the local search. If the required test data to execute this branch is found, node 2 is reached and finally the condition on node 4 is also evaluated as true.

In the case where a node leading to a last definition node also became problematic, the chaining approach procedure would recurse to find the last definition nodes for the variables used at these new problem nodes. This led to the generation of longer and longer sequences until some satisfying sequence was found or some pre-imposed length limit was reached.

The use of the chaining approach to change execution at problem nodes is a useful concept that seems applicable to ET and the state problem. In the chaining approach, problem nodes are those for which input data could not be found with local search methods. With the state problem, difficult nodes are those that are not easily executed with ET, due to the use of internal state variables. The chaining approach found previous nodes that could change the outcome of the target node. For the ET state problem; transitional statements need to be found and executed that manipulate the state, as such guidance is not already provided by the fitness function.

5 Applying the Chaining Approach

Our system (from here on referred to as ET-State) aims to deal with test objects exhibiting state behavior whose structural elements could not be covered using traditional ET. The system identifies potential event sequences. Traditional ET then attempts to find the input sequence that will lead to both execution of the event sequence, and if the identified event sequence leads to the desired state, the structural target also. This is performed using path-oriented fitness functions, so that each transitional statement receives the required guidance that leads to its execution.

Event sequences for state problems will be potentially much longer than those for problem nodes for functions solely dependent on the current input, as identified by the original chaining approach. As will be seen, our system also performs data dependency analysis that is more extensive than the original chaining approach, meaning that the number of nodes available to add to each sequence at each step of its construction is also greater. For complicated examples an exhaustive search of the chaining tree may not always be tractable. We plan to accommodate this by using a further stage of optimization to heuristically build and evaluate sequences – namely the ant colony system.

5.1 Ant Colony System (ACS)

Ant colony system (ACS) [2] is an optimizing technique inspired by the foraging behavior of real ants [7]. In ACS the problem is represented by a graph. Ants then incrementally construct solutions using this graph by applying problem-specific heuristics and by taking into account artificial pheromone deposited by previous ants on graph edges. This informs the ant of the previous performance of edges in previous solutions, since the edges belonging to the better solutions receive more pheromone.

ACS is an attractive search technique to use for building event sequences for the state problem, since the incremental solution construction process allows for straightforward incorporation of data dependency procedures for identifying possible transitional program nodes. The space of viable event sequences for a state problem is underpinned by the control and data dependencies of the underlying code structure, and as these factors can be taken into account, ants are prevented from exploring solutions that are unintelligent or infeasible from the outset.

Dorigo et al. [4] originally devised ant systems for the traveling salesman problem (TSP). Here, graph nodes correspond to cities in the problem. Initially, graph edges are initialized to some initial pheromone amount τ_0. Then, in t_{max} cycles, m ants are placed on a random city and progressively construct tours through the use of a probabilistic transition rule. In this rule, a random number, q, $0 \leq q \leq 1$ is selected and compared with a tunable exploration parameter q_0. If $q \leq q_0$, the most appealing edge in terms of heuristic desirability and pheromone concentration is always chosen. However if $q > q_0$ a transition probability rule is used. The probability of ant k to go from node i to node j whilst building its tour in cycle t, is given by:

$$p_{ij}^k(t) = \frac{[\tau_{ij}(t)] \cdot [\eta_{ij}]^\beta}{\sum_{l \in J_i^k} [\tau_{il}(t)] \cdot [\eta_{il}]^\beta} \quad (1)$$

where $\tau_{ij}(t)$ is the amount of pheromone on edge (i,j), β is an adjustable parameter that controls the relative importance of pheromone on edges, and η_{ij} is a desirability heuristic value from city i to j. In the TSP, the desirability heuristic is simply the inverse of the distance from node i to j, making nearer cities more attractive than those further away. Ants keep track of towns they have visited, and exclude these cities from future choices.

Every time an edge is selected a local update is performed whereby some of the pheromone on that edge evaporates. This has the effect of making that edge slightly less attractive to other ants, so as to encourage exploration of other edges of the graph. A local update for an edge *(i,j)* is computed using the formula:

$$\tau_{ij}(t) \leftarrow (1-\rho) \cdot \tau_{ij}(t) + \rho \cdot \tau_0 \qquad (2)$$

where ρ is the pheromone decay coefficient $0 < \rho \leq 1$.

After a cycle has finished, a global update of the graph takes place on the basis of the best solution found so far (i.e. the shortest tour). This encourages ants to search around the vicinity of the best solution in building future tours. The global update is performed for every edge *(i,j)* in the best tour using the formula:

$$\tau_{ij}(t) \leftarrow (1-\rho) \cdot \tau_{ij}(t) + \rho \cdot \Delta \tau_{ij}(t) \qquad (3)$$

where $\Delta \tau_{ij}(t)$ is computed as the inverse of the tour distance.

At the end of the last cycle the algorithm terminates and delivers the shortest tour found.

In the following sections we outline how the basic ACS algorithm is adapted for solving state problems.

Problem Representation. The problem representation for the state problem is simply a graph of all program nodes linked to one another.

However, in the state problem program nodes can be visited more than once in a sequence. This means that arcs belonging to good solutions can also be reinforced more than once. In order to prevent a situation where a sub-path is reinforced to the point where ants begin to travel around the graph in infinite loops, the search space is extrapolated so that nodes in the graph correspond to the *n*th use of some program node in building an event sequence. This expansion of the search space can take place dynamically, so there is no need to pre-impose limits on the number of times each individual program node can be visited.

Solution Construction. At each stage of event sequence construction, each ant chooses from a subset of all program nodes, as identified by the data dependency analysis procedure.

The original chaining approach built event sequences by working backwards from the target node. Data dependency analysis for a problem node ended at the last definitions of variables used at that node. For state problems, we extend the data dependency analysis by considering all nodes that could potentially affect the problem node (for example by analyzing variables used in assignments at last definitions, and so on). The algorithms for doing this are similar to those used to construct backward program slices [8].

Evaluation of Event Sequences. In the ET-State ant algorithm, the heuristic used to evaluate the desirability of the inclusion of a node into the event sequence is also equivalent to the "goodness" of the entire event sequence including that node. In order to evaluate event sequences, they are first executed by traditional ET using path-oriented fitness functions. The fitness information from the best individual is then fed into ET-State. In most cases, the best individual found would likely have a series of diverge points corresponding to nodes depending on states.

When adding a new node to an event sequence, ants use inputs used from the previous sequence to seed the first generation of the evolutionary algorithm for finding inputs leading to the execution of the new sequence.

Pheromone Updates. Local and global pheromone updates are handled in a similar fashion to ACS for the TSP, however $\Delta \tau_{ij}(t)$ for global updates is computed using the fitness of the event sequence as evaluated with ET.

Termination Criteria. As the length of solutions for the state problem is not fixed as they are for tours in the TSP, termination criteria are required to stop ants building infinitely long sequences. Two rules include a) the ant stops if it discovers a solution better than the current best solution found so far, and b) ants can only explore sequences up to a certain number of nodes longer than the current best so far, or the ant fails to improve its solution over the last n nodes.

5.2 Simple Initial Experiment

A simple initial experiment was run with a preliminary version of the system. The code used for this study can be seen in figure 4. The aim is for the ants to find an event sequence to get into the true branch from the if statement in fn_test (part (a)). The outcome of this branch predicate is dependent on functions in the module shown in part (b), which depend on internal state variables.

Four ants were used in ten cycles, the desirability heuristic exponent β was set to 2, the exploration parameter q_0 was set to 0.75, and a pheromone decay ρ of 0.25 was used. Sequences were built using backward dependency analysis. Therefore when adding the first node to its sequence, the ant could only choose to execute fn_c. For the second node the ant could execute fn_c or fn_b, since the assignment statement for k in fn_c uses the variable j that in turn is defined in fn_b. For every node after fn_b was called, any function from the dependent module could be used, since fn_b uses the variable i which in turn is defined in fn_a.

In the first cycle ants were prohibited from using any node more than once, so that the full definition-use chain from k through to i could potentially be explored. The greedy desirability heuristic for the addition of nodes, and for evaluating entire sequences, was simply based on the branch distance for the true branch in fn_test. In this case, and in order to make the search more complicated, the ants were directed to find the shortest satisfying sequence once a satisfying sequence had been found, by simply rewarding shorter sequences (of course, shorter sequences ultimately require less mental effort on behalf of the human checking the results of the structural tests).

Ants finished building their sequence if they had found a new best, or they had made no improvement on their sequence for the last five nodes. The results showed promise, as seen in table 1. In ten runs of the experiment, ants found a satisfying sequence in 2-3 cycles. The shortest function call sequence found (<fn_a, fn_a,

fn_b, fn_c, fn_c, fn_c, fn_test> (or similar sequence of the same length)) was ascertained in an average of 4.9 cycles.

// ...
const int target = 32;
void fn_test()
{
   if (fn_c() == target)
   { /* target branch */ }
}
// ...
	static int i=0, j=0, k=0;
	void fn_a()
	{ i ++; }
	void fn_b()
	{ j += i * 2; }
	int fn_c()
	{ k += j * 2; return k; }

(a) Function under test	(b) Dependent module

Fig. 4. C code for the initial experiment

Table 1. Results for the initial experiment

	Average	Best	Worst
Satisfying Event Sequence Found	2.1	2	3
Shortest Satisfying Event Sequence Found	4.9	3	7

No of ants: 4   No of trials: 10

6 Conclusions and Future Work

This paper has introduced the state problem for evolutionary testing; showing that state variables can hinder or render impossible the search for test data. State variables can be dependent on the input history to the test object, as well as just the current input. They do not form part of the input domain to the test object, and therefore cannot be optimized by the ET search process. The only way to control these variables is to attempt to execute the statements that perform assignments to them, statements that form the transitions of the underlying state machine of the system. Such statements can be identified by data dependency analysis. The use of ant colony algorithms is proposed as a means of heuristic search and evaluation of sequences of transitional statements, in order to find one that makes the current test goal possible.

The next step in this work is the complete implementation of the system, which will then be tried on a set of industrial examples. These experiments will be considered successful if coverage of structural elements involving states are achieved which were previously unobtainable by the use of ET or even random testing alone. If not all structural elements involving states can be covered then further analysis will take place to understand the features of these programs that are still causing problems.

Acknowledgements. This work is sponsored by DaimlerChrysler Research and Technology. The authors have benefited greatly from the discussion of this work with Joachim Wegener, André Baresel and other members of the ET group at DaimlerChrysler, along with members of the EPSRC-funded FORTEST and SEMINAL networks.

References

1. Aho A., Sethi R., Ullman J. D.: Compilers: Principles, Techniques and Tools. Addison-Wesley (1986)
2. Bonabeau E., Dorigo M., Theraulaz G.: Swarm Intelligence. Oxford University Press (1999)
3. Bottaci, L.: Instrumenting Programs with Flag Variables for Test Data Search by Genetic Algorithm, Proceedings of the Genetic and Evolutionary Computation Conference, New York, USA (2002)
4. Dorigo M., Maniezzo, V., Colorni A.: Ant System: An Autocatalytic Optimizing Process. Technical report, Politechnico di Milano, Italy, No. 91-016 (1991)
5. Ferguson R., Korel B.: The Chaining Approach for Software Test Data Generation. ACM Transactions on Software Engineering and Methodology, Vol. 5, No. 1, pp. 63–86 (1996)
6. Harman M., Hu L., Hierons R., Baresel A., Sthamer H: Improving Evolutionary Testing by Flag Removal. Proceedings of the Genetic and Evolutionary Computation Conference, New York, USA (2002)
7. Goss S., Aron S., Denenubourg J. L., Pasteels J. M.: Self Organized Shortcuts in the Argentine Ant. Naturwissenschaften, Vol. 76, pp. 579–581 (1989)
8. Tip F., A Survey of Program Slicing Techniques. Journal of Programming Languages, Vol.3, No.3, pp.121–189 (1995)
9. Tracey N., Clark J., Mander K.: Automated Flaw Finding using Simulated Annealing. International Symposium on Software Testing and Analysis, pp. 73–81 (1998).
10. Wegener J., Baresel A. Sthamer H.: Evolutionary Test Environment for Automatic Structural Testing. Information and Software Technology, Vol. 43, pp. 841–854 (2001)
11. Wegener J., Buhr K., Pohlheim H.: Automatic Test Data Generation for Structural Testing of Embedded Software Systems by Evolutionary Testing. Proceedings of the Genetic and Evolutionary Computation Conference, New York, USA (2002)
12. Wegener J., Grochtmann M.: Verifying Timing Constraints of Real-Time Systems by Means of Evolutionary Testing. Real-Time Systems, Vol. 15, pp. 275–298 (1998)

Modeling the Search Landscape of Metaheuristic Software Clustering Algorithms

Brian S. Mitchell and Spiros Mancoridis

Department of Computer Science
Drexel University, Philadelphia PA 19104, USA
{bmitchell, spiros.mancoridis}@drexel.edu
http://www.mcs.drexel.edu/~{bmitchel,smancori}

Abstract. Software clustering techniques are useful for extracting architectural information about a system directly from its source code structure. This paper starts by examining the Bunch clustering system, which uses metaheuristic search techniques to perform clustering. Bunch produces a subsystem decomposition by partitioning a graph formed from the entities (e.g., modules) and relations (e.g., function calls) in the source code, and then uses a fitness function to evaluate the quality of the graph partition. Finding the best graph partition has been shown to be a NP-hard problem, thus Bunch attempts to find a sub-optimal result that is "good enough" using search algorithms. Since the validation of software clustering results often is overlooked, we propose an evaluation technique based on the search landscape of the graph being clustered. By gaining insight into the search landscape, we can determine the quality of a typical clustering result. This paper defines how the search landscape is modeled and how it can be used for evaluation. A case study that examines a number of open source systems is presented.

1 Introduction and Background

Since many software systems are large and complex, appropriate abstractions of their structure are needed to simplify program understanding. Because the structure of software systems is usually not documented accurately, researchers have expended a great deal of effort studying ways to recover design artifacts from source code.

For small systems, source code analysis tools [3,9] can extract the source-level components (*e.g.*, modules, classes, functions) and relations (*e.g.*, method invocation, function calls, inheritance) of a software system. However, for large systems, these tools are, at best, useful for studying only specific areas of the system.

For large systems there is significant value in identifying the abstract (high-level) entities, and then modeling them using architectural components such as subsystems and their relations. Subsystems provide developers with structural information about the numerous software components, their interfaces, and their interconnections. Subsystems generally consist of a collection of collaborating source code resources that implement a feature or provide a service to the rest of the system. Typical resources found in subsystems include modules, classes, and possibly other subsystems.

The entities and relations needed to represent software architectures are not found in the source code. Thus, without external documentation, we seek other techniques to recover a reasonable approximation of the software architecture using source code information. Researchers in the reverse engineering community have applied a variety of software clustering approaches to address this problem. These techniques determine clusters (subsystems) using source code component similarity [18,4,17], concept analysis [10,5,1], or information available from the system implementation such as module, directory, and/or package names [2].

In this paper we examine the Bunch software clustering system [12,11,13]. Unlike the other software clustering techniques described in the literature, Bunch uses search techniques to determine the subsystem hierarchy of a software system, and has been shown to produce good results for many different types of systems. Additionally, Bunch works well on large systems [16] such as **swing** and **linux**, and some work has been done on evaluating the effectiveness of the tool [14,15].

Bunch starts by generating a random solution from the search space, and then improves it using search algorithms. Bunch rarely produces the exact same result on repeated runs of the tool. However, its results are very similar – this similarity can be validated by inspection and by using similarity measurements [14]. Furthermore, as we will demonstrate later in this paper, we validated that the probability of finding a good solution by random selection is extremely small, thus Bunch appears to be producing a family of related solutions.

The above observations intrigued us to answer why Bunch produces similar results so consistently. The key to our approach will be to model the search landscape of each system undergoing clustering, and then to analyze how Bunch produces results within this landscape.

2 Software Clustering with Bunch

The goal of the software clustering process is to partition a graph of the source-level entities and relations into a set of clusters such that the clusters represent subsystems. Since graph partitioning is known to be NP-hard [7], obtaining a good solution by random selection or exhaustive exploration of the search space is unlikely. Bunch overcomes this problem by using metaheuristic-search techniques.

Figure 1 illustrates the clustering process used by Bunch. In the preparation phase, source code analysis tools are used to parse the code and build a repository of information about the entities and relations in the system. A series of scripts are then executed to query the repository and create the *Module Dependency Graph* (*MDG*). The *MDG* is a graph where the source code components are modeled as nodes, and the source code relations are modeled as edges.

Once the *MDG* is created, Bunch generates a random partition of the *MDG* and evaluates the "quality" of this partition using a fitness function that is called

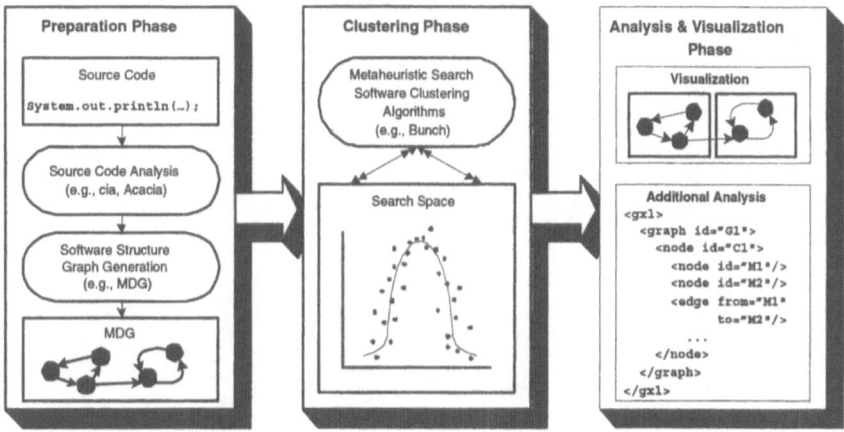

Fig. 1. Bunch's Software Clustering Process

Modularization Quality (MQ) [16]. *MQ* is designed to reward cohesive clusters and penalize excessive inter-cluster coupling. *MQ* increases as the number of *intraedges* (*i.e.*, internal edges contained within a cluster) increases and the number of *interedges* (*i.e.*, external edges that cross cluster boundaries) decreases.

Given that the fitness of an individual partition can be measured, metaheuristic search algorithms are used in the clustering phase of Figure 1 in an attempt to improve the *MQ* of the randomly generated partition. Bunch implements several hill-climbing algorithms [12,16] and a genetic algorithm [12].

Once Bunch's search algorithms converge, a result can be viewed as XML [8] or using the dotty [6] graph visualization tool.

2.1 Clustering Example with Bunch

This section presents an example illustrating how Bunch can be used to cluster the JUnit system. JUnit is an open-source unit testing tool for Java, and can be obtained online from http://www.junit.org. JUnit contains four main packages: the framework itself, the user interface, a test execution engine, and various extensions for integrating with databases and J2EE. The JUnit system contains 32 classes and has 55 dependencies between the classes.

For the purpose of this example, we limit our focus to the framework package. This package contains 7 classes, and 9 inter-class relations. All remaining dependencies to and from the other packages have been collapsed into a single relation to simplify the visualization.

Figure 2 animates the process that Bunch used to partition the *MDG* of JUnit into subsystems. In the left corner of this figure we show the *MDG* of JUnit.

Step 2 illustrates the random partition generated by Bunch as a starting point in the search for a solution. Since the probability of generating a "good" random partition is small, we expect the random partition to be a low quality

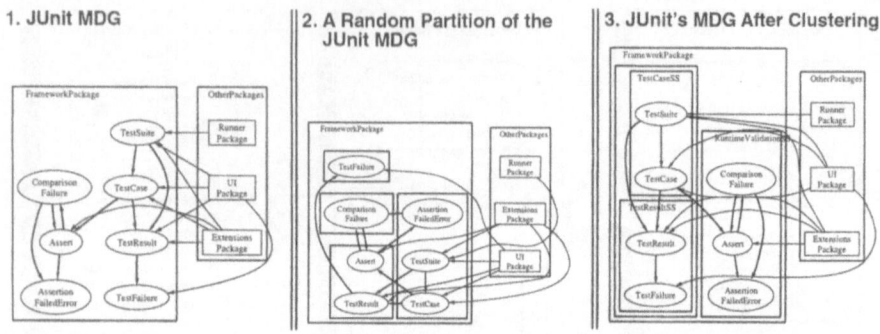

Fig. 2. JUnit Example

solution. This intuition is validated by inspecting the random partition, which contains 2 singleton clusters and a disproportionately large number of interedges (*i.e.*, edges that cross subsystem boundaries).

The random partition shown in step 2 is converted by Bunch into the final result shown in step 3 of Figure 2. The solution proposed by Bunch has many good characteristics as it groups the test case modules, the test result modules, and the runtime validation modules into clusters.

The overall result shown in step 3 is a good result, but Bunch's search algorithms are not guaranteed to produce exactly the same solution for every run. Thus, we would like to gain confidence in the *stability* of Bunch's clustering algorithms by analyzing the search landscape associated with each *MDG*.

3 Modeling the Search Landscape

This section presents an approach to modeling the search landscape of software clustering algorithms. The search landscape is modeled using a series of views, because software clustering results have many dimensions, and combining too much detail into a single view is confusing.

The search landscape is examined from two different perspectives. The first perspective examines the structural aspects of the search landscape, and the second perspective focuses on the similarity aspects of the landscape.

3.1 The Structural Search Landscape

The structural search landscape highlights similarities and differences from a collection of clustering results by identifying trends in the structure of graph partitions. Thus, the goal of the structural search landscape is to validate the following hypotheses:

- We expect to see a relationship between *MQ* and the number of clusters. Both *MQ* and the number of clusters in the partitioned *MDG* should not vary widely across the clustering runs.

- We expect a good result to produce a high percentage of intraedges (edges that start and end in the same cluster) consistently.
- We expect repeated clustering runs to produce similar MQ results.
- We expect that the number of clusters remains relatively consistent across multiple clustering runs.

3.2 The Similarity Search Landscape

The similarity search landscape focuses on modeling the extent of similarity across all of the clustering results. For example, given an edge $<u,v>$ from the MDG, we can determine, for a given clustering run, if modules u and v are in the same or different clusters. If we expand this analysis to look across many clustering runs we would like to see modules u and v consistently appearing (or not appearing) in the same cluster for most of the clustering runs. The other possible relationship between modules u and v is that they sometimes appear in the same cluster, and other times appear in different clusters. This result would convey a sense of dissimilarity with respect to these modules, as the $<u,v>$ edge drifts between being an interdedge and an intraedge.

4 Case Study

This section describes a case study illustrating the effectiveness of using the search landscape to evaluate Bunch's software clustering results. Table 1 describes the systems used in this case study, which consist of 7 open source systems and 6 randomly generated MDGs. Mixing the random graphs with real software graphs enables us to compare how Bunch handles real versus randomly generated graphs.

Table 1. Application descriptions

Application Name	Modules in MDG	Relations in MDG	Application Description
Telnet	28	81	Terminal emulator
PHP	62	191	Internet scripting language
Bash	92	901	Unix terminal environment
Lynx	148	1,745	Text-based HTML browser
Bunch	220	764	Software clustering tool
Swing	413	1,513	Standard Java user interface framework
Kerberos5	558	3,793	Security services infrastructure
Rnd5	100	247	Random graph with 5% edge density
Rnd50	100	2,475	Random graph with 50% edge density
Rnd75	100	3,712	Random graph with 75% edge density
Bip5	100	247	Random bipartite graph with 5% edge density
Bip50	100	2,475	Random bipartite graph with 50% edge density
Bip75	100	3,712	Random bipartite graph with 75% edge density

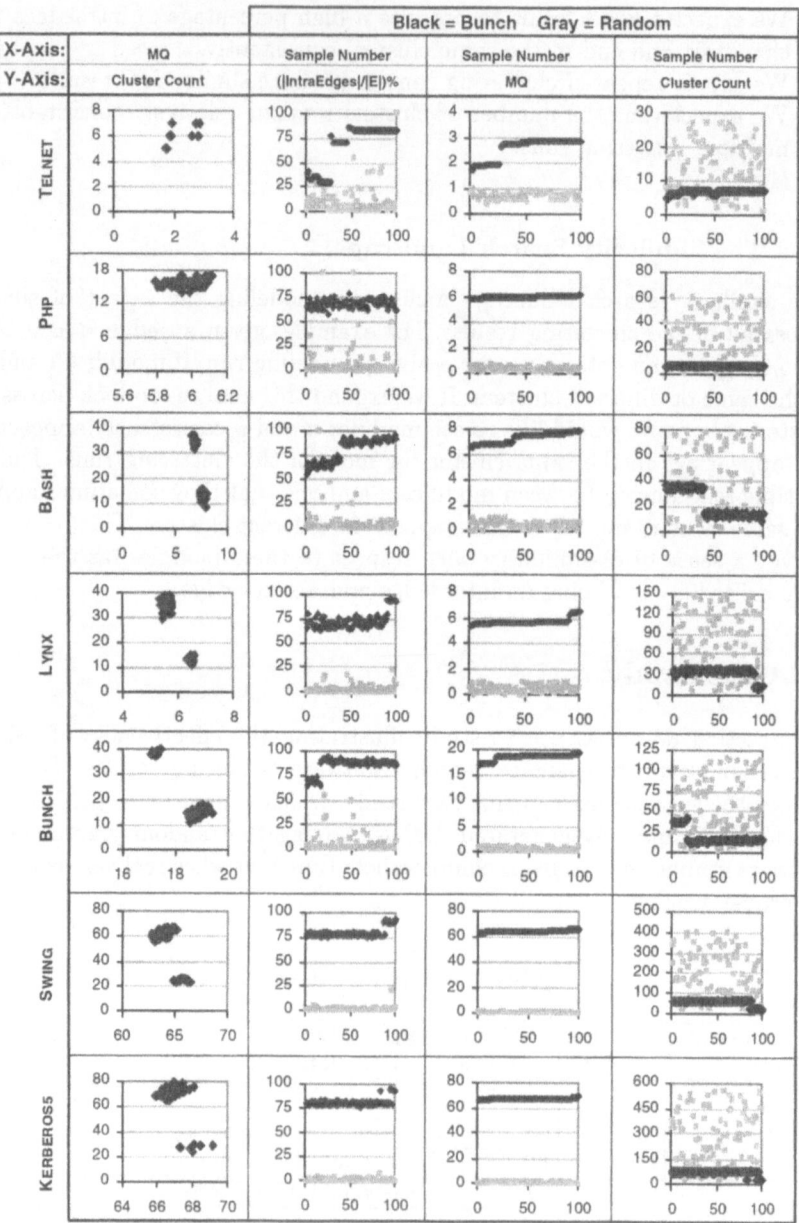

Fig. 3. The Structural Search Landscape for the Open Source Systems

Some explanation is necessary to describe how the random *MDG*s were generated. Each *MDG* used in this case study consists of 100 modules, and has a name that ends with a number. This number is the edge density of the *MDG*.

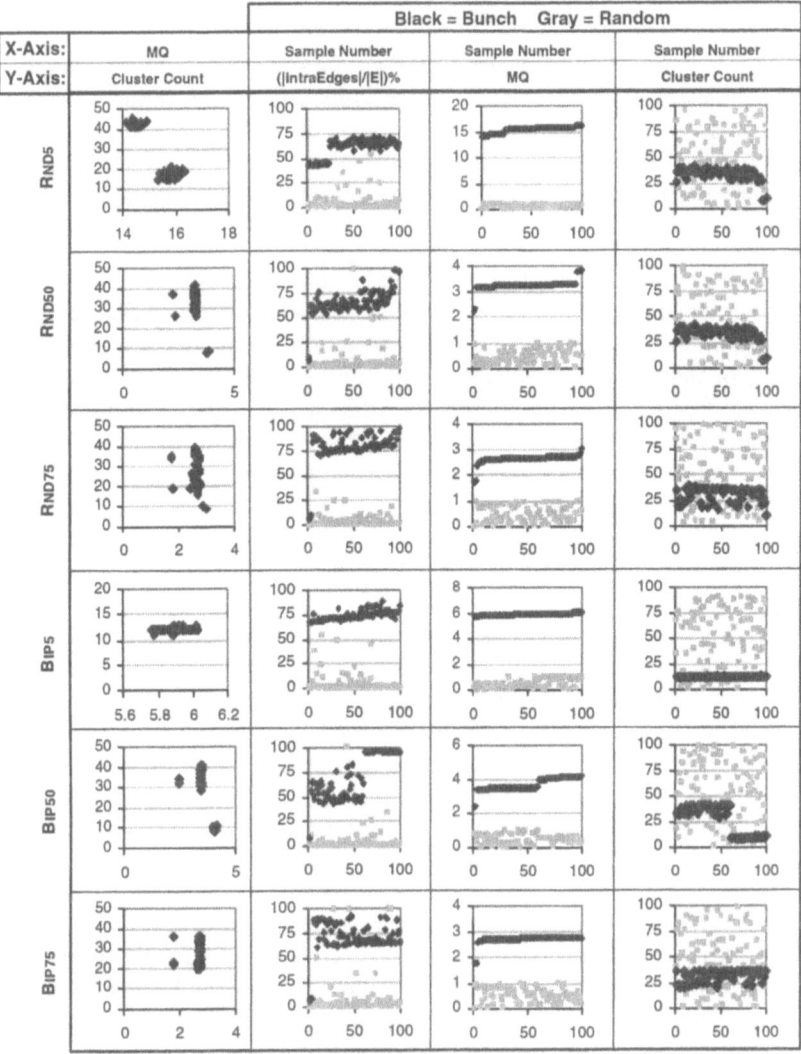

Fig. 4. The Structural Search Landscape for the Random MDGs

For example a 5% edge density means that the graph will have an edge count of $0.05 * (n(n-1)/2)$, which is 5% of the total possible number of edges that can exist for a graph containing n modules.

Each system in Table 1 was clustered 100 times using Bunch's default settings.[1] It should also be noted that although the individual clustering runs are independent, the landscapes are plotted in order of increasing MQ. This sorting highlights some results that would not be obvious otherwise.

---

[1] An analysis on the impact of altering Bunch's clustering parameters was done in a 2002 GECCO paper [16].

4.1 The Structural Landscape

Figure 3 shows the structural search landscape of the open source systems, and Figure 4 illustrates the structural search landscape of the random graphs used in this study.

The results produced by Bunch appear to have many consistent properties. This observation can be supported by examining the results shown in Figures 3 and 4:

- By examining the views that compare the cluster counts (*i.e.*, the number of clusters in the result) to the MQ values (far left) we notice that Bunch tends to converge to one or two "basins of attraction" for all of the systems studied. Also, for the real software systems, these attraction areas appear to be tightly packed. For example, the php system has a point of concentration were all of the clustering results are packed between MQ values of 5.79 and 6.11. The number of clusters in the results are also tightly packed ranging from a minimum of 14 to a maximum of 17 clusters. An interesting observation can be made when examining the random systems with a higher edge density (*i.e.*, rnd50, rnd75, bip50, bip75). Although these systems converged to a consistent MQ, the number of clusters varied significantly over all of the clustering runs. We observe these wide ranges in the number of clusters, with little change in MQ for most of the random systems, but do not see this characteristic in the real systems.
- The view that shows the percentage of intraedges in the clustering results (second column from the left) indicates that Bunch produces consistent solutions that have a relatively large percentage of intraedges. Also, since the 100 samples were sorted by MQ, we observe that the intraedge percentage increases as the MQ values increase. By inspecting all of the graphs for this view it appears that the probability of selecting a random partition (gray data points) with a high intraedge percentage is rare.
- The third view from the left shows the MQ value for the initial random partition, and the MQ value of the partition produced by Bunch. The samples are sorted and displayed by increasing MQ value. Interestingly, the clustered results produce a relatively smooth line, with points of discontinuity corresponding to different "basins of attraction". Another observation is that Bunch generally improves MQ much more for the real software systems, when compared to the random systems with a high edge density (rnd50, rnd75, bip50, bip75).
- The final view (far right column) compares the number of clusters produced by Bunch (black) with the number of clusters in the random starting point (gray). This view indicates that the random starting points appear to have a uniform distribution with respect to the number of clusters. We expect the random graphs to have from 1 (*i.e.*, a single cluster containing all nodes) to N (*i.e.*, each node is placed in its own cluster) clusters. The y-axis is scaled to the total number of modules in the system (see Table 1). This view shows that Bunch always converges to a "basin of attraction" regardless of the number of clusters in the random starting point. This view also supports

the claim made in the first view where the standard deviation for the cluster counts appears to be smaller for the real systems than they are for the random systems.

When examining the structural views collectively, the degree of commonality between the landscapes for the systems in the case study is surprisingly similar. We do not know exactly why Bunch converges this way, although we speculate that this positive result may be based on a good design property of the MQ fitness function. Section 2 described that the MQ function works on the premise of maximizing intraedges, while at the same time, minimizing interedges. Since the results converge to similar MQ values, we speculate that the search space contains a large number of isomorphic configurations that produce similar MQ values. Once Bunch encounters one of these areas, it's search algorithms cannot find a way to transform the current partition into a new partition with higher MQ.

4.2 The Similarity Landscape

The main observation from analyzing the structural search landscapes is that the results produced by Bunch are stable. For all of the MDGs we observe similar characteristics, but we were troubled by the similarity of the search landscapes for real software systems when compared to the landscapes of the random graphs. We expected that the random graphs would produce different results when compared to the real software systems.

In order to investigate the search landscape further we measured the degree of similarity of the placement of nodes into clusters across all of the clustering runs to see if there were any differences between random graphs and real software systems. Bunch creates a subsystem hierarchy, where the lower levels contain detailed clusters, and higher levels contain clusters of clusters. Because each subsystem hierarchy produced by Bunch may have a different height,[2] we decided to measure the similarity between multiple clustering runs using the initial (most detailed) clustering level.

The procedure used to determine the similarity between a series of clustering runs works as follows:

1. Create a counter $C_{<u,v>}$ for each edge $<u,v>$ in the MDG, initialize the counter to zero for all edges.
2. For each clustering run, take the lowest level of the clustering hierarchy and traverse the set of edges in the MDG:
   - If the edge $<u,v>$ is an intraedge increment the counter $C_{<u,v>}$ associated with that edge.

---

[2] The height of the subsystem hierarchy produced by Bunch generally does not differ by more than 3 levels for a given system, but the heights of the hierarchy for different systems may vary dramatically.

3. Given that there are N clustering runs, each counter $C_{<u,v>}$ will have a final value in the range of $0 \leq C_{<u,v>} \leq N$. We then normalize the $C_{<u,v>}$, by dividing by N, which provides the percentage $P_{<u,v>}$ of the number of times that edge $<u,v>$ appears in the same cluster across all clustering runs.
4. The frequency of the $P_{<u,v>}$ is then aggregated into the ranges: $\{[0,0], (0,10], (10,75), [75,100]\}$. These ranges correspond to **no (zero), low, medium** and **high** similarity, respectively. In order to compare results across different systems, the frequencies are normalized into percentages by dividing each value in the range set by the total number of edges in the system ($|E|$).

Using the above process across multiple clustering runs enables the overall similarity of the results to be studied. For example, having a large *zero* and *high* similarity is good, as these values highlight edges that either never or always appear in the same cluster. The *low* similarity measure captures the percentage of edges that appear together in a cluster less than 10% of the time, which is desirable. However, having a large *medium* similarity measure is undesirable, since this result indicates that many of the edges in the system appear as both inter- and intraedges in the clustering results.

Table 2. The Similarity Landscape of the Case Study Systems

Application Name	Edge Density Percent	Similarity Percentage (S)			
		Zero (%) ($S=0\%$)	Low (%) ($0\% < S \leq 10\%$)	Medium (%) ($10\% < S < 75\%$)	High (%) ($S \geq 75\%$)
Telnet	21.42	34.56	27.16	13.58	24.70
PHP	10.10	48.16	19.18	11.70	20.96
Bash	21.52	58.15	22.86	6.32	12.67
Lynx	16.04	49.28	30.64	8.99	11.09
Bunch	3.17	48.70	20.23	9.36	21.71
Swing	1.77	61.53	13.81	9.84	14.82
Kerberos5	2.44	57.55	19.06	9.75	13.64
Rnd5	5.00	12.14	54.65	24.71	8.50
Rnd50	50.00	32.46	37.97	29.49	0.08
Rnd75	75.00	33.80	30.39	35.81	0.00
Bip5	5.00	37.27	20.97	13.46	28.30
Bip50	50.00	47.21	23.89	28.60	0.30
Bip75	75.00	29.90	38.36	31.74	0.00

Now that the above approach for measuring similarity has been described, Table 2 presents the similarity distribution for the systems used in the case study. This table presents another interesting view of the search landscape, as it exhibits characteristics that differ for random and real software systems. Specifically:

- The real systems tend to have large values for the *zero* and *high* categories, while the random systems score lower in these categories. This indicates

that the results for the real software systems have more in common than the random systems do.
- The random systems tend to have much higher values for the *medium* category, further indicating that the similarity of the results produced for the real systems is better than for the random systems.
- The real systems have relatively low edge densities. The swing system is the most sparse (1.77%), and the bash system is the most dense (21.52%). When we compare the real systems to the random systems we observe a higher degree of similarity between the sparse random graphs and the real systems than we do between the real systems and the dense random graphs (rnd50, rnd75, bip50, bip75).
- It is noteworthy that the dense random graphs typically have very low numbers in the *high* category, indicating that it is very rare that the same edge appears as an intraedge from one run to another. The result is especially interesting considering that the *MQ* values presented in the structural landscape change very little for these random graphs. This outcome also supports the isomorphic "basin of attraction" conjecture proposed in the previous section, and the observation that the number of clusters in the random graphs vary significantly.

5 Conclusions

Ideally, the results produced by Bunch could be compared to an optimal solution, but this option is not possible since the graph partitioning problem is NP-hard. User feedback has shown that the results produced by Bunch are "good enough" for assisting developers performing program comprehension and software maintenance activities, however, work on investigating why Bunch produces consistent results had not been performed until now.

Through the use of a case study, we highlighted several aspects of Bunch's clustering results that would not have been obvious by examining individual clustering results. We also gained some intuition about why the results produced by Bunch have several common properties regardless of whether the *MDG*s were real or randomly generated. A final contribution of this paper is that it demonstrates that modeling the search landscape of metaheuristic search algorithms is a good technique for gaining insight into the solution quality of these types of algorithms.

Acknowledgements. This research is sponsored by grants CCR-9733569 and CISE-9986105 from the National Science Foundation (NSF). Any opinions, findings, and conclusions or recommendations expressed in this material are those of the authors and do not necessarily reflect the views of the NSF.

References

1. Anquetil, N.: A Comparison of Graphis of Concept for Reverse Engineering. In *Proc. Intl. Workshop on Program Comprehension*, June 2000.
2. Anquetil, N., Lethbridge, T.: Recovering Software Architecture from the Names of Source Files. In *Proc. Working Conf. on Reverse Engineering*, October 1999.
3. Chen, Y-F.: Reverse engineering. In B. Krishnamurthy, editor, *Practical Reusable UNIX Software*, chapter 6, pages 177–208. John Wiley & Sons, New York, 1995.
4. Choi, S., Scacchi, W.: Extracting and Restructuring the Design of Large Systems. In *IEEE Software*, pages 66–71, 1999.
5. van Deursen,A., Kuipers, T.: Identifying Objects using Cluster and Concept Analysis. In *International Conference on Software Engineering, ICSM'99*, pages 246–255. IEEE Computer Society, May 1999.
6. Gansner, E.R., Koutsofios, E., North, S.C., Vo, K.P: A Technique for Drawing Directed Graphs. *IEEE Transactions on Software Engineering*, 19(3):214–230, March 1993.
7. Garey, M.R., Johnson, D.S: *Computers and Intractability*. W.H. Freeman, 1979.
8. GXL: Graph eXchange Language: Online Guide. http://www.gupro.de/GXL/.
9. Korn, J., Chen, Y-F., Koutsofios, E.: Chava: Reverse Engineering and Tracking of Java Applets. In *Proc. Working Conference on Reverse Engineering*, October 1999.
10. Lindig, C., Snelting, G.: Assessing Modular Structure of Legacy Code Based on Mathematical Concept Analysis. In *Proc. International Conference on Software Engineering*, May 1997.
11. Mancoridis, S., Mitchell, B.S., Chen, Y-F., Gansner, E.R.: Bunch: A Clustering Tool for the Recovery and Maintenance of Software System Structures. In *Proceedings of International Conference of Software Maintenance*, pages 50–59, August 1999.
12. Mancoridis, S., Mitchell, B.S., Rorres, C., Chen, Y-F., Gansner, E.R.: Using Automatic Clustering to Produce High-Level System Organizations of Source Code. In *Proc. 6th Intl. Workshop on Program Comprehension*, June 1998.
13. Mitchell, B.: *A Heuristic Search Approach to Solving the Software Clustering Problem*. PhD thesis, Drexel University, Philadelphia, PA, USA, 2002.
14. Mitchell, B.S., Mancoridis, S.: Comparing the Decompositions Produced by Software Clustering Algorithms using Similarity Measurements. In *Proceedings of International Conference of Software Maintenance*, November 2001.
15. Mitchell, B.S., Mancoridis, S.: CRAFT: A Framework for Evaluating Software Clustering Results in the Absence of Benchmark Decompositions. In *Proc. Working Conference on Reverse Engineering*, October 2001.
16. Mitchell, B.S., Mancoridis, S.: Using Heuristic Search Techniques to Extract Design Abstractions from Source Code. In *Proceedings of Genetic and Evolutionary Computation Conference*, 2002.
17. Müller, H., Orgun, M., Tilley, S., Uhl., J.: A Reverse Engineering Approach to Subsystem Structure Identification. *Journal of Software Maintenance: Research and Practice*, 5:181–204, 1993.
18. Schwanke, R., Hanson, S.: Using Neural Networks to Modularize Software. *Machine Learning*, 15:137–168, 1998.

Search Based Transformations

Deji Fatiregun, Mark Harman, and Robert Hierons

Department of Information Systems and Computing
Brunel University
Uxbridge, Middlesex, UB8 3PH
ayodeji.fatiregun@brunel.ac.uk

1 Introduction

Program transformation [1,2,3] can be described as the act of changing one program to another. The aim being an alteration of the program syntax but not its semantics, hence leaving both source and target programs functionally equivalent. Consider as examples of transformations the following program fragments:

T_1: if (e1) s1; else s2; $\Rightarrow$ if (!e1) s2; else s1;
T_2: if (true) s1; else s2; $\Rightarrow$ s1;
T_3: x = 2; x = x - 1; y = 10; x = x + 1; $\Rightarrow$ x = 2; y = 10;
T_4: for (s1; e2; s2) s3; $\Rightarrow$ s1; while (e2) s3; s2;

Program Transformations are generally written in order to generate *better* programs. In transformations, we apply a number of simple transformation axioms to parts of a program source code to obtain a functionally equivalent program. The application of these axioms is treated as a search problem and we apply a meta–heuristic search algorithm such as hill climbing to guide the direction of the search.

1.1 The Transformation Problem

An overall program transformation from one program p to an improved version p' typically consists of many many smaller transformation tactics [1,2]. Each tactic consists of the application of a set of transformation rules. A transformation rule is an atomic transformation capable of performing the simple alterations like those captured in the examples $T_1 \ldots T_4$. At each stage in the application of these simple rules, there are many points in the program at which a chosen transformation rule could be applied.

There are many points in a program; typically one per node of the Control Flow Graph. The set of pairs of possible transformation rules and their corresponding application point is therefore large. Furthermore, to achieve an effective overall program transformation tactic, many rules may need to be applied, and each will have to be applied in the correct order to achieve the desired result.

2 Local Search-Based Transformation

Meta-heuristic search algorithms such as hill-climbing may be applied to arrive at an optimum result, or at least, a locally optimal result. Rather than apply the transformations manually, one after the other, we allow the algorithm to pick the best transformations to apply from a given set. We assume a search space containing all the possible allowable transformation rules and define our fitness function using an existing software metric [5,4], for example to be size of the program in Lines of Code (LoC).

An optimum solution would be the sequence of transformations that results in an equivalent program with the fewest possible number of statements. For instance, in the examples used in the introduction, transformation T_3 clearly shows a reduction in the size of the program from 4 nodes to 2 nodes and so would be selected as one which returned a better program, than, for example, the identity transformation.

Using a simple hill-climbing search algorithm and a size-based metric such as LoC, after a particular transformation is applied, the size of the new program is compared with that of the previous program. Any transformation which reduces size, is retained and the search continues from the new smaller program found. When no smaller program is found by the application of a rule, the search terminates.

3 Evolutionary Search-Based Transformation

Our approach is to use the *transformation sequence* to be applied to the program as the individual to be optimised. Using the transformation sequence as the individual makes it possible to define crossover relatively easily. Two sequences of transformations can be combined to change information, using single point, multiple point and uniform crossover. The result is a valid transformation sequence and since all transformation rules are meaning preserving, so are all sequences of transformation rules.

References

1. I. D. Baxter. Transformation systems: Domain-oriented component and implementation knowledge. In *Proceedings of the Ninth Workshop on Institutionalizing Software Reuse*, Austin, TX, USA, January 1999.
2. Keith H. Bennett. Do program transformations help reverse engineering? In *IEEE International Conference on Software Maintenance (ICSM'98)*, pages 247–254, Bethesda, Maryland, USA, November 1998. IEEE Computer Society Press, Los Alamitos, California, USA.
3. John Darlington and Rod M. Burstall. A tranformation system for developing recursive programs. *Journal of the ACM*, 24(1):44–67, 1977.
4. Norman E. Fenton. *Software Metrics: A Rigorous Approach*. Chapman and Hall, 1990.
5. Martin J. Shepperd. *Foundations of software measurement*. Prentice Hall, 1995.

Finding Building Blocks for Software Clustering

Kiarash Mahdavi, Mark Harman, and Robert Hierons

Department of Information Systems and Computing
Brunel University
Uxbridge, Middlesex, UB8 3PH
kiarash.mahdavi@brunel.ac.uk

1 Introduction

It is generally believed that good modularization of software leads to systems which are easier to design, develop, test, maintain and evolve [1].

Software clustering using search–based techniques has been well studied using a hill climbing approach [2,4,5,6]. Hill climbing suffers from the problem of local optima, so some improvement may be expected by by considering more sophisticated search-based techniques. However, hitherto, the use of other techniques to overcome this problem such as Genetic Algorithms (GA) [3] and Simulated Annealing [7] have been disappointing.

This poster paper looks at the possibility of using results from multiple hill climbs to form a basis for subsequent search. The findings will be presented in the poster created for the poster session in GECCO 2003.

2 Multiple Hill Climbing to Identify Building Blocks

The goal of module clustering is to arrive at a graph partition in which each cluster maximizes the number of internal edges and minimizes the number of external edges [1]. To capture this in our approach, we use the "Basic MQ" fitness function [4].

The multiple hill climbing process consists of two stages. In the initial stage a set of hill climbs are carried out to produce a set of clusterings. These clusterings are then used to create building blocks for the final stage of the process.

The hill climbers use a nearest neighbor approach used in Bunch [4,6]. In this approach, the nearest neighbors of a clustering are constructed by moving a node (or module) to a new cluster or into an existing cluster. In our approach however, the nodes correspond to building blocks (sets of modules) not individual modules.

The clusterings from the initial stage of the process are compared to identify groups of nodes that are placed in the same cluster across clusterings. To reduce the disruptive effect of highly unfit solutions on the creation of the initial building blocks, cut off points based on the quality of initial hill climbs are used. These are made from the best 10% to the best 100% of the hill climbed, in increments of 10%, resulting in 10 sets of building blocks. These building blocks are then used in the second stage to improve the results of subsequent hill climbing and GAs.

3 Using Building Blocks to Reduce the Search Space for Repeated Hill Climbing

Simply reducing the number of nodes by grouping nodes together as building blocks reduces the search space. Furthermore, created building blocks provide a way to identify and preserve useful structures within the solution landscape and remove the destructive effects of mutation on these structures.

4 Using Building Blocks to Seed Genetic Algorithms

GAs can find clustering problems difficult [2,3,6]. This could be attributed to the difficulty they face in preserving building blocks from the destructive effects of the genetic operators. To combat this, our approach is to use the identified building blocks as atomic units by the GA. This reduces the size of the search space and potentially assists the GA in forming solutions at the meta level of combining building blocks.

References

1. Larry L. Constantine and Edward Yourdon. *Structured Design*. Prentice Hall, 1979.
2. D. Doval, S. Mancoridis, and B. S. Mitchell. Automatic clustering of software systems using a genetic algorithm. In *International Conference on Software Tools and Engineering Practice (STEP'99)*, Pittsburgh, PA, 30 August – 2 September 1999.
3. Mark Harman, Robert Hierons, and Mark Proctor. A new representation and crossover operator for search-based optimization of software modularization. In *GECCO 2002: Proceedings of the Genetic and Evolutionary Computation Conference*, pages 1351–1358, New York, 9–13 July 2002. Morgan Kaufmann Publishers.
4. Spiros Mancoridis, Brian S. Mitchell, Yih-Farn Chen, and Emden R. Gansner. Bunch: A clustering tool for the recovery and maintenance of software system structures. In *Proceedings; IEEE International Conference on Software Maintenance*, pages 50–59. IEEE Computer Society Press, 1999.
5. Spiros Mancoridis, Brian S. Mitchell, C. Rorres, Yih-Farn Chen, and Emden R. Gansner. Using automatic clustering to produce high-level system organizations of source code. In *International Workshop on Program Comprehension (IWPC'98)*, pages 45–53, Ischia, Italy, 1998. IEEE Computer Society Press, Los Alamitos, California, USA.
6. Brain S. Mitchell. *A Heuristic Search Approach to Solving the Software Clustering Problem*. PhD Thesis, Drexel University, Philadelphia, PA, January 2002.
7. Brian S. Mitchell and Spiros Mancoridis. Using heuristic search techniques to extract design abstractions from source code. In *GECCO 2002: Proceedings of the Genetic and Evolutionary Computation Conference*, pages 1375–1382, New York, 9–13 July 2002. Morgan Kaufmann Publishers.

Author Index

Abbass, Hussein A. 483, 1612
Acan, Adnan 695
Ackley, David H. 150
Aguilar Contreras, Andrés 874
Aguilar-Ruiz, Jesús S. 979
Aguirre, Hernán 2264
Ahn, Byung-Ha 1610
Akama, Kiyoshi 1222, 1616
Alba, Enrique 955
Almeida, Jonas 1776
Ando, Shin 1926
Angelov, P.P. 1938
Aporntewan, Chatchawit 1566
Araujo, Lourdes 1951
Armstrong, Marc P. 1530
Arnold, Dirk V. 525
Arslan, Tughrul 1614
Asai, Kiyoshi 2288
Auger, Anne 512
Aune, Thomas 2277
Aupetit, S. 140
Azad, R. Muhammad Atif 1626

Bacardit, Jaume 1818
Balan, Gabriel Catalin 1729
Balan, M. Sakthi 425
Ballester, Pedro J. 706
Banerjee, Nilanjan 2179
Banzhaf, Wolfgang 390
Baraglia, R. 2109
Barbosa, Helio J.C. 718
Baresel, André 2428, 2442
Barfoot, Timothy D. 377
Barkaoui, Mohamed 646
Barry, Alwyn 1832
Barz, Christiane 754
Bates Congdon, Clare 2034
Baugh Jr., John W. 730
Behrens, Ivesa 754
Beielstein, Thomas 1963
Belle, Terry Van 150
Benekohal, Rahim F. 2420
Benson, Karl 1690
Berger, Jean 646
Beyer, Hans-Georg 525

Bhanu, Bir 332, 2227
Blackwell, T.M. 1
Blain, Derrel 413
Blindauer, Emmanuel 369
Bonacina, Claudio 1382
Boryczka, Mariusz 142
Botello Rionda, Salvador 573
Bottaci, Leonardo 2455
Bouvry, Pascal 1369
Bouzerdoum, Abdesselam 742
Bowersox, Rodney 2157
Branke, Jürgen 537, 754, 766, 1568
Brown, Martin 778
Buason, Gunnar 458
Bucci, Anthony 250
Buckles, Bill P. 1624
Bull, Larry 1924
Burke, Edmund 1800
Buswell, R.A. 1938
Butz, Martin V. 1844, 1857, 1906

Camens, Doug 1590
Cantú-Paz, Erick 790, 801
Cardona, Cesar 219
Carter, Jonathan N. 706
Carvalho, Andre de 634
Castillo, Flor 1975
Cattron, James 298
Chafekar, Deepti 813
Cheang, Sin Man 1802, 1918
Chen, Hui 379
Chen, Ping 1986
Chen, Ying-Ping 825, 837, 1620
Chen, Yu-Hung 681
Choe, Heemahn 850
Choi, Sung-Soon 850, 862, 1998
Choi, Yoon-Seok 2010
Chong, H.W. 2396
Chongstitvatana, Prabhas 1566
Chuang, Yue-Ru 681
Clark, John A. 146, 2022
Clergue, Manuel 1788
Coello, Carlos A. 158, 573, 640
Collard, Philippe 1788
Collins, J.J. 1320

Cruz Cortés, Nareli 158
Culberson, Joseph 948
Cutello, Vincenzo 171, 1570
Czech, Zbigniew J. 142

D'Eleuterio, Gabriele M.T. 377
Daida, Jason M. 1639, 1652, 1665
Dasgupta, Dipankar 183, 195, 219, 1580
Dawson, Devon 1870
Delbem, A.C.B. 634
Dick, Grant 1572
Dijk, Steven van 886
Divina, Federico 898, 1574
Doerner, Karl 2465
Dongarra, Jack 1015
Doya, Kenji 507
Dozier, Gerry 561
Drake, Stephen 1576
Droste, Stefan 909
Dubreuil, Marc 1578

Eikelder, Huub M.M. ten 344
Elfwing, Stefan 507
Elliott, Lionel 2046
Eppstein, Maggie 967
Espinoza, Felipe P. 922
Esterline, Albert 657
Ewald, Claus-Peter 1963

Falke II, William Joseph 1920
Fan, Zhun 1029, 1764, 2058
Fatiregun, Deji 2511
Feinstein, Mark 61
Ferguson, M.I. 442
Fernández, Francisco 1812, 2109
Ficici, Sevan G. 286
Fischer, Marco 2398
Forman, Sean L. 2072
Foster, James A. 2313, 456
Fry, Rodney 1804
Fu, Zhaohui 1986
Fukami, Kiichiro 2134
Furutani, Hiroshi 934

Gaag, Linda C. van der 886
Gagné, Christian 1578
Galeano, Germán 1812
Gallagher, John C. 431, 454
Galos, Peter 1806
Gao, Yong 948

Garnica, O. 2109
Garrell, Josep Maria 1818
Garrett, Deon 1469
Garway-Heath, David 2360
Garzon, Max H. 379, 413
Gérard, Pierre 1882
Giacobini, Mario 955
Gilbert, Joshua 967
Giráldez, Raúl 979
Goldberg, David E. 801, 825, 837, 922, 1172, 1271, 1332, 1554, 1620, 1844, 1857, 1906
Gómez, D. 243
Gomez, Faustino J. 2084
Gómez, Jonatan 195, 1580
González, Fabio 195, 219, 1580
Goodman, Erik D. 1029, 1764, 2058, 2121
Green, James 1975
Greene, William A. 1582
Guinot, C. 140
Guo, Haipeng 1584
Guo, Pei F. 322
Gustafson, Steven 1800
Gutjahr, Walter J. 2465
Gwaltney, D.A. 442

Hahn, Lance W. 2412
Hallam, Jonathan 1586
Hamza, Karim 2096
Han, Kuk-Hyun 427, 2147
Hanby, V.I. 1938
Handa, Hisashi 991
Hang, Dehua 13
Harman, Mark 2511, 2513
Hart, Emma 1295
Hayasaka, Taichi 1600
Heckendorn, Robert B. 1003
Hermida, R. 2109
Hernández Aguirre, Arturo 573
Heywood, Malcolm I. 2325
Hidalgo, J.I. 2109
Hierons, Robert 2511, 2513
Hilbers, Peter A.J. 344
Hilss, Adam M. 1639, 1652, 1665
Hiroshima, Koji 2134
Hiroyasu, Tomoyuki 1015
Holcombe, Mike 2488
Holder, Mike 2400
Holifield, Gregory A. 1588

Homaifar, Abdollah 657
Horn, Jeffrey 298
Hornby, Gregory S. 1678
Howard, Daniel 1690
Hsu, William H. 1584
Hu, Jianjun 1029, 1764, 2058
Hu, Shangxu 134
Huang, Chien-Feng 1041
Huang, Gaofeng 1053
Huang, Yong 2121
Huang, Zhijian 2121
Hussain, Mudassar 657

Iba, Hitoshi 470, 1259, 1715, 1816, 1926, 2288
Ichikawa, Kazuo 2134
Ichikawa, Manabu 2134
Ingham, Derek B. 2046
Ishibuchi, Hisao 1065, 1077, 1234
Isono, Katsunori 2288

Jacob, Jeremy L. 2022
Jähn, Hendrik 2398
Jang, Jun-Su 2147
Jansen, Thomas 310
Jin, Yaochu 636
Johnson, Paul 156
Jong, Edwin D. de 262, 274
Jong, Kenneth De 1604
Julstrom, Bryant A. 2402
Just, Winfried 154

Kaegi, Simon 122
Kaige, Shiori 1234
Kalisker, Tom 1590
Kamio, Shotaro 470
Karr, Charles L. 2157, 2400, 2404, 2406
Kavka, Carlos 1089
Keijzer, Maarten 898, 1574, 1752
Kelsey, Johnny 207
Kendall, Graham 1800
Kharma, Nawwaf 322
Khoshgoftaar, Taghi M. 1808
Kim, Jong-Hwan 427, 638, 2147
Kim, Jong-Pil 2408
Kim, Jung-Hwan 1101
Kim, Keum Joo 642
Kim, Yong-Geon 2426
Kim, Yong-Hyuk 1112, 1123, 1136, 1345, 2168, 2215, 2410

Kimiaghalam, Bahram 657
Klein, Jon 61
Klinkenberg, Ralf 1606
Kohl, Nate 356
Kondo, Shoji 2134
Korczak, Jerzy 369
Kordon, Arthur 1975
Kramer, Gregory R. 454
Krawiec, Krzysztof 332
Kumar, Rajeev 1592, 2179
Kumar, Sujay V. 730
Kwan, Raymond S.K. 693
Kwok, N.M. 2191
Kwon, Yung-Keun 1112, 2203, 2426
Kwong, Sam 2191, 2396
Kyne, Adrian G. 2046

Labroche, Nicolas 25
Lai, Eugene 681
Lanchares, J. 2109
Langdon, W.B. 1702
Lanzi, Pier Luca 1894, 1922
Lattaud, Claude 144
Le Bris, Claude 512
Lee, Chi-Ho 638
Lee, Kin Hong 1802, 1918
Lee, Seung-Kyu 2168, 2215
Lefley, Martin 2477
Leier, André 390
Lemonge, Afonso C.C. 718
Leung, Kwong-Sak 585, 1160, 1802, 1918
Li, Gaoping 2121
Li, Hsiaolei 1665
Li, Xiaodong 37
Liang, Yong 585, 1160
Liekens, Anthony M.L. 344
Lim, Andrew 1053, 1594, 1596, 1986
Lin, Kuo-Chi 1541
Lin, Yingqiang 2227
Lipson, Hod 1518
Liszkai, Tamás 2418
Liu, Hongwei 1715
Liu, Xiaohui 2360
Liu, Yi 1808
Lizárraga, Giovanni 573
Llorà, Xavier 1172
Long, Stephen L. 1652
Louis, Sushil J. 2424
Lu, Ming 1148
Luke, Sean 1729, 1740

Mahdavi, Kiarash 2513
Majumdar, Nivedita Sumi 183
Mancoridis, Spiros 2499
Marchiori, Elena 898, 1574
Mărginean, Flaviu Adrian 1184
Marín-Blázquez, Javier G. 1295
Markon, Sandor 1963
Marshall, Kenric 1975
Martín-Vide, Carlos 401
Masahiro, Hiji 2337
Masaru, Tezuka 2337
Matsui, Shouichi 1598, 2240
Mauch, Holger 1810
Mazza, Raymond H. 2034
McMinn, Phil 2488
McQuay, Brian N. 2384
Mera, Nicolae S. 2046
Messimer, Sherri 2406
Mezura-Montes, Efrén 640
Miikkulainen, Risto 356, 2084
Miki, Mitsunori 1015
Minaei-Bidgoli, Behrouz 2252
Minsker, Barbara S. 922
Misevicius, Alfonsas 598
Mistry, Paavan 693
Mitavskiy, Boris 1196
Mitchell, Brian S. 2499
Mitrana, Victor 401
Mitsuhashi, Hideyuki 470
Mohan, Chilukuri 110
Monmarché, Nicolas 25, 140
Moon, Byung-Ro 669, 850, 862, 1101, 1112, 1123, 1136, 1345, 1357, 1998, 2010, 2168, 2203, 2215, 2408, 2410, 2426
Moore, Jason H. 2277, 2412
Morrison, Ronald W. 1210
Moura Oliveira, P.B. de 510
Mueller, Rainer 742
Munetomo, Masaharu 1222, 1616
Murao, Naoya 1222
Murata, Tadahiko 1234
Myodo, Emi 2264
Myung, Hyun 638

Nasraoui, Olfa 219
Neel, Andrew 379
Nguyen, Minh Ha 483
Nicolau, Miguel 1752
Nicosia, Giuseppe 171
Niño, F. 243

Nordin, Peter 495, 1806
Northern, James III 2414

Ocenasek, Jiri 1247
Oda, Terri 122, 231
Ofria, Charles 13
Oka, Kazuhiro 1814
Okabe, Tatsuya 636
Olsen, Anne 2416
Olsén, Joel 1806
Olsen, Nancy 2277
Oppacher, Franz 1481, 1493
Orla Kimbrough, Steven 1148
Osadciw, Lisa Ann 110

Palacios-Durazo, Ramón Alfonso 371
Palmes, Paulito P. 1600
Panait, Liviu 1729, 1740
Pappalardo, Francesco 1570
Parizeau, Marc 1578
Park, Sung-Joon 1602
Paul, Topon Kumar 1259
Pauw, Guy De 549
Pavone, Mario 171
Pei, Min 2121
Pelikan, Martin 1247, 1271
Peram, Thanmaya 110
Perego, R. 2109
Pérez-Jiménez, Mario J. 401
Perry, Chris 61
Pohlheim, Hartmut 2428
Pollack, Jordan B. 250, 274, 286
Popovici, Elena 1604
Pourkashanian, Mohamed 2046
Powell, David J. 2347
Prügel-Bennett, Adam 1586
Punch, William F. 1431, 2252

Raich, Anne 2418
Ramos, Fernando 429
Ramos, Fernando Manuel 375
Ranji Ranjithan, S. 1622
Rasheed, Khaled 813
Reif, David M. 2277
Reyes-Luna, Juan F. 2096
Ringnér, Kristofer Sundén 1806
Riquelme, José C. 979
Ritthoff, Oliver 1606
Rockett, Peter 1592
Rodebaugh, Brandon 148

Rodrigues, Brian 1594, 1986
Rodriguez-Tello, Eduardo 1283
Rojas, Carlos 219
Rosenberg, Ronald C. 1029, 1764, 2058
Ross, Peter 1295, 1920
Rothlauf, Franz 1307, 1608
Rowe, Jonathan E. 874, 1505
Russell, Matthew 146
Ruwei, Dai 152
Ryan, Conor 1320, 1626, 1752

Sadeghipour, Sadegh 2428
Saeki, Shusuke 2288
Saitou, Kazuhiro 2096
Sánchez, J.M. 2109
Sancho-Caparrini, Fernando 401
Sano, Masaki 1015
Santos, Eugene Jr. 642
Santos, Eunice E. 642
Sarafis, Ioannis 2301
Sastry, Kumara 1332, 1554, 1857
Sayyarodsari, Bijan 657
Schiavone, Guy 1541
Schmeck, Hartmut 1568
Schmidt, Christian 766
Schmidt, Thomas M. 13
Schmitt, Lothar M. 373
Schoenauer, Marc 512, 1089
Schulenburg, Sonia 1295
Schwarz, Josef 1247
Sciortino, John C. Jr. 2384
Scott, Douglas A. 2404
Sendhoff, Bernhard 636
Seo, Dong-Il 669, 1345, 1357
Seo, Kisung 1029, 1764, 2058
Seredyński, Franciszek 1369
Settles, Matthew 148
Shanblatt, Michael 2414
Shepherd, Rob 456
Shepperd, Martin J. 2477
Sheu, Shiann-Tsong 681
Shibata, Youhei 1065
Shimohara, Katsunori 74
Shimosaka, Hisashi 1015
Shyu, Conrad 2313
Sigaud, Olivier 1882
Silva, Sara 1776
Singh, Vishnu 2157
Slimane, M. 140
Smith, R.E. 1382

Smith, Robert E. 778
Soak, Sang-Moon 1610
Socha, Krzysztof 49
Solteiro Pires, E.J. 510
Song, Dong 2325
Soule, Terence 148
Sousa, Fabiano Luis de 375
Spector, Lee 61
Stacey, A. 2422
Stein, Gunnar 644
Stein, Michael 1568
Stephens, Christopher R. 874, 1394
Stepney, Susan 146, 2022
Sthamer, Harmen 2442
Stone, Christopher 1924
Stone, Peter 356
Storch, Tobias 1406
Streeter, Matthew J. 1418
Streichert, Felix 610, 644
Suen, Ching Y. 322
Sun, Dazhi 2420
Suzuki, Tomoya 1814

Takahashi, Katsutoshi 2288
Takama, Yasufumi 246
Tanabe, Shoko 2134
Tanaka, Kiyoshi 2134, 2264
Tanev, Ivan 74
Tang, Ricky 1665
Tarakanov, Alexander O. 248
Teich, Tobias 2398
Tekol, Yüce 695
Tenreiro Machado, J.A. 510
Teo, Jason 483, 1612
Thangavelautham, Jekanthan 377
Tharakunnel, Kurian K. 1906
Thierens, Dirk 886
Tian, Lirong 1614
Timmis, Jon 207
Tokoro, Ken-ichi 1598, 2240
Tomassini, Marco 955, 1788, 1812, 2109
Tominaga, Kazuto 1814
Tong, Siu 2347
Topchy, Alexander 1431
Torng, Eric 13
Torres-Jimenez, Jose 1283
Toussaint, Marc 86, 1444
Trinder, Phil 2301
Truong, T.Q.S. 2422
Tsuji, Miwako 1616

Tsutsui, Shigeyoshi 1015
Tucker, Allan 2360
Tyrrell, Andy 1804

Uchibe, Eiji 507
Ueno, Yutaka 2288
Ulmer, Holger 610, 644
Upal, M. Afzal 98
Usui, Shiro 1600

Valenzuela-Rendón, Manuel 371, 1457
Vallejo, Edgar E. 429
Vanneschi, Leonardo 1788, 1812
Veeramachaneni, Kalyan 110
Vejar, R. 243
Venturini, Gilles 25, 140
Vigraham, Saranyan 431
Vlassov, Valeri 375
Vose, Michael D. 1505

Wakaki, Hiromi 1816
Wallace, Jeff 2424
Waller, S. Travis 2420
Wallin, David 1320
Wang, Wei 537
Ward, David J. 1652
Watanabe, Isamu 1598, 2240
Watson, Jean-Paul 1469
Wegener, Ingo 622, 1406
West, Michael 413
White, Bill C. 2277
White, Tony 122, 231
Whiteson, Shimon 356
Whitley, Darrell 1469
Wieczorek, Wojciech 142
Wiegand, R. Paul 310
Wilson, Chritopher W. 2046
Wilson, Eric 2406
Wineberg, Mark 1481, 1493
Witt, Carsten 622
Wolff, Krister 495

Wood, David Harlan 1148
Wright, Alden H. 1003, 1505
Wright, J.A. 1938
Wu, Annie S. 1541, 1588, 2384
Wu, D.J. 1148
Wyatt, Danica 1518

Xianghui, Dong 152
Xiao, Fei 1594
Xiao, Ningchuan 1530
Xiao, Ying 154
Xu, Zhou 1596
Xuan, Jiang 813

Yamamoto, Takashi 1077
Yamamura, Masayuki 1602
Yang, Jinn-Moon 2372
Yang, Seung-Jin 2426
Yang, Shengxiang 1618
Yassine, Ali 1620
Yilmaz, Ayse S. 2384
Yu, Han 1541, 2384
Yu, Huanjun 134
Yu, Senhua 183
Yu, Tian-Li 1554, 1620
Yu, Tina 156
Yuan, Xiaohui 1624
Yuchi, Ming 638
Yue, Yading 2072

Zalzala, Ali 2301
Zamora, Adolfo 1394
Zechman, Emily M. 1622
Zell, Andreas 610, 644
Zhang, Jian 1624
Zhang, Liping 134
Zhang, Y. 1938
Ziemke, Tom 458
Zincir-Heywood, A. Nur 2325
Zomaya, Albert Y. 1369

Lecture Notes in Computer Science

For information about Vols. 1–2646
please contact your bookseller or Springer-Verlag

Vol. 2647: K.Jansen, M. Margraf, M. Mastrolli, J.D.P. Rolim (Eds.), Experimental and Efficient Algorithms. Proceedings, 2003. VIII, 267 pages. 2003.

Vol. 2648: T. Ball, S.K. Rajamani (Eds.), Model Checking Software. Proceedings, 2003. VIII, 241 pages. 2003.

Vol. 2649: B. Westfechtel, A. van der Hoek (Eds.), Software Configuration Management. Proceedings, 2003. VIII, 241 pages. 2003.

Vol. 2650: M.-P. Huget (Ed.), Communication in Multiagent Systems. VIII, 323 pages. 2003. (Subseries LNAI).

Vol. 2651: D. Bert, J.P. Bowen, S. King, M, Waldén (Eds.), ZB 2003: Formal Specification and Development in Z and B. Proceedings, 2003. XIII, 547 pages. 2003.

Vol. 2652: F.J. Perales, A.J.C. Campilho, N. Pérez de la Blanca, A. Sanfeliu (Eds.), Pattern Recognition and Image Analysis. Proceedings, 2003. XIX, 1142 pages. 2003.

Vol. 2653: R. Petreschi, Giuseppe Persiano, R. Silvestri (Eds.), Algorithms and Complexity. Proceedings, 2003. XI, 289 pages. 2003.

Vol. 2655: J.-P. Rosen, A. Strohmeier (Eds.), Reliable Software Technologies – Ada-Europe 2003. Proceedings, 2003. XIII, 489 pages. 2003.

Vol. 2656: E. Biham (Ed.), Advances in Cryptology – EUROCRPYT 2003. Proceedings, 2003. XIV, 429 pages. 2003.

Vol. 2657: P.M.A. Sloot, D. Abramson, A.V. Bogdanov, J.J. Dongarra, A.Y. Zomaya, Y.E. Gorbachev (Eds.), Computational Science – ICCS 2003. Proceedings, Part I. 2003. LV, 1095 pages. 2003.

Vol. 2658: P.M.A. Sloot, D. Abramson, A.V. Bogdanov, J.J. Dongarra, A.Y. Zomaya, Y.E. Gorbachev (Eds.), Computational Science – ICCS 2003. Proceedings, Part II. 2003. LV, 1129 pages. 2003.

Vol. 2659: P.M.A. Sloot, D. Abramson, A.V. Bogdanov, J.J. Dongarra, A.Y. Zomaya, Y.E. Gorbachev (Eds.), Computational Science – ICCS 2003. Proceedings, Part III. 2003. LV, 1165 pages. 2003.

Vol. 2660: P.M.A. Sloot, D. Abramson, A.V. Bogdanov, J.J. Dongarra, A.Y. Zomaya, Y.E. Gorbachev (Eds.), Computational Science – ICCS 2003. Proceedings, Part IV. 2003. LVI, 1161 pages. 2003.

Vol. 2663: E. Menasalvas, J. Segovia, P.S. Szczepaniak (Eds.), Advances in Web Intelligence. Proceedings, 2003. XII, 350 pages. 2003. (Subseries LNAI).

Vol. 2664: M. Leuschel (Ed.), Logic Based Program Synthesis and Transformation. Proceedings, 2002. X, 281 pages. 2003.

Vol. 2665: H. Chen, R. Miranda, D.D. Zeng, C. Demchak, J. Schroeder, T. Madhusudan (Eds.), Intelligence and Security Informatics. Proceedings, 2003. XIV, 392 pages. 2003.

Vol. 2666: C. Guerra, S. Istrail (Eds.), Mathematical Methods for Protein Structure Analysis and Design. Proceedings, 2000. XI, 157 pages. 2003. (Subseries LNBI).

Vol. 2667: V. Kumar, M.L. Gavrilova, C.J.K. Tan, P. L'Ecuyer (Eds.), Computational Science and Its Applications – ICCSA 2003. Proceedings, Part I. 2003. XXXIV, 1060 pages. 2003.

Vol. 2668: V. Kumar, M.L. Gavrilova, C.J.K. Tan, P. L'Ecuyer (Eds.), Computational Science and Its Applications – ICCSA 2003. Proceedings, Part II. 2003. XXXIV, 942 pages. 2003.

Vol. 2669: V. Kumar, M.L. Gavrilova, C.J.K. Tan, P. L'Ecuyer (Eds.), Computational Science and Its Applications – ICCSA 2003. Proceedings, Part III. 2003. XXXIV, 948 pages. 2003.

Vol. 2670: R. Peña, T. Arts (Eds.), Implementation of Functional Languages. Proceedings, 2002. X, 249 pages. 2003.

Vol. 2671: Y. Xiang, B. Chaib-draa (Eds.), Advances in Artificial Intelligence. Proceedings, 2003. XIV, 642 pages. 2003. (Subseries LNAI).

Vol. 2672: M. Endler, D. Schmidt (Eds.), Middleware 2003. Proceedings, 2003. XIII, 513 pages. 2003.

Vol. 2673: N. Ayache, H. Delingette (Eds.), Surgery Simulation and Soft Tissue Modeling. Proceedings, 2003. XII, 386 pages. 2003.

Vol. 2674: I.E. Magnin, J. Montagnat, P. Clarysse, J. Nenonen, T. Katila (Eds.), Functional Imaging and Modeling of the Heart. Proceedings, 2003. XI, 308 pages. 2003.

Vol. 2675: M. Marchesi, G. Succi (Eds.), Extreme Programming and Agile Processes in Software Engineering. Proceedings, 2003. XV, 464 pages. 2003.

Vol. 2676: R. Baeza-Yates, E. Chávez, M. Crochemore (Eds.), Combinatorial Pattern Matching. Proceedings, 2003. XI, 403 pages. 2003.

Vol. 2678: W. van der Aalst, A. ter Hofstede, M. Weske (Eds.), Business Process Management. Proceedings, 2003. XI, 391 pages. 2003.

Vol. 2679: W. van der Aalst, E. Best (Eds.), Applications and Theory of Petri Nets 2003. Proceedings, 2003. XI, 508 pages. 2003.

Vol. 2680: P. Blackburn, C. Ghidini, R.M. Turner, F. Giunchiglia (Eds.), Modeling and Using Context. Proceedings, 2003. XII, 525 pages. 2003. (Subseries LNAI).

Vol. 2681: J. Eder, M. Missikoff (Eds.), Advanced Information Systems Engineering. Proceedings, 2003. XV, 740 pages. 2003.

Vol. 2684: M.V. Butz, O. Sigaud, P. Gérard (Eds.), Anticipatory Behavior in Adaptive Learning Systems. X, 303 pages. 2003. (Subseries LNAI).

Vol. 2685: C. Freksa, W. Brauer, C. Habel, K.F. Wender (Eds.), Spatial Cognition III. X, 415 pages. 2003. (Subseries LNAI).

Vol. 2686: J. Mira, J.R. Álvarez (Eds.), Computational Methods in Neural Modeling. Proceedings, Part I. 2003. XXVII, 764 pages. 2003.

Vol. 2687: J. Mira, J.R. Álvarez (Eds.), Artificial Neural Nets Problem Solving Methods. Proceedings, Part II. 2003. XXVII, 820 pages. 2003.

Vol. 2688: J. Kittler, M.S. Nixon (Eds.), Audio- and Video-Based Biometric Person Authentication. Proceedings, 2003. XVII, 978 pages. 2003.

Vol. 2689: K.D. Ashley, D.G. Bridge (Eds.), Case-Based Reasoning Research and Development. Proceedings, 2003. XV, 734 pages. 2003. (Subseries LNAI).

Vol. 2691: V. Mařík, J. Müller, M. Pěchouček (Eds.), Multi-Agent Systems and Applications III. Proceedings, 2003. XIV, 660 pages. 2003. (Subseries LNAI).

Vol. 2692: P. Nixon, S. Terzis (Eds.), Trust Management. Proceedings, 2003. X, 349 pages. 2003.

Vol. 2693: A. Cechich, M. Piattini, A. Vallecillo (Eds.), Component-Based Software Quality. X, 403 pages. 2003.

Vol. 2694: R. Cousot (Ed.), Static Analysis. Proceedings, 2003. XIV, 505 pages. 2003.

Vol. 2695: L.D. Griffin, M. Lillholm (Eds.), Scale Space Methods in Computer Vision. Proceedings, 2003. XII, 816 pages. 2003.

Vol. 2697: T. Warnow, B. Zhu (Eds.), Computing and Combinatorics. Proceedings, 2003. XIII, 560 pages. 2003.

Vol. 2698: W. Burakowski, B. Koch, A. Bęben (Eds.), Architectures for Quality of Service in the Internet. Proceedings, 2003. XI, 305 pages. 2003.

Vol. 2701: M. Hofmann (Ed.), Typed Lambda Calculi and Applications. Proceedings, 2003. VIII, 317 pages. 2003.

Vol. 2702: P. Brusilovsky, A. Corbett, F. de Rosis (Eds.), User Modeling 2003. Proceedings, 2003. XIV, 436 pages. 2003. (Subseries LNAI).

Vol. 2704: S.-T. Huang, T. Herman (Eds.), Self-Stabilizing Systems. Proceedings, 2003. X, 215 pages. 2003.

Vol. 2706: R. Nieuwenhuis (Ed.), Rewriting Techniques and Applications. Proceedings, 2003. XI, 515 pages. 2003.

Vol. 2707: K. Jeffay, I. Stoica, K. Wehrle (Eds.), Quality of Service - IWQoS 2003. Proceedings, 2003. XI, 517 pages. 2003.

Vol. 2709: T. Windeatt, F. Roli (Eds.), Multiple Classifier Systems. Proceedings, 2003. X, 406 pages. 2003.

Vol. 2710: Z. Ésik, Z, Fülöp (Eds.), Developments in Language Theory. Proceedings, 2003. XI, 437 pages. 2003.

Vol. 2711: T.D. Nielsen, N.L. Zhang (Eds.), Symbolic and Quantitative Approaches to Reasoning with Uncertainty. Proceedings, 2003. XII, 608 pages. 2003. (Subseries LNAI).

Vol. 2712: A. James, B. Lings, M. Younas (Eds.), New Horizons in Information Management. Proceedings, 2003. XII, 281 pages. 2003.

Vol. 2713: C.-W. Chung, C.-K. Kim, W. Kim, T.-W. Ling, K.-H. Song (Eds.), Web and Communication Technologies and Internet-Related Social Issues - HSI 2003. Proceedings, 2003. XXII, 773 pages. 2003.

Vol. 2714: O. Kaynak, E. Alpaydin, E. Oja, L. Xu (Eds.), Artificial Neural Networks and Neural Information Processing - ICANN/ICONIP 2003. Proceedings, 2003. XXII, 1188 pages. 2003.

Vol. 2715: T. Bilgiç, B. De Baets, O. Kaynak (Eds.), Fuzzy Sets and Systems - IFSA 2003. Proceedings, 2003. XV, 735 pages. 2003. (Subseries LNAI).

Vol. 2716: M.J. Voss (Ed.), OpenMP Shared Memory Parallel Programming. Proceedings, 2003. VIII, 271 pages. 2003.

Vol. 2718: P. W. H. Chung, C. Hinde, M. Ali (Eds.), Developments in Applied Artificial Intelligence. Proceedings, 2003. XIV, 817 pages. 2003. (Subseries LNAI).

Vol. 2719: J.C.M. Baeten, J.K. Lenstra, J. Parrow, G.J. Woeginger (Eds.), Automata, Languages and Programming. Proceedings, 2003. XVIII, 1199 pages. 2003.

Vol. 2720: M. Marques Freire, P. Lorenz, M.M.-O. Lee (Eds.), High-Speed Networks and Multimedia Communications. Proceedings, 2003. XIII, 582 pages. 2003.

Vol. 2721: N.J. Mamede, J. Baptista, I. Trancoso, M. das Graças Volpe Nunes (Eds.), Computational Processing of the Portuguese Language. Proceedings, 2003. XIV, 268 pages. 2003. (Subseries LNAI).

Vol. 2722: J.M. Cueva Lovelle, B.M. González Rodríguez, L. Joyanes Aguilar, J.E. Labra Gayo, M. del Puerto Paule Ruiz (Eds.), Web Engineering. Proceedings, 2003. XIX, 554 pages. 2003.

Vol. 2723: E. Cantú-Paz, J.A. Foster, K. Deb, L.D. Davis, R. Roy, U.-M. O'Reilly, H.-G. Beyer, R. Standish, G. Kendall, S. Wilson, M. Harman, J. Wegener, D. Dasgupta, M.A. Potter, A.C. Schultz, K.A. Dowsland, N. Jonoska, J. Miller (Eds.), Genetic and Evolutionary Computation – GECCO 2003. Proceedings, Part I. 2003. XLVII, 1252 pages. 2003.

Vol. 2724: E. Cantú-Paz, J.A. Foster, K. Deb, L.D. Davis, R. Roy, U.-M. O'Reilly, H.-G. Beyer, R. Standish, G. Kendall, S. Wilson, M. Harman, J. Wegener, D. Dasgupta, M.A. Potter, A.C. Schultz, K.A. Dowsland, N. Jonoska, J. Miller (Eds.), Genetic and Evolutionary Computation – GECCO 2003. Proceedings, Part II. 2003. XLVII, 1274 pages. 2003.

Vol. 2725: W.A. Hunt, Jr., F. Somenzi (Eds.), Computer Aided Verification. Proceedings, 2003. XII, 462 pages. 2003.

Vol. 2726: E. Hancock, M. Vento (Eds.), Graph Based Representations in Pattern Recognition. Proceedings, 2003. VIII, 271 pages. 2003.

Vol. 2727: R. Safavi-Naini, J. Seberry (Eds.), Information Security and Privacy. Proceedings, 2003. XII, 534 pages. 2003.

Vol. 2731: C.S. Calude, M.J. Dinneen, V. Vajnovszki (Eds.), Discrete Mathematics and Theoretical Computer Science. Proceedings, 2003. VIII, 301 pages. 2003.

Vol. 2733: A. Butz, A. Krüger, P. Olivier (Eds.), Smart Graphics. Proceedings, 2003. XI, 261 pages. 2003.

Vol. 2734: P. Perner, A. Rosenfeld (Eds.), Machine Learning and Data Mining in Pattern Recognition. Proceedings, 2003. XII, 440 pages. 2003. (Subseries LNAI).

Vol. 2743: L. Cardelli (Ed.), ECOOP 2003 – Object-Oriented Programming. Proceedings, 2003. X, 501 pages. 2003.

Vol. 2745: M. Guo, L.T. Yang (Eds.), Parallel and Distributed Processing and Applications. Proceedings, 2003. XII, 450 pages. 2003.

Vol. 2749: J. Bigun, T. Gustavsson (Eds.), Image Analysis. Proceedings, 2003. XXII, 1174 pages. 2003.

Vol. 2750: T. Hadzilacos, Y. Manolopoulos, J.F. Roddick, Y. Theodoridis (Eds.), Advances in Spatial and Temporal Databases. Proceedings, 2003. XIII, 525 pages. 2003.